ELECTRICAL TRANSPORT IN NANOSCALE SYSTEMS

In recent years there has been a huge increase in the research and development of nanoscale science and technology, with electrical transport playing a central role. This graduate textbook provides an in-depth description of the transport phenomena relevant to systems of nanoscale dimensions.

In this textbook the different theoretical approaches are critically discussed, with emphasis on their basic assumptions and approximations. The book also covers information content in the measurement of currents, the role of initial conditions in establishing a steady state, and the modern use of density-functional theory. Topics are introduced by simple physical arguments, with particular attention to the non-equilibrium statistical nature of electrical conduction, and followed by a detailed formal derivation. This textbook is ideal for graduate students in physics, chemistry, and electrical engineering.

MASSIMILIANO DI VENTRA is Professor of Physics at the University of California, San Diego. He has published over 70 papers in refereed journals, co-edited the textbook *Introduction to Nanoscale Science and Technology* (Springer, 2004), and has delivered more than 100 invited talks worldwide on the subject of this book.

ELECTRICAL TRANSPORT IN NANOSCALE SYSTEMS

MASSIMILIANO DI VENTRA
University of California, San Diego

CAMBRIDGE
UNIVERSITY PRESS

University Printing House, Cambridge CB2 8BS, United Kingdom

Cambridge University Press is part of the University of Cambridge.

It furthers the University's mission by disseminating knowledge in the pursuit of education, learning and research at the highest international levels of excellence.

www.cambridge.org
Information on this title: www.cambridge.org/9780521896344

© M. Di Ventra 2008

This publication is in copyright. Subject to statutory exception and to the provisions of relevant collective licensing agreements, no reproduction of any part may take place without the written permission of Cambridge University Press.

First published 2008

A catalogue record for this publication is available from the British Library

ISBN 978-0-521-89634-4 Hardback

Cambridge University Press has no responsibility for the persistence or accuracy of URLs for external or third-party internet websites referred to in this publication, and does not guarantee that any content on such websites is, or will remain, accurate or appropriate.

To
Elena, Matteo and Francesca

Contents

1	**A primer on electron transport**		*page* 1
	1.1	Nanoscale systems	1
	1.2	Generating currents	3
		1.2.1 Finite versus infinite systems	8
		1.2.2 Electron sources	9
		1.2.3 Intrinsic nature of the transport problem	10
	1.3	Measuring currents	11
		1.3.1 Microscopic states	12
		1.3.2 The current operator	13
		1.3.3 The measurement process	16
		1.3.4 Complete measurement and pure states	17
	1.4	The statistical operator and macro-states	19
		1.4.1 Pure and mixed states	21
		1.4.2 Quantum correlations	22
		1.4.3 Time evolution of the statistical operator	23
		1.4.4 Random or partially specified Hamiltonians	24
		1.4.5 Open quantum systems	25
		1.4.6 Equilibrium statistical operators	29
	1.5	Current measurement and statistical operator truncation	32
	1.6	One current, different viewpoints	34
	Summary and open questions		36
	Exercises		36
2	**Drude model, Kubo formalism and Boltzmann equation**		39
	2.1	Drude model	39
	2.2	Resistance, coherent and incoherent transport	42
		2.2.1 Relaxation vs. dephasing	44

		2.2.2 Mean-free path	48
		2.2.3 The meaning of momentum relaxation time	49
	2.3	Kubo formalism	50
		2.3.1 The current-current response function	55
		2.3.2 The use of Density-Functional Theory in the Kubo approach	57
		2.3.3 The fluctuation-dissipation theorem	60
		2.3.4 Ohmic vs. ballistic regimes	66
	2.4	Chemical, electrochemical and electrostatic potentials	68
	2.5	Drift-diffusion equations	72
		2.5.1 Diffusion coefficient of an ideal electron gas in the non-degenerate limit	73
		2.5.2 Generalization to spin-dependent transport	75
	2.6	Distribution functions	77
	2.7	Boltzmann equation	79
		2.7.1 Approach to local equilibrium	82
	2.8	Entropy, loss of information, and macroscopic irreversibility	83
		2.8.1 The classical statistical entropy	85
		2.8.2 Quantum statistical entropy	86
		2.8.3 Information content of the N- and one-particle statistical operators	89
		2.8.4 Entropy of open quantum systems	90
		2.8.5 Loss of information in the Kubo formalism	91
		2.8.6 Loss of information with stochastic Hamiltonians	92
		2.8.7 Entropy associated with the measurement of currents	93
	Summary and open questions		94
	Exercises		95
3	**Landauer approach**		101
	3.1	Formulation of the problem	102
	3.2	Local resistivity dipoles and the "field response"	113
	3.3	Conduction from transmission	115
		3.3.1 Scattering boundary conditions	115
		3.3.2 Transmission and reflection probabilities	119
		3.3.3 Total current	123
		3.3.4 Two-probe conductance	128
	3.4	The Lippmann–Schwinger equation	132
		3.4.1 Time-dependent Lippmann–Schwinger equation	132
		3.4.2 Time-independent Lippmann–Schwinger equation	140
	3.5	Green's functions and self-energy	145

		3.5.1	Relation to scattering theory	154
	3.6	The S matrix		159
		3.6.1	Relation between the total Green's function and the S matrix	162
	3.7	The transfer matrix		167
		3.7.1	Coherent scattering of two resistors in series	169
		3.7.2	Incoherent scattering of two resistors in series	171
		3.7.3	Relation between the conductance and the transfer matrix	173
		3.7.4	Localization, ohmic and ballistic regimes	174
	3.8	Four-probe conductance in the non-invasive limit		178
		3.8.1	Single-channel case	179
		3.8.2	Geometrical "dilution"	181
		3.8.3	Multi-channel case	182
	3.9	Multi-probe conductance in the invasive limit		185
		3.9.1	Floating probes and dephasing	187
	3.10	Generalization to spin-dependent transport		190
		3.10.1	Spin-dependent transmission functions	194
		3.10.2	Multi-probe conductance in the presence of a magnetic field	195
		3.10.3	Local resistivity spin dipoles and dynamical effects	196
	3.11	The use of Density-Functional Theory in the Landauer approach		198
	Summary and open questions			202
	Exercises			203
4	**Non-equilibrium Green's function formalism**			**209**
	4.1	Formulation of the problem		211
		4.1.1	Contour ordering	215
	4.2	Equilibrium Green's functions		217
		4.2.1	Time-ordered Green's functions	218
		4.2.2	Dyson's equation for interacting particles	221
		4.2.3	More Green's functions	223
		4.2.4	The spectral function	225
	4.3	Contour-ordered Green's functions		231
		4.3.1	Equations of motion for non-equilibrium Green's functions	233
	4.4	Application to steady-state transport		236
	4.5	Coulomb blockade		244
	4.6	Quantum kinetic equations		250

	Summary and open questions	255
	Exercises	257
5	**Noise**	**258**
5.1	The moments of the current	261
5.2	Shot noise	263
	5.2.1 The classical (Poisson) limit	264
	5.2.2 Quantum theory of shot noise	266
5.3	Counting statistics	274
5.4	Thermal noise	275
	Summary and open questions	277
	Exercises	277
6	**Electron-ion interaction**	**280**
6.1	The many-body electron-ion Hamiltonian	281
	6.1.1 The adiabatic approximation for a current-carrying system	282
	6.1.2 The phonon subsystem	284
	6.1.3 Electron-phonon coupling in the presence of current	288
6.2	Inelastic current	290
	6.2.1 Inelastic current from standard perturbation theory	291
	6.2.2 Inelastic current from the NEGF	296
6.3	Local ionic heating	312
	6.3.1 Lattice heat conduction	319
6.4	Thermopower	323
6.5	Current-induced forces	328
	6.5.1 Elastic vs. inelastic contribution to electro-migration	328
	6.5.2 One force, different definitions	330
	6.5.3 Local resistivity dipoles and the force sign	333
	6.5.4 Forces at equilibrium	333
	6.5.5 Forces out of equilibrium	335
	6.5.6 Are current-induced forces conservative?	340
6.6	Local ionic heating vs. current-induced forces	343
	Summary and open questions	344
	Exercises	344
7	**The micro-canonical picture of transport**	**346**
7.1	Formulation of the problem	347
	7.1.1 Transport from a finite-system point of view	347
	7.1.2 Initial conditions and dynamics	349
7.2	Electrical current theorems within dynamical DFTs	351
	7.2.1 Closed and finite quantum systems in a pure state	351

	7.2.2	Closed quantum systems in a pure state with arbitrary boundary conditions	353
	7.2.3	Current in open quantum systems	354
	7.2.4	Closure of the BBGKY hierarchy	356
	7.2.5	Functional approximations and loss of information	357
7.3	Transient dynamics		358
7.4	Properties of quasi-steady states		360
	7.4.1	Variational definition of quasi-steady states	360
	7.4.2	Dependence of quasi-steady states on initial conditions	364
7.5	A non-equilibrium entropy principle		365
7.6	Approach to steady state in nanoscale systems		369
7.7	Definition of conductance in the micro-canonical picture		374
Summary and open questions			375

8 Hydrodynamics of the electron liquid — 376

8.1	The Madelung equations for a single particle		378
8.2	Hydrodynamic form of the Schrödinger equation		380
	8.2.1	Quantum Navier–Stokes equations	382
8.3	Conductance quantization from hydrodynamics		388
8.4	Viscosity from Time-Dependent Current Density-Functional Theory		391
	8.4.1	Functional approximation, loss of information, and dissipative dynamics	394
	8.4.2	Effect of viscosity on resistance	395
8.5	Turbulent transport		397
8.6	Local electron heating		403
	8.6.1	Electron heat conduction	405
	8.6.2	Hydrodynamics of heat transfer	406
	8.6.3	Effect of local electron heating on ionic heating	410
Summary and open questions			412
Exercises			413

Appendices

Appendix A **A primer on second quantization** 415

Appendix B **The quantum BBGKY hierarchy** 420

Appendix C **The Lindblad equation** 423

C.1	The Lindblad theorem	424
C.2	Derivation of the Lindblad equation	426
C.3	Steady-state solutions	430

Appendix D Ground-state Density-Functional Theory   431
- D.1 The Hohenberg–Kohn theorem   431
- D.2 The Kohn–Sham equations   432
- D.3 Generalization to grand-canonical equilibrium   434
- D.4 The local density approximation and beyond   434

Appendix E Time-Dependent DFT   436
- E.1 The Runge–Gross theorem   436
- E.2 The time-dependent Kohn–Sham equations   437
- E.3 The adiabatic local density approximation   437

Appendix F Time-Dependent Current DFT   439
- F.1 The current density as the main variable   439
- F.2 The exchange-correlation electric field   440
- F.3 Approximate formulas for the viscosity   442

Appendix G Stochastic Time-Dependent Current DFT   444
- G.1 The stochastic Schrödinger equation   444
- G.2 Derivation of the quantum master equation   446
- G.3 The theorem of Stochastic TD-CDFT   449

Appendix H Inelastic corrections to current and shot noise   451

Appendix I Hydrodynamic form of the Schrödinger equation   454

Appendix J Equation of motion for the stress tensor   458

Appendix K Cut-off of the viscosity divergence   461

Appendix L Bernoulli's equation   463

References   464

Index   470

Preface

"The important thing is not to stop questioning.
Curiosity has its own reason for existing."
Albert Einstein

About ten years ago I was resting between session breaks of a busy American Physical Society March meeting. A colleague, whom I had not seen in years, was with me and inquired about my work. I told him I was working on understanding transport in nanoscale systems. He replied, "Aren't the most important facts already understood?"

As unsettling as that question was, I realized he was simply echoing a sentiment in the community: the field of mesoscopic systems – larger "cousins" of nanoscale systems – had provided us with a wealth of experimental results, and a theoretical construct – known as the single-particle scattering approach to conduction – that had almost assumed the characteristics of a "dogma". Many transport properties of mesoscopic systems could be understood in terms of this approach. Books on the subject had appeared which enumerated the successes of this theory. Nanoscale systems were nothing else than smaller versions of mesoscopic systems. All we needed to do was transfer the established experimental knowledge – and proven theoretical and computational techniques – to this new length scale. Or so it seemed.

The past decade has shown that the field of transport in nanoscale systems is *not* a simple extension of mesoscopic physics. Thanks to improved experimental capabilities and new theoretical approaches and viewpoints, it has become clear that novel transport properties emerge at the nanometer scale. In addition, many physical assumptions and approximations we reasonably make to describe mesoscopic systems may not hold for nanoscale structures. Most importantly, it is now starting to sink in that we need to treat the many-body transport problem for what it truly is: a *non-equilibrium statis-*

tical problem. Conducting electrons – and the background ionic structure – are in a state of non-equilibrium, whose properties are known, at best, statistically, even at steady state. By neglecting the true non-equilibrium statistical nature of this problem, we may neglect important dynamical phenomena of particular relevance in nanostructures.

This book attempts to reframe the transport problem with this perspective in mind. Therefore, attention is given to questions that are often overlooked in the literature, e.g., how electrical current is generated, what do we measure when we measure currents, what is the role of initial conditions in establishing a steady state, etc. The language of information theory is used throughout the book to quantify the amount of information one can gather from either the measurement of the current, or the various descriptions of electrical conduction. In addition, I have tried to critically point out the underlying physical assumptions and approximations of the different approaches to transport. It is my opinion that some of the concepts we generally take for granted need to be applied with more care in the case of nanoscale systems, and novel physics may emerge if some of these approximations/assumptions are lifted.

Transport theories belong to the field of non-equilibrium statistical mechanics and are, first and foremost, based on *viewpoints*, not sets of equations. Each of these viewpoints contributes bits of information to our understanding of electrical conduction. The book is thus roughly divided into the description of these viewpoints, the similarities and differences among them, and the physical phenomena one can predict from them. In addition, due to the growing importance of density-functional theory (DFT) – in both its ground-state and dynamical formulations – in transport, the book contains a description of DFT so that it is as self-contained as possible. In particular, the fundamental limitations of ground-state DFT in approaches to electrical conduction are highlighted, and several theorems on the total current are formulated and demonstrated within dynamical density-functional theories. The inclusion of these theorems is not a tribute to mathematics. Rather, it shows the conceptual and formal strengths of these theories in describing electrical transport.

A colleague of mine, who has written textbooks, once told me: "A book is useful to at least one person: its own author." While this statement is definitely true in my case – in the sense that by writing it I have deepened my knowledge and understanding of the subject beyond what everyday research would have probably allowed me – I truly hope this textbook will be of use to its readership, especially those students and researchers who approach the subject for the first time. I have tried to write it at a level accessible to graduate students with a good background in quantum mechanics and statistical mechanics. Some knowledge of solid state physics may help but is not necessary. The derivations of almost all the main results are written explicitly. When this would have resulted in an unnecessary increase in length, I left them as exercises for the reader, or in few cases referred to other textbooks. In this respect, the most difficult topic is probably the non-equilibrium Green's function formalism of Chapter 4. As a compromise between synthesis and clarity, I have written enough about this many-body technique for the reader to follow its basic tenets and results, and referred to other textbooks whenever the level of details seemed to overshadow the main physics. Finally, some of the exercises add to the topics discussed in the main text, or provide useful reference to some mathematical statements I use but have no space to prove.

I have left out topics like superconductivity, the Kondo problem, Luttinger liquid, weak localization, and universal conductance fluctuations. This is not because I believe they are not important but because a comprehensive description of these phenomena would have resulted in a very lengthy extension of the manuscript, with the addition of advanced mathematical formalisms. There are other excellent textbooks that cover these topics. I also apologize in advance to all the authors who feel their work has not been properly credited, and remind them that this is intended as a textbook not a review.

Despite all my efforts and the amount of energy spent to write this book with as much care as I could possibly muster, it would be foolish of me to think that with more than 1200 equations – and a comparable number of concepts – this manuscript would be free of errors. I will therefore post any correction I uncover after its publication on a link to my website **http://physics.ucsd.edu/~diventra/**, and take comfort in the old saying: "Those who never make mistakes make the biggest mistake of all: they never try anything new."

There are too many people who directly or indirectly have contributed to my personal understanding of the subject, and have helped me in this endeavor. I particularly wish to thank Norton Lang, who introduced me to the topic of transport in nanoscale systems. My gratitude also goes to Hardy Gross, Doug Natelson, Nongjian Tao, Tchavdar Todorov, Jan van Ruitenbeek, and Giovanni Vignale, for enlightening discussions, and to Congjun Wu for taking time off his busy schedule to read most of the manuscript. His suggestions and criticisms have helped me improve it. I am also indebted to Dan Arovas who has shared with me his lecture notes on mesoscopic physics. Some topics of Chapter 3 have been inspired by these notes.

I feel fortunate to have worked over the years with talented students and post-doctoral associates. They have shared with me the difficulties and excitement of some of the research that has made it through the pages of this book. Those directly involved in the topics presented here are Neil Bushong, Yu-Chang Chen, Roberto D'Agosta, Yonatan Dubi, John Gamble, Matt Krems, Johan Lagerqvist, Yuriy Pershin, Na Sai, Eric Wright, Zhongqin Yang, and Mike Zwolak. Many of them have also read parts of the manuscript, found several misprints, and made valuable suggestions. Needless to say, any remaining error is due solely to the author.

I also wish to thank the National Science Foundation, the Department of Energy, the National Institutes of Health, and the Petroleum Research Fund for generously funding my research over the years. These funding agencies are, however, not responsible for the ideas expressed in this manuscript.

Finally, the writing of a book takes an enormous amount of energy and time at the expense of the relationships that are most dear to one's life. It is the loving support, understanding, and patience of my wife, Elena, and of my two children, Francesca and Matteo, that have made this project possible. Their presence and encouragement have sustained me during the most difficult times, when it would have been so easy to simply give up. A thank you does not make full justice of my feelings of gratitude and love towards them.

Massimiliano Di Ventra
  La Jolla, San Diego

1
A primer on electron transport

1.1 Nanoscale systems

Let us briefly discuss the systems I will consider in this book, those of *nanoscale* dimensions (1 nm = 10^{-9} m). The phenomena and theoretical approaches I will present are particularly relevant for these structures rather than those with much larger dimensions.

So, what is a nanoscale system? The simplest – and most natural – answer is that it is a structure with at least one dimension at the nanoscale, meaning that such dimension is anywhere in between a few tens of nanometers and the size of an atom (Di Ventra *et al.*, 2004a). One can then define structures with larger – but still not yet macroscopic – dimensions as *mesoscopic*. This separation of scales is arguably fuzzy. Mesoscopic structures share some of the transport properties of nanostructures; the theoretical description of both classes of systems is often similar; and in certain literature no distinction between them is indeed made.

Is there then, in the context of electrical conduction, another key quantity that characterizes nanoscale systems? As I will emphasize several times in this book, this key quantity is the *current density* – current per unit area – they can carry. This can be extremely large.

As an example, consider a wire made using a mechanically controllable break junction (Muller *et al.*, 1992), a junction that is created by mechanically breaking a metal wire. Such a structure – a type of metallic *quantum point contact* – may result in a single atom in between two large chunks of the same material (see schematic in Fig. 1.1). If a typical current of 1μA is set to flow across the system, at the atom position, considering a cross section of 10 Å^2, we would expect a current density of about 1×10^9 A/cm^2! These current densities are typically orders of magnitude larger than those found in mesoscopic/macroscopic systems.

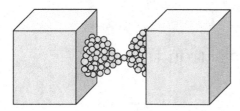

Fig. 1.1. Schematic of an atomic metallic quantum point contact.

A large current density implies a large number of scattering events per unit time and unit volume. This means that interactions among electrons, or among electrons and ions are particularly important.

Note that I have said nothing about how fast a single electron "crosses" a nanoscale structure. This *transit time* may be extremely short. However, due to the large current density the *cumulative* effect of all electrons is to amplify electron-electron and electron-ion interactions locally in the nanostructure. For instance, as I will discuss later in the book, both ions and electrons heat up locally in the junction above their nominal background temperature, thus affecting its structural stability under current flow.

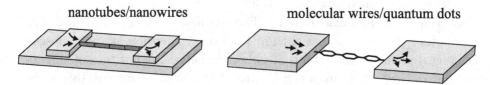

Fig. 1.2. Left panel: A quasi-1D wire laid on top of a surface and in between two bulk electrodes. Right panel: A single molecule between bulk electrodes. The arrows indicate the direction of charge flow.

I mention here a few other nanoscale systems of present interest. These include *nanotubes* or long *atomic wires* in contact with metal electrodes (see Fig. 1.2, left panel), and small *molecules* sandwiched between bulk metals (Fig. 1.2, right panel). These systems may sometimes be referred to as *quantum dots* if their bonding to the electrodes is very weak (these concepts will become clearer as we go along with the book).

Many other structures – and their combinations – that confine electrons in one or more dimensions can be fabricated. These systems represent ideal test beds to understand electron and ion dynamics at these length scales, and may find application in the broadly defined field of optoelectronics, or even in biotechnology and medicine. The latter point is particularly relevant nowadays as the conducting properties of DNA and its single units – called nucleotides – are being studied for possible use in sequencing technology (Zwolak and Di Ventra, 2008).

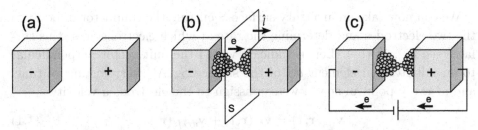

Fig. 1.3. (a) Two finite electrodes charged differently. (b) The electrodes are connected via a junction. (c) Symbol used to represent a battery.

The above list of examples can go on and on. To illustrate the phenomena discussed in this book, I will refer to selected nanoscale systems whose properties have been studied experimentally. The choice of these specific examples reflects both their pedagogical appeal and, of course, the author's taste.

1.2 Generating currents

Before considering the different approaches one can employ to formulate the transport problem in nanoscale systems, let us ask a basic question that will guide us in developing theories of charge transport, namely

How do we generate electrical currents?

There are several answers to this question suggesting different ways to describe the corresponding transport problem.

Let us start from the simplest experimental realization possible. Consider two large but *finite* electrodes and charge them differently: one has more electrical charge than the other, or equivalently one is charged negatively, the other positively. How we charge these electrodes is irrelevant. For instance, we can do it by simply rubbing them with some other material!

I label the one with more electrons with a "−" sign, the other with a "+" sign (see Fig. 1.3(a)). At equilibrium, and due to the conducting nature of each electrode, the extra charge (whether negative or positive) is found on the surface of the electrodes. The electrons in the "−" electrode have higher potential energy (and lower electrostatic potential) than the electrons in the "+" electrode. A potential difference, or *bias*, between the two electrodes has been created so that if we put them in contact by means of some other conducting material, electrons will flow from the "−" region to the "+" region (see Fig. 1.3(b)). *Electrical current* has been thus established.

We can now take an arbitrary surface S cutting the conductor in between the two electrodes and determine the rate at which electrons cross this surface (see Fig. 1.3(b)). Let us indicate with $\hat{\mathbf{l}}$ the unit vector perpendicular to an infinitesimal element dS of the surface S. At every instant of time, and at every point in space, we can assign to the electrons a velocity

$$\mathbf{v}_{tot}(\mathbf{r},t) = \mathbf{v}_{th}(\mathbf{r},t) + \mathbf{v}_{drift}(\mathbf{r},t) \quad (1.1)$$

which is sum of a randomly oriented[1] velocity, $\mathbf{v}_{th}$, due to thermal fluctuations, and a component, $\mathbf{v}_{drift}$, we call *drift velocity*, that, on average, points in the direction of global electron flow.[2] If we average the above velocity over all particles, only the average drift velocity is different from zero,

$$\mathbf{v}(\mathbf{r},t) \equiv \langle \mathbf{v}_{th}(\mathbf{r},t) \rangle + \langle \mathbf{v}_{drift}(\mathbf{r},t) \rangle = \langle \mathbf{v}_{drift}(\mathbf{r},t) \rangle, \quad (1.2)$$

where the bracket operation $\langle \cdots \rangle$ means average over the *ensemble* of particles. If $n(\mathbf{r},t)$ is the *number density* of carriers with charge e at any given point in space and any given instant of time,[3] the amount of charge dQ that crosses the surface dS in an infinitesimal time dt is

$$dQ = en\mathbf{v} \cdot \hat{\mathbf{l}} \, dS \, dt. \quad (1.3)$$

The current across the surface dS is thus

$$dI = \frac{dQ}{dt} = en\mathbf{v} \cdot \hat{\mathbf{l}} \, dS \equiv \mathbf{j} \cdot d\mathbf{S}, \quad (1.4)$$

where I have defined the *current density* vector

$$\mathbf{j}(\mathbf{r},t) = en(\mathbf{r},t)\mathbf{v}(\mathbf{r},t), \quad (1.5)$$

and the surface vector $d\mathbf{S} = \hat{\mathbf{l}} dS$. The total average current across the surface S is then the integral

$$I_S(t) = \int_S d\mathbf{S} \cdot \mathbf{j}(\mathbf{r},t), \quad (1.6)$$

where the subscript S is to remind us that, in general, the total current depends on the chosen surface.[4]

---

[1] Randomly oriented means that the thermal velocity has an isotropic spatial distribution.
[2] In the next chapter I will discuss physical reasons why the drift velocity does not increase indefinitely in time.
[3] For electrons, I choose the convention $e = -|e|$. Note that the standard convention for the direction of current is opposite to the electron flow direction.
[4] In strictly 2D systems the surface integral in the definition 1.6 is replaced by a line integral, and the current density 1.5 is defined with the 2D number density. In 1D the current and the current density are the same quantity, with the definition 1.5 containing the 1D number density.

Polarization and magnetization

In addition to "free" charges, which are able to move across macroscopic regions of the system, there may be bound charges and localized currents, i.e., charges and currents localized to microscopic regions of the sample. Bound charges give rise to *polarization*, and localized currents to *magnetization*. Let us call q_i these bound charges at position $\mathbf{r}_i$, and $\mathbf{v}_i$ their velocities. From classical electrodynamics we know that the polarization is (Jackson, 1975)

$$\mathbf{P}(\mathbf{r}) = \frac{1}{V} \sum_i q_i \, \mathbf{r}_i, \tag{1.7}$$

i.e., the average dipole moment per unit volume, and the magnetization

$$\mathbf{M}(\mathbf{r}) = \frac{1}{2cV} \sum_i q_i \, (\mathbf{r}_i \times \mathbf{v}_i), \tag{1.8}$$

is the average magnetic moment per unit volume, where the sums extend over all charges in a volume V centered at position $\mathbf{r}$, and c is the speed of light. The total current density is (Jackson, 1975)

$$\mathbf{j}_{\text{tot}}(\mathbf{r}, t) = en(\mathbf{r}, t)\mathbf{v}(\mathbf{r}, t) + \frac{\partial \mathbf{P}}{\partial t} + c\nabla \times \mathbf{M}, \tag{1.9}$$

and the total density

$$en_{\text{tot}} = en - \nabla \cdot \mathbf{P}. \tag{1.10}$$

The total current is given again by Eq. 1.6, the integral of the total current density over a surface S. In the following, when discussing densities and current densities, I will always refer to the total density 1.10 and current density 1.9, even though the contributions to the current from polarization and magnetization are, in most of the systems and conditions discussed in this book, small.[5]

In addition, the current density 1.9 generates a magnetic field, which "acts back" on the current, and thus modifies it. This *self-consistent* effect is known as *magnetic screening*. It is generally a small effect for the systems I consider in this book, for the same reasons that the magnetic current

---

[5] For instance, at steady state the polarization current is zero; for non-magnetic materials and small magnetic fields, the magnetization is small. Indeed, for paramagnetic and diamagnetic materials $|\mathbf{M}| \propto \chi|\mathbf{B}|$, with $\mathbf{B}$ an external magnetic field, and χ is the magnetic susceptibility. χ is of the order of $10^{-3} - 10^{-5}$ cm^3/mol (Ashcroft and Mermin, 1975) so that the magnetization is a very small fraction of the field. We will see in Chapter 7 that stochastic time-dependent current density functional theory provides a formally – in principle – exact way to calculate all contributions (from free and bound charges and localized currents) to the current density of a many-body system using effective single-particle equations.

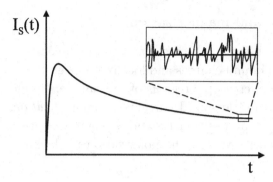

Fig. 1.4. The current as a function of time during the discharge of a capacitor. The current fluctuates around some average value as shown schematically in the inset.

contribution in Eq. 1.9 is small. I thus either neglect it altogether,[6] or assume the current has been determined self-consistently with this magnetic screening effect included.

To the above we need to add the *continuity equation*[7]

$$e\frac{\partial n(\mathbf{r},t)}{\partial t} = -\nabla \cdot \mathbf{j}(\mathbf{r},t) \qquad (1.11)$$

that states the conservation of charge.

Fluctuations and reservoirs

The *gedanken* experiment I have outlined describes the discharge of a capacitor across a conductor. We then know from experiments that as time goes on the current $I_S(t)$ decays (see Fig. 1.4) so that, if we wait enough time, no current (in a time-averaged sense) will flow across the surface S: all negative charges on the "−" electrode have neutralized all positive charges on the "+" electrode, and the electrode-conductor-electrode system ends up in *global equilibrium*.

We can however make this decay time longer and longer by increasing the size of the electrodes while keeping the conductor in between them unchanged. This corresponds to increasing the number of carriers that are stored in the two electrodes. In classical circuit theory this corresponds to increasing the *capacitance* of the capacitor while keeping the electron *resis-*

---

[6] This is clearly a theoretical statement: in experiments one cannot eliminate this screening effect.

[7] Since the divergence of the curl of a vector is zero, the continuity equation is identically satisfied between the total current 1.9 and the total density 1.10:

$$e\frac{\partial n_{\text{tot}}(\mathbf{r},t)}{\partial t} = -\nabla \cdot \mathbf{j}_{\text{tot}}(\mathbf{r},t).$$

1.2 Generating currents

tance to flow from one plate of the capacitor to the other constant. Intuitively, one may guess that if the dimensions of the electrodes were set to go to infinity while keeping the conductor and its contact with the electrodes unchanged, the average current would not change in time, and its value would be independent of the chosen surface S used to evaluate Eq. 1.6.[8]

In reality, the electrical current continually *fluctuates* about an average value (see inset of Fig. 1.4). If we take a time interval T large enough that many current fluctuations occur in that interval, but not so large that the decay of the current is appreciable, we can define

$$\langle I_S \rangle = \frac{1}{T} \int_{-T/2}^{+T/2} dt\, I_S(t) \qquad (1.12)$$

as the average current in that interval.[9] If we take the limit of infinite electrodes first, we can then take the limit of $T \to +\infty$. We thus realize the *current-carrying steady state* condition

$$\frac{d\langle I_S \rangle}{dt} = 0 \quad \text{stationarity condition}, \qquad (1.13)$$

in which the time derivative of the *average* total current is zero.[10]

The above limit of infinite electrodes is also the first encounter with the theoretical concept of *reservoir*, which I will develop further in Chapter 3. Here, I give the following definition

Reservoir: *An ideal "system" that can supply and receive an arbitrary amount of carriers and energy without changing its internal state.*

If this "ideal system" exchanges only energy, and not particles, with some other physical system – whose dynamics we are interested in – I shall call it **bath** and not reservoir. Both the bath and the reservoir embody the notion that physical systems are never truly *closed*; they always have some degree of interaction with an external *environment*.[11]

---

[8] Note, however, that the infinite electrode size is no guarantee that the stationary condition 1.13 is satisfied (see discussion in Sec. 7.6).

[9] Note that an analogous averaging process is always performed by the actual apparatus measuring the current. This is due to the fact that such apparatus has a finite frequency bandwidth $\Delta \nu$. Here I am assuming that $1/\Delta\nu \ll T$.

[10] I anticipate here that the time average 1.12 in the limit of $T \to +\infty$ may be replaced by the average over all possible electron state configurations (*ensemble-averaged* current). This is known as the *ergodic hypothesis*. I will come back to it in Chapter 2 where I will also discuss the conditions for its validity.

[11] Consider for instance electrons in a solid kept at a finite temperature. The finite temperature is that of an external bath. For this bath to be called as such, the dynamics of the electrons cannot affect its properties, namely its temperature.

1.2.1 Finite versus infinite systems

Indeed, the experimental fact that the discharge of a *finite* capacitor ends up in a state of global equilibrium (no average current flows) is a consequence of the fact that the electrons in the capacitor are coupled to some external environment.[12] They thus lose energy during the discharge. If the electron system is truly *closed*, namely electrons do not interact with any external environment – and they are not subject to any external field[13] – their total energy is conserved, and their dynamics is *deterministic* in the sense that we can (in principle) solve exactly their equation of motion at all times – the Schrödinger equation 1.16 in the quantum case. We then know that the following *Poincaré recurrence theorem* applies: there exists a time – called the *recurrence time* – after which the system will return to the neighborhood of its initial conditions. "Neighborhood" means as close as possible to the initial conditions.

For the case at hand, this implies that a finite, closed and isolated capacitor will eventually return to its initial state: it will recharge itself! This apparent paradox, however, does not pose any practical worry. Even if the electron system were truly closed, its recurrence time would be extremely long.

For clarity, consider the quantum case. The many-body states of the capacitor are eigenstates of some Hamiltonian $\hat{H}$. Since we are dealing with a large number N of *interacting* electrons (of the order of an Avogadro's number $N_A = 6 \times 10^{23}$ particles/mol) the spectrum of $\hat{H}$ is extremely *dense*, namely the separation between the many-body energy states dE is extremely small. If ΔE is a representative energy interval of the system,[14] and M the number of states in that interval, it is indeed easy to argue that the separation between many-body energy states is of the order of (Landau and Lifshitz, 1959a)

$$dE \sim \frac{\Delta E}{M} \sim \Delta E \, e^{-S/k_B} \sim \Delta E \, e^{-N}, \qquad (1.14)$$

where S is the statistical entropy I will discuss in Sec. 2.8, $k_B = 1.38 \times 10^{-23}$ J/K is the Boltzmann constant. In the above I have used the fact that the

---

[12] The ionic vibrations of the underlying lattice may be such an environment if they form a *dense* energy spectrum, and no correlated state between electrons and vibrations forms. Otherwise, energy will flow back and forth between electrons and ionic vibrations, and the following considerations apply to this combined system.

[13] Which means the system is also *isolated*. In this book, I will choose the convention of calling a system *closed* when it is not interacting *dynamically* with an environment, and *isolated* when it is not subject to external *deterministic* forces, like a classical electromagnetic field.

[14] Statistically, this interval represents the width of the energy distribution of the system, like the bandwidth in the single-particle case.

number of states is exponentially related to the statistical entropy[15] and the latter is an *extensive variable*, namely it increases with the dimensions of the system, or, equivalently, with the number of particles.

The recurrence time is thus of the order of

$$\Delta t_{rec} \sim \frac{\hbar}{dE} \sim \frac{\hbar}{\Delta E} e^N, \qquad (1.15)$$

which for a macroscopically large number of particles may be much larger than the age of the universe![16] For all practical purposes, there is no way for us to observe the reverse process, the recharge of a closed capacitor, and the dynamics of the system is thus practically *irreversible*.[17]

The above point is interesting for another reason. If I keep the electron system closed, but I make it truly *infinite*, namely with an infinite number of particles, the spectrum of its Hamiltonian becomes a continuum, $dE \to 0$, and the recurrence time tends to infinity, $\Delta t_{rec} \to +\infty$. This means that an infinite system will never return to its initial state, and therefore we do not need to worry about recurrence times: *the dynamics of an infinite system is intrinsically irreversible*. In this case, the energy is *diluted* into an infinite number of *degrees of freedom* (property of a bath, as I have discussed above), and if we added a particle to it we would not be able to modify the internal state of the system (property of a reservoir), and that particle would be "lost" into the reservoir, in the sense that, almost immediately, we would not be able to follow its dynamics.[18]

1.2.2 Electron sources

In real life there is clearly no such a thing as an infinite system and in order to set current flowing one generally relies on batteries attached to the electrodes. The batteries are devices that, via internal chemical processes, continually charge the surfaces of the electrodes so that a constant potential difference can be applied at the two contacts between the electrodes and

---

[15] I will demonstrate this explicitly in Sec. 2.8, Eq. 2.144. Here I just need to use this property to show that an infinite capacitor has an infinite recurrence time.

[16] As an example, consider a system with only 100 electrons, whose many-body spectrum is distributed over $\Delta E = 1$ eV of energy. From Eq. 1.14, $dE \approx 10^{-44}$ eV, and the recurrence time, Eq. 1.15, is $\Delta t_{rec} \sim \hbar/dE \approx 10^{28}$ s, or about 10^{20} years! This number has to be compared with the age of the universe which is about 10^{18} s.

[17] This result applies even if we could solve *exactly* the equation of motion of a macroscopic number of particles. Needless to say, this is a practically impossible task. Approximations to the exact dynamics introduce *loss of information*, and thus provide another source of irreversibility of dynamical systems (see Sec. 2.8).

[18] This statement is obvious in the quantum case where electrons are *indistinguishable*. It is also true in classical mechanics where, for all practical purposes, it is impossible to follow the dynamics of a single particle in an ensemble of infinite identical particles.

the battery, with consequent current flow across the electrode-conductor-electrode structure. In other words, the battery creates a capacitor-like situation as discussed above, and, in addition, *maintains* it till the chemical processes run out. I will use the conventional symbol for a battery as represented in Fig. 1.3(c).

Electrons from the negative side of the battery cross the whole length of the structure to be collected at the other positive end of the battery. On average, at the same time electrons are injected from the negative side into the structure thanks to the electrochemical reaction inside the battery. If for a moment we put quantum mechanics aside and could label single electrons, we could easily argue that the ones that get into the positive side of the battery are not necessarily the same ones that get out of the battery from its negative side. Quite generally, we would need *inelastic* processes to carry electrons *inside* the battery from its positive terminal to the negative one. Inelastic means that any one electron changes its energy in going from the positive terminal to the negative terminal inside the battery.

Most importantly, once an electron is collected at the positive end of the battery, its subsequent dynamics is practically impossible to follow, and this electron is "lost" into the battery, most likely without changing the internal state of the latter. Similarly, we can assume that an electron leaving the battery does not modify its state considerably. In other words, the battery effectively acts as a reservoir of electrons.

1.2.3 Intrinsic nature of the transport problem

Are these the only ways we can generate current? Not quite. For instance, we can go back to our finite electrode-conductor-electrode system of Fig. 1.3(b) after the discharge is complete, i.e., the system is at equilibrium. We can then immerse it in an oscillating electric field. Electric charges would then respond to the field and would start moving from one electrode to the other, generating current. One can also generate current in a metallic ring threaded by a magnetic field (see, e.g., Kamenev and Kohn, 2001), and so on and so forth.

All this seems awfully complicated to describe theoretically. There appear to be so many ways we can generate currents, and so many processes we need to take into account (e.g, the chemical processes in the battery). And I have not even discussed what happens in the conductor itself, or how we *probe* these currents!

This leads us to the following question: What is our main concern when we want to study transport properties of a given system? For instance, are we really interested in what happens inside a battery or at the battery-electrodes interface? The answer is clearly no. Indeed, what we are really interested in are the conducting properties of the nanoscale structure in between the two electrodes. We are less concerned with the way electrons come out of the source or behave very far away from the conductor.

If so, can we simply ignore the electron source or, generally, forget about the way electron motion is created – its initial conditions, the immediate ensuing dynamics, its history dependence, etc. – and focus on the conductor alone? Unfortunately, as we will see in the following chapters, the answer to this question is generally a sound no. Indeed, ignoring such issues may simply mean neglecting important physical processes of particular relevance in nanostructures.

The above details would not bother us if we were simply to study the equilibrium properties of the conductor. However, the forced attention to these details reflects the one basic issue we will have to deal with in this whole book:

> *Electrical transport is a non-equilibrium statistical problem.*

Conducting electrons are in a state of non-equilibrium whose properties, quite generally, are at best known statistically. Under certain experimental conditions, electrons may be very far from their equilibrium state. And as we all know, many concepts, theorems and results that make equilibrium statistical mechanics a well-accepted theoretical framework cannot be straightforwardly carried over to the non-equilibrium case. Nonetheless, it is the combination of statistical mechanics ideas with quantum mechanics that will be the basis of the transport theories I will discuss in the following chapters; the ultimate test of these theories being their experimental verification.

1.3 Measuring currents

I have discussed how we generate currents. A natural question is now

What do we measure when we measure currents?

The answer is non-trivial since we are dealing with a quantum-mechanical process.

Let us then stop for a moment and familiarize ourselves a bit more with the quantum mechanics of the transport problem. This will be necessary to understand what type of information we can extract from a measurement of the current.

1.3.1 Microscopic states

As stated previously, electrons flowing in a conductor are in a state of non-equilibrium. Quantum mechanics tells us that, given an initial condition $|\Psi_0\rangle$ on the many-body state of the system, its time evolution is described by a many-body state vector $|\Psi(t)\rangle$, governed by the time-dependent Schrödinger equation

$$i\hbar \frac{d|\Psi(t)\rangle}{dt} = \hat{H}(t)|\Psi(t)\rangle , \quad |\Psi(t_0)\rangle = |\Psi_0\rangle \quad (1.16)$$

where $\hat{H}$ is the many-body electron Hamiltonian, sum of the kinetic energy of all electrons, all interaction energies between them and the surrounding ions, and any other interaction energy, for instance, the one due to external fields. A formal solution of this equation is

$$|\Psi(t)\rangle = \hat{U}(t,t_0)|\Psi_0\rangle, \quad (1.17)$$

with the *time-evolution* unitary operator

$$\hat{U}(t,t_0) = \mathrm{T}\left\{\exp\left[-\frac{i}{\hbar}\int_{t_0}^{t} dt'\,\hat{H}(t')\right]\right\} \quad (1.18)$$

that *propagates* the initial-time state $|\Psi_0\rangle$ to the state $|\Psi(t)\rangle$ at time t.

The symbol T stands for the time-ordering operator which orders products so that "past" times appear to the right.[19] From the Schrödinger equation 1.16 and Eq. 1.17, we find that the equation of motion of the time-evolution operator is

$$i\hbar \frac{\partial \hat{U}(t,t_0)}{\partial t} = \hat{H}(t)\hat{U}(t,t_0), \quad (1.19)$$

with boundary condition

$$U(t_0,t_0) = 1. \quad (1.20)$$

The formal solution of Eq. 1.19 is again Eq. 1.18. If the Hamiltonian $\hat{H}$ does not depend on time, from 1.18 we find

$$\hat{U}(t,t_0) \equiv \hat{U}(t-t_0) = e^{-i\hat{H}(t-t_0)/\hbar}. \quad (1.21)$$

The state vector $|\Psi(t)\rangle$ represents a *micro-state* of the system at time t. Similarly, the state $|\Psi(t_0)\rangle$ represents the micro-state of the system at initial time $t = t_0$. In order to prepare the system in such a state one generally measures one observable $\hat{A}$, or many commuting observables.

---

[19] For two times t_1 and t_2, $\mathrm{T}[\hat{H}(t_1)\hat{H}(t_2)] = \hat{H}(t_1)\hat{H}(t_2)$ if $t_1 > t_2$, and $\mathrm{T}[\hat{H}(t_1)\hat{H}(t_2)] = \hat{H}(t_2)\hat{H}(t_1)$ if $t_2 > t_1$.

1.3.2 The current operator

The observable we are interested in is the current. We thus need to define its quantum-mechanical operator. The current operator can be constructed from the current density operator. For a system of electrons interacting with an arbitrary classical electromagnetic field with vector potential $\mathbf{A}(\mathbf{r},t)$, the current density operator is[20]

$$\hat{\mathbf{j}}(\mathbf{r},t) = \frac{1}{2}\sum_i \{\delta(\mathbf{r}-\hat{\mathbf{r}}_i), \hat{\mathbf{v}}_i\} \tag{1.22}$$

where the sum extends over all particles with position operator $\hat{\mathbf{r}}_i$ and velocity operator

$$\hat{\mathbf{v}}_i = \frac{\hat{\mathbf{p}}_i - e\mathbf{A}(\hat{\mathbf{r}}_i, t)/c}{m}, \tag{1.23}$$

with the symbol $\{\hat{A}, \hat{B}\} \equiv (\hat{A}\hat{B} + \hat{B}\hat{A})$ indicating the anti-commutator of any two operators $\hat{A}$ and $\hat{B}$; δ is the Dirac delta function; and m is the electron mass.

We can rewrite Eq. 1.22 in another form which will be convenient in the following chapters. By replacing 1.23 into 1.22, and defining the *number density operator*

$$\hat{n}(\mathbf{r}) = \sum_i \delta(\mathbf{r}-\hat{\mathbf{r}}_i), \tag{1.24}$$

we get

$$\hat{\mathbf{j}}(\mathbf{r},t) = \hat{\mathbf{j}}_p(\mathbf{r}) - \frac{e}{mc}\hat{n}(\mathbf{r})\mathbf{A}(\hat{\mathbf{r}}_i, t), \tag{1.25}$$

where

$$\hat{\mathbf{j}}_p(\mathbf{r},t) = \frac{1}{2m}\sum_i \{\delta(\mathbf{r}-\hat{\mathbf{r}}_i), \hat{\mathbf{p}}_i\} \tag{1.26}$$

is the *paramagnetic current density* operator, and the term $-e\hat{n}(\mathbf{r})\mathbf{A}(\hat{\mathbf{r}}_i, t)/mc$ is called the *diamagnetic current density* operator.[21]

---

[20] The operator $\hat{\mathbf{j}}(\mathbf{r},t)$ has three components, one for each cartesian coordinate.
[21] Note that the paramagnetic current density operator is Hermitian but not gauge invariant. Therefore, it is not an observable. Only the total current density 1.22 is an observable.

Given a surface S, the current operator is defined as[22]

$$\hat{I} = \int_S d\mathbf{S} \cdot \hat{\mathbf{j}}(\mathbf{r},t), \qquad (1.27)$$

where $d\mathbf{S}$ is again the surface vector $\hat{\mathbf{I}}dS$ we have defined in Sec. 1.2.[23] This operator has, in general, a *spectrum* of eigensolutions with a *discrete* component

$$\hat{I}|\Psi_{I_0}\rangle = I_0|\Psi_{I_0}\rangle, \qquad (1.28)$$

with the states $|\Psi_{I_0}\rangle$ belonging to the Hilbert space of the system, and satisfying the orthonormality condition

$$\langle \Psi_{I_0}|\Psi_{I_0'}\rangle = \delta_{I_0,I_0'}. \qquad (1.29)$$

These discrete states correspond to *confined* currents such as those set by a magnetic field threading a conducting ring.[24] But most often, and particularly for all the cases I will discuss in this book, the operator $\hat{I}$ has mainly a *continuum* spectrum

$$\hat{I}|\Psi_I\rangle = I|\Psi_I\rangle. \qquad (1.30)$$

These states are not square-integrable – their norm is not finite – and thus they do not belong to the Hilbert space of the system. Their "orthonormality" condition is instead

$$\langle \Psi_I|\Psi_{I'}\rangle = \delta(I - I'), \qquad (1.31)$$

and their statistical interpretation is a bit trickier than for states that belong to a discrete spectrum. Indeed, only the *eigendifferentials* one can construct

---

[22] Alternatively, we can close the surface S so that it encloses one of the electrodes. We can then define an operator $\hat{N}_V$ that counts the number of electrons in the volume V enclosed by the surface S. The current operator can then be equivalently defined as the rate of change of charges in that volume

$$\hat{I}_V = -\frac{d\hat{N}_V}{dt}.$$

We will use this definition in Chapter 4.

[23] The current operator 1.27 clearly depends on the chosen surface S, and, as we are considering here, may have an explicit dependence on time due to the presence of a time-dependent field. Even if it does not depend on time explicitly, it may not commute with the Hamiltonian of the system. The current is thus not necessarily a constant of motion. This may be true even if the system is in a current-carrying steady state, because in that case, it is truly the time average 1.12 that is constant in time.

[24] These are also called confined or *persistent* currents (Büttiker et al., 1983; Levý et al., 1990). In this case, we say that the current is a property of the "ground state" of the system in the presence of a magnetic field since it can be obtained from the eigenstates of the Hamiltonian in the presence of the field.

from these states, namely the states

$$|\Psi_{\delta I}\rangle = \frac{1}{\sqrt{\delta I}} \int_{I}^{I+\delta I} dI |\Psi_I\rangle, \qquad (1.32)$$

with δI a small interval of current in the continuum, are square-integrable and thus belong to the Hilbert space. They indeed satisfy the orthonormality condition (Messiah, 1961)

$$\lim_{\substack{\delta I \to 0 \\ \delta I' \to 0}} \langle \Psi_{\delta I} | \Psi_{\delta I'} \rangle = \delta_{I,I'}. \qquad (1.33)$$

Strictly speaking, whenever we construct averages of operators over states in the continuum, we implicitly assume that we are dealing with eigendifferentials of the type 1.32, corresponding to the states of that observable, and not the continuum states themselves. We will see in Sec. 6.5.5 that this has implications on the definition of current-induced forces.

Here I show that as a consequence of what I have just discussed, the measurement of the current, which provides a value in the continuum, is *always* affected by an intrinsic error – even if the measurement apparatus itself does not introduce any other error in the measurement.[25]

In order to see the fundamental difference between a measurement of the current that provides a value in the discrete part of the spectrum and the measurement of a value in the continuum, let us write the state of the system $|\Psi(t)\rangle$ in terms of the eigenvectors of the current operator,[26]

$$|\Psi(t)\rangle = \sum_{I_0} c_{I_0}(t) |\Psi_{I_0}\rangle + \int dI\, c_I(t) |\Psi_I\rangle, \qquad (1.34)$$

where the coefficients $c_{I_0}(t)$ and $c_I(t)$ are some complex numbers. The probability that a measurement of the current $\hat{I}$ gives a discrete value I_0 at time t is then

$$P(\hat{I} \to I_0 | t) = |c_{I_0}(t)|^2 = |\langle \Psi_{I_0} | \Psi(t) \rangle|^2. \qquad (1.35)$$

On the other hand, for a measurement in the continuum, we can only ask what the probability is that a measurement of the current $\hat{I}$ provides a value in an interval $(I, I + \delta I)$ at time t. This probability is

$$P(\hat{I} \to (I, I+\delta I)|t) = \int_{I}^{I+\delta I} dI\, |c_I(t)|^2 = \int_{I}^{I+\delta I} dI\, |\langle \Psi_I | \Psi(t) \rangle|^2. \qquad (1.36)$$

---

[25] Obviously, a measuring apparatus is affected by its own intrinsic error, in addition to the bandwidth I discussed in Footnote [9].

[26] Equation 1.34 assumes that there are no other quantum numbers associated with the state of the system. As I will discuss in a moment, this is generally not true.

If the interval δI is infinitesimal, the probability is simply $P(\hat{I} \to (I, I + dI)|t) = |c_I(t)|^2 dI$.

We can rewrite the above probability in an alternative way by introducing the *projection operator* onto the subspace of the eigenvalues of the current. For the discrete spectrum I define

$$\hat{P}_{I_0} = |\Psi_{I_0}\rangle\langle\Psi_{I_0}|, \qquad (1.37)$$

which has eigenvalue 1 for the state $|\Psi_{I_0}\rangle$ corresponding to current I_0, and 0 for any other state $|\Psi_{I_0'}\rangle$, with $I_0' \neq I_0$. With this definition, the probability 1.36 may be written as

$$P(\hat{I} \to I_0|t) = \langle\Psi(t)|\hat{P}_{I_0}|\Psi(t)\rangle. \qquad (1.38)$$

For the continuum, I define the projection operator onto the subspace of the current eigenvalues in the interval $(I, I + \delta I)$ as

$$\hat{P}_I = \int_I^{I+\delta I} dI |\Psi_I\rangle\langle\Psi_I|, \qquad (1.39)$$

which acquires the value 1 when operating on the eigenstates $|\Psi_I\rangle$ in the interval $(I, I + \delta I)$, and 0 for any eigenstate $|\Psi_{I'}\rangle$, with $I' \neq (I, I + \delta I)$. Equation 1.36 can then be written as

$$P(\hat{I} \to (I, I + \delta I)|t) = \langle\Psi(t)|\hat{P}_I|\Psi(t)\rangle. \qquad (1.40)$$

1.3.3 The measurement process

If the state of the system is given by Eq. 1.34 *before* the measurement of the current, and by measuring the latter at time t we observe a discrete value I_0, the system is in the state[27]

$$|\Psi(t+dt)\rangle = |\Psi_{I_0}\rangle \qquad (1.41)$$

immediately *after* the measurement. An immediate repetition of the same measurement would lead to exactly the same result. This is known as the *postulate of wave-packet reduction*.

For a measurement in the continuum instead, if we measure a value of the current in the interval $(I, I + \delta I)$, the system is in the state

$$|\Psi(t+dt)\rangle = \frac{1}{\sqrt{\int_I^{I+\delta I} dI |c_I(t)|^2}} \int_I^{I+\delta I} dI\, c_I(t) |\Psi_I\rangle, \qquad (1.42)$$

---

[27] Here, as in the rest of the book, I will assume that the system is not appreciably perturbed by the measurement of the current, except for the wave-packet collapse. This is called an *ideal measurement*.

immediately after the measurement.

Using the definition 1.39 of the projection operator, with the results 1.36 and 1.40 the postulate of wave-packet reduction becomes

$$|\Psi(t+dt)\rangle = \frac{\hat{P}_I|\Psi(t)\rangle}{\sqrt{\langle\Psi(t)|\hat{P}_I|\Psi(t)\rangle}} = \frac{\hat{P}_I|\Psi(t)\rangle}{\sqrt{P(\hat{I} \to (I, I+\delta I)|t)}}, \quad (1.43)$$

immediately after the measurement, which reduces to 1.32 if δI is infinitesimal.

As anticipated, unlike for the discrete spectrum, a measurement in the continuum does *not* specify the state of the system with absolute precision, but provides a linear combination of states in a given interval of values of the spectrum. This interval can be made very small, but never exactly zero. This means that a measurement of the current for the majority of systems I consider in this book is always affected by an intrinsic (unavoidable) error of quantum-mechanical origin.

1.3.4 Complete measurement and pure states

There is yet another point I need to make regarding the measurement of currents. In the above discussion, I have implicitly assumed that by measuring the current I can completely specify the micro-state of the system at any time, i.e., I can write $|\Psi(t)\rangle$ as in Eq. 1.34. In other words, I have assumed that the current provides *maximum information* on the state of the system. For the majority of physical systems I consider in this book, this is not the case. In quantum mechanics, specifying $|\Psi(t)\rangle$ at any given time generally requires measuring more than one observable. To make this clearer, let us start from classical mechanics.

In classical mechanics, we completely know the state of N particles if we know their $3N$ coordinates and their $3N$ momenta at any time. In quantum mechanics, if two observables do not commute (for instance, the conjugated position and momentum operators), we cannot measure their values simultaneously: the measurement of one of them introduces an uncertainty in the other.

Therefore, the best one can do is to measure the maximum number of observables $\hat{A}_1$, $\hat{A}_2$, ..., $\hat{A}_l$, that commute with each other (i.e., $[\hat{A}_i, \hat{A}_j] = \hat{A}_i\hat{A}_j - \hat{A}_j\hat{A}_i = 0, \forall i,j = 1,\ldots,l$), so that the result of these observations provides a set of eigenvalues to which is associated a single eigenstate, i.e.,

$$\hat{A}_1|\Psi_{a_1,\ldots,a_l}\rangle = a_1|\Psi_{a_1,\ldots,a_l}\rangle$$
$$\vdots$$
$$\hat{A}_l|\Psi_{a_1,\ldots,a_l}\rangle = a_l|\Psi_{a_1,\ldots,a_l}\rangle. \tag{1.44}$$

This is called a *complete set of commuting* or *compatible observables*, and the corresponding state (of which we know *all* possible quantum numbers) is also known as *pure state*. As I will discuss in Sec. 2.8 this is a state with "maximum knowledge", in the sense that it provides *maximum information* on the system.

An example

Consider the simplest case possible: a free particle, one whose dynamics is determined by the Hamiltonian $\hat{H} = \hat{\mathbf{p}}^2/2m$. For this case, a complete set of observables is provided by the three components of the momentum $\hat{\mathbf{p}}$. Let us indicate with $|\Psi_{p_x,p_y,p_z}\rangle$ the common eigenstates of these operators, which in the position representation are

$$\langle\mathbf{r}|\Psi_{p_x,p_y,p_z}\rangle = \frac{1}{(2\pi\hbar)^{3/2}}\exp\left[\frac{i}{\hbar}(p_x x + p_y y + p_z z)\right] = \frac{1}{(2\pi\hbar)^{3/2}}e^{\frac{i}{\hbar}\mathbf{p}\cdot\mathbf{r}}, \tag{1.45}$$

with eigenvalues p_x, p_y and p_z for the three coordinates of the momentum (see Eq. 1.44). For a free particle, the current density is constant in space, so that the current operator 1.27 defined across a given surface is simply the area of that surface times the current density operator. From Eq. 1.22 we then see that the current operator is a function of the momentum components, commutes with them, and thus shares with them the same complete set of eigenstates. Therefore, in the case of a free particle, the measurement of the current *and of any two* of its momenta specifies completely the simultaneous state of the three components of the momentum that leads to that current.

However, if we simply measure a current I, we do not know all degrees of freedom of the particle (its momentum), and its state after the measurement is simply

$$|\Psi(t+dt)\rangle = \mathcal{A}\int_{-\infty}^{+\infty}d\mathbf{p}\,c_{p_x,p_y,p_z}(t)|\Psi_{p_x,p_y,p_z}\rangle\bigg|_I, \tag{1.46}$$

a linear combination of momentum states of the particle, *constrained* to give the measured current I, with $\mathcal{A}$ a normalization constant.[28] This is even more so for a system of N interacting particles: *the measurement of the current is generally not enough to specify the micro-state of the system.*

Since in transport experiments one generally measures only the current, and not all other compatible degrees of freedom, the state of the system is not completely known. The current measurement provides only an average *macroscopic* quantity without information on the *microscopic* degrees of freedom.

1.4 The statistical operator and macro-states

From the previous discussion I conclude that the state of a current-carrying system is only known statistically: one does not know the time evolution of all degrees of freedom, only of some average macroscopic quantities. In this case, the quantum-mechanical evolution of a single state vector $|\Psi(t)\rangle$ does not provide full information on the state of the system, and one has to resort to a different formalism. The latter needs to take into account both the "intrinsic" statistical nature of quantum mechanics, and a classical type of statistics. Here I briefly introduce the mathematical concept that accomplishes this task, known as the *statistical operator* or *density operator*. I will use it in several parts of this book, whenever I need to consider explicitly the statistical nature of the transport problem.

If the quantum-mechanical state of the system is not known exactly, we can only discuss what is the probability that our system is in the state $|\Psi_i(t)\rangle$ at any given time t. It is thus convenient to envision an *ensemble* of identical copies of our system, all prepared with similar initial conditions, and attribute at any given time t, a state vector $|\Psi_i(t)\rangle$ to *each* element i of the ensemble.[29] This state vector occurs with probability p_i in the ensemble (see Fig. 1.5). These classical probabilities do not depend on time,[30] as they simply describe how the ensemble is distributed over the individual elements; and in order to represent a probability they have to add to unity

$$\sum_i p_i = 1. \qquad (1.47)$$

---

[28] Clearly, for a free particle in one dimension, the measurement of the current specifies its momentum in the sense 1.36 of a measurement in the continuum.

[29] Of course, even if we knew the state of the system exactly (that is we have prepared it via a complete measurement), due to the statistical interpretation of quantum mechanics, we have to deal with an ensemble of systems, each prepared with exact precision. As I will show in a moment, this is a particular case (a *pure state*) of the more general statistical operator formalism I am discussing here. Therefore, to be precise, one should consider the elements i as *sub-ensembles* of the whole ensemble of copies of the system.

[30] This is true for closed quantum systems. For open quantum systems these probabilities may vary in time (see Appendix G).

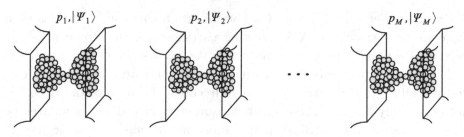

Fig. 1.5. An ensemble of identical copies of a system, each described by a state vector $|\Psi_i\rangle$ occurring with probability p_i. The whole set $\{|\Psi_i\rangle, p_i\}$ describes a macro-state of the system with M accessible micro-states.

Let us then define the statistical operator

$$\hat{\rho}(t) = \sum_i p_i |\Psi_i(t)\rangle \langle \Psi_i(t)|. \qquad (1.48)$$

With this definition, the probabilities p_i are simply

$$p_i = \langle \Psi_i | \hat{\rho} | \Psi_i \rangle, \qquad (1.49)$$

where I have used the orthonormality of the states $|\Psi_i\rangle$.

The set $\{|\Psi_i\rangle, p_i\}$ of all possible micro-states $|\Psi_i\rangle$ and their probability p_i of occurring in the ensemble is called a *macro-state* of the system, or a *mixed state*.[31]

Given an operator $\hat{A}$ we then define its expectation value

$$\langle \hat{A} \rangle_t = \sum_i p_i \langle \Psi_i(t) | \hat{A} | \Psi_i(t) \rangle = \text{Tr}\{\hat{\rho}(t)\hat{A}\}, \qquad (1.50)$$

by performing both a quantum-mechanical average over the states of the system, and a classical average over the ensemble of replicas of the system. The symbol Tr indicates the trace of the matrix obtained from a given operator in a specified basis set (note that the trace is independent of the chosen basis set), and the last equality of Eq. 1.50 comes from the definition of the trace of an operator, or equivalently, from the fact that given any two states $|\phi\rangle$ and $|\psi\rangle$ the trace satisfies the following property (see Exercise 1.1)

$$\text{Tr}\{|\phi\rangle\langle\psi|\} = \langle\psi|\phi\rangle. \qquad (1.51)$$

---

[31] The representation of a given macro-state is not necessarily unique. There may indeed be two (or possibly many) distinct macro-states $\{|\Psi_i\rangle, p_i\}$, $\{|\tilde{\Psi}_i\rangle, \tilde{p}_i\}$ that give the *same* statistical operator 1.48. Since only the averages 1.50 are physically measurable, these macro-states represent the same dynamical properties of the same physical system, and cannot be thus distinguished (see Exercise 1.3).

Equation 1.50 is written in the *Schrödinger picture* where the time evolution is attributed to the state vectors, and the operators are independent of time, apart from a possible explicit time dependence. In the *Heisenberg picture* the operators evolve in time, while the state vectors are constant in time.[32] The average 1.50 must be identical in the two representations, and in the Heisenberg picture we write it as

$$\langle \hat{A} \rangle_t = \langle \hat{A}_H(t) \rangle \equiv \text{Tr}\{\hat{\rho}(t=0)\hat{A}_H(t)\}, \qquad (1.52)$$

where $\hat{\rho}(t=0)$ is the initial-time statistical operator, and $A_H(t)$ is the observable in the Heisenberg picture.

In any given basis $\{|\Phi\rangle\}$ of the system's Hilbert space the statistical operator 1.48 is represented by the *density matrix* elements $\langle \Phi | \hat{\rho} | \Phi' \rangle$. In addition,

$$\text{Tr}\{\hat{\rho}(t)\} = \sum_i p_i = 1, \qquad (1.53)$$

i.e., the statistical operator has unit trace at all times. I also mention the following important properties of the statistical operator (Exercise 1.2), namely $\hat{\rho}$ is Hermitian, $\hat{\rho} = \hat{\rho}^\dagger$, and non-negative, $\langle \Psi | \hat{\rho} | \Psi \rangle \geq 0$, $\forall \, |\Psi\rangle$ in the Hilbert space, which guarantees that the expectation value 1.50 of a positive operator, and the variance of any operator $\hat{A}$, $\langle \hat{A}^2 \rangle_t - \langle \hat{A} \rangle_t^2$, are positive.

Finally, due to the properties of the trace, the average of any observable 1.50 is independent of the chosen basis. Instead of using the exact many-body states $|\Psi_i(t)\rangle$ one can then change basis and use the eigenstates of the system whose interactions among particles have been switched off. We will see in the following chapters that with suitable approximations on the Hamiltonian of the system this leads to enormous simplifications in the calculations.

1.4.1 Pure and mixed states

When the macro-state has probability distribution $\{1, 0, \ldots, 0\}$, the statistical operator is

$$\hat{\rho}(t) = |\Psi(t)\rangle\langle\Psi(t)|, \quad \text{pure state}, \qquad (1.54)$$

and the system is in a well-defined micro-state, of which we have *full information*. It is thus a state that can be prepared by the measurement of a complete set of commuting observables, and as we have discussed in

---

[32] I recall that given an operator $\hat{A}$ in the Schrödinger picture, its Heisenberg representation is $\hat{A}_H(t) = \hat{U}^\dagger(t,t_0)\,\hat{A}\,\hat{U}(t,t_0) = \hat{U}(t_0,t)\,\hat{A}\,\hat{U}(t,t_0)$, where the unitary operator $\hat{U}(t,t_0)$ is given in Eq. 1.18. In this book I will make use of both representations. I will specify when I switch from the Schrödinger to the Heisenberg picture.

Sec. 1.3.4 we call it a *pure state*. Otherwise, the statistical operator is given by Eq. 1.48, and, as previously discussed, we say the system is in a *mixed state*.

1.4.2 Quantum correlations

What is the meaning of the matrix elements of the statistical operator in a given basis set? The answer to this question is important to understand the phenomenon of *decoherence* I will discuss in the next chapter.

Consider a system that can exist in only two states (for instance the two states of an electron spin). Call these two "spin" states $|\uparrow\rangle$ and $|\downarrow\rangle$. Assume the system is in a pure state $|\Psi\rangle$, which is a linear combination of these two spin states

$$|\Psi\rangle = \alpha|\uparrow\rangle + \beta|\downarrow\rangle, \quad (1.55)$$

with α and β two complex numbers such that $|\alpha|^2 + |\beta|^2 = 1$. The statistical operator is

$$\hat{\rho} = |\Psi\rangle\langle\Psi| = |\alpha|^2|\uparrow\rangle\langle\uparrow| + \alpha\beta^*|\uparrow\rangle\langle\downarrow| + \alpha^*\beta|\downarrow\rangle\langle\uparrow| + |\beta|^2|\downarrow\rangle\langle\downarrow|, \quad (1.56)$$

or in matrix notation (in the basis $|\uparrow\rangle, |\downarrow\rangle$)

$$\rho = \begin{pmatrix} |\alpha|^2 & \alpha\beta^* \\ \alpha^*\beta & |\beta|^2 \end{pmatrix}. \quad (1.57)$$

From the above we immediately see that we can attribute a classical interpretation to the diagonal elements of the density matrix: they represent the probability of finding the system in either the state $|\uparrow\rangle$ or $|\downarrow\rangle$. The off-diagonal terms, on the other hand, relate to the *quantum correlations* between the two spin states, and originate precisely from the superposition (or *quantum coherence*) of quantum-mechanical states (our equation 1.55). These correlations do not have a classical counterpart and therefore they do not allow for a strictly classical interpretation. They quantify the degree of "quantumness" of a physical system.[33]

As I will discuss in the next chapter, when the system is in interaction with an environment (or with a measuring apparatus), the effect of the environment is to transform the density matrix into a diagonal one as time elapses, namely

$$\rho = \begin{pmatrix} |\alpha|^2 & \alpha\beta^* \\ \alpha^*\beta & |\beta|^2 \end{pmatrix} \xrightarrow[\text{environment}]{t \to +\infty} \begin{pmatrix} |\alpha'|^2 & 0 \\ 0 & |\beta'|^2 \end{pmatrix}, \quad (1.58)$$

---

[33] Note that the term "quantum correlations" is not limited to the off-diagonal elements of the density matrix. We may indeed work in a basis in which the density matrix is diagonal to begin with. The diagonal terms would then contain quantum correlations.

with α' and β' two complex numbers such that $|\alpha'|^2 + |\beta'|^2 = 1$. The *correlation degrees of freedom* embodied in the off-diagonal matrix elements of the density matrix are "lost" in the environment. We thus say that the latter has created *decoherence* or *dephasing*, and the system ends up in an *incoherent* state, which, as I have discussed above, allows for an easy classical interpretation.

1.4.3 Time evolution of the statistical operator

The equation of motion for the statistical operator 1.48 can be easily obtained from the time-dependent Schrödinger equation 1.16. I find

$$i\hbar \frac{d\hat{\rho}(t)}{dt} = i\hbar \sum_i p_i \frac{d|\Psi_i(t)\rangle\langle\Psi_i(t)|}{dt}$$

$$= \sum_i p_i [\hat{H}(t)|\Psi_i(t)\rangle\langle\Psi_i(t)| - |\Psi_i(t)\rangle\langle\Psi_i(t)|\hat{H}(t)]$$

$$= \hat{H}(t)\left(\sum_i p_i |\Psi_i(t)\rangle\langle\Psi_i(t)|\right) - \left(\sum_i p_i |\Psi_i(t)\rangle\langle\Psi_i(t)|\right)\hat{H}(t)$$

$$= \hat{H}(t)\hat{\rho}(t) - \hat{\rho}(t)\hat{H}(t). \tag{1.59}$$

Using the definition of the commutator of any two operators $\hat{A}$ and $\hat{B}$, $[\hat{A}, \hat{B}] = \hat{A}\hat{B} - \hat{B}\hat{A}$, I can write the above equation as

$$\boxed{i\hbar \frac{d\hat{\rho}(t)}{dt} = [\hat{H}, \hat{\rho}(t)], \quad \text{closed quantum system.}} \tag{1.60}$$

The above equation, known as the *Liouville–von Neumann equation*, describes the dynamics of a *closed quantum system* whose state may or may not be completely specified.[34] It must be solved with a given initial condition $\hat{\rho}(t_0)$.

Using the definition 1.18 of the time-evolution operator $U(t, t_0)$, a formal solution to Eq. 1.60 is

$$\hat{\rho}(t) = U(t, t_0)\hat{\rho}(t_0)U^\dagger(t, t_0), \tag{1.61}$$

which can be checked by direct substitution into 1.60.

---

[34] I remind the reader that I call *closed* a system that is *not* in dynamical interaction with an environment with which it may exchange energy and/or particles. However, Eq. 1.60 is also valid for systems subject to deterministic external forces, such as an electromagnetic force. Therefore, the system may not be *isolated*.

1.4.4 Random or partially specified Hamiltonians

Note that in order to derive 1.60 I have implicitly assumed that the Hamiltonian $\hat{H}$ *does not* depend on the micro-states $|\Psi_i(t)\rangle$,[35] namely $\hat{H} \neq \hat{H}[\{|\Psi_i\rangle\}]$ – otherwise the third equality in Eq. 1.59 does not hold. What if this is not the case, or more generally, the system is so complicated that we do not even know the Hamiltonian completely? This issue will appear in Chapter 7 when we will make use of approximate Hamiltonians derived from density-functional theory. It is, however, a general problem when the number of degrees of freedom is macroscopic (as in a transport problem), and by making a series of approximations to render the problem analytically or numerically tractable we are left with some degrees of freedom that we know only statistically.[36]

The above results can also be extended to the case in which the Hamiltonian itself is not known exactly, i.e., when the Hamiltonian $\hat{H}$ contains some unknown components, or components that may vary *randomly*. Let us call $\hat{H}_j$ each Hamiltonian in this ensemble of *random* or *stochastic Hamiltonians*, and $\tilde{p}_j$ the probability that this Hamiltonian appears in the ensemble.[37] A statistical operator $\hat{\rho}_j(t)$ associated with a given Hamiltonian $\hat{H}_j$ satisfies an equation of motion 1.61 with a given $U_j(t, t_0)$. For each of these statistical operators we can calculate the average of any observable $\hat{A}$ via Eq. 1.50. The total average over *all* statistical operators is simply weighted by the probabilities $\tilde{p}_j$ that these statistical operators occur. The definition 1.50 can be thus generalized as

$$\langle \hat{A} \rangle_t = \mathrm{Tr}\left\{\hat{\rho}(t)\hat{A}\right\} = \sum_j \tilde{p}_j \mathrm{Tr}\left\{\hat{\rho}_j(t)\hat{A}\right\} = \mathrm{Tr}\left\{\sum_j \tilde{p}_j \hat{\rho}_j(t)\hat{A}\right\}$$

$$= \mathrm{Tr}\left\{\sum_j \tilde{p}_j U_j(t,t_0)\hat{\rho}_j(t_0) U_j^\dagger(t,t_0)\hat{A}\right\}. \qquad (1.62)$$

Since this must be true for any observable $\hat{A}$, from the second and last

---

[35] If it does depend, like the Hamiltonian operator of the Hartree–Fock equations (Ashcroft and Mermin, 1975), we say that it is a *functional* of the wave-functions. This is also the case, for instance, if the Hamiltonian is a functional of the density and/or a functional of the current density.

[36] For instance, we may neglect some correlations among electrons that we know constrain the current to a certain *average* value. These correlation are then known only statistically and we may treat them as random variables.

[37] This probability is not necessarily the probability that, given a Hamiltonian, a micro-state $|\Psi_i\rangle$ appears in the ensemble of micro-states corresponding to that Hamiltonian.

equalities we see that the statistical operator $\hat{\rho}$ at time t is simply

$$\hat{\rho}(t) = \sum_j \tilde{p}_j \hat{\rho}_j(t) = \sum_j \tilde{p}_j U_j(t, t_0) \hat{\rho}_j(t_0) U_j^\dagger(t, t_0). \qquad (1.63)$$

Note that this statistical operator does *not* satisfy Eq. 1.60 with a given Hamiltonian, i.e., its time evolution is not known in a closed form, and in order to find it, one needs to evaluate the average 1.63, at every time step, over all stochastic evolutions of each element of the ensemble of Hamiltonians.

1.4.5 Open quantum systems

Consider now the case in which our system can be separated into two subsystems. For instance, I will show in Chapter 6 that the many-body Hamiltonian of electrons and ions of a given material, can, under specific approximations, be written as the sum of an electronic Hamiltonian $\hat{H}_e$, plus a Hamiltonian describing ionic vibrations (whose quanta are called *phonons*) $\hat{H}_{ph}$, and a Hamiltonian describing their mutual interaction $\hat{H}_{e-ph}$. The total Hamiltonian is thus[38]

$$\hat{H} = \hat{H}_e \otimes \hat{1}_e + \hat{1}_{ph} \otimes \hat{H}_{ph} + \hat{H}_{e-ph}, \qquad (1.64)$$

where $\hat{1}_e$ is the identity operator on the Hilbert space $\mathcal{H}_e$ of the electrons only, and $\hat{1}_{ph}$ is the identity operator on the Hilbert space $\mathcal{H}_{ph}$ of the phonons. The symbol $\otimes$ means tensor product of the two Hilbert spaces.

If we are only interested in the dynamics of the electrons alone, i.e., we are interested in the current carried by the electrons in the presence of the *bath of phonons* without concern for the microscopic dynamics of the latter, we can proceed as follows.

Let $\{|\Psi_e\rangle\}$ be a complete set of eigenstates in the Hilbert space $\mathcal{H}_e$ of the electrons, and $\{|\Psi_{ph}\rangle\}$ a complete set of eigenstates in the Hilbert space $\mathcal{H}_{ph}$ of the phonons. The states

$$|\Psi_{e,ph}\rangle = |\Psi_e\rangle|\Psi_{ph}\rangle \qquad (1.65)$$

span the composite Hilbert space of the total system, electrons plus phonons, $\mathcal{H} = \mathcal{H}_e \otimes \mathcal{H}_{ph}$.

The macro-state of the total system is described by some statistical operator $\hat{\rho}_{e-ph}(t)$ whose matrix elements in the basis 1.65 are $\langle \Psi_{e,ph}|\hat{\rho}_{e-ph}|\Psi_{e',ph'}\rangle$, with the symbols $\{e, e', ph, ph'\}$ indicating all possible electron and phonon

---

[38] I will derive the explicit form of these Hamiltonians in Chapter 6. Here it suffices to know that they describe different sub-systems, and their mutual interaction.

quantum numbers. The current operator on the composite Hilbert space, $\mathcal{H}$, is the tensor product of the current operator 1.27 on the Hilbert space of the electrons, and the identity operator on the Hilbert space of the phonons, $\hat{1}_{ph}$, i.e.,

$$\hat{I}_{e-ph} \equiv \hat{I} \otimes \hat{1}_{ph}. \tag{1.66}$$

According to Eq. 1.50 the expectation value of this current operator (that represents the expectation value of the current operator of the electron subsystem) is then

$$\begin{aligned}\langle \hat{I} \rangle_t &\equiv \langle \hat{I}_{e-ph} \rangle_t = \text{Tr}\{\hat{\rho}_{e-ph}(t)\hat{I}_{e-ph}\} = \text{Tr}\{\hat{\rho}_{e-ph}(t)[\hat{I} \otimes \hat{1}_{ph}]\} \\ &= \sum_{e,e',ph,ph'} \langle \Psi_{e,ph}|\hat{\rho}_{e-ph}|\Psi_{e',ph'}\rangle\langle \Psi_{e',ph'}|\hat{I} \otimes \hat{1}_{ph}|\Psi_{e,ph}\rangle \\ &= \sum_{e,e',ph} \langle \Psi_{e,ph}|\hat{\rho}_{e-ph}|\Psi_{e',ph}\rangle\langle \Psi_{e'}|\hat{I}|\Psi_e\rangle,\end{aligned} \tag{1.67}$$

where in the above, the symbol $\{e, e', ph, ph'\}$ means that the sum extends over all states $|\Psi_{e,ph}\rangle$ of the composite system, and in the last equality I have used the orthonormality condition of the phonon states $\langle \Psi_{ph}|\hat{1}_{ph}|\Psi_{ph'}\rangle = \delta_{ph,ph'}$.

If we define the matrix elements of the *reduced* statistical operator acting only on the Hilbert space $\mathcal{H}_e$ of the electrons[39]

$$\langle \Psi_e|\hat{\rho}_e|\Psi_{e'}\rangle = \sum_{ph}\langle \Psi_{e,ph}|\hat{\rho}_{e-ph}|\Psi_{e',ph}\rangle \equiv \langle \Psi_e|\text{Tr}_{ph}\{\hat{\rho}_{e-ph}(t)\}|\Psi_{e'}\rangle, \tag{1.68}$$

we can rewrite Eq. 1.67 as

$$\langle \hat{I} \rangle_t = \text{Tr}_e\{\hat{\rho}_e(t)\hat{I}\} \tag{1.69}$$

where the symbol Tr_e means that we are integrating over the electronic degrees of freedom, and I have defined the statistical operator

$$\hat{\rho}_e(t) = \text{Tr}_{ph}\{\hat{\rho}_{e-ph}(t)\}, \tag{1.70}$$

constructed by *integrating* or *tracing out* the phonon degrees of freedom. In the rest of the book we will be concerned mostly with traces over the electronic degrees of freedom, so to simplify the notation, I will use the symbol Tr to mean Tr_e.

Note that after this integration, the state of the electronic subsystem is *not* necessarily in a pure state, even if it was prepared as such at the initial time.

---

[39] This is the quantum-mechanical analogue of the classical reduced densities I will define in Sec. 2.6.

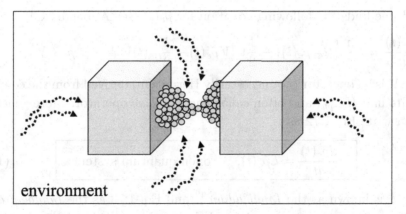

Fig. 1.6. A physical system is, in general, always in interaction with an environment. For electrons in a conductor this environment may be provided by the ionic vibrations, or the ensemble of electrons themselves.

This is a general property of composite quantum systems: *a subcomponent of the whole quantum system cannot be generally described by pure states.*

In addition, unlike the total statistical operator $\hat{\rho}_{e-ph}(t)$ whose dynamics is given by Eq. 1.60 with Hamiltonian 1.64, the dynamics of the reduced statistical operator $\hat{\rho}_e(t)$ does *not* generally obey a closed equation of motion, i.e., a simple equation of the type 1.60.[40] This is precisely due to the interaction $\hat{H}_{e-ph}$ between electrons and phonons. Nonetheless, we can still derive a closed equation under specific conditions.

Let us assume that at the initial time $t = t_0$ the total statistical operator can be separated into

$$\hat{\rho}_{e-ph}(t_0) = \hat{\rho}_e(t_0) \otimes \hat{\rho}_{ph}(t_0), \qquad (1.71)$$

the electrons and phonons are initially *uncorrelated*. One can then switch on their mutual interaction for $t > t_0$ (provided the interaction is weak), and by assuming that the phonons "lose" memory of their internal correlations

---

[40] In fact, from Eq. 1.60

$$i\hbar \frac{d(\text{Tr}_{ph}\{\hat{\rho}_{e-ph}(t)\})}{dt} = \text{Tr}_{ph}\{[\hat{H}, \hat{\rho}_{e-ph}(t)]\} \neq [\hat{H}, \text{Tr}_{ph}\{\hat{\rho}_{e-ph}(t)\}].$$

fast,[41] one finds the following equation for $\hat{\rho}_e(t)$ (see Appendix C)

$$\frac{d\hat{\rho}_e(t)}{dt} = -\frac{i}{\hbar}\left[\hat{H}_e, \hat{\rho}_e(t)\right] - \frac{1}{2}\hat{V}^\dagger \hat{V}\hat{\rho}_e(t) - \frac{1}{2}\hat{\rho}_e(t)\hat{V}^\dagger \hat{V} + \hat{V}\hat{\rho}_e(t)\hat{V}^\dagger, \quad (1.72)$$

where V is an operator (not necessarily Hermitian) derived from the coupling Hamiltonian $\hat{H}_{e-ph}$, and often called the *Lindblad* operator. More generally we can write Eq. 1.72 as

$$\boxed{\frac{d\hat{\rho}_e(t)}{dt} = \mathcal{L}\hat{\rho}_e(t), \quad \text{open quantum system}} \quad (1.73)$$

where $\mathcal{L}$ is known as the *Lindbladian*,[42] and Eq. 1.73 as the *Lindblad equation*, or *quantum master equation*. I derive this equation in Appendix C, where I also discuss some of its general properties. In particular, I stress here that the Hamiltonian appearing in Eq. 1.72 may not necessarily coincide with the Hamiltonian of the electrons in the absence of the bath. The bath, via interaction with the system, may also contribute to the unitary part of the system evolution by "shifting" (or *renormalizing*) its energy states (see Appendix C).

In Appendix G I will re-derive Eq. 1.73 from the *stochastic Schrödinger equation* G.1, which will be more useful when we transform the many-body electron problem into an effective single-particle one using density-functional methods.

Equation 1.73 describes the dynamics of the electrons *open* to a bath of phonons. This dynamics is *not* unitary like the time-dependent Schrödinger equation 1.16, and we thus say it describes electron evolution "beyond Hamiltonian dynamics".

Equation 1.73 is not limited to electrons in interaction with phonons. It generally describes the interaction of a physical system in interaction with (open to) an environment whose strength is represented by some operator $\hat{V}$ (Fig. 1.6). If one considers single electrons, this environment may be due to the presence of the other electrons in the conductor, which produce fluctuations in the electron dynamics via electron-electron scattering.

I also note that for an arbitrary operator $\mathcal{L}$ Eq. 1.73 does not necessarily

---

[41] In other words, the electron dynamics is slower than any correlation time of the phonons (see definition in Sec. 5.1). As I have discussed in Sec. 1.2, this is precisely the condition for a bath to behave as such: its interaction with the system of interest – in this case the electrons – does not modify its internal state. This is also called the *Markovian* approximation. However, in the case of phonons, by doing so we neglect any correlated feedback the electrons may provide to the phonon dynamics. We will see in Sec. 8.6 that this feedback is not necessarily negligible locally in a nanoscale junction. If we relaxed the Markov approximation, we would still obtain a closed equation of motion for the statistical operator but with "memory" effects (see Appendix C).

[42] The operator $\mathcal{L}$ is a type of *super-operator* because it transforms an operator into another.

admit a stationary solution in the limit of $t \to +\infty$, namely a statistical operator $\hat{\rho}_e^{ss}$ such that

$$\lim_{t \to +\infty} \frac{d\hat{\rho}_e^{ss}(t)}{dt} = 0. \tag{1.74}$$

If it does admit a solution for a given $\mathcal{L}$, this statistical operator may not be unique, i.e., there may be many (possibly a macroscopic number of) solutions to Eq. 1.72 satisfying the condition 1.74 (see Appendix C).

1.4.6 Equilibrium statistical operators

The statistical operators of electrons and phonons at the initial time may be arbitrary. Indeed, at the initial time both subsystems may be in a state of non-equilibrium according to how the system is prepared – for instance by the battery that drives the current. In many instances, it will be however convenient, or necessary, to assume that both electrons and phonons are in an initial state of equilibrium. This state is easy to calculate from equilibrium statistical mechanics.

When a system is in *global* equilibrium with a reservoir at temperature θ with which it may exchange particles,[43] the probability p_i that it can be found in a micro-state $|\Psi_i\rangle_N$, that contains a number N of particles, and has energy E_i^N (solution of $\hat{H}_N|\Psi_i\rangle_N = E_i^N|\Psi_i\rangle_N$, with $\hat{H}_N$ the N-particle Hamiltonian) is given by the equilibrium *grand-canonical* distribution (see, e.g., Balian, 1991)

$$p_i^{eq} = \frac{e^{-\beta(E_i^N - \bar{\mu}N)}}{\sum_M \sum_j e^{-\beta(E_j^M - \bar{\mu}M)}}, \tag{1.75}$$

where $\bar{\mu}$ is the energy change due to the addition of an extra particle to the system (also known as *chemical potential* that I discuss in detail in the next chapter), and I have defined $\beta = 1/k_B\theta$. Using the energy eigenstates $|\Psi_i\rangle_N$ and the definition 1.48 of statistical operator, the latter can be written as (see also 1.49)

$$\hat{\rho}_G^{eq} = \frac{e^{-\beta(\hat{H} - \bar{\mu}\hat{N})}}{\text{Tr}\{e^{-\beta(\hat{H} - \bar{\mu}\hat{N})}\}} \equiv \frac{e^{-\beta(\hat{H} - \bar{\mu}\hat{N})}}{\mathcal{Z}_G}, \tag{1.76}$$

with the grand-canonical partition function $\mathcal{Z}_G = \text{Tr}\{e^{-\beta(\hat{H} - \bar{\mu}\hat{N})}\}$,[44] and $\hat{N}$

---

[43] This equilibrium state can be thought of as the long-time stationary solution of Eq. 1.73 for some super-operator $\mathcal{L}$.

[44] The sums in both the canonical and grand-canonical partition functions run over all degeneracies of the energy states.

an operator that counts the number of particles in a given many-body state (see also Appendix A and Chapter 4)

$$\hat{N}|\Psi_i\rangle_N = \left(\sum_{M,j} M|\Psi_j\rangle_{M\ M}\langle\Psi_j|\right)|\Psi_i\rangle_N = N|\Psi_i\rangle_N. \quad (1.77)$$

If the number of particles in the system is fixed (not just its average) then the equilibrium statistical operator is the *canonical*[45]

$$\hat{\rho}_C^{eq} = \frac{e^{-\beta\hat{H}}}{\text{Tr}\{e^{-\beta\hat{H}}\}} \equiv \frac{e^{-\beta\hat{H}}}{\mathcal{Z}_C}, \quad (1.78)$$

with the canonical partition function $\mathcal{Z}_C = \text{Tr}\{e^{-\beta\hat{H}}\}$. The canonical probabilities are

$$p_i^{eq} = \frac{e^{-\beta E_i}}{\sum_j e^{-\beta E_j}}. \quad (1.79)$$

The form of the above statistical operators is the same for both electrons and phonons, the difference being clearly in the quantities appearing in them, such as the Hamiltonian and chemical potential.

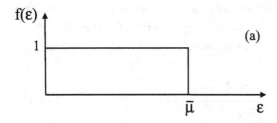

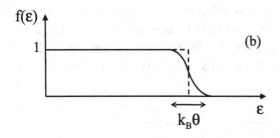

Fig. 1.7. (a) Degenerate limit of the Fermi–Dirac distribution for independent electrons. (b) Broadening of the Fermi–Dirac distribution at a finite temperature θ. $\bar{\mu}$ is the chemical potential.

When the N particles are *non-interacting*, or weakly interacting,[46] and *indistinguishable* then the total energy of the system E is simply the sum

---

[45] In the limit of zero temperature, $\beta \to +\infty$, this reduces to the statistical operator of the ground state $\hat{\rho}_0 = |\Psi_0\rangle\langle\Psi_0|$.
[46] This is called an *ideal gas* and I will employ its properties in several parts of this book.

over the single-particle energies $\mathcal{E}_j$ of its constituents, $E = \sum_j N_j \mathcal{E}_j$, where N_j is now the average number of particles that occupy the single-particle level with energy $\mathcal{E}_j$, with $\sum_j N_j = N$. We then find that the average *occupation number* or *distribution* of these single-particle levels weighted according to 1.75 is

$$f_j \equiv N_j = \frac{1}{e^{(\mathcal{E}_j - \bar{\mu})/k_B \theta} + 1}, \qquad \text{Fermi–Dirac distribution}, \qquad (1.80)$$

for electrons, and

$$N_j = \frac{1}{e^{\mathcal{E}_j/k_B \theta} - 1}, \qquad \text{Bose–Einstein distribution}, \qquad (1.81)$$

for phonons.[47]

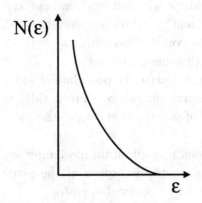

Fig. 1.8. Bose–Einstein distribution as a function of energy for independent phonons.

In the limit of zero temperature the Fermi–Dirac distribution tends to a step function with unit height, with the chemical potential delimiting the edge of the step function:

$$f(\epsilon) \to \Theta(\bar{\mu} - \epsilon); \qquad \theta \to 0, \qquad (1.82)$$

where $\Theta(\bar{\mu} - \epsilon)$ is the Heaviside step function, which is zero for $\bar{\mu} - \epsilon < 0$, and takes the value 1 for $\bar{\mu} - \epsilon > 0$ (Fig. 1.7(a)). In this limit we say the non-interacting electrons form a *degenerate Fermi gas*, and the chemical potential coincides with the Fermi energy, E_F, the energy of the highest occupied level.[48]

At finite temperatures the Fermi–Dirac distribution broadens due to the

---

[47] The number of phonons (like the number of photons, quanta of the electromagnetic field) is not conserved. At thermal equilibrium it cannot be modified independently of the temperature. Therefore, the chemical potential for phonons is taken to be zero.

[48] More generally, the degenerate limit of the Fermi gas occurs when the thermal energy $k_B \theta$ is much smaller than the Fermi energy E_F.

fact that some electrons are *excited* to states above the chemical potential, with the broadening of the order of $k_B\theta$ (Fig. 1.7(b)).[49]

The Bose–Einstein distribution instead tends to zero with increasing energy, and diverges as the energy goes to zero (see Fig. 1.8). We refer the reader to other textbooks for a derivation of these well-known results (see for instance Pathria, 1972).

1.5 Current measurement and statistical operator truncation

As a particular case of open quantum systems, let us consider our electron system "open" to an apparatus measuring the current. Indeed, the coupling of a measurement apparatus with the system of interest can be interpreted as the interaction of the latter with other degrees of freedom external to it. And similarly, the interaction of a system with an environment can be interpreted as the continuous "measurement" of the open system (with the environment carrying information on the correlations with the system), without actually observing the outcome of that measurement.

This particular "open" case allows us to reformulate the postulate of wave-packet reduction in the language of statistical operators. From this, in Sec. 2.8 I will quantify the amount of *loss of information* during the measurement of currents.

As I have previously discussed, measurements in quantum mechanics are associated with an irreversible projection onto the subspace of the given observable. This has an important effect on the statistical operator.

To see this, let us assume for simplicity that before the current measurement the system is prepared in a pure state $|\Psi(t)\rangle$. Its statistical operator is then given by Eq. 1.54, i.e., $\hat{\rho}(t) = |\Psi(t)\rangle\langle\Psi(t)|$. Let us also assume that the current measurement provides a complete measurement, namely, we do not need to measure any other observable to know the state of the system.

Let us now couple the system to an apparatus that measures currents. Before the coupling, the statistical operator of the combined object, system plus apparatus, is the tensor product of the statistical operator of the closed system and that of the apparatus (similar to the product 1.71 between the statistical operators of electrons and phonons before their interaction). We then allow for coupling, and immediately separate the measurement appa-

---

[49] When the temperature is large enough that the thermal energy $k_B\theta$ is a large fraction of the Fermi energy E_F, we say the electrons form a *non-degenerate* Fermi gas. In metallic systems the Fermi energy is of the order of eVs. The temperatures considered in this book are close to room temperature or much smaller. Therefore, for the metallic systems considered here, the electron gas is practically degenerate. In doped semiconductors the Fermi energy (whether above the conduction band minimum, or below the valence band maximum) is of the order of few $k_B\theta$s. For this type of systems the non-degenerate limit is easily attainable.

1.5 Current measurement and statistical operator truncation

ratus from the system.[50] We know that such a measurement process can provide *any* of the current eigenvalues I, with probabilities

$$p_I = P(\hat{I} \to (I, I + \delta I)|t) = \mathrm{Tr}\{\hat{\rho}\hat{P}_I\}, \quad (1.83)$$

given by 1.40, associated with the states $|\Psi_{\delta I}\rangle$.[51] This is equivalent to the definition of a statistical mixture to which corresponds a statistical operator of the form 1.48. We thus see that, even *before any observation is made*, i.e., before we actually know the outcome of the measurement, by simply coupling the system to the measurement apparatus (and its subsequent separation), the statistical operator of the system transforms into that representing the statistical mixture

$$\hat{\rho}_\mathcal{I}(t) = \sum_I p_I |\Psi_I\rangle\langle\Psi_I| = \sum_I p_I \frac{\hat{P}_I|\Psi(t)\rangle}{\sqrt{p_I}} \frac{\langle\Psi(t)|\hat{P}_I}{\sqrt{p_I}}$$

$$= \sum_I \hat{P}_I|\Psi(t)\rangle\langle\Psi(t)|\hat{P}_I = \sum_I \hat{P}_I \hat{\rho}(t) \hat{P}_I, \quad (1.84)$$

where in the first line I have used 1.43.

The above result is true even if the statistical operator of the system is not initially in a pure state. It amounts to *decoupling* any two different subspaces corresponding to any two eigenvalues of the current operator. In other words, the off-diagonal elements of the statistical operator corresponding to the coupling between any two eigenvalues of the current operator are set to zero. Therefore, this stage of the measurement corresponds to the *truncation* of the initial statistical operator $\hat{\rho}$.

The final stage of the measurement is to observe a specific outcome I which, according to Eq. 1.43 transforms the statistical operator into $\hat{\rho}_I = \hat{P}_I \hat{\rho} \hat{P}_I / \mathrm{Tr}\{\hat{\rho}\hat{P}_I\}$. The full measurement process, from the coupling of the system with the apparatus, to the final observation of a specific value of the current (wave-packet reduction) can then be summarized in the two-step process

$$\hat{\rho} \longrightarrow \hat{\rho}_\mathcal{I} = \sum_I \hat{P}_I \hat{\rho} \hat{P}_I \longrightarrow \hat{\rho}_I = \frac{\hat{P}_I \hat{\rho} \hat{P}_I}{\mathrm{Tr}\{\hat{\rho}\hat{P}_I\}}. \quad (1.85)$$

We will see in Sec. 2.8 that the first stage of the current measurement actually leads to a loss of information.[52]

---

[50] This "thought" separation is because we are interested in the properties of the system alone, without concern about the apparatus.

[51] I work here only with the continuum spectrum of the current operator. Extension to the discrete component is straightforward. I also implicitly assume that I have divided the continuum spectrum of the current operator into infinitesimally small intervals δI. The summation 1.84 runs over these intervals.

[52] We also note that, quite generally, the interaction of the system with the apparatus is not

1.6 One current, different viewpoints

I have discussed the different ways we can generate current, anticipated some of the processes we need to take into account to have a comprehensive understanding of its properties, and analyzed the amount of information one can extract from a current measurement.

At this point we may ask the question: isn't this enough to calculate electrical currents? After all, if the system is in a pure state we can solve the time-dependent Schrödinger equation 1.16, and calculate, at any time, the expectation value of the current operator 1.27 over this state.

If the system is in a mixed state but closed, we can solve Eq. 1.60 for the statistical operator $\hat{\rho}(t)$ (or calculate the average 1.63 if the Hamiltonian is known only statistically) from which the current can be calculated according to 1.50 as

$$\boxed{I(t) = e\langle \hat{I} \rangle_t = e\text{Tr}\{\hat{\rho}(t)\hat{I}\}\,,} \qquad (1.86)$$

the pure state being a particular case.

Finally, if the system is open – for instance, if the electrons interact with a given bath or a reservoir – and we know how to determine the statistical operator $\hat{\rho}(t)$ (e.g., if its dynamics can be described by the Lindblad equation 1.73), once again we can calculate the expectation value of the current operator according to 1.86. From this, we can also calculate all possible fluctuations of the current as I will discuss in Chapter 5.

As a matter of principle, we may indeed do all of the above, and our book would be complete at this point (it would be a very short book after all!). There are, however, two issues I need to expand on.

(1) For all practical purposes the procedures I have outlined above, while formally correct, are unfeasible: the non-equilibrium many-body problem we are dealing with is too complicated, and thus requires some simpler mathematical procedure, or outright approximations, in order to have tractable equations we can either solve analytically, or (as is often the case) numerically. In the following chapters, we will see that we have a fundamentally exact way to transform the intractable time-dependent many-body problem into a much simpler effective single-particle problem for the various time-dependent cases considered above. The most important aspect of this trans-

---

limited to the current only, but may be extended to the coupling with any other type of observable, or also to the interaction with a different type of particles. In Sec. 1.4.5 this interaction led us to consider a new equation of motion for the statistical operator of the system coupled to a bath, the Lindblad equation 1.73. We can thus re-interpret that equation as describing the continuous "measurement" of the state of the system via interaction with a bath, without ever actually "observing" the outcome of this measurement (Dalibard *et al.*, 1992; Bushong and Di Ventra, 2007).

1.6 One current, different viewpoints

formation is that it is based on exact theorems that provide solid ground for further approximations.

(2) Approximations and assumptions one reasonably makes to solve the transport problem may also suggest new ways of interpreting electrical conduction, and even provide clues to predict novel phenomena that are "hidden" in the average current 1.86.[53]

Therefore, I now move on to describe the different transport theories that have been suggested over the years, based on definite physical assumptions and approximations. These theories will reveal important basic concepts and properties of electrical conduction, and suggest simpler ways to calculate not only the electrical current, but all other related phenomena as well.

Before doing this, however, let us briefly summarize the *viewpoints* underlying these theories. These viewpoints can be grouped into two main lines of thought:

> **Viewpoint 1.** *The electrical current is a consequence of an applied electric field: the field is the cause, the current is the response to this field.*

> **Viewpoint 2.** *The current flux is determined by the boundary conditions at the surfaces of the sample whose properties we are interested in. Carrier flow incident on the sample boundaries generates, self-consistently, a pileup of carriers at the surfaces of the sample or in its interior. An inhomogeneous electric field is thus generated across the sample. The field is a consequence of carrier flow.*

To Viewpoint 1 belong the classical *Drude conductivity model* (Drude, 1900), the *Kubo formalism* (Kubo, 1959), the *Boltzmann equation* (Boltzmann, 1872), and variations of these theories. Viewpoint 2 is generally referred to as the *Landauer approach* (Landauer, 1957).

What about the discharge of a closed capacitor across a nanojunction I have discussed at the beginning of this chapter? This alternative viewpoint to transport – I call it *micro-canonical* (Di Ventra and Todorov, 2004) for

---

[53] Indeed, the knowledge of the *exact* many-body state of the system is not, by itself, necessarily that useful. This state contains a huge amount of information. Therefore, extracting from it the relevant macroscopic properties of the system would be a formidable task. No less than finding the many-body state itself.

reasons I will discuss in Chapter 7 – falls in between the above two categories. The current discharge is the response to the electric field between the two electrodes. The magnitude of the current flux, however, is determined by the charge pileup and consequent local field at the nanoscale junction.

Summary and open questions

In this chapter I have briefly anticipated the type of physical systems we will be mainly interested in: those of nanoscale dimensions. A key quantity relevant to these structures is the large current density they can carry as opposed to their mesoscopic/macroscopic counterparts. I have also described different ways electrical current can be generated, and discussed – from the point of view of quantum measurement theory – what is measured when one measures currents.

In addition, I have stressed the most important message of this book: *electrical transport is a non-equilibrium statistical problem*. The reason is that electrons are driven out of their equilibrium state, and our experimental knowledge of their time evolution is generally incomplete. For this reason, I have introduced the statistical operator formalism, which will be central in discussing the information content during a measurement of the current, and, in general, the loss of information in the description of transport phenomena.

Finally, I have summarized the main *viewpoints* upon which the transport theories I will discuss in the next chapters are based. They are all equally valid in describing electrical conduction, and each one of them will provide important bits of understanding into the transport problem.

Exercises

1.1 **Properties of the trace.** Given an operator $\hat{A}$, and a basis set $|\phi_k\rangle$, the trace of the operator $\hat{A}$ is defined as

$$\mathrm{Tr}\{\hat{A}\} \equiv \sum_k \langle \phi_k | \hat{A} | \phi_k \rangle, \quad (\text{E1.1})$$

and is independent of the basis set. Prove the following properties of the trace that we make use of in this book:

(a) The trace of a dyadic $|\phi\rangle\langle\psi|$ is

$$\mathrm{Tr}\{|\phi\rangle\langle\psi|\} = \langle\psi|\phi\rangle. \quad (\text{E1.2})$$

(b) Given two operators $\hat{A}$ and $\hat{B}$, the trace is invariant under a cyclic permutation:

$$\mathrm{Tr}\{\hat{A}\hat{B}\} = \mathrm{Tr}\{\hat{B}\hat{A}\}, \quad \text{cyclic property of the trace.} \quad \text{(E1.3)}$$

(c) The trace of a commutator of two operators $\hat{A}$ and $\hat{B}$ is zero:

$$\mathrm{Tr}\{[\hat{A}, \hat{B}]\} = 0. \quad \text{(E1.4)}$$

(d) Given a general operator $\hat{A}$

$$\mathrm{Tr}\{\hat{A}^\dagger\} = [\mathrm{Tr}\{\hat{A}\}]^*. \quad \text{(E1.5)}$$

1.2 **Properties of the statistical operator.** Prove that the statistical operator $\hat{\rho}(t)$ (Eq. 1.48) satisfies the following properties:

(a) It is Hermitian:

$$\hat{\rho}^\dagger = \hat{\rho}. \quad \text{(E1.6)}$$

Using the properties of the trace, prove that this guarantees that the expectation value of a general observable (Eq. 1.50) is a *real* number.

(b) The trace of the statistical operator is unit:

$$\mathrm{Tr}\{\hat{\rho}\} = 1, \quad \text{(E1.7)}$$

and it is preserved at all times by the the Liouville–von Neumann equation 1.60.

(c) The statistical operator is non-negative:

$$\langle \Psi | \hat{\rho} | \Psi \rangle \geq 0, \quad \forall \, |\Psi\rangle \in \mathcal{H}, \quad \text{(E1.8)}$$

where $\mathcal{H}$ is the Hilbert space of the system. Prove that this property implies that the variance of a general operator $\hat{A}$, $(\hat{A} - \langle \hat{A} \rangle)^2$, is positive.

Prove that an operator $\hat{\rho}$ that satisfies properties (a), (b) and (c) can always be written as in Eq. 1.48, and thus represents a density operator.

1.3 **Multiplicity of the statistical operator.** A given statistical operator may be represented by many different macro-states (see Footnote [31]). All of these different representations cannot be distinguished because only the average 1.50 is accessible experimentally.

Consider a spin 1/2 particle prepared in an *unpolarized* state. Write down different microscopic descriptions of its statistical operator. What is the general transformation that relates all these macro-states which produce the *same* statistical operator?

1.4 **Idempotency and pure states.** Show that a statistical operator representing a pure state is *idempotent* at any time

$$\hat{\rho}^2(t) = \hat{\rho}(t), \quad \text{pure state,} \quad (E1.9)$$

which implies that $\text{Tr}\{\hat{\rho}^2(t)\} = 1$. Show that this latter condition is *necessary* and *sufficient* for a statistical operator to represent a pure state. Prove that for a mixed state

$$\text{Tr}\{\hat{\rho}^2(t)\} < 1, \quad \text{mixed state.} \quad (E1.10)$$

2
Drude model, Kubo formalism and Boltzmann equation

In this chapter I discuss the transport theories belonging to **Viewpoint 1**: the current is the response to an electric field (Sec. 1.6). They are ideal to introduce several important concepts I will be making use of in the rest of the book.

2.1 Drude model

In the spirit of **Viewpoint 1**, let us consider an electric field $\mathbf{E}$ that drives a current density $\mathbf{j}$ inside a wire. Let us assume that these two quantities are linearly related. We call the proportionality constant between the field and the current density, *resistivity*, and write[1]

$$\mathbf{E} = \rho \mathbf{j} \qquad (2.1)$$

which is the geometry-independent version of *Ohm's law*

$$V = RI, \quad \text{Ohm's law}, \qquad (2.2)$$

where V is the potential drop along the wire (also called *bias*), I the current that flows across it, and R is its *resistance*. The current density in Eq. 2.1 is the *response* of the system to an external field. Equivalently, we could have written

$$\mathbf{j} = \sigma \mathbf{E}, \qquad (2.3)$$

where $\sigma = 1/\rho$ is the *conductivity*, the inverse of the resistivity, and

$$I = GV \qquad (2.4)$$

with $G = 1/R$ the *conductance*.

---
[1] We will see in Sec. 2.3 that this local relation between the field and the current density is not necessarily valid.

Ohm's law is an example of a *linear-response* approach to conductance: a small field induces a current response in the system that is linearly related to the field itself. The term "small" strictly means that the system can be considered very close to equilibrium. In Sec. 2.3, when introducing the Kubo formalism, all these terms will become clearer, and we will make a connection between linear-response theory and the fluctuations of the system close to equilibrium.

Let us now assume that the conductor is uniform with resistivity ρ and has a constant cross section S and length L. We also consider the field static and uniform. The potential drop along the wire is then $V = EL$. From 1.6 (for steady-state currents) we find $I = Sj$, the current density j being constant everywhere in the wire. From Eqs. 2.1 and 2.2 we finally find

$$R = \frac{\rho L}{S}, \tag{2.5}$$

or for the conductance

$$G = \frac{1}{R} = \frac{1}{\rho}\frac{S}{L} = \frac{\sigma S}{L}. \tag{2.6}$$

We have thus obtained a known classical result: for *ohmic conductors* – those following Ohm's law 2.1 – the resistance increases with the length of the conductor. We will see in Chapter 3 that for mesoscopic and nanoscopic systems this is not necessarily true.

In general, for 2D and 3D systems, the current density may not be parallel to the direction of the field. The resistivity would then be represented by a tensor, i.e., an $N \times N$ matrix with $N = 1, 2, 3$ according to the dimensionality. In this case, if a field component $E_\nu(\mathbf{r})$ is applied in the ν direction and induces a current density j_μ in the μ direction, the linear response relation 2.3 becomes

$$j_\mu = \sum_\nu \sigma_{\mu\nu} E_\nu, \tag{2.7}$$

with $\sigma_{\mu\nu}$ the elements of the conductivity *tensor*. In addition, all the above quantities may depend on time if the driving field depends on time.[2]

For simplicity, let us still assume the electric field static and uniform,

---

[2] For a small, arbitrary electric field, the electrical response may also be *non-local* in space (for an inhomogeneous system) and *history* dependent: the current density at point $\mathbf{r}$ and time t may depend on the conductivity at every other point in space and earlier times. The general current response is thus

$$j_\mu(\mathbf{r}, t) = \sum_\nu \int d\mathbf{r}' \int_{-\infty}^{t} dt' \sigma_{\mu\nu}(\mathbf{r}, \mathbf{r}'; t - t') E_\nu(\mathbf{r}', t').$$

I will discuss this general expression in Sec. 2.3.

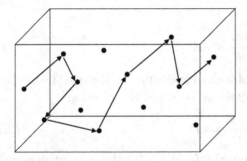

Fig. 2.1. The semiclassical electron path due to scattering at different points in space. The scattering centers are represented by dots.

so that the resistivity and conductivity are independent of time. We also assume the material composing the wire to be isotropic so that the conductivity tensor is diagonal, and we can work with the relation 2.1 (or equivalently 2.3). Let us now relate the resistivity to microscopic quantities, such as the mass m of the electrons, their density n, etc.[3]

Electrons moving under the action of the electric field will experience collisions that will change their momentum $\mathbf{p}$. These collisions could be due to the scattering with ionic vibrations, or collisions with other electrons or impurities (see Fig. 2.1). For the time being the nature of the collisions is irrelevant. However, what is important for the considerations we make here is that all collisions are *independent events*, namely, electrons experience a collision independently of the outcome of previous collisions.[4]

Electrons have an average velocity $\mathbf{v}$. Let us now define the *relaxation time* τ as the average time between two successive collisions. As a consequence of these collisions, in the time interval τ, electrons change, on average, their momentum by the amount

$$\frac{d\mathbf{p}}{dt} = \frac{m\mathbf{v}}{\tau}, \qquad (2.8)$$

where m is the effective mass of the electron in the given material.

We assume that after a collision the electron emerges with a velocity $\mathbf{v}_0$ that is randomly oriented. This implies that on average (over all particles) the randomly oriented velocity immediately after the collision does not contribute to the total average velocity $\mathbf{v}$, i.e., $\langle \mathbf{v}_0 \rangle = 0$.[5] After the collision, and before the next collision occurs, the electron is accelerated by the driving field. Between collisions, the change of momentum must then be

---

[3] Since we are working in linear response, the density here is the density of the system at equilibrium, as will become clear in Sec. 2.3.
[4] We say that a collision at time t is *uncorrelated* with a collision at time t'.
[5] Similar to the zero average of the thermal velocity (see Eq. 1.1 in Sec. 1.2)

proportional to the field

$$\frac{d\mathbf{p}}{dt} = e\mathbf{E}. \tag{2.9}$$

At steady state, if all the momentum change is equal to the one the electrons experience during collisions (Eq. 2.8) we then get the relation[6]

$$\frac{m\mathbf{v}}{\tau} = e\mathbf{E}. \tag{2.10}$$

Comparing 2.10 with 2.1, and using the form 1.5 for the current density we find that the resistivity is

$$\rho = \frac{m}{ne^2\tau}. \tag{2.11}$$

In this model, the conductivity is given by

$$\boxed{\sigma = \frac{1}{\rho} = \frac{ne^2\tau}{m}, \quad \text{Drude conductivity.}} \tag{2.12}$$

This is the Drude relation between the conductivity and the microscopic properties of the conducting electrons. An equivalent relation holds for a frequency-dependent field. I leave the derivation of this case as Exercise 2.1.

The relaxation time is generally the only unknown and can be determined experimentally precisely from the relation 2.12, once the conductivity is measured.[7]

Relaxation times in bulk metallic wires are typically of the order of 10^{-14} s at room temperature and reflect mainly inelastic scattering of electrons by ionic vibrations. The detailed discussion of this type of scattering mechanism in nanoscale systems is discussed in Chapter 5.

2.2 Resistance, coherent and incoherent transport

The above analysis tells us the real physical origin of resistance: *Momentum change*. We can thus give the following physical definition

---

[6] As we will see in Chapter 3, at steady state the momentum *gain* due to the field may not necessarily equal the momentum *loss* due to scattering. For instance, this is the case when charge piles up at defects due to scattering.

[7] The ratio n/m can be obtained from a well-known *sum rule* satisfied by the symmetric part of the Fourier transform in frequency of the conductivity tensor $\sigma^s_{\mu\nu}(\omega) = [\sigma_{\mu\nu}(\omega) + \sigma^T_{\mu\nu}(\omega)]/2$ (where the superscript T means transpose of the tensor) that one measures by applying an electric field $\mathbf{E}(\omega)$ (Kubo, 1959)

$$\frac{2}{\pi}\int_0^{+\infty} d\omega\, \sigma^s_{\mu\nu}(\omega) = \frac{ne^2}{m}\delta_{\mu\nu}.$$

2.2 Resistance, coherent and incoherent transport

Resistance: *Amount of momentum change during collisions.*

In this context, momentum change is also referred to as momentum *relaxation*. The collisions can then be ideally grouped into

- **Elastic:** Momentum relaxes but single-particle energy is conserved.
- **Inelastic:** Momentum *and* energy relax.

Relaxation times can be accordingly defined. For instance, an *elastic relaxation time* is the time between two successive collisions that change electron momentum, but conserve single-particle energy. This is the case, for instance, for scattering off a rigid impurity.

Following the distinction between elastic and inelastic resistance, it is also customary to name the electron dynamics in which the system experiences only elastic scattering, *phase coherent transport*, from the fact that if the single-particle energy is conserved, the phase of the corresponding time-dependent wave-function has a simple oscillatory behavior, with constant frequency, over the whole time evolution. In other words, no phase is "lost" or "gained" during the collision.

This is clearly an idealization, as scattering processes that change single-particle energy are always present to some degree. Electron transport is thus truly *incoherent*. We then define an *energy* or *inelastic relaxation time*

$$\tau_E = \text{average time between two successive scattering events that change energy.} \quad (2.13)$$

In certain literature it is also customary to define as *dephasing processes* those that change the phase without, however, changing the energy considerably. This is assumed to be the case, for instance, for *voltage probes*: devices that measure a voltage difference, without carrying overall current. (I will come back to these probes in the next chapter.)

In this case we define a *coherence time* as

$$\tau_\phi = \text{average time between two successive scattering events that change wave-function phase.} \quad (2.14)$$

If this time is longer than the time it takes electrons to go from one side of the sample to the other then electron transport may, to some degree, be approximated as coherent. As I will discuss in the next chapter this is one of the underlying assumptions of the Landauer approach.

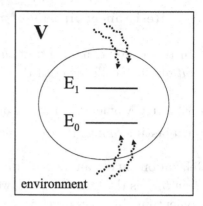

Fig. 2.2. A system with two allowed energy states in interaction with an environment via an operator $\hat{V}$.

2.2.1 Relaxation vs. dephasing

Here I want to provide a somewhat more rigorous definition of dephasing and its relation to energy relaxation. This can be done by means of the statistical operator I have introduced in Sec. 1.4.

Let us start by saying that, in general, dephasing and energy relaxation go hand in hand: usually, one cannot have only dephasing and no energy relaxation at all.[8] To see this, let us consider the following simple example. Let us assume the many-body electron system has only two current-carrying available states, – e.g., two states of a conducting ring threaded by a magnetic field – one with lower energy E_0, and the other with energy E_1, with $E_1 > E_0$.

Let us also assume that the electrons are coupled to a phonon bath, whose only purpose is to provide a finite temperature θ to the electrons. This could be, e.g., a very *dense* (in energy) set of phonons, so that the interaction with the environment is nothing other than the scattering with phonons at all possible energies (van Kampen, 1992). We are interested in the dynamics of the electrons in the presence of this bath (Fig. 2.2). We can therefore use the Lindblad equation 1.72 to study the time evolution of the statistical operator of the electrons. In the energy basis the Hamiltonian is simply

$$H = \begin{pmatrix} E_0 & 0 \\ 0 & E_1 \end{pmatrix}. \qquad (2.15)$$

For a two-level system, the operator $\hat{V}$ appearing in the Lindblad equation 1.72, and describing the bath at temperature θ can be written in the

---

[8] As I will discuss in a moment the time scales associated with dephasing and energy relaxation may be so different that we may, for all practical purposes, assume one of the two processes negligible. For instance, it is usually assumed that inelastic scattering by low-energy phonons in bulk is a very effective dephasing process, thus the corresponding relaxation time is much shorter than the energy relaxation time (Altshuler et al., 1982).

2.2 Resistance, coherent and incoherent transport

energy basis as (van Kampen, 1992)

$$V = \frac{1}{\sqrt{\tilde{\tau}}} \begin{pmatrix} 0 & e^{-\beta E_0/2} \\ e^{-\beta E_1/2} & 0 \end{pmatrix}, \quad (2.16)$$

where $\tilde{\tau}$ is a quantity with dimension of time. That the operator 2.16 describes the effect of a bath at temperature θ will become clearer in a moment.

The statistical operator in the same representation is a 2×2 matrix

$$\rho_e(t) = \begin{pmatrix} \rho_{00}(t) & \rho_{01}(t) \\ \rho_{10}(t) & \rho_{11}(t) \end{pmatrix}. \quad (2.17)$$

The electrons are prepared (for instance by the battery that initiates the current) in some initial statistical mixture defined by some density matrix $\rho_e(t=0)$. We can now solve Eq. 1.72 with Hamiltonian 2.15 and bath operator 2.16. One finds for the diagonal elements of the density matrix (I leave this to the reader as Exercise 2.2)

$$\rho_{00}(t) = \frac{e^{-\beta E_0}}{\mathcal{Z}_C} + \left(\rho_{00}(t=0) - \frac{e^{-\beta E_0}}{\mathcal{Z}_C} \right) e^{-t\mathcal{Z}_C/\tilde{\tau}}, \quad (2.18)$$

$$\rho_{11}(t) = \frac{e^{-\beta E_1}}{\mathcal{Z}_C} + \left(\rho_{11}(t=0) - \frac{e^{-\beta E_1}}{\mathcal{Z}_C} \right) e^{-t\mathcal{Z}_C/\tilde{\tau}}, \quad (2.19)$$

where $\mathcal{Z}_C = e^{-\beta E_0} + e^{-\beta E_1}$ is the canonical partition function (see Eq. 1.78). In the limit of $t \to +\infty$ these diagonal elements tend to

$$\lim_{t \to +\infty} \rho_{00}(t) = \frac{e^{-\beta E_0}}{\mathcal{Z}_C} \equiv \rho_{00}^{eq}, \quad (2.20)$$

$$\lim_{t \to +\infty} \rho_{11}(t) = \frac{e^{-\beta E_1}}{\mathcal{Z}_C} \equiv \rho_{11}^{eq}. \quad (2.21)$$

These are precisely the elements of the canonical statistical operator 1.78 in the energy basis, for a system at equilibrium with a bath at temperature θ. We thus see that the electron system evolves from the initial-time mixture to the canonical equilibrium (Fig. 2.3). In doing so it exchanges energy with the bath: it *relaxes* energy. To see this, let us calculate the expectation value of the energy. From Eq. 1.50 we get

$$\langle \hat{H} \rangle_t = \text{Tr}\{\hat{\rho}(t)\hat{H}\} = \text{Tr} \left\{ \begin{pmatrix} E_0 \rho_{00}(t) & E_1 \rho_{01}(t) \\ E_0 \rho_{10}(t) & E_1 \rho_{11}(t) \end{pmatrix} \right\} = E_0 \rho_{00}(t) + E_1 \rho_{11}(t). \quad (2.22)$$

From this and Eqs. 2.18, 2.19, we see that the average energy relaxation time is $\tau_E = \tilde{\tau}/\mathcal{Z}_C$, and this is the characteristic time scale over which equilibrium is reached.

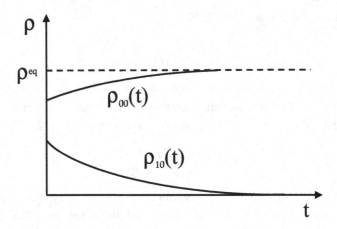

Fig. 2.3. Time dependence of the diagonal $\rho_{00}(t)$ and off-diagonal $\rho_{10}(t)$ matrix elements of the statistical operator.

What happens to the off-diagonal elements of the density matrix? Their expression is a bit more complicated and I leave the full result as an exercise (Exercise 2.2). Here I am interested only in their main time dependence at large times. This is

$$\rho_{01}(t) \propto \rho_{01}(t=0)e^{-t\mathcal{Z}_C/2\tilde{\tau}} + \rho_{10}(t=0)e^{-t\mathcal{Z}_C/2\tilde{\tau}}, \qquad (2.23)$$

$$\rho_{10}(t) \propto \rho_{10}(t=0)e^{-t\mathcal{Z}_C/2\tilde{\tau}} + \rho_{01}(t=0)e^{-t\mathcal{Z}_C/2\tilde{\tau}}. \qquad (2.24)$$

This result is quite different than the one for the diagonal elements of the density matrix. First of all, these off-diagonal elements both tend to zero as $t \to +\infty$ (Fig. 2.3). This is precisely what we have anticipated in Sec. 1.4: the effect of the environment is to introduce *decoherence* or *dephasing* in the system by reducing the density matrix to a diagonal one. We also note that these off-diagonal elements tend to zero with a *different* time constant than the one responsible for energy relaxation (Fig. 2.2). In this model, this relaxation time is $\tau_\phi = 2\tilde{\tau}/\mathcal{Z}_C = 2\tau_E$, and is precisely the dephasing (or decoherence) time I have discussed previously.

In general, however, the decoherence time may be smaller than the energy relaxation time, or, put differently, the ratio τ_ϕ/τ_E depends on the nature of the environment or scattering process, and the system at hand.[9]

From the above discussion, we can conclude that the dephasing time is

---

[9] Macroscopic systems are generally characterized by decoherence times very short compared to relaxation times. It is indeed this property that allows us to describe them classically (Zurek, 1991).

the average time it takes for the off-diagonal elements of the density matrix to decay. This may not necessarily occur exponentially as in the above model. Also, according to the properties of the environment and its interaction with the system, it may be different for the different available states. As we have discussed in Sec. 1.4, when the density matrix evolves into a diagonal one, its elements have a well-defined classical interpretation. We can thus say that the dephasing time is the average time it takes the system to lose quantum correlations (entanglement) and thus tend to behave like a "classical" system.[10]

Finally, what happens to the average current? In the energy basis the current operator $\hat{I}$ (Eq. 1.27) is the 2×2 matrix

$$I = \begin{pmatrix} \langle E_0|\hat{I}|E_0\rangle & \langle E_0|\hat{I}|E_1\rangle \\ \langle E_1|\hat{I}|E_0\rangle & \langle E_1|\hat{I}|E_1\rangle \end{pmatrix} \equiv \begin{pmatrix} I_{00} & I_{01} \\ I_{10} & I_{11} \end{pmatrix}. \qquad (2.25)$$

We can use Eq. 1.86 to get the average current

$$\langle \hat{I} \rangle_t = \mathrm{Tr}\{\hat{\rho}(t)\hat{I}\} = I_{00}\rho_{00}(t) + I_{01}\rho_{10}(t) + I_{10}\rho_{01}(t) + I_{11}\rho_{11}(t). \qquad (2.26)$$

This current thus contains correlations – or *mixing* – between the two many-particle states, represented by the terms $I_{01}\rho_{10}(t)$ and $I_{10}\rho_{01}(t)$. It is only in the limit $t \to +\infty$ that these terms do not contribute, and the average current becomes

$$\lim_{t \to +\infty} \langle \hat{I} \rangle_t = I_{00}\rho_{00}^{eq} + I_{11}\rho_{11}^{eq}, \qquad (2.27)$$

which is simply the sum of *independent* (or *incoherent*) contributions from the currents corresponding to the two states.[11]

If these states are single-particle states we can similarly argue that the effect of the environment is to eliminate their correlated contribution to the current as time evolves. If this is the case, the current is, in the long-time limit, a linear combination of independent currents, one for *each* single-particle state.

In the next chapter I will come back to the concept of mixing of states when discussing the Landauer approach. Here, I conclude this part by stressing once more that in a single-particle picture the "environment" could also be due to the presence of all other electrons. These "other" electrons introduce fluctuations – via electron-electron scattering – in the dynamics of single electrons, and contribute to single-particle wave-function dephasing.

---

[10] This is only an intuitive picture. Due to the interaction with the environment, the system can still evolve/dephase into final states which contain quantum correlations.
[11] Note that if the two states are not current-carrying, $I_{00} = I_{11} = 0$, and the system ends up in a global equilibrium state with no current.

2.2.2 Mean-free path

Another concept (derived from the relaxation time) that will be useful in the following discussions is that of *mean-free path*. The latter is defined as the average distance traveled by electrons between two successive collisions. If the average electron velocity is **v**, and τ is the average time between collisions, then the mean-free path is

$$\lambda = |\mathbf{v}|\tau. \tag{2.28}$$

According to the type of relaxation time we introduce in 2.28 we can define the corresponding mean-free path. For instance, if the above τ is the energy relaxation time τ_E, we call λ the *inelastic* (or *energy*) mean-free path: the average distance traveled by electrons between two successive collisions that change energy.

What velocity **v** should we introduce in 2.28? For electrons in a bulk material, the most natural choice is the *Fermi velocity*, $\mathbf{v}_F$, that corresponding to the highest-occupied single-particle energy level (Ashcroft and Mermin, 1975). If the collisions are uncorrelated from each other, and we assume the electrons form an *ideal gas* (independent electrons, also known as *Fermi gas*) one can indeed show that such a velocity enters in 2.28 (see Sec. 2.3.4).

Considering a typical Fermi velocity of the order of 10^8 cm/s, and an inelastic relaxation time $\tau = 10^{-14}$ s due to collisions with phonons, we get an inelastic mean-free path of 100 Å, larger than most nanoscale structures we consider in this book. By lowering the temperature, the relaxation time due to scattering with ionic vibrations increases, so that the corresponding mean-free path increases accordingly.

The *coherence length* is similarly defined as $\lambda_\phi = |\mathbf{v}|\tau_\phi$. We can thus restate our previous point: If the coherence length is much longer than the length of the sample, then electron transport may, to some degree, be approximated as coherent.

All this seems quite simple and intuitive. The above concepts of relaxation time, mean-free path, etc. are very useful as a starting point to discuss transport in nanoscale systems, and we will indeed expand on them in the following chapters.

However, as we have analyzed above, the "exact" way to derive their value is via the time evolution of the statistical operator. Unfortunately, this procedure is impossible to follow most of the time for the full many-body non-equilibrium problem we are interested in. The reader thus needs to keep in mind that these parameters can, at best, be estimated only in simple cases or under specific physical assumptions (see, e.g., Chapter 6 and

Altshuler et al., 1982; Stern et al., 1990, for some examples of calculations of relaxation times in specific cases).

In addition, they are not the only parameters of interest in describing electron dynamics at nanometer length scales. For instance, I have just discussed that nanostructures are typically much smaller than inelastic meanfree paths. This would naively suggest that all the inelastic effects associated with the loss of energy of the electrons occur away from the nanojuction, deep inside the electrodes.

As I will discuss in Chapters 6 and 8, this is not necessarily the case. In Chapter 1 I have anticipated that the large current density these systems carry is another fundamental quantity to consider. This large current density leads to enhanced inelastic scattering locally in the junction.

2.2.3 The meaning of momentum relaxation time

I have discussed the physical meaning of dephasing and energy relaxation times, but I have not actually discussed our initial (and basic) time, the one related to momentum relaxation, and thus directly to resistance. For inelastic scattering this is the same as the energy relaxation time: both energy and momentum change during the same collision.

We have seen in the above simple example that this is the time it takes the system to reach equilibrium. Since the above reasoning can be generalized to any region of the sample, we can say that this is the time it takes the system to reach *local equilibrium*. I will show this explicitly when I derive the Boltzmann equation (Sec. 2.7).

What if the scattering is elastic? We note first that in this case the momentum relaxation time is *not* the dephasing time. Instead, it is again the time it takes the system to reach local equilibrium. We can see this within the simple Drude picture.

Between collisions the electrons are accelerated by the electric field, and are thus driven away from equilibrium. The elastic collisions stop the otherwise indefinite acceleration, randomizing the electron velocity, and restoring a local constant velocity.

This is an important point, especially for electrons moving into a nanojunction. While in bulk electrodes electrons may avoid isolated impurities by simply moving "around" them, a nanoscale structure between bulk electrodes, like the ones pictured in Figs. 1.1 and 1.2, is *unavoidable*.

Electrons necessarily have to go through the nanostructure, experiencing at least elastic (and in most instances also inelastic) scattering. In the process, electrons have to adjust *dynamically* and *continually* their motion

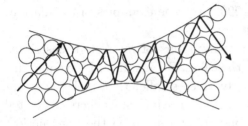

Fig. 2.4. Schematic of the semi-classical trajectory of an electron moving in a nanojunction. The velocity field is the average over all particles.

according to the junction geometry (see Fig. 2.4). This adjustment is the main source of resistance that the nanostructure produces, and, as discussed above, also forces the electron system *continually* into local equilibrium (Di Ventra and Todorov, 2004).

The interesting point is that this relaxation time can be extremely short for nanoscale junctions. We can estimate it as follows.

Let us assume that the nanojunction has a cross section w and an electron wave-packet with a given average momentum of magnitude p moves into it. The wave-packet has to adjust its motion to the given junction geometry in a time $\Delta t \sim \hbar/\Delta E$. Since $\Delta E \approx v\Delta p \sim v\hbar/\sqrt{w}$, where $v = p/m$, then the time it takes for the wave-packet to change momentum in going into the junction is simply $\Delta t \sim \sqrt{w}/v$. For a nanojunction of linear width $\sqrt{w} = 1$ nm, assuming the electron wave-packet moves at a typical velocity of 10^8 cm/s, Δt is of the order of 10^{-15} s, a very fast time scale.

In other words, even in the absence of other inelastic effects, the mere presence of the nanojunction would contribute to fast relaxation of electron momentum, and thus to fast approach to local equilibrium. In Chapter 8 we will see that this effect may lead to interesting phenomena.

2.3 Kubo formalism

Let us now expand on the Drude approach with a much more rigorous formalism, still belonging to **Viewpoint 1** (see Sec. 1.6). This approach goes under the name of *Kubo*, or *linear response formalism* (Kubo, 1959). My goal here, as in the Drude model, is to determine the current response to a weak electric field.

The experimental situation I have in mind is a system (a sample) initially in *global* equilibrium subject to an external perturbation – typically an electromagnetic field. I want to describe the *deviations* from equilibrium – electrical current formation, in particular – induced by the external perturbation by taking full advantage of quantum mechanics.

2.3 Kubo formalism

For this purpose, the formalism of statistical operators I have described in Sec. 1.4 comes in really handy. Let us then outline the assumptions underlying the Kubo approach. Some of them also form the basis of the non-equilibrium Green's function formalism I describe in Chapter 4.

Assumption 1: *Closed quantum systems*

The Kubo formalism I develop here refers to systems evolving under Hamiltonian dynamics, namely closed quantum systems, possibly in a mixed state (see Sec. 1.4.3). However, these systems are not isolated: they are subject to external *deterministic* forces.

I therefore solve (albeit approximately) the Schrödinger equation 1.16 – or its mixed-state version, the Liouville–von Neumann equation 1.60 – and not an equation of motion for open systems like, e.g., the Lindblad equation 3.4.

I work out the general theory, and at the same time refer, as an example, to the specific transport problem. Let us then consider a system with Hamiltonian $\hat{H}_0$. For electrons in a conductor we may take their many-body Hamiltonian

$$\hat{H}_0 = \frac{1}{2m}\sum_i \hat{p}_i^2 + \frac{1}{2}\sum_{i\neq j} W(\hat{r}_i - \hat{r}_j) + \sum_i \sum_j V_{el-ion}(\hat{r}_i - \mathbf{R}_j), \quad (2.29)$$

where $W(\hat{r}_i - \hat{r}_j)$ is the Coulomb potential describing two-particle interactions (see also Eq. 2.96), and $V_{el-ion}(\hat{r}_i - \mathbf{R}_j)$ describes the electron interaction with a static ion at position $\mathbf{R}_j$ (see also Chapter 6).

Assumption 2: *Initial global equilibrium*

The system is at the initial time[12] $t = t_0$ in *global* canonical equilibrium with a bath at temperature θ, namely it is described by the canonical statistical operator (see Eq. 1.78)[13]

$$\hat{\rho}(t \leq t_0) = \frac{e^{-\beta \hat{H}_0}}{\mathcal{Z}_C} \equiv \hat{\rho}_C^{eq}. \quad (2.30)$$

---

[12] It is convenient to take $t_0 \to -\infty$ but this is not a requirement.
[13] From a formal point of view, this is not really a necessary assumption: a different choice of initial conditions may be possible. However, many simplifications that practically help in the determination of the conductivity cannot be made if the system is not initially in global equilibrium. If this were not the case, one may also run into serious conceptual problems. For instance, if the initial state is an unstable steady state, even a small perturbation may drive the system out of the steady state.

Assumption 3: *Small perturbations*

At time t_0, we apply a time-dependent perturbation $\hat{H}'(t)$. The statistical operator $\hat{\rho}(t)$ thus changes upon this perturbation. We nonetheless assume that the perturbation is so weak that the statistical operator $\hat{\rho}(t)$ differs only slightly from the canonical equilibrium 2.30. The total Hamiltonian is for $t > t_0$

$$\hat{H}(t) = \hat{H}_0 + \hat{H}'(t). \tag{2.31}$$

Let us also assume this perturbation can be written in the form[14]

$$\hat{H}'(t) = \sum_i \hat{B}_i \lambda_i(t). \tag{2.32}$$

In the specific case of electrical transport, this perturbation can be obtained as follows. We apply an electromagnetic field with vector potential $\mathbf{A}(\hat{\mathbf{r}}, t)$ so that the Hamiltonian 2.29 is modified into

$$\hat{H}_{tot}(t) = \frac{1}{2m} \sum_i \left[\hat{p}_i - \frac{e}{c}\mathbf{A}(\hat{r}_i, t)\right]^2 + \frac{1}{2} \sum_{i \neq j} W(\hat{r}_i - \hat{r}_j) + \sum_{i,j} V_{el-ion}(\hat{r}_i - \mathbf{R}_j). \tag{2.33}$$

Since, by assumption, the field is small we can expand 2.33 and keep only the linear terms in $\mathbf{A}(\hat{r}, t)$. By making use of the definition 1.26 of the paramagnetic current density operator we can finally write[15]

$$\hat{H}_{tot}(t) \simeq H_0 - \frac{e}{c} \int d\mathbf{r}\, \mathbf{j}_p(\hat{\mathbf{r}}) \cdot \mathbf{A}(\hat{\mathbf{r}}, t) \equiv \hat{H}_0 + \hat{H}'(t). \tag{2.34}$$

The perturbation term in the above equation is of the same form as in 2.32, where $\hat{B}$ corresponds to the paramagnetic current density operator, and $\lambda(t)$ to the "field" $-e\mathbf{A}(\hat{\mathbf{r}}, t)/c$.

Assumption 4: *Adiabatic approximation*

As anticipated in Assumption 1, even if we allow for the presence of a bath, we work with the Liouville equation 1.60 for closed quantum systems, and not with equations of motions for open quantum systems (such as the Lindblad equation 1.72). This is quite a strong physical approximation.

It means that the time evolution of the system induced by the perturbation $\hat{H}'(t)$ is *fast* compared to the energy relaxation time τ_E that drives the system to global equilibrium, and no transition between energy states of the unperturbed Hamiltonian $\hat{H}_0$ is thus allowed by the presence of the bath.

---

[14] Since we are working in linear response, the contributions from the different perturbations add independently and it is thus enough to work with just one term in the sum 2.32.
[15] We work in a gauge in which the scalar potential is zero at any time (see Appendix F).

2.3 Kubo formalism

This is called the *adiabatic approximation* and we will make such an approximation also when discussing the non-equilibrium Green's function approach in Chapter 4. This approximation also entails an *adiabatic switch-on* of the perturbation $\hat{H}'(t)$ at time t_0 so that, once again, no transition between the energy states of the unperturbed Hamiltonian $\hat{H}_0$ is induced by the interaction with the bath at this initial time.

In large parts of this book we will be concerned with the limit $\omega \to 0$, the *static* (also called d.c. – for direct current) limit, that corresponds to the long-time limit of the perturbation. The longer this time, the easier it is for the bath to induce transitions between states of $\hat{H}_0$, thus invalidating one of the basic assumptions of the formalism. An important point to keep in mind when working with the Kubo formalism and comparing its results with experiments in the d.c. limit.

Under the above assumptions, I now want to determine the change of the expectation value of any observable $\hat{A}$ – in our case, the total current density $\hat{\mathbf{j}}$, Eq. 1.22 – from its equilibrium value. Using the definition 1.50 our goal is to find

$$\Delta \langle \hat{A} \rangle_t \equiv \langle \hat{A} \rangle_t - \langle \hat{A} \rangle_{eq} = \mathrm{Tr}\{\hat{\rho}(t)\hat{A}\} - \mathrm{Tr}\{\hat{\rho}_C^{eq}\hat{A}\}, \qquad (2.35)$$

which for the current density reads

$$\Delta \langle \hat{\mathbf{j}} \rangle_t \equiv \langle \hat{\mathbf{j}} \rangle_t - \langle \hat{\mathbf{j}} \rangle_{eq} = \mathrm{Tr}\{\hat{\rho}(t)\hat{\mathbf{j}}\} - \mathrm{Tr}\{\hat{\rho}_C^{eq}\hat{\mathbf{j}}\} = \mathrm{Tr}\{\hat{\rho}(t)\hat{\mathbf{j}}\}, \qquad (2.36)$$

where the last equality simply states that the average current is zero at equilibrium.

Let us then write the statistical operator in the *interaction picture*[16]

$$\hat{\rho}(t)_I \equiv e^{\frac{i}{\hbar}\hat{H}_0 t} \hat{\rho}(t) e^{-\frac{i}{\hbar}\hat{H}_0 t}. \qquad (2.37)$$

The reverse equation is

$$\hat{\rho}(t) = e^{-\frac{i}{\hbar}\hat{H}_0 t} \hat{\rho}(t)_I e^{\frac{i}{\hbar}\hat{H}_0 t}. \qquad (2.38)$$

We also write the perturbation $\hat{H}'$ in the same representation

$$\hat{H}'(t)_I \equiv e^{\frac{i}{\hbar}\hat{H}_0 t} \hat{H}'(t) e^{-\frac{i}{\hbar}\hat{H}_0 t}. \qquad (2.39)$$

If we replace 2.38 into the Liouville equation 1.60 we easily get the equation of motion for the statistical operator in the interaction picture

$$i\hbar \frac{d\hat{\rho}(t)_I}{dt} = [\hat{H}'(t)_I, \hat{\rho}(t)_I]. \qquad (2.40)$$

---

[16] Not to be confused with the statistical operator 1.84 I introduced to discuss the current measurement. Together with the Schrödinger and Heisenberg pictures, this is the third picture used in this book (there are no more!).

A formal solution of this equation is

$$\hat{\rho}(t)_I = \rho(t_0)_I - \frac{i}{\hbar} \int_{t_0}^{t} dt' \, [\hat{H}'(t')_I, \hat{\rho}(t')_I], \qquad (2.41)$$

which can be checked by direct differentiation, and recalling that the equilibrium statistical operator does not change in time.

Since we are assuming the statistical operator differs only slightly from its global equilibrium value we can replace in the integrand of 2.41 $\hat{\rho}(t)_I$ with $\hat{\rho}(t_0)_I$ (*first Born approximation*). The solution of the Liouville equation 2.40 to *first order* in the perturbation is thus[17]

$$\hat{\rho}(t)_I \simeq \rho(t_0)_I - \frac{i}{\hbar} \int_{t_0}^{t} dt' \, [\hat{H}'(t')_I, \hat{\rho}(t_0)_I]. \qquad (2.42)$$

Using the reverse relation 2.38, taking into account that $\hat{H}_0$ commutes with the equilibrium statistical operator $\hat{\rho}_C^{eq}$, and writing $\hat{H}'(t) = \hat{B}\lambda(t)$, this reads

$$\hat{\rho}_{\text{Kubo}}(t) \simeq \rho(t_0) - \frac{i}{\hbar} \int_{t_0}^{t} dt' \, e^{\frac{i}{\hbar}\hat{H}_0(t'-t)} [\hat{B}, \hat{\rho}(t_0)] e^{-\frac{i}{\hbar}\hat{H}_0(t'-t)} \lambda(t'). \qquad (2.43)$$

If we replace this result into 2.35 we get

$$\Delta\langle\hat{A}\rangle_t = -\frac{i}{\hbar} \text{Tr} \left\{ \int_{t_0}^{t} dt' \, e^{\frac{i}{\hbar}\hat{H}_0(t'-t)} [\hat{B}, \hat{\rho}(t_0)] e^{-\frac{i}{\hbar}\hat{H}_0(t'-t)} \, \hat{A} \right\} \lambda(t')$$

$$= -\frac{i}{\hbar} \int_{t_0}^{t} dt' \, \text{Tr} \left\{ [\hat{B}, \hat{\rho}_C^{eq}] e^{-\frac{i}{\hbar}\hat{H}_0(t'-t)} \, \hat{A} \, e^{\frac{i}{\hbar}\hat{H}_0(t'-t)} \right\} \lambda(t'), \qquad (2.44)$$

where in the last equality I have used the cyclic property of the trace (Eq. E1.3 in Exercise 1.1).

Let us define the interaction representation of the operator $\hat{A}$

$$\hat{A}(t - t') \equiv e^{\frac{i}{\hbar}\hat{H}_0(t-t')} \hat{A} e^{-\frac{i}{\hbar}\hat{H}_0(t-t')}, \qquad (2.45)$$

and replace it into 2.44. By making use of the cyclic property of the trace once more, we finally get

$$\Delta\langle\hat{A}\rangle_t = -\frac{i}{\hbar} \int_{t_0}^{t} dt' \, \text{Tr} \left\{ \hat{\rho}_C^{eq} [\hat{A}(t - t'), \hat{B}] \right\} \lambda(t')$$

$$= -\frac{i}{\hbar} \int_{t_0}^{t} dt' \, \text{Tr} \left\{ \hat{\rho}_C^{eq} [\hat{A}(t), \hat{B}(t')] \right\} \lambda(t'), \qquad (2.46)$$

---

[17] Note that for an arbitrary perturbation $\hat{H}'(t)$, the approximate statistical operator 2.42 may lose at a given time its fundamental property of positivity (see Sec. 1.4). We exclude here such cases.

with $\hat{B}(t)$ and $\hat{A}(t)$ the interaction representation 2.45 of the respective operators.

We can extend the above formalism to the initial time $t_0 \to -\infty$, provided the perturbation tends to zero in the same limit. We can also extend the upper limit of integration to $+\infty$ by defining the *retarded response function*

$$\chi_{AB}(t-t') \equiv -\frac{i}{\hbar}\Theta(t-t')\text{Tr}\left\{\hat{\rho}_C^{eq}[\hat{A}(t-t'),\hat{B}]\right\} \quad (2.47)$$

with $\Theta(t-t')$ the step function, which vanishes for $t < t'$, and is equal to 1 for $t > t'$. We finally obtain the linear-response relation

$$\boxed{\Delta\langle\hat{A}\rangle_t = \int_{-\infty}^{+\infty} dt' \, \chi_{AB}(t-t')\lambda(t').} \quad (2.48)$$

We thus see that the perturbation $\hat{H}'(t) = \hat{B}\lambda(t)$ modifies the expectation value of the observable $\hat{A}$ at time t by taking into account the whole *history* of the effect of that perturbation on the observable $\hat{A}$ at previous times. That is the reason the response function 2.47 is called retarded.

2.3.1 The current-current response function

We can now apply what we have just obtained to the transport problem. The perturbation is given in Eq. 2.34, and we are interested in the change of the current density, $\hat{j} = \hat{A}$. As shown in Eq. 1.25, this can be written as the sum of paramagnetic and diamagnetic current densities. The latter is already linear in the vector potential, so it is enough to calculate the response of the paramagnetic current, and sum the two terms at the end. For the paramagnetic current density $\hat{\mathbf{j}}_p$ we thus obtain from 2.48 the response in the direction μ induced by the field in the direction ν

$$j_{p\mu}(\mathbf{r},t) \equiv \Delta\langle\hat{j}_{p\mu}(\mathbf{r})\rangle_t = -\frac{e}{c}\sum_\nu \int d\mathbf{r}' \int_{-\infty}^{+\infty} dt' \, \chi_{\hat{j}_{p\mu},\hat{j}_{p\nu}}(\mathbf{r},\mathbf{r}',t-t')A_\nu(\mathbf{r}',t'), \quad (2.49)$$

where

$$\chi_{\hat{j}_{p\mu},\hat{j}_{p\nu}}(\mathbf{r},\mathbf{r}',t-t') = -\frac{i}{\hbar}\Theta(t-t')\text{Tr}\left\{\hat{\rho}_C^{eq}[\hat{j}_{p\mu}(\mathbf{r},t-t'),\hat{j}_{p\nu}(\mathbf{r}')]\right\} \quad (2.50)$$

is the paramagnetic *current-current response function*. By adding the diamagnetic contribution we obtain the total response function

$$\chi^j_{\mu\nu}(\mathbf{r},\mathbf{r}',t-t') = \chi_{\hat{j}_{p\mu},\hat{j}_{p\nu}}(\mathbf{r},\mathbf{r}',t-t') + \frac{n_{eq}}{m}\delta(\mathbf{r}-\mathbf{r}')\delta_{\mu\nu}, \quad (2.51)$$

with $n_{eq} = \text{Tr}\{\hat{\rho}_C^{eq}\hat{n}\}$ the density of the system at equilibrium without the perturbation.

We thus confirm one of the points I have raised in Chapter 1: in general, in order to calculate transport properties of a physical system, we cannot ignore its past dynamics! The current response at a given time t contains information on the previous evolution.

The frequency-dependent conductivity

The conductivity can be read directly from Eqs. 2.50 and 2.51, by taking into account the relation between the electric field and vector potential, which in the present gauge is (the vector potential is, by assumption, zero at $t = t_0$)

$$\mathbf{E}(\mathbf{r},t) = -\frac{1}{c}\frac{\partial \mathbf{A}(\mathbf{r},t)}{\partial t} \iff \mathbf{A}(\mathbf{r},t) = -c\int_{t_0}^{t} dt'\, \mathbf{E}(\mathbf{r},t'). \tag{2.52}$$

The expression simplifies enormously in the case of a periodic perturbation[18]

$$\mathbf{A}(\mathbf{r},t) = \mathbf{A}(\mathbf{k},\omega)e^{i(\mathbf{k}\cdot\mathbf{r}-\omega t)} + \text{c.c.}, \tag{2.53}$$

where c.c. means "complex conjugate" of the previous term. In this case, we can define the Fourier transform of the terms appearing in Eqs. 2.49 and 2.51. For instance, the Fourier transform of the response function 2.51 is

$$\chi_{\mu\nu}^{j}(\mathbf{r},\mathbf{r}',t-t') = \int_{-\infty}^{+\infty} \frac{d\omega}{2\pi} e^{-i\omega(t-t')} \sum_{\mathbf{k}} e^{i\mathbf{k}\cdot\mathbf{r}} \sum_{\mathbf{k}'} e^{-i\mathbf{k}'\cdot\mathbf{r}'} \chi_{\mu\nu}^{j}(\mathbf{k},\mathbf{k}',\omega). \tag{2.54}$$

The Fourier transform of the vector potential 2.52 is simply related to the Fourier transform of the electric field via

$$\mathbf{A}(\mathbf{k},\omega) = -\frac{ic}{\omega}\mathbf{E}(\mathbf{k},\omega), \tag{2.55}$$

so that the total response of the physical current in Fourier space is from 2.49 and 2.51

$$ej_\mu(\mathbf{k},\omega) = \frac{ie^2}{\omega}\sum_\nu\sum_{\mathbf{k}'} \chi_{\mu,\nu}^{j}(\mathbf{k},\mathbf{k}',\omega)E_\nu(\mathbf{k}',\omega)$$

$$\equiv \sum_\nu\sum_{\mathbf{k}'} \sigma_{\mu,\nu}(\mathbf{k},\mathbf{k}',\omega)E_\nu(\mathbf{k}',\omega), \tag{2.56}$$

---

[18] A periodic perturbation is indeed what is generally realized in experiments, so this case is the most important.

where we have defined the conductivity tensor[19]

$$\sigma_{\mu,\nu}(\mathbf{k},\mathbf{k}',\omega) = \frac{ie^2}{\omega}\chi^j_{\mu,\nu}(\mathbf{k},\mathbf{k}',\omega) = \frac{ie^2}{\omega}\left[\chi_{j_{p\mu},j_{p\nu}}(\mathbf{k},\mathbf{k}',\omega) + \frac{n_{eq}}{m}\delta_{\mu\nu}\right].$$
(2.57)

This is the quantum-mechanical generalization of the Drude conductivity 2.12. It reduces to that result if one assumes independent electrons interacting with a set of random impurities whose effect on each electron is independent of the electron scattering with the other impurities.[20]

2.3.2 The use of Density-Functional Theory in the Kubo approach

Let us stop for a moment and discuss here how one could actually *calculate* the above current response. (It is always easier to write down an equation, like the Schrödinger equation 1.16, than actually solve it for a real material!) In the next section, I will show, using as example the response to a static uniform electric field, that in the case of independent electrons the calculation of the current response can be done analytically. On the other hand, for an arbitrary interacting many-body system it still remains an outstanding problem, and we refer the reader to other books for some of the analytical techniques used to tackle this problem (Giuliani and Vignale, 2005).

Due to the growing importance of density-functional theory (DFT) in transport theories, here I want to discuss the possible use of DFT in combination with the Kubo formalism.[21] In Appendix D I provide a primer of the main concepts of ground-state DFT. In Appendices E and F I discuss the ideas behind its generalization to the time-dependent case for closed quantum systems, and in Appendix G its generalization to dynamical open quantum systems. I really urge the reader not familiar with the basic tenets of DFT (and interested in this section) to stop here for a moment, and take a look at these appendices.

If we use ground-state DFT we know that we can, in principle, find the *exact* ground-state density associated with the Hamiltonian $\hat{H}_0$ (Eq. 2.29) of the full Hamiltonian $\hat{H}(t) = \hat{H}_0 + \hat{H}'(t)$ (Eq. 2.34) using effective single-particle equations (Eqs. D.10) known as the *Kohn–Sham equations*. This

---

[19] It is only when the system is homogeneous (i.e., translationally invariant) that the conductivity tensor depends only on one $\mathbf{k}$ vector. This includes also the case in which the system is disordered, but the disorder does not introduce a dependence on the position of the impurities.
[20] For an explicit derivation of this result, see, e.g., Giuliani and Vignale, 2005.
[21] I will discuss DFT theorems on the total current in Chapter 7.

system of *auxiliary* independent particles is known as the *Kohn–Sham system*.

The above result is also valid for a system in (grand-)canonical equilibrium, namely we can, in principle, find a statistical operator $\hat{\rho}_{gs}^{KS}$ (which is a functional of the equilibrium electronic density) that gives the correct density of the interacting many-body system (see Appendix D)[22]

$$n_{eq} = \text{Tr}\{\hat{\rho}_C^{eq}\hat{n}\} = \text{Tr}\{\hat{\rho}_{gs}^{KS}\hat{n}\}. \tag{2.58}$$

This auxiliary non-interacting statistical operator would also provide the correct average energy at equilibrium

$$\langle \hat{H}_0 \rangle_{eq} = \text{Tr}\{\hat{\rho}_C^{eq}\hat{H}_0\} = \text{Tr}\{\hat{\rho}_{gs}^{KS}\hat{H}_0\}. \tag{2.59}$$

The result 2.58 would take care of the density response component of the full current response 2.57. However, this is the best we can do with ground-state DFT: the paramagnetic current-current response function cannot be calculated with the equilibrium statistical operator $\hat{\rho}_{gs}^{KS}$, even if we knew it exactly, namely

$$\mathbf{j}_{\text{Kubo}}(\mathbf{r}, t) = \text{Tr}\{\hat{\rho}_{\text{Kubo}}(t)\hat{\mathbf{j}}_p\} - \frac{e}{mc}\text{Tr}\{\hat{\rho}_C^{eq}\hat{n}\}\mathbf{A}(\mathbf{r}, t)$$
$$\neq \text{Tr}\{\hat{\rho}_{gs}^{KS}\hat{\mathbf{j}}_p\} - \frac{e}{mc}\text{Tr}\{\hat{\rho}_{gs}^{KS}\hat{n}\}\mathbf{A}(\mathbf{r}, t), \tag{2.60}$$

showing once more that we need a time-dependent approach to evaluate transport properties.

What if we use time-dependent current density-functional theory (Appendix F)? This theory guarantees that, given an initial condition on the state of the system and the external vector potential $\mathbf{A}(\mathbf{r}, t)$, the current density $\mathbf{j}(\mathbf{r}, t)$ of the many-body system is given *exactly* by the corresponding quantity in the Kohn–Sham system calculated using effective time-dependent single-particle equations (Eqs. F.1 of Appendix F), if one knows the exact functional:

$$\mathbf{j}^{KS}(\mathbf{r}, t) = \sum_{\alpha=1}^{N} \left[\frac{\hbar}{2im} \left([\phi_\alpha^{KS}(\mathbf{r})]^* \nabla_\mathbf{r} \phi_\alpha^{KS}(\mathbf{r}) - \phi_\alpha^{KS}(\mathbf{r}) \nabla_\mathbf{r}[\phi_\alpha^{KS}(\mathbf{r})]^* \right) \right.$$
$$\left. - \frac{e}{mc}|\phi_\alpha^{KS}(\mathbf{r}, t)|^2 \mathbf{A}_{eff}(\mathbf{r}, t) \right]$$
$$= \mathbf{j}(\mathbf{r}, t), \tag{2.61}$$

where the sum runs over the N orthonormal occupied one-electron Kohn–Sham

---

[22] I work here in the canonical equilibrium but these results are valid also in the grand-canonical equilibrium. I also assume to know the *exact* ground-state functional of DFT.

orbitals ϕ_α^{KS}, and

$$\mathbf{A}_{eff}(\mathbf{r},t) = \mathbf{A}(\mathbf{r},t) + \mathbf{A}_{xc}(\mathbf{r},t) \qquad (2.62)$$

is an effective vector potential, sum of the external and exchange-correlation vector potentials, with the latter including all many-body effects due to the Pauli exclusion principle (exchange part) and quantum correlations among electrons (correlation part). The exchange-correlation vector potential depends on the current density $\mathbf{j}(\mathbf{r},t')$ at times $t' \leq t$ (i.e., it is *history* dependent).

The Kubo formalism relates to the first-order expansion of the statistical operator (approximation 2.42). One then needs to do the same in the auxiliary Kohn–Sham system, i.e., we need to do a perturbation expansion of the exact Kohn–Sham density matrix and stop at the first order. The zero-order term of this expansion provides the current-current response function, $\chi_{KS}(\omega)$, of the *static* Kohn–Sham system, namely the one that yields the *exact* ground-state density.

This, however, neglects *dynamical* correlations pertaining to the linear response of the true many-body system. Therefore, the relation between the true interacting current-current response function and the equivalent Kohn–Sham quantity needs to take into account these extra correlations.

I state here this relation without demonstrating it for a homogeneous electron gas in the presence of a periodic perturbation 2.53. For a homogeneous system we can work in momentum $\mathbf{k}$ space, and we can separate the response into a *longitudinal* component χ_L (parallel to $\mathbf{k}$) and a *transverse* component χ_T (perpendicular to $\mathbf{k}$). One then finds that the current-current response function, χ, of the true many-body system is related to that in the Kohn–Sham system, χ_{KS}, via the relations (Ullrich and Vignale, 2002)

$$\chi_L^{-1}(k,\omega) = \chi_{KS,L}^{-1}(k,\omega) - \frac{k^2}{\omega}[v_H(k) + f_{xc,L}(k,\omega)], \qquad (2.63)$$

and

$$\chi_T^{-1}(k,\omega) = \chi_{KS,T}^{-1}(k,\omega) - \frac{k^2}{\omega}f_{xc,T}(k,\omega), \qquad (2.64)$$

where $v_H(k)$ is proportional to the Fourier transform of the direct Coulomb interaction potential $v_H(\mathbf{r})$ I will define in Sec. 2.4 (Eq. 2.95), and $f_{xc,L(T)}$ are the longitudinal and transverse components of a quantity known as the *exchange-correlation kernel*, whose role is precisely to compensate for the correlations missing in the Kohn–Sham response function χ_{KS} with respect to the many-body response function χ.[23]

---

[23] Note that an extra complication arises in the case of a finite system subject to weak *periodic*

Without this kernel term, the use of DFT in combination with the Kubo formalism for the current response must be considered, at best, a type of *mean-field approximation*,[24] where single electrons experience a mean field (albeit with some quantum correlations) from all other electrons.[25]

I will show in Sec. 8.4.2 that the terms related to the exchange-correlation kernel give rise to important *dynamical* effects in the resistance of nanoscale systems. These effects are related to the *viscous* nature of the electron liquid.

2.3.3 The fluctuation-dissipation theorem

I now want to discuss a very important result of the Kubo formalism that once again shows that electrical transport is not an equilibrium problem, and thus cannot be derived from ground-state calculations.

In classical statistical mechanics there exists an important relation known as the *fluctuation-dissipation theorem* that states the following (Kubo et al., 1985; van Kampen, 1992)

Theorem: *The response of a system close to global equilibrium, subject to a small external perturbation, is equivalent to its spontaneous fluctuations.*

Put differently, the dissipation that drives the system *towards* global equilibrium is accompanied by fluctuations that tend to displace the system *away* from its equilibrium state. The two effects combine to give the correct expectation value of the given observable (in our case, the current).

We can derive a similar relation for a quantum mechanical system using the Kubo formalism we have just derived. I leave the general result as an exercise (Exercise 2.10), and I will come back to it in Chapter 5 when discussing thermal noise. Here I want to derive it for the electrical current in the presence of a static and uniform electric field.

First of all, it is easy to show the following general commutation relation between the canonical statistical operator 2.30 and any observable $\hat{B}$

---

external perturbations (those usually employed in linear-response theory). The reason is that one can find examples in which two *different* such perturbations produce the *same* linear response, thus violating the theorems of dynamical DFTs (Gross at al., 1988).

[24] I will use mean-field approximations several times in this book. Indeed, when discussing *independent* electrons I will always consider electrons that are either truly independent, or interacting via a mean field provided by the other electrons. In Sec. 4.2.4 of Chapter 4 I will give a formal definition of mean-field approximation.

[25] Note that I am not saying that DFT (in all its formulations: ground-state, dynamic and stochastic) is a mean-field theory (it is not!). I am just saying that its use in the context of the Kubo formalism for conduction, when the exchange-correlation kernel in 2.63 and 2.64 is neglected, can be thought of as a type of mean-field approximation.

(Exercise 2.4)

$$\left[\hat{\rho}_C^{eq}, \hat{B}\right] = i\hbar \int_0^\beta d\tau\, \hat{\rho}_C^{eq} \frac{d\hat{B}(-i\hbar\tau)}{dt}, \qquad (2.65)$$

where

$$\frac{d\hat{B}(-i\hbar\tau)}{dt} \equiv e^{\tau \hat{H}_0} \frac{d\hat{B}(t)}{dt} e^{-\tau \hat{H}_0} \qquad (2.66)$$

with τ a real number, and $\hat{B}(t)$ the interaction representation 2.45 of the operator $\hat{B}$.[26] From Eq. 2.46 we thus find ($t_0 \to -\infty$)

$$\begin{aligned}
\Delta \langle \hat{A} \rangle_t &= -\frac{i}{\hbar} \int_{-\infty}^t dt'\, \text{Tr}\left\{[\hat{\rho}_C^{eq}, \hat{A}(t-t')]\hat{B}\right\} \lambda(t') \\
&= \frac{i}{\hbar} \int_{-\infty}^t dt'\, \text{Tr}\left\{\hat{A}(t-t')[\hat{\rho}_C^{eq}, \hat{B}]\right\} \lambda(t') \\
&= \frac{i}{\hbar} \int_0^{+\infty} dt''\, \text{Tr}\left\{\hat{A}(t'')[\hat{\rho}_C^{eq}, \hat{B}]\right\} \lambda(t-t'') \\
&= -\int_0^{+\infty} dt'' \int_0^\beta d\tau\, \text{Tr}\left\{\hat{\rho}_C^{eq} \frac{d\hat{B}(-i\hbar\tau)}{dt''} \hat{A}(t'')\right\} \lambda(t-t''),
\end{aligned}$$
$$(2.67)$$

where in the third equality I have made the change of variables $t'' = t - t'$, and I have used the identity 2.65 in the last step. By comparing this result with Eq. 2.48 we thus see that we can write the response function as

$$\chi_{AB}(t-t') \equiv -\int_0^\beta d\tau\, \text{Tr}\left\{\hat{\rho}_C^{eq} \frac{d\hat{B}(-i\hbar\tau)}{dt'} \hat{A}(t-t')\right\}. \qquad (2.68)$$

Response to a uniform and static electric field

The above result does not seem very transparent. To clarify it let us apply it to the response to a uniform and static electric field $\mathbf{E}$. Here, we need to determine only the paramagnetic current response. The perturbation to $\hat{H}_0$ is thus

$$\hat{H}' = -e \sum_i \int d\mathbf{r}\, \delta(\mathbf{r} - \hat{\mathbf{r}}_i) \mathbf{r} \cdot \mathbf{E} = -e \int d\mathbf{r}\, \hat{n}(\mathbf{r}) \mathbf{r} \cdot \mathbf{E}, \qquad (2.69)$$

where $\mathbf{r}_i$ are the positions of the particles, and I have used the definition 1.24

---

[26] I am thus working with "complex" times up to a value $i\hbar\beta$. I will be using this "trick" again in Chapter 4 when discussing the non-equilibrium Green's function formalism.

of number density operator. The operator $\hat{B}$ is now the vector $\int d\mathbf{r}\, \hat{n}(\mathbf{r})\mathbf{r}$, and $\lambda = -e\mathbf{E}$.[27]

The time derivative of $\hat{B}$ can be calculated from the continuity equation 1.11, which in operator form reads[28]

$$\frac{\partial \hat{n}(\mathbf{r},t)}{\partial t} = -\nabla \cdot \hat{\mathbf{j}}(\mathbf{r},t). \tag{2.70}$$

By assuming that the current density vanishes at infinity (finite-system assumption[27]) and integrating by parts we find

$$\frac{d\hat{B}}{dt} = \frac{\partial}{\partial t}\int d\mathbf{r}\, \hat{n}(\mathbf{r},t)\mathbf{r} = -\int d\mathbf{r}\, \nabla \cdot \hat{\mathbf{j}}(\mathbf{r},t)\mathbf{r} = \int d\mathbf{r}\, \hat{\mathbf{j}}(\mathbf{r},t) \equiv \hat{\mathbf{j}}(t). \tag{2.71}$$

Finally, from Eq. 2.67 the change of the expectation value of the physical current density in the direction μ to a uniform and static field in direction ν is

$$ej_\mu = e^2 \sum_\nu \int_0^{+\infty} dt' \int_0^\beta d\tau\, \mathrm{Tr}\left\{\hat{\rho}_C^{eq}\, \hat{j}_{p\nu}(-i\hbar\tau)\, \hat{j}_{p\mu}(t')\right\} E_\nu, \tag{2.72}$$

from which the conductivity tensor is

$$\sigma_{\mu\nu}(0) = e^2 \int_0^{+\infty} dt' \int_0^\beta d\tau\, \mathrm{Tr}\left\{\hat{\rho}_C^{eq}\, \hat{j}_{p\nu}(-i\hbar\tau)\, \hat{j}_{p\mu}(t')\right\}$$

$$= e^2 \int_0^{+\infty} dt' \int_0^\beta d\tau\, \langle \hat{j}_{p\nu}(-i\hbar\tau)\, \hat{j}_{p\mu}(t') \rangle_{eq}. \tag{2.73}$$

The symbol "0" is to remind us that this is the conductivity for a static and uniform field, and indeed it can be obtained from the limits $\{\mathbf{k} \to 0, \mathbf{k}' \to 0\}$ and $\omega \to 0$[29] of Eq. 2.57 (I leave this as Exercise 2.3). The last symbol I used $\langle \cdots \rangle_{eq}$ represents the *current-current correlation function* or *current density autocorrelation* evaluated at equilibrium.[30] These are

---

[27] For an infinite system the dipole moment $\int d\mathbf{r}\, \hat{n}(\mathbf{r})\mathbf{r}$ is ill defined. So, here, I am actually working with a finite system. This requirement would not be necessary if I followed a different route, by taking the limits $\{\mathbf{k} \to 0, \mathbf{k}' \to 0\}$ and $\omega \to 0$ of Eq. 2.57 (see Footnote [29] and Exercise 2.3.)

[28] Here, all operators are in the interaction representation.

[29] In this precise order. The opposite order of limits (first $\omega \to 0$, then $\{\mathbf{k} \to 0, \mathbf{k}' \to 0\}$) corresponds to the case in which the electrons first find their "ground state", and then respond to a long-wavelength perturbation. Unless there is long-range coherence (like in the case of a superconductor), this cannot lead to conductivity, and the response is thus zero.

[30] Strictly speaking this correlation is in the complex time, while the "standard" correlations are defined on the real time axis. Also, we generally define the correlation between two observables $\hat{A}$ and $\hat{B}$ over a statistical operator $\hat{\rho}$ as

$$S_{AB}(t,t') = \frac{1}{2}\mathrm{Tr}\left\{\hat{\rho}(t=0)[\hat{A}(t)\hat{B}(t') + \hat{B}(t')\hat{A}(t)]\right\},$$

(with both operators in the Heisenberg representation) to take into account the fact that these operators may not commute (see also Chapter 5).

the correlations between the *spontaneously* fluctuating current density at a given time and the current density at a different time, when the statistical ensemble is the canonical equilibrium one.

We thus see that the response to the electric field, embodied in the conductivity tensor 2.73, can be written in terms of the *spontaneous fluctuations* of the current itself. This the core of the fluctuation-dissipation theorem I have stated above.

This result also shows another important fact: even in the static limit of the external field, the current *cannot*, in principle, be evaluated as an equilibrium property of the system. Instead, its value is related to its own fluctuations at equilibrium. No degree of self-consistency in the charge density would compensate for this fact.

2.3.3.1 Static conductivity of an ideal gas

We can apply the above result to the case of non-interacting electrons. This will provide a well-known form of the fluctuation-dissipation theorem.

For non-interacting electrons the correlation is the same for every complex time and we can choose it to be the time $t = 0$.[31] Using the definition 1.5 of current density we get from Eq. 2.73 ($\mathcal{V}$ is the volume of the system)

$$\sigma_{\mu\nu}^{ideal}(0) = e^2 \beta \mathcal{V} \int_0^{+\infty} dt' \, \langle n(0) v_\nu(0) \, n(t') v_\mu(t') \rangle_{eq}. \qquad (2.74)$$

If we assume the density of the electrons is uncorrelated with their velocity we can write the above as

$$\sigma_{\mu\nu}^{ideal}(0) = e^2 \beta \mathcal{V} \int_0^{+\infty} dt' \, \langle n(0) n(t') \rangle_{eq} \langle v_\nu(0) \, v_\mu(t') \rangle_{eq}$$

$$= n e^2 \beta \int_0^{+\infty} dt' \, \langle v_\nu(0) \, v_\mu(t') \rangle_{eq}, \qquad (2.75)$$

where I have used the fact that in linear response the density is independent of time, and the well-known result that the variance of density fluctuations of an ideal gas in canonical equilibrium is related to its average density, namely $\langle (\Delta n)^2 \rangle_{eq} = \langle n^2 \rangle_{eq} - \langle n \rangle_{eq}^2 = \langle n \rangle_{eq} / \mathcal{V} \equiv n/\mathcal{V}$.

Let us now assume that in the long-time limit the autocorrelation of the velocities satisfies the relation

$$\lim_{t \to +\infty} \langle v_\nu(0) \, v_\mu(t) \rangle_{eq} = \langle v_\nu \rangle_{eq} \langle v_\mu \rangle_{eq}, \qquad (2.76)$$

---

[31] I am working here in the non-degenerate limit of the Fermi gas (see Sec. 1.4.6).

where

$$\langle v_\nu \rangle_{eq} = \lim_{T \to +\infty} \frac{1}{T} \int_0^T dt\, v_\nu(t). \qquad (2.77)$$

The above two relations constitute what we call *ergodicity*, and are based on the (see, e.g., Pathria, 1972)[32]

Ergodic hypothesis: *The trajectory of a representative point in the system's phase space traverses, during time evolution, any neighborhood of any point of the relevant phase space.*

In other words, during time evolution the system "explores" all trajectories in the phase space of all positions and momenta of the particles. If this is true, we can replace time averages with ensemble averages (which are easier to compute). Therefore, instead of following the dynamics of the system in time, and then performing a *time average* of the observable of interest (right-hand side of Eq. 2.77), we can compute the average value of that observable over an ensemble of replicas of the system at *any* given time (left-hand side of Eq. 2.77).[33]

In the quantum case one can prove that a closed, finite and isolated system is ergodic if and only if there are no other constants of motion other than the energy (see, e.g., Reichl, 1998).[34]

The above also states that the time average 2.77 is, for $T \to \infty$, independent of the initial conditions. While this is a reasonable assumption, it may not always be true for all initial conditions. In addition, one can show that the ergodicity relation 2.77 holds if Eq. 2.76 is satisfied, and vice versa, Eq. 2.76 is true if Eq. 2.77 is valid (Khinchine, 1949).

Within this ergodic assumption we can compute the time integral in Eq. 2.75 as

$$\int_0^{+\infty} dt'\, \langle v_\nu(0)\, v_\mu(t') \rangle_{eq} = \lim_{T \to +\infty} \frac{1}{T} \int_0^T dt \int_0^T dt'\, \langle v_\nu(t)\, v_\mu(t') \rangle_{eq}$$

$$= \lim_{T \to +\infty} \frac{1}{T} \langle [r_\nu(T) - r_\nu(0)]\, [r_\mu(T) - r_\mu(0)] \rangle_{eq}$$

$$\equiv D_{\mu\nu}, \qquad (2.78)$$

---

[32] Strictly speaking the average on the left-hand side of Eq. 2.77 (which appears also on the right-hand side of Eq. 2.76) is performed over a *micro-canonical* ensemble, namely an ensemble of systems with constant total energy (see Sec. 2.8.2).

[33] Note that this hypothesis may not necessarily hold in nanoscale systems at steady state (see also Chapter 5).

[34] In this case, all possible micro-states of the system are equally probable (Eq. 2.143) and the appropriate ensemble is the micro-canonical one with entropy given by Eq. 2.144.

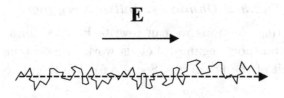

Fig. 2.5. Semiclassical picture of brownian motion of an electron in an electron gas driven by a small electric field.

where the quantity $D_{\mu\nu}$ is called the *diffusion coefficient*. It allows us to re-interpret this result as follows.

From classical statistical mechanics we know that if a particle performs a *random walk* (or *brownian motion*) close to equilibrium, its mean-square displacement $\langle(\Delta x)^2\rangle_{eq}$ during an interval of time T is related to this time as $\langle(\Delta x)^2\rangle_{eq} = DT$, where D is the diffusion coefficient (van Kampen, 1992).

Comparing this result with Eq. 2.78 we see that if we assume the electrons do not interact with each other (or interact via the *mean field* of other electrons) they perform, under the influence of a (small) static and uniform electric field, a random walk (or brownian motion). The fluctuations of this random walk tend to drive them away from equilibrium. On the other hand, the processes that ensure a finite conductivity tend to drive the system back to equilibrium (Fig. 2.5). From 2.75 and 2.78 we thus get the fluctuation-dissipation relation

$$\sigma_{\mu\nu}^{ideal}(0) = \frac{e^2 n D_{\mu\nu}}{k_B \theta}, \qquad (2.79)$$

also known as the *Einstein relation*.

The interacting case

Finally, if the particles do interact, then the density variance is not simply equal to the average density. Instead, we have the relation (Pathria, 1972)

$$\langle n^2 \rangle_{eq} - \langle n \rangle_{eq}^2 = \frac{k_B \theta}{\mathcal{V}} \left. \frac{\partial n}{\partial \bar{\mu}} \right|_\theta, \qquad (2.80)$$

where $\bar{\mu}$ is the chemical potential (see Sec. 2.4), and $\partial n/\partial \bar{\mu}|_\theta$ is proportional to the *isothermal compressibility*, $K = \frac{1}{n^2}\partial n/\partial \bar{\mu}|_\theta$, of the electron liquid. In this case, the above Einstein relation becomes

$$\sigma_{\mu\nu}^{int}(0) = e^2 D_{\mu\nu} \left. \frac{\partial n}{\partial \bar{\mu}} \right|_\theta. \qquad (2.81)$$

2.3.4 Ohmic vs. ballistic regimes

For non-interacting electrons we can rewrite Eq. 2.81 in a form that suggests different transport regimes. Let us work at zero temperature $\theta = 0$ (degenerate limit of the Fermi gas, Sec. 1.4.6). We then have

$$\left.\frac{\partial n}{\partial \bar{\mu}}\right|_{\theta=0} = 2D_\sigma(E_F), \qquad (2.82)$$

where $D_\sigma(E_F)$ is the *density of states* per unit volume per spin evaluated at the Fermi energy E_F. The factor of 2 in 2.82 takes into account the spin degeneracy. The density of states $D_\sigma(E_F)$ for non-interacting electrons in d dimensions is (Giuliani and Vignale, 2005)

$$D_\sigma(E_F) = \frac{\Omega_d}{(2\pi\hbar)^d} m^{d/2} (2E_F)^{(d-2)/2}, \qquad (2.83)$$

where Ω_d is the solid angle in d dimensions ($\Omega_d = 4\pi, 2\pi, 2$ in $d = 3, 2, 1$, respectively). The density at equilibrium of non-interacting electrons is related to the density of states via

$$n = \frac{4 E_F D_\sigma(E_F)}{d}. \qquad (2.84)$$

If we now assume that the system is isotropic, we can use the Drude result 2.12 and Eqs. 2.81 and 2.82 to write the diffusion coefficient in d dimensions (and degenerate limit) as

$$D = \frac{2E_F \tau}{dm} = \frac{p_F^2 \tau}{dm^2} = \frac{v_F^2 \tau}{d}, \qquad (2.85)$$

where $p_F = \sqrt{2mE_F}$ and $v_F = p_F/m$ are the Fermi momentum and velocity, respectively, and τ is the relaxation time.

Let us now consider the conductance 2.6 of a wire of constant cross section S and length L. Using the above results we can rewrite Eq. 2.6 as

$$G = \frac{2e^2}{h} \frac{\Omega_d}{d(2\pi)^{d-1}} \left(\frac{p_F}{\hbar}\right)^{d-2} \left(\frac{2E_F \tau}{\hbar}\right) \frac{S}{L}, \qquad (2.86)$$

which shows that in the linear response and at zero electronic temperature *the conductance is determined by properties of the electron gas at the Fermi level*. This is consistent with the fact that the only possible scattering events, that satisfy the Pauli exclusion principle, involve electrons at the Fermi surface.[35]

---

[35] Electrons cannot scatter into states that are already occupied, like those below the Fermi surface (Ashcroft and Mermin, 1975).

If we replace the relaxation time with the corresponding mean-free path, $\lambda = v_F \tau$, we get

$$G = \frac{2e^2}{h} \frac{\Omega_d}{d(2\pi)^{d-1}} \left(\frac{p_F}{\hbar}\right)^{d-1} \lambda \frac{S}{L} = \frac{2e^2}{h} \frac{\Omega_d}{d(2\pi)^{d-1}} (k_F)^{d-1} \lambda \frac{S}{L}. \qquad (2.87)$$

Let us now assume that the wire is strictly one-dimensional (the area S is a point). For $d = 1$ we get

$$G_{1D} = \frac{2e^2}{h} \frac{2\lambda}{L}. \qquad (2.88)$$

Ohmic regime: $\lambda \gg L$ or $L \gg \lambda$. The conductance is dependent on the length of the wire as in the classical regime. That is the reason we call it the *ohmic regime*. In the limit in which $\lambda \to +\infty$ first, we can then set $L \to +\infty$. The conductance is thus infinite, as one would expect of a defect-free wire (see also Exercise 2.5).

If λ is finite, the conductance tends to zero for $L \to \infty$: the system becomes an *insulator*.[36] This is a well-known result: an infinite one-dimensional system becomes an insulator if one introduces an infinite series of disordered regions.

Ballistic regime: $2\lambda = L$. The conductance becomes

$$G_{1D} = \frac{2e^2}{h} \equiv G_0 = 77.5\,\mu\text{S}, \qquad (2.89)$$

which is known as the *quantum of conductance* (it includes spin degeneracy), and is independent of the length of the wire. Its inverse is the *quantized resistance*

$$R_0 = \frac{1}{G_0} = \frac{h}{2e^2} = 12.9\,\text{k}\Omega. \qquad (2.90)$$

This regime is known as *ballistic*.

This "residual" quantized resistance can be interpreted as follows. It is the resistance that an electron wave-packet leaving from the *center* of the wire would experience when it reached the two opposite *ends* of the wire, traveling a distance $\lambda = L/2$ in both directions (Fig. 2.6). At these ends, it cannot be reflected back to the center, or the wire resistance would increase. Instead, we can *assume* the wave-packet is transported *out* of the wire without further resistance into a region where it is collected. This region can be thought of as a *reservoir*.

[36] Actually, as I will show in Sec. 3.7.4, when the length of the wire is larger than a quantity we call the localization length, the conductance depends exponentially on the length of the wire.

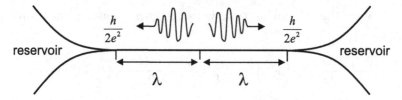

Fig. 2.6. The quantized conductance can be *interpreted* as originating at the "contact" between a wire of length $L = 2\lambda$, with λ the scattering length, and two reservoirs, one at each end of the wire, when a wave-packet travels from the center of the wire towards both ends without reflection. For each spin component, each contact provides a resistance of $R_{\text{contact}} = h/2e^2$.

The quantized resistance can be thus interpreted as the resistance of the "contact" between the wire and two reservoirs (one at each end of the wire) where electrons get "absorbed" without being reflected, as if the wire is attached to the reservoirs *adiabatically* (Imry, 1986). This contact resistance is $R_{\text{contact}} = h/2e^2$, for *each spin* component (Fig. 2.6).

I will return to these concepts in the next chapter when developing the Landauer approach. Here I wanted to show how they can be derived under specific conditions from the Kubo formalism applied to non-interacting electrons.

2.4 Chemical, electrochemical and electrostatic potentials

Let us now turn to another important concept I have already mentioned several times. As anticipated, the quantity $\bar{\mu}$ appearing, for instance, in the equilibrium grand-canonical distribution 1.75, is called the *chemical potential*. The latter is defined as the amount of energy change of the system if an extra particle is added to it, while keeping its volume and thermodynamic entropy – Eq. 2.146 – unchanged. More precisely, it is defined through the Helmholtz free energy of a system with N particles at temperature θ, $F_N = U - \theta S$ (with U the internal energy and S the thermodynamic entropy) via the relation[37]

$$\bar{\mu} = F_{N+1} - F_N. \qquad (2.91)$$

---

[37] This is equivalent to the definition

$$\bar{\mu} = \left(\frac{\partial F}{\partial N}\right)_{\theta, V},$$

the partial derivative of the free energy with respect to the total number of particles N, at fixed temperature θ and constant volume V.

2.4 Chemical, electrochemical and electrostatic potentials

In a metallic system at equilibrium (and zero temperature), the chemical potential corresponds precisely to the *Fermi energy*, i.e., the energy of the highest occupied energy level. The larger the chemical potential, the larger the number of electrons in the system, so that the latter can also be interpreted as a measure of energy level filling.

The above definitions are strictly valid for systems in thermodynamic equilibrium.[38] As we have discussed before, electron transport is truly a non-equilibrium process. How can we then extend such concepts to this case? The answer to this question is not unique, and we generally adopt a pragmatic view that allows us to make sense of these quantities using our knowledge of equilibrium statistical mechanics.

Let us first rely on this experimental fact. Consider again two separate metallic electrodes at equilibrium, each with its own chemical potential. Call these potentials $\bar{\mu}_1$ and $\bar{\mu}_2$. For instance, these electrodes could be two different types of metal, i.e., metals with different density. If we put together these two metallic systems, and there is no infinite barrier between them, carriers will flow from the metal with higher chemical potential to the one with lower chemical potential. This *diffusion* process is a natural tendency of all physical systems to approach thermodynamic equilibrium by canceling any existing chemical potential difference. Put differently, electrons respond to a *gradient* of the chemical potential as if a "field" of magnitude $|\nabla \bar{\mu}/e|$ is present.

As we have seen in the discussion of the Drude and Kubo approaches, in addition to this chemical potential difference electrons respond *mechanically* to an electric field $\mathbf{E}$ and *drift* accordingly. It is then the *sum* of the chemical potential "field" $-\nabla \bar{\mu}/e$ and electric field $\mathbf{E}$ that drives electrons in the system: the carriers respond to

$$\mathcal{E} = \mathbf{E} - \nabla \bar{\mu}/e. \tag{2.92}$$

The integral of this quantity along the circuit that connects the two end points $\mathbf{r}_1$ and $\mathbf{r}_2$ of a *voltmeter* (a voltage probe) is precisely what the latter measures, i.e.,

$$V \equiv \frac{\mu(\mathbf{r}_2) - \mu(\mathbf{r}_1)}{e} = -\int_{\mathbf{r}_1}^{\mathbf{r}_2} \mathcal{E} \cdot d\mathbf{r} = \phi(\mathbf{r}_2) - \phi(\mathbf{r}_1) + \frac{\bar{\mu}(\mathbf{r}_2)}{e} - \frac{\bar{\mu}(\mathbf{r}_1)}{e}. \tag{2.93}$$

---

[38] Or close to equilibrium in a *quasi-static* sense: the time evolution of the system is so slow that the latter has time to equilibrate with a succession of reservoirs, each with a different temperature and chemical potential.

In Eq. 2.93 we have introduced the *electrostatic potential* ϕ defined as

$$\phi(\mathbf{r}) = \sum_i \frac{|e|Z_i}{|\mathbf{r} - \mathbf{R}_i|} + \int d\mathbf{r}' \, \frac{e\,n(\mathbf{r}')}{|\mathbf{r} - \mathbf{r}'|}, \qquad (2.94)$$

where $Z_i|e|$ and $\mathbf{R}_i$ are the charge and position of ion i, respectively, of the given material. The second term on the right-hand side of Eq. 2.94 is known as the *Hartree potential*,

$$v_H(\mathbf{r}) = \int d\mathbf{r}' \, \frac{e\,n(\mathbf{r}')}{|\mathbf{r} - \mathbf{r}'|}, \qquad (2.95)$$

and it is the *electrostatic potential* created by a distribution of charges of density $n(\mathbf{r})$.[39]

In this book, we will sometimes use the Hartree potential energy $V_H \equiv e v_H(\mathbf{r})$ as an approximation to the true many-body Coulomb interaction energy

$$W = \frac{1}{2} \sum_{i \neq j} \frac{e^2}{|\mathbf{r}_i - \mathbf{r}_j|}, \qquad (2.96)$$

of all electrons interacting with each other, namely

$$W \simeq \sum_i e v_H(\mathbf{r}_i) = \sum_i V_H(\mathbf{r}_i). \qquad (2.97)$$

This is one form of mean-field approximation: electrons can be considered as moving independently of each other, and the effect of the other electrons is only to provide the *mean field* 2.95.

We call V the *bias* or, equivalently, the *electrochemical potential difference* per unit charge, where, as discussed in Sec. 1.2, the bias can be realized in various ways, such as via the introduction of a battery or the electrostatic potential difference of a capacitor.

We then generalize the above concepts by defining the space- and time-dependent electrochemical potential as

$$\mu(\mathbf{r}, t) = e\phi(\mathbf{r}, t) + \bar{\mu}(\mathbf{r}, t). \qquad (2.98)$$

At the microscopic level, due to the presence of the ions, this quantity oscillates strongly at the atomic positions (see Eq. 2.94). In the bulk, if we are interested in its value in the direction of current flow (call this direction x), we can get rid of these oscillations by performing first a *planar* average $\langle \mu(x,t) \rangle = S^{-1} \int_S dy\, dz\, \mu(\mathbf{r}, t)$ over a surface S, and then a *macroscopic*

---

[39] Here, $n(\mathbf{r})$ is the *self-consistent* charge density of the global current-carrying system. The Hartree potential may also depend on time if the driving field depends on time.

2.4 Chemical, electrochemical and electrostatic potentials

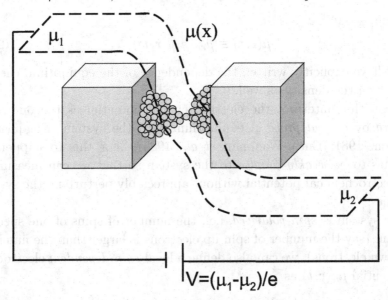

Fig. 2.7. Macroscopic average of the local electrochemical potential along a biased nanostructure.

average over the distance a between two successive planes of the material

$$\bar{\mu}(x,t) = \frac{1}{a} \int_{-a/2}^{a/2} dx' \, \langle \mu(x+x',t) \rangle. \tag{2.99}$$

A schematic of a possible electrochemical potential (averaged as above) along a nanostructure kept at a (time-independent) bias V is shown in Fig. 2.7.

In the following, I will always assume this spatial averaging has been performed, but to lighten the notation, I will still employ the symbol $\mu(\mathbf{r},t)$ even for this averaged electrochemical potential.

The electrochemical potential is a key quantity in transport theories and I will come back to it in the next chapters as well. Here I make a few general remarks. First of all, I have defined a *local* (in space) time-dependent chemical potential $\bar{\mu}(\mathbf{r},t)$. Since the chemical potential has a well-defined meaning only in thermodynamic equilibrium and in a global sense (see definition 2.91), what is the meaning of this quantity?

Here, again, I give a pragmatic reply. If we look back at the definition 2.91, an $\mathbf{r}$- and t-dependent chemical potential requires us to *postulate* that a *local* temperature $\theta(\mathbf{r},t)$ and a *local* entropy function $S(\mathbf{r},t)$ can be defined such that, at any given time, the system is, at position r, in *local equilibrium* with a reservoir at that same temperature $\theta(\mathbf{r},t)$. In mathematical terms we define

$$\bar{\mu}(\mathbf{r},t) \equiv \bar{\mu}_{eq}(n, \theta(\mathbf{r},t)), \qquad (2.100)$$

where I have explicitly written the dependence of the equilibrium chemical potential on the density as well.

We can thus interpret the electrochemical potential as the one *ideally measured* by a *local probe* in equilibrium with the system (Engquist and Anderson, 1981; Entin-Wohlman et al., 1986). For this to happen, the probe has to be *weakly coupled* to the system so that we can measure the local electrochemical potential without appreciably perturbing the system's properties.[40]

If the system is *spin-polarized*, i.e., the number of spins of one species is dominant (say the number of spin-up electrons is larger than the number of spin-down electrons), we can also define a local *spin-dependent* electrochemical potential $\mu_\alpha(\mathbf{r},t)$ as

$$\mu_\alpha(\mathbf{r},t) \equiv e\phi(\mathbf{r},t) + \bar{\mu}_{\alpha,eq}(n_\alpha, \theta(\mathbf{r},t)), \qquad (2.101)$$

where the definition of equilibrium chemical potential now applies to a given species α of spins.[41]

Alternative definitions of electrochemical potential can be provided (see, for instance, Büttiker et al., 1985). All these definitions – which are not all necessarily equivalent, both conceptually and practically – try to attribute a non-equilibrium (and local) meaning to the equilibrium chemical potential, either as a measure of the energy level filling of independent electrons, or as the energy required to add electrons locally into the system.

The above conceptual differences show that while *local* chemical potential, temperature and entropy are useful theoretical constructs in the description of transport properties (we will indeed make heavy use of them in the next chapters), they do not necessarily have a unique definition for a system out of equilibrium.[42]

2.5 Drift-diffusion equations

If we use the concepts developed above, we can calculate the total current density due *both* to an electric field $\mathbf{E}$ that creates a *drift current*, and to a chemical potential gradient $\nabla\mu$ that creates a *diffusion current*. In order to proceed in determining this current density, let us assume the temperature

---

[40] Whether this can be *practically* done is clearly a different story.
[41] The spin species with the largest number of electrons is called *majority spins*, the one with the lowest, *minority spins*.
[42] We will come back to the concept of entropy and its various definitions in Sec. 2.8.

θ is uniform across the whole system, so that the only variation of the chemical potential is due to variations of the density. Let us also work in linear response and with an isotropic material. We thus know from the analysis of the Drude model (Sec. 2.1) and the Kubo formalism (Sec. 2.3) that an electric field induces a conductivity response, σ, related to the drift current. A chemical potential gradient must also induce some other response, associated with the diffusion current. Let us call L_e this positive *diffusion response coefficient*.

Taking into account both the response to an electric field and the response to a chemical potential difference we thus write the total current density as

$$\mathbf{j} = \sigma \mathbf{E} - e\frac{L_e}{\theta}\nabla\bar{\mu} = \sigma\mathbf{E} - e\left.\frac{L_e}{\theta}\frac{\partial\bar{\mu}}{\partial n}\right|_\theta \nabla n, \qquad (2.102)$$

where σ is the Drude conductivity 2.12.[43]

We call *diffusion coefficient* the quantity

$$D \equiv \left.\frac{L_e}{\theta}\frac{\partial\bar{\mu}}{\partial n}\right|_\theta, \qquad (2.103)$$

so that Eq. 2.102 can be written as

$$\boxed{\mathbf{j} = \sigma\mathbf{E} - eD\nabla n.} \qquad (2.104)$$

This *drift-diffusion* equation must be supplied by the equation that relates the self-consistent charge density and the current density, i.e., the continuity equation 1.11.

2.5.1 Diffusion coefficient of an ideal electron gas in the non-degenerate limit

We can relate the diffusion coefficient 2.103 to the conductivity σ under the following conditions. Let us assume that the electrons constitute an *ideal gas*, i.e., they are non-interacting, or at most, interact via the mean-field potential $\phi(\mathbf{r})$ (Eq. 2.94). At global equilibrium the chemical potential must be uniform in space. On the other hand, the electrostatic potential may

---

[43] The temperature appearing in the denominator of the second term on the right-hand side of Eq. 2.102 comes from the fact that, quite generally, in the presence of a position-dependent temperature, a chemical potential difference induces an electrical current of the type (Balian, 1991)

$$\mathbf{j}_D = eL_{eE}\nabla\left(\frac{1}{\theta}\right) - eL_e\nabla\left(\frac{\bar{\mu}}{\theta}\right).$$

The response coefficient L_{eE} is related to the fact that a temperature gradient can induce *both* an energy flow *and* a particle flow, even when $\bar{\mu}/\theta$ is constant (see Sec. 6.4).

change in space due to local variations of the electron density. From equilibrium statistical mechanics of ideal gases we then know that the particle density must have the form

$$n(\mathbf{r}) = n_0 \exp\left(\frac{\bar{\mu} - e\phi(\mathbf{r})}{k_B \theta}\right), \tag{2.105}$$

where here $\bar{\mu}$ is constant, and n_0 is some constant density. We can apply this approximate form to the non-degenerate Fermi gas.[44]

At thermal equilibrium the total current density must vanish so that from 2.102 we find

$$0 = -\sigma \nabla \phi(\mathbf{r}) + \frac{e^2 D n(\mathbf{r})}{k_B \theta} \nabla \phi(\mathbf{r}), \tag{2.106}$$

where we have used the electrostatic relation $\mathbf{E} = -\nabla \phi$.

From 2.106 we then obtain the relation for an ideal gas in the non-degenerate limit

$$D = \frac{\sigma k_B \theta}{e^2 n}. \tag{2.107}$$

We define *mobility* (a concept most useful for transport in doped semiconductors) as the mean drift velocity per unit field

$$\mu_{el} = \frac{|\mathbf{v}|}{|\mathbf{E}|}. \tag{2.108}$$

In the Drude model we have the relation 2.10, so that the mobility is

$$\mu_{el} = \frac{e\tau}{m}, \tag{2.109}$$

and the Drude conductivity (from Eq. 2.12)

$$\sigma = e n \mu_{el}. \tag{2.110}$$

Using these definitions we thus get

$$D = \frac{k_B \theta}{e} \mu_{el}, \tag{2.111}$$

or

$$\sigma = \frac{e^2 n D}{k_B \theta}, \tag{2.112}$$

---

[44] This approximation is well suited for *non-degenerate* semiconductors, i.e., semiconductors in which, if E_c is the conduction band minimum, $E_c - \bar{\mu} \gg k_B \theta$ (or if one considers holes, $\bar{\mu} - E_v \gg k_B \theta$, with E_v the top of the valence band), see, e.g., Ashcroft and Mermin, 1975. This is the regime in which the drift-diffusion equation 2.104 is mostly applied.

2.5.2 Generalization to spin-dependent transport

Let us generalize Eq. 2.104 to spin-dependent transport.[45] Assume there are two species of spin, which we label spin up ↑, and spin down ↓. For clarity, let us assume these are the eigenstates of the Pauli operator $\hat{\sigma}_z$.[46] The spin-up electrons have density $n_\uparrow(\mathbf{r})$, the spin-down electrons have density $n_\downarrow(\mathbf{r})$, with $n_\uparrow(\mathbf{r}) + n_\downarrow(\mathbf{r}) = n(\mathbf{r})$, the total density. If there is no mechanism that destroys spin polarization, such as spin-orbit coupling, Eq. 2.104 can be simply generalized to the two equations (one for each spin component)

$$\mathbf{j}_{\uparrow(\downarrow)} = \sigma_{\uparrow(\downarrow)}\mathbf{E} - eD_{\uparrow(\downarrow)}\nabla n_{\uparrow(\downarrow)}, \qquad (2.113)$$

where we have also introduced the spin-dependent diffusion coefficients $D_{\uparrow(\downarrow)}$.[47] The current densities $\mathbf{j}_{\uparrow(\downarrow)}$ are the two distinct components of the electrical current density associated with the two different spins. The total current density is simply $\mathbf{j} = \mathbf{j}_\uparrow + \mathbf{j}_\downarrow$.

These equations need to be supplemented by the continuity equations

$$e\frac{\partial n_{\uparrow(\downarrow)}(\mathbf{r},t)}{\partial t} = -\nabla \cdot \mathbf{j}_{\uparrow(\downarrow)}(\mathbf{r},t), \qquad (2.114)$$

which now state the conservation of charge for *each* spin component.

In the presence of spin-orbit coupling, or, in general, of any mechanism such that $\hat{S}_z$ is not a constant of motion,[48] one cannot separately define the two spin components of the current densities, and the only meaningful quantity is the total current density.

Nonetheless, if these *spin-relaxation* effects act on the spin very "slowly", i.e., the time it takes for an electron initially prepared into one spin state (say spin up) to completely relax into the opposite spin state (spin flip),

---

[45] Once again, the drift-diffusion equations we are going to derive find their most natural application in non-degenerate Fermi systems (see Footnote [44]). For a recent review on spin-dependent phenomena see, e.g., Žutić et al., 2004.

[46] In this basis the Pauli operator $\hat{\sigma}_z$ is

$$\hat{\sigma}_z = \begin{pmatrix} 1 & 0 \\ 0 & -1 \end{pmatrix}.$$

The z-component of the spin angular momentum $\hat{S}_z$ is simply $\hat{S}_z = \hbar\hat{\sigma}_z/2$.

[47] Coulomb interactions create an effect known as spin Coulomb drag (D'Amico and Vignale, 2002) that decreases the diffusion coefficient $D_{\uparrow(\downarrow)}$ with respect to its non-interacting value 2.107.

[48] One such mechanism may be the coupling of the electron spin with the nuclear spins.

is longer than any other relaxation time, one may approximately distinguish the current components $\mathbf{j}_{\uparrow(\downarrow)}$ and introduce these relaxation effects phenomenologically (Yu and Flatté, 2002; Pershin and Di Ventra, 2007). Typically, *spin coherence* – the analogue of the *orbital* wave-function coherence we have discussed in Sec. 2.2 – can be maintained for distances of the order of 100 μm or larger, and times of the order of $10^{-9} - 10^{-8}$ s in both bulk metals (Johnson and Silsbee, 1985) and semiconductors (Kikkawa and Awshalom, 1999), i.e., orders of magnitude longer than mean-free paths and relaxation times due to inelastic effects associated with the charge (see Sec 2.1). (Long spin relaxation times are one of the reasons spin transport is so attractive for applications.)

For clarity, let us focus on the current density $\mathbf{j}_{\uparrow}$. With respect to this current, spin-flip events that transform a spin-up electron into a spin-down electron can be viewed as a "sink" of spin-up electrons: the population of spin-up electrons decreases in time in favor of the spin-up population. And vice versa, these mechanisms act as a "source" for the spin-down electron population: their number increases with time. As a first approximation we can thus say that the equation for the current density 2.113 remains unchanged, while the continuity equation *for each spin component* is not satisfied, i.e., a source/sink term must be included.

If we call τ_{sf} the average relaxation time for spin flip, the number density of spin-up electrons that are transformed into spin-down electrons per unit time must be proportional to $(n_{\uparrow} - n_{\downarrow})/\tau_{sf}$, i.e., to the number of spin-up electrons available per unit time. We can, however, do the same analysis for the spin-down electrons and conclude that the number density of spin-down electrons that are transformed into spin-up electrons per unit time must be proportional to $(n_{\downarrow} - n_{\uparrow})/\tau_{sf}$. We can thus modify the continuity equations 2.114 as follows

$$e\frac{\partial n_{\uparrow}(\mathbf{r}, t)}{\partial t} = -\nabla \cdot \mathbf{j}_{\uparrow}(\mathbf{r}, t) + \frac{e}{2\tau_{sf}}(n_{\uparrow} - n_{\downarrow}), \quad (2.115)$$

$$e\frac{\partial n_{\downarrow}(\mathbf{r}, t)}{\partial t} = -\nabla \cdot \mathbf{j}_{\downarrow}(\mathbf{r}, t) + \frac{e}{2\tau_{sf}}(n_{\downarrow} - n_{\uparrow}). \quad (2.116)$$

This is the simplest form of continuity equation with a source/sink term that guarantees that $n_{\uparrow}(\mathbf{r}) + n_{\downarrow}(\mathbf{r}) = n(\mathbf{r})$ and satisfies the continuity equation 1.11 with $\mathbf{j} = \mathbf{j}_{\uparrow} + \mathbf{j}_{\downarrow}$ (no charge is lost due to spin-relaxation effects).

From all this discussion, we see that the drift-diffusion equations 2.104 and 2.113 are most useful in discussing transport in linear response for a non-degenerate Fermi gas, and when self-consistent charge redistributions due to quantum-mechanical tunneling effects are of less importance, or can be somehow lumped into either the value of the conductivity σ or the diffusion coefficient D.

2.6 Distribution functions

Apart from the discussion about the electrochemical potential, I have so far assumed that the system is very close to equilibrium. What if the electrons are far from their equilibrium state? We can proceed using different approaches. Here I will describe the one that introduces the concept of *distribution functions*.

In the Drude model we have assumed that electrons emerge from a collision with an average velocity that is randomly oriented. The *magnitude* of such velocity is the same before and after the collision. In reality, some electrons may emerge from a collision with a given velocity, and another set of electrons with a different velocity. This same set of electrons may emerge from another collision at a later time and different point in space with yet another velocity, and so on. A better description of the transport problem would then involve *distribution functions*, i.e., functions that describe how electrons are distributed in momentum space, as well as in real space in the course of time.

At equilibrium we are well familiar with one such distribution in the case of non-interacting electrons. As we have discussed in Sec. 1.4.5, non-interacting electrons at equilibrium with an external bath at temperature θ are distributed in momentum space according to the Fermi–Dirac distribution (see Eq. 1.80)

$$f^{eq}(\mathbf{p}) = \frac{1}{e^{(E(\mathbf{p})-\bar{\mu})/k_B\theta} + 1}, \tag{2.117}$$

with $\bar{\mu}$ the chemical potential, and $E(\mathbf{p})$ is the relation between momentum and energy. For instance, for a free particle of mass m it is simply $E(\mathbf{p}) = \mathbf{p}^2/2m$.

We want to extend this definition to the non-equilibrium problem. Let us proceed as follows.

Let us consider N *interacting* and *indistinguishable* particles and define a *phase density* $D(\mathbf{r}_1, \mathbf{p}_1; \ldots; \mathbf{r}_N, \mathbf{p}_N, t)$ such that, if $d\Omega$ is an infinitesimal element of phase space spanned by the coordinates and momenta of all particles, the quantity

$$D(\mathbf{r}_1, \mathbf{p}_1; \ldots; \mathbf{r}_N, \mathbf{p}_N, t)\, d\Omega \tag{2.118}$$

is the probability that, at a time t, the particles can be found in the small volume $d\Omega$ centered at $(\mathbf{r}_1, \mathbf{p}_1; \ldots; \mathbf{r}_N, \mathbf{p}_N)$.[49] The phase density is the classical counterpart of the statistical operator I have defined in Sec. 1.4.

If $d\mathbf{r}_i\, d\mathbf{p}_i$ is the infinitesimal phase-space volume associated with particle i, the volume $d\Omega$ is

$$d\Omega = \frac{1}{N!} \prod_{i=1}^{N} d\mathbf{r}_i\, d\mathbf{p}_i, \qquad (2.119)$$

where the factor $1/N!$ is to account for the indistinguishability of the particles.

For most of the systems of interest, the phase density is too difficult to determine since it contains all possible *correlations* among particles, i.e., how each particle motion correlates with the motion of the other particles.[50]

We are thus content with studying a simpler quantity: the *one-particle reduced density*, or one-particle *non-equilibrium distribution function*

$$f(\mathbf{r},\mathbf{p},t) = \int \frac{1}{(N-1)!} \prod_{i=2}^{N} d\mathbf{r}_i\, d\mathbf{p}_i\, D(\mathbf{r},\mathbf{p};\ldots;\mathbf{r}_N,\mathbf{p}_N,t) \qquad (2.120)$$

such that, given the phase-space volume $d\mathbf{r}\, d\mathbf{p}$, the quantity[51]

$$\begin{aligned}f(\mathbf{r},\mathbf{p},t)\, d\mathbf{r}\, d\mathbf{p} = \;& \text{average number of particles that at time } t\\ & \text{is found in a phase-space volume } d\mathbf{r}\, d\mathbf{p},\\ & \text{around the phase-space point } \mathbf{r},\mathbf{p}.\end{aligned} \qquad (2.121)$$

The total number of particles N is therefore

$$N = \int f(\mathbf{r},\mathbf{p},t)\, d\mathbf{r}\, d\mathbf{p}, \qquad (2.122)$$

where the integral is over all allowed space and momenta.

Similarly, one can define the *two-particle distribution function*

$$f_2(\mathbf{r},\mathbf{p};\mathbf{r}',\mathbf{p}',t) = \int \frac{1}{(N-2)!} \prod_{i=3}^{N} d\mathbf{r}_i\, d\mathbf{p}_i\, D(\mathbf{r},\mathbf{p};\mathbf{r}',\mathbf{p}';\ldots;\mathbf{r}_N,\mathbf{p}_N,t), \qquad (2.123)$$

---

[49] Since D is a probability density it satisfies $\int D(\mathbf{r}_1,\mathbf{p}_1;\ldots;\mathbf{r}_N,\mathbf{p}_N,t)\, d\Omega = 1$ at all times.

[50] A notable exception is the phase density of uncorrelated particles (such as those of an ideal gas). In this case, the phase density factorizes into the product of a common function

$$D(\mathbf{r}_1,\mathbf{p}_1;\ldots;\mathbf{r}_N,\mathbf{p}_N,t) = N! \prod_{i=1}^{N} f(\mathbf{r}_i\, \mathbf{p}_i,t).$$

[51] We implicitly assume that in the element of phase space $d\mathbf{r}\, d\mathbf{p}$ there are many one-particle states so that a continuous distribution function can be defined.

and, in general, the M-particle distribution function f_M by integrating out all but M degrees of freedom.

The calculation of the one-particle non-equilibrium distribution function 2.121 for a general many-body interacting system is still a difficult task (let alone the calculation of the two-particle distribution). In the next section I will discuss the Boltzmann equation, which is precisely a semiclassical equation of motion for the non-equilibrium distribution function. In Chapter 4 I will introduce the quantum analogue of the Boltzmann equation and discuss the non-equilibrium distribution functions in that context.

2.7 Boltzmann equation

In a semiclassical theory of transport one can derive an equation of motion, known as the Boltzmann equation, for the non-equilibrium distribution function $f(\mathbf{r}, \mathbf{p}, t)$. To do this consider first a set of N *non*-interacting particles subject to an external potential $V_{ext}(\mathbf{r}, t)$, and thus evolving according to the Hamiltonian

$$H = \sum_{i=1}^{N} \left(\frac{\mathbf{p}_i^2}{2m} + V_{ext}(\mathbf{r}_i, t) \right). \tag{2.124}$$

By "non-interacting" I mean that they do not experience any type of scattering. Hamiltonian 2.124 describes the dynamics of independent particles each moving with the *same* equations of motion. The phase density $D(\mathbf{r}_1, \mathbf{p}_1; \ldots; \mathbf{r}_N, \mathbf{p}_N, t)$ we have defined in Eq. 2.118 thus factorizes into the product of N one-particle reduced densities $f(\mathbf{r}, \mathbf{p}, t)$ (see Footnote [50]). Instead of working with the N-particle density we can thus work with the quantity $f(\mathbf{r}, \mathbf{p}, t) d\mathbf{r} \, d\mathbf{p}$, that gives the number of particles in $d\mathbf{r} \, d\mathbf{p}$.

Since the particles evolve under Hamiltonian dynamics the *Liouville theorem* guarantees that the phase-space volume is conserved during time evolution (Goldstein, 1950). In addition, no interactions are present among particles, therefore the number of particles in volume $d\mathbf{r} \, d\mathbf{p}$ remains constant during time evolution under Hamiltonian 2.124. All this implies that the single-particle density $f(\mathbf{r}, \mathbf{p}, t)$ must be conserved, namely

$$\frac{d}{dt} f(\mathbf{r}, \mathbf{p}, t) = 0. \tag{2.125}$$

By differentiating the above equation we get

$$\frac{\partial f(\mathbf{r}, \mathbf{p}, t)}{\partial t} + \frac{d\mathbf{r}}{dt} \cdot \nabla_\mathbf{r} f(\mathbf{r}, \mathbf{p}, t) + \frac{d\mathbf{p}}{dt} \cdot \nabla_\mathbf{p} f(\mathbf{r}, \mathbf{p}, t) = 0. \tag{2.126}$$

We can now use the Hamilton canonical equations for the particles' conjugated variables $\{r_j, p_j\}$ (i.e., the j-th component of the particle position and momentum, respectively)

$$\frac{dr_j}{dt} = \frac{\partial H}{\partial p_j}; \quad \frac{dp_j}{dt} = -\frac{\partial H}{\partial r_j}, \qquad (2.127)$$

with the single-particle Hamiltonian $H = \mathbf{p}^2/2m + V_{ext}(\mathbf{r}, t)$. We obtain

$$\frac{\partial f(\mathbf{r}, \mathbf{p}, t)}{\partial t} + \frac{\mathbf{p}}{m} \cdot \nabla_\mathbf{r} f(\mathbf{r}, \mathbf{p}, t) - \nabla_\mathbf{r} V_{ext}(\mathbf{r}, t) \cdot \nabla_\mathbf{p} f(\mathbf{r}, \mathbf{p}, t) = 0. \qquad (2.128)$$

Let us now allow the particles to interact via, say, a two-particle potential $W(|\mathbf{r} - \mathbf{r}'|)$ that depends on their relative distance (this could be, e.g., the mutual Coulomb interaction for charged particles), or via some other general potential that scatters particles. The presence of this interaction changes the particles' momenta via collisions, and the particles can scatter "in" and "out" of the phase-space volume $d\mathbf{r}\, d\mathbf{p}$. The distribution function f is no longer a conserved quantity and condition 2.125 is not satisfied. The change in time of the distribution function must thus be balanced by an equal amount of change due to collisions. We write this as

$$\frac{d}{dt} f(\mathbf{r}, \mathbf{p}, t) = \left(\frac{\partial f(\mathbf{r}, \mathbf{p}, t)}{\partial t} \right)_{\text{coll}} \equiv \mathbf{I}[f]. \qquad (2.129)$$

The quantity $\mathbf{I}[f]$ is called the *collision integral* (or *scattering operator*) and is a *functional* of the distribution function. It gives the net rate of change of particles with momentum $\mathbf{p}$, at position $\mathbf{r}$ and time t. For arbitrary interaction potentials, it contains information on the type of processes that drive the system to local equilibrium. These processes may be elastic or inelastic, i.e., they may or may not change the energy of the single particles.

Equating Eq. 2.129 with Eq. 2.128 we finally get the Boltzmann equation[52]

$$\boxed{\frac{\partial f(\mathbf{r}, \mathbf{p}, t)}{\partial t} + \frac{\mathbf{p}}{m} \cdot \nabla_\mathbf{r} f(\mathbf{r}, \mathbf{p}, t) - \nabla_\mathbf{r} V_{ext}(\mathbf{r}, t) \cdot \nabla_\mathbf{p} f(\mathbf{r}, \mathbf{p}, t) = \mathbf{I}[f].}$$

(2.130)

Given an interaction potential, the collision integral can be formally calculated *exactly*. For instance, for a two-body potential $W(|\mathbf{r} - \mathbf{r}'|)$ one finds (Exercise 2.11)

$$\mathbf{I}[f] = \int d\mathbf{r}'\, d\mathbf{p}'\, \nabla_\mathbf{r} W(|\mathbf{r} - \mathbf{r}'|) \cdot \nabla_\mathbf{p} f_2(\mathbf{r}, \mathbf{p}; \mathbf{r}', \mathbf{p}', t). \qquad (2.131)$$

---

[52] Note that here V_{ext} may be the sum of the potential energy $e\phi(\mathbf{r})$ and any other external potential energy. It thus depends, self-consistently, on the charge density (see Eq. 2.94) and therefore on the distribution function (Eq. 2.136).

2.7 Boltzmann equation

Unfortunately, the above expression contains the two-particle reduced density f_2. We thus need an equation of motion for this quantity.

Carrying out a derivation similar to that leading to Eq. 2.131, one can show that the equation of motion for f_2 contains the three-particle reduced density f_3. In turn, the equation of motion for f_3 depends on the four-particle reduced density f_4, and so on. This generates an infinite *hierarchy* of coupled equations, known as the *BBGKY hierarchy*,[53] thus making the problem practically unsolvable.

Approximations to the collision integral are thus required to solve the problem either analytically or numerically. For electrons, a common approximation to the collision integral is (Ashcroft and Mermin, 1975)

$$\mathbf{I}[f] = -\int d\mathbf{p}' \left\{ W_{\mathbf{p},\mathbf{p}'} f(\mathbf{r},\mathbf{p},t) \left[1 - f(\mathbf{r},\mathbf{p}',t)\right] \right. \\ \left. - W_{\mathbf{p}',\mathbf{p}} f(\mathbf{r},\mathbf{p}',t) \left[1 - f(\mathbf{r},\mathbf{p},t)\right] \right\}, \quad (2.132)$$

where the quantity $W_{\mathbf{p},\mathbf{p}'}$ is the *transition probability density* per unit time that a particle with momentum $\mathbf{p}$ is scattered into a state with momentum $\mathbf{p}'$. This probability density may depend on the distribution function itself. The term $f(\mathbf{r},\mathbf{p},t)$ that appears in 2.132 counts how many electrons are in the initial state $\mathbf{p}$, while the term $1 - f(\mathbf{r},\mathbf{p}',t)$ counts the available states with momentum $\mathbf{p}'$ into which the electrons can scatter (due to the exclusion principle, electrons can only scatter into states that are unoccupied). With this choice of collision integral, equation 2.130 is a non-linear integro-differential equation for the distribution function.

According to the type and strength of the scattering potential, the transition probability $W_{\mathbf{p},\mathbf{p}'}$ can be calculated with standard techniques of quantum mechanics, such as perturbation theory (see Sec. 6.2.1 for one such calculation). A further approximation, known as the *relaxation-time approximation* assumes the identity

$$\frac{1}{\tau(\mathbf{p})} = \int d\mathbf{p}' W_{\mathbf{p},\mathbf{p}'} \left[1 - f(\mathbf{r},\mathbf{p}',t)\right], \quad (2.133)$$

where, as in the Drude model (Sec. 2.1), $\tau(\mathbf{p})$ is the *relaxation time*, i.e., the average time between successive collisions.[54] With this approximation we can rewrite the collision integral as

$$\mathbf{I}[f] = -\frac{[f(\mathbf{r},\mathbf{p},t) - f^{eq}(\mathbf{r},\mathbf{p})]}{\tau}. \quad (2.134)$$

---

[53] From the names of Born, Bogoliubov, Green, Kirkwood, and Yvon. We will come back to its quantum analogue in Chapter 4.
[54] The relaxation time may also depend on energy and position.

Here $f^{eq}(\mathbf{r},\mathbf{p})$ is the local equilibrium Fermi distribution function

$$f^{eq}(\mathbf{r},\mathbf{p}) = \frac{1}{e^{(E(\mathbf{p})-\bar{\mu}(\mathbf{r}))/k_B\theta(\mathbf{r})}+1}, \qquad (2.135)$$

with position-dependent chemical potential and temperature.

Equation 2.134 quantifies deviations from equilibrium, and contains the physical notion that once electrons reach local equilibrium, then further scattering will not modify their distribution ($\mathbf{I}[f] = 0$ at local equilibrium).

Irrespective of the approximations employed for the collision integral, once the distribution function is known we can calculate the electron number density

$$n(\mathbf{r},t) = \int d\mathbf{p}\, f(\mathbf{r},\mathbf{p},t), \qquad (2.136)$$

and current density

$$\mathbf{j}(\mathbf{r},t) = e \int d\mathbf{p}\, \frac{\mathbf{p}}{m} f(\mathbf{r},\mathbf{p},t). \qquad (2.137)$$

2.7.1 Approach to local equilibrium

Let us assume that at $t = 0$ we prepare the system so that the distribution function is far from local equilibrium, i.e., $f(\mathbf{r},\mathbf{p},t=0)$ is very different from $f^{eq}(\mathbf{r},\mathbf{p})$. If the distribution f and the external potential do not have large spatial variations, at the initial times the collision integral dominates the dynamics of the system as is evident from 2.130. During the transient time, therefore, the other terms in Eq. 2.130 can be neglected and the Boltzmann equation reduces to

$$\frac{\partial f(\mathbf{r},\mathbf{p},t)}{\partial t} \approx \mathbf{I}[f]. \qquad (2.138)$$

If we make the relaxation-time approximation 2.134 to the collision integral, this equation becomes

$$\frac{\partial f(\mathbf{r},\mathbf{p},t)}{\partial t} \approx -\frac{[f(\mathbf{r},\mathbf{p},t)-f^{eq}(\mathbf{r},\mathbf{p})]}{\tau}, \qquad (2.139)$$

whose solution is

$$f(\mathbf{r},\mathbf{p},t) = f^{eq}(\mathbf{r},\mathbf{p}) + [f(\mathbf{r},\mathbf{p},t=0)-f^{eq}(\mathbf{r},\mathbf{p})]e^{-t/\tau}. \qquad (2.140)$$

We thus see that the system approaches local equilibrium in a time interval of the order of τ, i.e., of the order of the collision time. The smaller this time, the faster local equilibrium can be reached. After this initial transient, the collision integral becomes comparable to the other terms in the Boltzmann equation, and the dynamics proceeds in such a way that the particles lose memory of their initial momentum. The particles' momentum is then continually *randomized* by collisions.[55]

The subsequent dynamics is determined by the two terms $\mathbf{p}\cdot\nabla_{\mathbf{r}}f(\mathbf{r},\mathbf{p},t)/m$, and $-\nabla_{\mathbf{r}}V_{ext}(\mathbf{r},t)\cdot\nabla_{\mathbf{p}}f(\mathbf{r},\mathbf{p},t)$ in Eq. 2.130. This does *not* mean that the local velocity of the particles is zero, and that the collision integral does not play any role. It simply means that the *global* evolution of the system is driven by these terms, while the collision integral forces the system to a *local* equilibrium.[56] This regime is called local equilibrium, or the *hydrodynamic regime*, since one can derive from the Boltzmann equation 2.130 equations of motion for the electron liquid similar to the hydrodynamic equations for a classical fluid.

In Chapter 8 I will show that hydrodynamic equations can indeed be derived more "naturally" in quantum mechanics than in classical physics: the time-dependent Schrödinger equation 1.16 can be equivalently written *exactly* in terms of the single-particle density and current density. In the same chapter I will also show that under appropriate conditions transport in nanostructures can be described by simple hydrodynamic equations.

I conclude by noting that, while I have discussed the Boltzmann equation for electrons, a similar equation may be derived to study the non-equilibrium properties of bosons, such as phonons. In this case, different approximations to the collision integral can be derived that do not require the Pauli exclusion principle.

2.8 Entropy, loss of information, and macroscopic irreversibility

The Boltzmann equation 2.130 has been derived to describe the dynamics of irreversible statistical processes of a large number of particles, i.e., processes that are not invariant under *time-reversal symmetry*. However, if we look back at its derivation, we have only made use of *microscopic* equations of motion – namely the Hamilton equations 2.127 – that are indeed indifferent to the arrow of time (in the sense that they are symmetric under time-reversal symmetry $t \to -t$, $\mathbf{r}_j \to \mathbf{r}_j$, $\mathbf{p}_j \to -\mathbf{p}_j$ if the Hamiltonian is an even function of the momenta[57]). This means that if the set $\{\mathbf{r}_j(t),\mathbf{p}_j(t)\}$ is a solution of the equations 2.127, $\{\mathbf{r}_j(-t),-\mathbf{p}_j(-t)\}$ is also a solution of

---

[55] Note that this is true for elastic collisions as well, i.e., collisions that conserve energy.
[56] Or *steady state* if we have an infinite system or continuous supply of charges.
[57] In the presence of a magnetic field the equations of motion are invariant with respect to time-reversal symmetry if the sign of the field is also changed.

the same equations. This is called *micro-reversibility* of the equations of motion.[58]

How is it then possible that one can derive an equation of motion that describes *macroscopic* irreversible processes from microscopic equations of motion that are time-reversible? The answer is *loss of information*. Let us expand on this.[59]

If we could solve the Boltzmann equation exactly, namely if we could solve the BBGKY hierarchy to all orders, we would know how the many-body phase density $D(\mathbf{r}_1, \mathbf{p}_1; \ldots; \mathbf{r}_N, \mathbf{p}_N, t)$ evolves in time. We would then know *all* correlations among particles, and how these correlations change in the course of time.

As we have discussed in the preceding section, this is clearly an impossible task – both experimentally and theoretically/numerically. We thus need approximations to the collision integral that generally entail neglecting correlations among particles. For instance, if we employ the approximation 2.132 for $\mathbf{I}[f]$, we account, via factors of the type $f(\mathbf{r}, \mathbf{p}', t)\left[1 - f(\mathbf{r}, \mathbf{p}, t)\right]$, for the fact that electrons cannot scatter into occupied single-particle states.[60] However, at the same time, we neglect how two, three or more electrons correlate with each other during scattering.

By making approximations to the collision integral we thus "lose information" on the degrees of freedom that correspond to particle correlations.[61] This information cannot be recovered during time evolution (since we do not know the exact collision integral), and thus the dynamics described by the Boltzmann equation 2.130, with an approximate $\mathbf{I}[f]$, is in fact *irreversible*. This seems an almost obvious statement, but is not the whole story.

Even if we could solve the Boltzmann equation exactly, by focusing on the single-particle degrees of freedom (that are described by the reduced density f), during time evolution we would nonetheless *transfer* information to the very many degrees of freedom related to the correlations of all particles in the system. The question is then how easy it is to recover such information after a long time has passed from the initial condition.

The answer to this question can only be given probabilistically, namely, the probability that we can recover information on the correlations, once

---

[58] Note that $\{\mathbf{r}_j(-t), -\mathbf{p}_j(-t)\}$ may represent a physically *different* trajectory in phase space than $\{\mathbf{r}_j(t), \mathbf{p}_j(t)\}$. In this case, we say that a *spontaneous symmetry breaking* has occurred: the solution of the equations of motion does not satisfy the same symmetry as the equations themselves (see, e.g., Gaspard, 2006).

[59] While the following discussion relies on the Boltzmann equation and classical concepts, the conclusions on macroscopic irreversibility are valid for quantum systems as well (see Sec. 2.8.3).

[60] This is also a type of "correlation" due to the Pauli exclusion principle.

[61] The approximation that correlations among particles can be neglected, i.e., one can consider only the one-particle reduced density f, is often known as *Stosszahlansatz*.

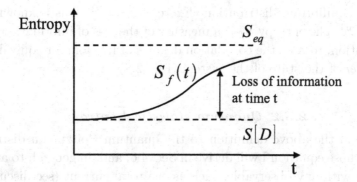

Fig. 2.8. The entropy $S_f(t)$ is associated with the single-particle distribution function f. At the initial time it is equal to the entropy of the total phase density $S[D]$, which remains constant in time. $S_f(t)$ instead increases with time. If the system is not subject to external time-dependent perturbations, $S_f(t)$ approaches the global equilibrium entropy S_{eq}.

a long time has passed from the initial time, and revert the dynamics to the initial condition is infinitesimally small. For all practical purposes, once this information is lost in the correlation degrees of freedom it cannot be recovered, and the dynamics becomes *practically irreversible*.

From the above discussion it is thus clear that macroscopic irreversibility of physical processes is related to loss of information during time evolution, or better still, to the *transfer of information* – in the form of classical or quantum correlations – from the system whose dynamics we are interested in to other degrees of freedom, that either we choose not to consider, or their number is so large that it is practically impossible to follow each and every one of them in time.

2.8.1 The classical statistical entropy

Let us try to quantify this loss of information by defining a *statistical entropy* S related to the quantity we are interested in. In the case of the Boltzmann equation we define

$$S_f(t) = -k_B \int d\mathbf{r}\, d\mathbf{p}\, f(\mathbf{r}, \mathbf{p}, t) \ln f(\mathbf{r}, \mathbf{p}, t) \equiv -\mathcal{H}(t). \qquad (2.141)$$

It can be shown that the entropy $S_f(t)$ *increases* with time from its initial value $S[D]$ ($\mathcal{H}(t)$ decreases), which corresponds to the entropy associated with the *exact* phase density D. It reaches a constant value only if f is the global equilibrium distribution (Exercise 2.13).[62] This is known as the *H-theorem*.[63] The entropy S_f is a measure of the loss of information, during time evolution, toward the correlation degrees of freedom, or equivalently of the *disorder* of the state of the system (Fig. 2.8).

2.8.2 Quantum statistical entropy

I can extend the above definition to the quantum Boltzmann distribution function whose equation I will derive in Sec. 4.6, and, in general, to any state associated with any observable, such as the total current (see discussion in Sec. 1.5).

Given a statistical operator $\hat{\rho}$ that represents the macro-state $\{|\Psi_i\rangle, p_i\}$ we define the *quantum statistical entropy* associated with such a statistical operator as

$$S[\hat{\rho}] = -k_B \operatorname{Tr}\{\hat{\rho} \ln \hat{\rho}\} = -k_B \sum_i p_i \ln p_i \qquad (2.142)$$

where the last equality comes from the definition of the trace (see Eq. 1.50), and the fact that $\hat{\rho}$ is diagonal in the basis $\{|\Psi_i\rangle\}$, with eigenvalues $\{p_i\}$.

Micro-canonical entropy

If $\{|\Psi_i\rangle\}$ is a complete set in the Hilbert space of the system, and the latter explores M available states, the *micro-canonical* statistical operator is simply

$$\hat{\rho} = \sum_{i=1}^{M} |\Psi_i\rangle \frac{1}{M} \langle\Psi_i|, \qquad (2.143)$$

namely the probability p_i of finding an element i of the ensemble in the state $|\Psi_i\rangle$ is constant and equal to $1/M$. From 2.142 we then find that the statistical entropy is a maximum and is equal to

$$S[\hat{\rho}] = k_B \ln M, \quad \text{for equiprobable states.} \qquad (2.144)$$

The above corresponds to the case of a closed system whose energy is well defined (not just on average) and no other constants of motion exist.

---

[62] The Boltzmann entropy reaches the global equilibrium entropy 2.146 only if the system is isolated. Also, its time derivative may be zero at some given time, if, at that time, it is equal to the local equilibrium distribution at every point in space (Exercise 2.13).

[63] There are exceptions to the H-theorem. For instance, if the initial state of the system is highly correlated (extremely disordered) compared to the final equilibrium state, one cannot simply neglect these initial correlations, and the approach to equilibrium may indeed occur via an increase of $\mathcal{H}(t)$. See Jaynes, 1971, for a discussion on H-violating initial conditions.

On the other hand, if the system is in a pure state the corresponding macro-state has probability distribution $\{1, 0, \ldots, 0\}$ (see Sec. 1.4) so that from 2.142 we find

$$S[\hat{\rho}] = 0, \quad \text{for a pure state.} \tag{2.145}$$

Canonical entropy

When the system is in *global equilibrium* with a bath at temperature θ, the probabilities p_i are given by their canonical value 1.79 and the associated entropy is

$$S[\hat{\rho}_C^{eq}] = -k_B \sum_i \frac{e^{-\beta E_i}}{\mathcal{Z}_C} \ln\left(\frac{e^{-\beta E_i}}{\mathcal{Z}_C}\right)$$

$$= k_B \ln \mathcal{Z}_C + k_B \beta \sum_i E_i \frac{e^{-\beta E_i}}{\mathcal{Z}_C}$$

$$= k_B \ln \mathcal{Z}_C + \frac{\langle \hat{H} \rangle_{eq}}{\theta} \equiv S_{eq}, \tag{2.146}$$

where $\langle \hat{H} \rangle_{eq} = \text{Tr}\{\hat{\rho}_C^{eq} \hat{H}\}$ is the average energy of the system at equilibrium, and $\mathcal{Z}_C$ is the canonical partition function (Eq. 1.78). Equation 2.146 is precisely the equilibrium entropy of the canonical ensemble from which one can derive the second law of thermodynamics (Exercise 2.14). It is also the maximum amount of disorder a system can reach when at equilibrium with a bath at temperature θ.

Since the partition function is an *extensive variable*, it is proportional to the number of particles N of the system. By comparing 2.146 with 2.144, we thus see that at equilibrium the number of microscopic states (or the entropy) is proportional to the number of particles; a fact I have used in Sec. 1.2 to show that the separation of many-body energy states of a capacitor decreases exponentially with the number of particles (Eq. 1.14).

Entropy as loss of information

That $S[\hat{\rho}]$ is a good measure of loss of information in the quantum case can be seen as follows. Let $\hat{\rho}$ be the statistical operator of a closed system that is not in a pure state but in a statistical mixture, evolving under Hamiltonian dynamics (Eq. 1.16). The statistical operator evolves according to Eq. 1.60.

The time derivative of $S[\hat{\rho}]$ is thus

$$\frac{dS[\hat{\rho}]}{dt} = \frac{\mathrm{i}k_B}{\hbar}\mathrm{Tr}\{[\hat{H},\hat{\rho}]\ln\hat{\rho}\} = \frac{\mathrm{i}k_B}{\hbar}\mathrm{Tr}\{\hat{H},[\hat{\rho},\ln\hat{\rho}]\} = 0, \qquad (2.147)$$

where in the last equality I have used the cyclic property of the trace (Eq. E1.3 in Exercise 1.1), and that an operator commutes with the logarithm of itself. We thus find that if we knew the dynamics of the system completely, even if its statistical operator varies in time, we would not lose any information.[64]

If, instead, we focus on the information we can extract from a *reduced* statistical operator, the amount of information one can gather from that statistical operator is limited, and the entropy must increase with time as in the Boltzmann equation case.

Finally, we can quantify the statement I have made before, namely that even if we knew exactly the single-particle distribution function f of the Boltzmann equation, the probability is negligible that we can recover full information on all degrees of freedom of the system, and thus revert its dynamics to the initial conditions.

Assume the system has reached, during its time evolution, some disordered macro-state. The number of micro-states associated with that macro-state is of the order of $\exp(S/k_B)$ (from Eq. 2.144). In macroscopic systems a typical equilibrium entropy is of the order of 1 J K^{-1} (Balian, 1991). The Boltzmann constant is $k_B = 1.38 \times 10^{-23}$ J K^{-1}. The number of microstates is thus of the order of $\exp(10^{23})$!

If the initial state is prepared in a pure state, then there is only one microstate associated with it. Even if this is not the case, due to the initial preparation of the system, its initial macro-state would, most of the time, have a different number of micro-states than macro-states at later times. In either case, we need to revert the dynamics of a macroscopic number of micro-states into one micro-state or another set of micro-states. For all practical purposes, this is clearly an impossible task, and reversibility is highly improbable.[65]

---

[64] In the classical case this statement would correspond to saying that the entropy $S[D] = -k_B \int d\Omega\, D\ln D$ associated with the phase density $D(\mathbf{r}_1,\mathbf{p}_1;\ldots;\mathbf{r}_N,\mathbf{p}_N,t)$ does not vary in time: if we could follow the time evolution of *all* degrees of freedom of the system, our knowledge of its evolution would be complete, and its entropy would remain constant.

[65] A statement known as the *fluctuation theorem* has been advanced (Evans et al., 1993) where it is argued that the ratio between the probability that the entropy decreases by an amount ΔS per unit time and the probability that it increases by the same amount decays exponentially with time as $\exp(-\Delta St/k_B)$. This shows again that reverting the microscopic dynamics is not impossible, simply very unlikely.

2.8.3 Information content of the N- and one-particle statistical operators

In the following chapters, most of the time we will make use of single-particle quantities to develop theories of transport (whether for non-interacting electrons, or for truly interacting electrons). This is, for instance, the case for the Landauer approach of Chapter 3, where we will consider single particles interacting at a mean-field level, or the non-equilibrium Green's function formalism of Chapter 4, where we will calculate *single-particle* Green's functions even if we consider interacting electrons. In other words, we focus on the degrees of freedom (and their time evolution) of single particles.

The discussion in the previous section points to the fact that, given an N-particle interacting system, the reduced single-particle statistical operator contains *less* information than the N-particle statistical operator. To be more specific, suppose the many-body system is in a pure state $|\Psi(t)\rangle$ (we can easily generalize the discussion to a mixed state), which in the position representation is $\langle \mathbf{r}_1, \mathbf{r}_2, \ldots, \mathbf{r}_N | \Psi(t) \rangle = \Psi(\mathbf{r}_1, \mathbf{r}_2, \ldots, \mathbf{r}_N, t)$. The N-particle density matrix is then, from Eq. 1.54

$$\rho_N(\{\mathbf{r}'\}, \{\mathbf{r}\}, t) = \langle \mathbf{r}'_1, \mathbf{r}'_2, \ldots, \mathbf{r}'_N | \Psi(t) \rangle \langle \Psi(t) | \mathbf{r}_1, \mathbf{r}_2, \ldots, \mathbf{r}_N \rangle, \qquad (2.148)$$

which contains *full information* on the state of the system: its entropy, $S[\hat{\rho}_N] = -k_B \text{Tr}\{\hat{\rho}_N \ln \hat{\rho}_N\}$, does not change in time – Eq. 2.147 (for the pure state case we consider here it is zero at all times).

From the above density matrix we can construct the reduced *single-particle* density matrix by integrating (tracing) out all degrees of freedom of all particles except one

$$\rho_1(\mathbf{r}'_1, \mathbf{r}_1, t) = \int d\mathbf{r}_2, \ldots, d\mathbf{r}_N \, \Psi^*(\mathbf{r}'_1, \mathbf{r}_2, \ldots, \mathbf{r}_N, t) \, \Psi(\mathbf{r}_1, \mathbf{r}_2, \ldots, \mathbf{r}_N, t). \qquad (2.149)$$

This density matrix does not necessarily describe a pure state of a single particle, and its entropy, $S[\hat{\rho}_1] = -k_B \text{Tr}\{\hat{\rho}_1 \ln \hat{\rho}_1\}$, is thus larger than the entropy of the N-particle density matrix. As in the Boltzmann approach, we have lost information by focusing on the single-particle degrees of freedom. Can we then recover this information as time evolves?

The answer to the above question seems a simple no, but density-functional theories allow us, at least in principle, to recover full information on the system dynamics. Here is how.

From the single-particle density matrix we can calculate the particle density of the system

$$n(\mathbf{r}, t) = N \rho_1(\mathbf{r}, \mathbf{r}, t), \qquad (2.150)$$

or the current density (see Eqs. 4.47 and 4.50 in Chapter 4)

$$\mathbf{j}(\mathbf{r},t) = \frac{i\hbar N}{2m} \lim_{\mathbf{r}'\to\mathbf{r}} (\nabla_{\mathbf{r}'} - \nabla_{\mathbf{r}}) \rho_1(\mathbf{r}, \mathbf{r}', t). \qquad (2.151)$$

Now, let us use density-functional theory (see Appendices E, F, and G).

For simplicity, let us focus only on the density, and work with time-dependent DFT (similar considerations apply to the current density if we work with time-dependent current-DFT or its stochastic version).

From the theorem of time-dependent DFT (see Appendix E), we then know that, for any initial condition, given the exact density (which we can obtain from the set of effective single-particle equations E.4 if we knew the exact functional) we can determine the scalar external potential $\hat{V}_{ext}(t)$ that generates such density (apart from a trivial time-dependent constant). If we know the external potential, we know the full many-body Hamiltonian $\hat{H}$ from which we can, in principle, determine the many-body wave-functions, and hence recover full information on the system at all times. Schematically

$$n(\mathbf{r},t) \longrightarrow \hat{V}_{ext}(t) \longrightarrow \hat{H} \longrightarrow \Psi(\mathbf{r}_1, \mathbf{r}_2, \ldots, \mathbf{r}_N, t) \to \hat{\rho}_N(t). \qquad (2.152)$$

Clearly, the step that goes from the many-body Hamiltonian to the many-body wave-function is purely formal,[66] since we do not generally have the means to calculate the many-body wave-functions directly from the N-particle Hamiltonian – otherwise we would have already solved all many-body problems! Due to this practical impossibility, if we focus only on the reduced single-particle density matrix, information is lost during time evolution, and the system dynamics becomes *practically irreversible*, even if we know the exact functional.

2.8.4 Entropy of open quantum systems

In the case of quantum systems open to an external environment (Sec. 1.4.5) the statistical entropy may either increase or decrease. For instance, in Sec. 2.2.1 I have shown that if we let the system interact with an environment via the operator $\hat{V}$ of Eq. 2.16 the statistical operator evolves into the canonical equilibrium one. In that model, the main time dependence of this evolution is given by the energy relaxation time, and has the simple exponential form $\exp(-t/\tau_E)$. Intuitively, we expect that the approach to equilibrium may occur via an *increase* of entropy if the system is initially prepared in a less disordered state (for instance, in a pure state, or a mixture of states at an initial temperature *lower* than the temperature of the

---

[66] The "reverse" problem of determining the potential that gives rise to a given density is not trivial to solve either.

environment). In this case, the information is lost irreversibly to the degrees of freedom of the bath.

The statistical entropy, however, can also *decrease* towards the value 2.146 if the initial state is more *disordered* than the final state (e.g., the initial state is at a *higher* temperature than the final one). The system becomes more *ordered* at the expense of an increase in disorder of the bath.

In general, one can prove that if the dynamics of the open system is described by the Lindblad equation 1.73, and the latter admits a stationary solution (satisfying Eq. 1.74), the *entropy production rate* (Eq. E2.23 of Exercise 2.15) increases with time, and its time derivative will vanish only when the statistical operator evolves into the stationary one. This means that the statistical entropy 2.142 may increase or decrease in time, with a corresponding *entropy flux* from or into the bath degrees of freedom (Exercise 2.15).

2.8.5 Loss of information in the Kubo formalism

Let us now discuss the information content in a calculation of the conductivity using the linear-response formalism of Sec. 2.3. In that approach, the system is prepared in a canonical equilibrium state whose entropy is given by 2.146. We then let the system evolve under the influence of a perturbation via Hamiltonian dynamics, i.e., we assume the system is *closed*.[67] If we know the *exact* statistical operator at any given time, from Eq. 2.147 we conclude that its entropy will not change in time,[68] and we would preserve all information on *all* the correlations degrees of freedom of the system.

However, in the Kubo formalism we solve the statistical operator to first order in perturbation theory (Eq. 2.42). If $\hat{\rho}_{\text{Kubo}}$ is the *approximate* statistical operator 2.42, its statistical entropy is

$$S[\hat{\rho}_{\text{Kubo}}(t)] = -k_B \text{Tr}\{\hat{\rho}_{\text{Kubo}}(t) \ln \hat{\rho}_{\text{Kubo}}(t)\}, \quad (2.153)$$

and its time derivative

$$\frac{dS[\hat{\rho}_{\text{Kubo}}]}{dt} = \frac{i k_B}{\hbar} \text{Tr}\{[\hat{H}'(t)_I, \hat{\rho}_C^{eq}] \ln \hat{\rho}_{\text{Kubo}}\}, \quad (2.154)$$

where I have used Eq. 2.42 for the approximate Kubo statistical operator. But the term on the right-hand side of Eq. 2.154 is not zero, since the equilibrium statistical operator $\hat{\rho}_C^{eq}$ does not generally commute with either the perturbation $\hat{H}'$, or the Kubo operator $\hat{\rho}_{\text{Kubo}}(t)$. We thus see that we

---

[67] In the sense that it is not interacting dynamically with any environment.
[68] However, it would not necessarily be the equilibrium entropy $S[\hat{\rho}_C^{eq}]$, Eq. 2.146.

lose information by approximating the exact statistical operator with the Kubo one.

2.8.6 Loss of information with stochastic Hamiltonians

I now discuss the case in which the system is closed but the Hamiltonian depends explicitly on the wave-functions or, in general, the system is so complicated that its Hamiltonian is not known exactly (see Sec. 1.4.3). An example of this is the use of approximate Hamiltonians from density-functional theory (Appendices D, E, F). These Hamiltonians depend on the density (or current density) and are thus different for each element of a statistical ensemble.

In all the above cases, we can define an entropy $S[\hat{\rho}_j]$ associated with the statistical operator evolving according to the Hamiltonian $\hat{H}_j$ in the ensemble of random Hamiltonians (see discussion preceding Eq. 1.62). Since the total statistical operator is the weighted sum of all these individual statistical operators with weights $\tilde{p}_j$ (Eq. 1.63), the total entropy is

$$S[\hat{\rho}(t)] = S\left[\sum_j \tilde{p}_j \hat{\rho}_j(t)\right] \geq \sum_j \tilde{p}_j S[\hat{\rho}_j(t)], \qquad (2.155)$$

where the last equality comes from a property of the entropy known as *concavity* which states that *the entropy (or disorder) of a mixture of states of the same system is larger than the sum of its parts*. I leave the demonstration of this result as an exercise for the reader (Exercise 2.17).

The property 2.155 shows that any Hamiltonian of which we do not know all the correlation degrees of freedom increases the disorder of the system. In fact, suppose that the single statistical operators $\hat{\rho}_j(t)$ *conserve* information: $S[\hat{\rho}_j(t)] = S[\hat{\rho}(t = t_0)]$. From Eq. 2.155 we get

$$S[\hat{\rho}(t)] \geq \sum_j \tilde{p}_j S[\hat{\rho}_j(t)] = \sum_j \tilde{p}_j S[\hat{\rho}(t = t_0)] = S[\hat{\rho}(t = t_0)], \qquad (2.156)$$

namely the disorder increases with time, and the associated loss of information cannot be retrieved even though the equations of motion are time-reversal invariant: *a partially known Hamiltonian explicitly introduces irreversibility in the system dynamics.*

2.8.7 Entropy associated with the measurement of currents

I finally quantify the amount of information one obtains from the measurement of currents (Sec. 1.5).[69] We have seen in Sec. 1.5 that the first step of a current measurement (the coupling with the apparatus) truncates the total statistical operator of the system $\hat{\rho}$ into the statistical operator $\hat{\rho}_\mathcal{I} = \sum_I \hat{P}_I \hat{\rho} \hat{P}_I$. The entropy associated with this new state is

$$S[\hat{\rho}_\mathcal{I}] \equiv -k_B \text{Tr}\{\hat{\rho}_\mathcal{I} \ln \hat{\rho}_\mathcal{I}\} = -k_B \text{Tr}\left\{\sum_I \hat{P}_I \hat{\rho} \hat{P}_I \ln \hat{\rho}_\mathcal{I}\right\}$$

$$= -k_B \text{Tr}\left\{\hat{\rho} \sum_I \hat{P}_I \ln \hat{\rho}_\mathcal{I} \hat{P}_I\right\} = -k_B \text{Tr}\{\hat{\rho} \ln \hat{\rho}_\mathcal{I}\}$$

$$\geq -k_B \text{Tr}\{\hat{\rho} \ln \hat{\rho}\} \equiv S[\hat{\rho}], \qquad (2.157)$$

where in going from the second to the third equality I have used the cyclic property of the trace (Eq. E1.3 in Exercise 1.1); from the third to the fourth, the fact that the projection over the current degrees of freedom does not affect the current statistical operator $\hat{\rho}_\mathcal{I}$; and finally I have used the inequality E2.26 of Exercise 2.16.

Therefore, the entropy associated with the first step of a current measurement is larger than the entropy associated with the total statistical operator: by merely coupling the system to the measurement apparatus we have lost information on other degrees of freedom that are "hidden" in the total statistical operator $\hat{\rho}$. This information is irreversibly lost into the apparatus degrees of freedom and thus cannot be recovered.

Finally, if we do *observe* a current value I, due to the postulate of wave-packet reduction, the statistical operator becomes $\hat{\rho}_I = \hat{P}_I \hat{\rho} \hat{P}_I / \text{Tr}\{\hat{\rho} \hat{P}_I\}$ (see Sec. 1.5). The information content of this operator,

$$S[\hat{\rho}_I] = -k_B \text{Tr}\{\hat{\rho}_I \ln \hat{\rho}_I\}, \qquad (2.158)$$

may be even *smaller* than the information contained in the state $\hat{\rho}_\mathcal{I}$ (once again, we have lost information). However, if we perform an ensemble of measurements on the system, and observe a distribution of values $\{I\}$, each with the probability $p_I = \text{Tr}\{\hat{\rho} \hat{P}_I\}$ (from Eq. 1.83), the *average* amount of information is *larger* than the one we obtain by simply coupling our system with the measurement apparatus. Indeed, from Eq. 2.142 we easily get (using the general property of a projector $\hat{P}_I^2 = \hat{P}_I$)

$$\bar{S}_I = -k_B \sum_I p_I \ln p_I = S[\hat{\rho}_\mathcal{I}] - \sum_I p_I S[\hat{\rho}_I]. \qquad (2.159)$$

---

[69] A similar discussion for a general observable can be found in Balian, 1991.

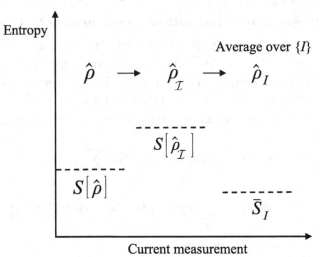

Fig. 2.9. Information content during an ideal measurement of the current. In the first step, we couple the system to an apparatus. The initial state of the system, represented by the statistical operator $\hat{\rho}$, irreversibly becomes $\hat{\rho}_\mathcal{I}$. We lose information during this step. If we observe one specific outcome I, the associated entropy may be even larger. However, on *average*, over the ensemble of measurements of the current $\{I\}$, we *gain* the average information $\bar{S}_I$, which may be smaller than the entropy $S[\hat{\rho}]$ before the measurement.

If the system ends up in a pure state this information gain is precisely $S[\hat{\rho}_\mathcal{I}]$, the entropy immediately after the coupling with the apparatus. The entropy $\bar{S}_I$ may be even smaller than the initial entropy $S[\hat{\rho}]$ *before* the measurement. We thus generally gain (on average) information during the reading of an ensemble of values of the current (Fig. 2.9).

From the above discussion we arrive at an important conclusion that I have already anticipated: unlike the equilibrium case, in a non-equilibrium problem there is *no unique* definition of entropy. We define an entropy according to the given observable or degrees of freedom we are really interested in. We then interpret such entropy as loss of information or amount of disorder associated with such degrees of freedom.

Summary and open questions

I have introduced the transport theories belonging to **Viewpoint 1**: the current is the response to an external electric field. These theories have allowed me to introduce several concepts I will use in the rest of the book, such as relaxation, dephasing, local electrochemical potential, temperature etc.

I have also explicitly shown that electrical conduction *cannot* be described using ground-state theories. This holds also in the limit of zero external field, where the conductivity is related to *spontaneous fluctuations* of the current close to equilibrium. In addition, I have proved that, in general, the conductivity at a given time requires knowledge of its value at previous times: the current is a *history-dependent* quantity.

I have also defined the statistical entropy which allows us to quantify the loss of information in the description of physical processes out of equilibrium. In particular, I have discussed the information content in the measurement of currents, and quantified the loss of information when we focus on a reduced description of a physical phenomenon. For instance, we generally lose information when describing the transport problem within a single-particle picture. Even if dynamical density-functional theories allow us – at least in principle – to recover such information at any given time, it is practically impossible to do so, and the system dynamics is *practically irreversible*.

Exercises

2.1 **Frequency-dependent Drude conductivity.** Consider independent electrons with average momentum $\mathbf{p}(t)$ and effective mass m^* in an oscillating electric field of amplitude $\mathbf{E}$ and frequency ω. The electrons experience collisions at an average rate of $1/\tau$, with τ the relaxation time. The equation of motion for the electron momentum is

$$\frac{d\mathbf{p}(t)}{dt} = -\frac{\mathbf{p}(t)}{\tau} + e\mathbf{E}e^{-i\omega t}. \quad (E2.1)$$

The current density is $\mathbf{j}(t) = en\mathbf{p}/m^*$, where n is the electron density. Using this model, show that in the long-time limit the *complex Drude conductivity* is

$$\sigma_{\alpha\beta}(\omega) = \frac{e^2 n\tau}{m^*} \frac{\delta_{\alpha\beta}}{1 - i\omega\tau}. \quad (E2.2)$$

The real part of the complex conductivity $\mathrm{Re}\{\sigma\}$ is related to dissipation and is

$$\sigma'(\omega) = \mathrm{Re}\{\sigma(\omega)\} = \frac{e^2 n\tau}{m^*} \frac{\delta_{\alpha\beta}}{1 + \omega^2\tau^2}. \quad (E2.3)$$

The value $\sigma'(\omega = 0)$ is known as the *Drude peak*. Using these results, check that the sum rule in Footnote [7] is verified.

2.2 Prove Eqs. 2.18 and 2.19 for the diagonal elements of the density matrix, and Eqs. 2.23 and 2.24 for the off-diagonal ones.

2.3 Prove that expression 2.57 is finite for $\omega \to 0$. Show that Eq. 2.73 can be obtained from Eq. 2.57 in the limits $\{\mathbf{k} \to 0, \mathbf{k}' \to 0\}$ and $\omega \to 0$, in this precise order.

2.4 Prove Eq. 2.65.

2.5 **Infinite conductivity of a homogenous system.** Show that the conductivity of a perfectly homogeneous system is infinite in the long-wavelength limit ($\mathbf{k} \to 0$ *before* $\omega \to 0$).

2.6 **Continuity.** From the continuity equation 2.70 show that

$$\sum_\mu k_\mu \chi^j_{\mu\nu} = 0, \tag{E2.4}$$

with $\chi^j_{\mu\nu}$ the Fourier transform of the full response function 2.54.

2.7 **Gauge invariance.** Using the general gauge invariance property of a vector potential

$$\mathbf{A}'(\mathbf{r}, t) = \mathbf{A}(\mathbf{r}, t) + \nabla \lambda(\mathbf{r}, t), \tag{E2.5}$$

with $\lambda(\mathbf{r}, t)$ an arbitrary real function, show that

$$\sum_\nu \chi^j_{\mu\nu} k_\nu = 0. \tag{E2.6}$$

2.8 **Real and imaginary parts of the response function.** Consider the Fourier transform of the response function 2.47

$$\chi_{AB}(\omega) = \int dt\, e^{i\omega t} \chi_{AB}(t - t'), \tag{E2.7}$$

and write it in terms of a real and an imaginary part

$$\chi_{AB}(\omega) = \mathrm{Re}\{\chi_{AB}(\omega)\} + i\mathrm{Im}\{\chi_{AB}(\omega)\}. \tag{E2.8}$$

Show that the following relations hold

$$\mathrm{Re}\{\chi_{AB}(\omega)\} = \mathrm{Re}\{\chi_{A^\dagger B^\dagger}(-\omega)\}$$
$$\mathrm{Im}\{\chi_{AB}(\omega)\} = -\mathrm{Im}\{\chi_{A^\dagger B^\dagger}(-\omega)\}. \tag{E2.9}$$

(*Hint:* write the response function in terms of a complete set of exact eigenstates of the unperturbed Hamiltonian.)

2.9 **Energy dissipation.** Define the average energy

$$U = \mathrm{Tr}\{\hat{\rho}(t)\hat{H}(t)\}, \tag{E2.10}$$

where $\hat{\rho}(t)$ is the statistical operator of the system and $\hat{H}(t)$ is Hamiltonian 2.31. Show that

$$\frac{dU}{dt} \propto \mathrm{Im}\{\chi_{AA^\dagger}(\omega)\}, \tag{E2.11}$$

namely the imaginary part of the response of an observable $\hat{A}$ to itself ($\hat{B} = \hat{A}^\dagger$) is related to energy dissipation.

2.10 **The fluctuation-dissipation theorem.** Consider the linear response of an observable $\hat{A}$ to a periodic perturbation (Sec. 2.3.1). Define the *correlation function*

$$S_{AA^\dagger}(\omega) = \frac{1}{2\pi}\int_{-\infty}^{\infty} dt\, e^{i\omega t} \langle \hat{A}(t)\hat{A}^\dagger \rangle_{eq}. \tag{E2.12}$$

Calculate the response function $\chi_{AA^\dagger}(\omega)$, namely the response of the observable $\hat{A}$ due to $\hat{B} = \hat{A}^\dagger$. Prove the following fluctuation-dissipation relation

$$\mathrm{Im}\{\chi_{AA^\dagger}(\omega)\} = -\frac{\pi}{\hbar}\left(1 - e^{-\beta\hbar\omega}\right) S_{AA^\dagger}(\omega), \tag{E2.13}$$

where $\beta = 1/k_B\theta$. Since the imaginary part of the response function describes dissipation (Exercise 2.9), the above shows that the latter is related to the correlations (spontaneous fluctuations) in the system, and generalizes the result 2.73 to an arbitrary observable.

2.11 **Exact collision integral.** Consider the classical phase density $D(\mathbf{r}_1, \mathbf{p}_1; \ldots; \mathbf{r}_N, \mathbf{p}_N, t)$ of N interacting and indistinguishable particles. The phase density satisfies the *Liouville equation*

$$\frac{\partial D}{\partial t} = \sum_k \left(\frac{\partial H}{\partial q_k}\frac{\partial D}{\partial p_k} - \frac{\partial D}{\partial q_k}\frac{\partial H}{\partial p_k}\right) \equiv \{H, D\}, \tag{E2.14}$$

where q_k and p_k are the canonical variables of the particles, and H is the system Hamiltonian. Consider the following Hamiltonian

$$H = \sum_{i=1}^{N}\left(\frac{\mathbf{p}_i^2}{2m} + V_{ext}(\mathbf{r}_i, t)\right) + \sum_{i,j=1; i<j}^{N} W(|\mathbf{r}_i - \mathbf{r}_j|), \tag{E2.15}$$

where W is a general two-body potential, and V_{ext} a one-body external potential. Using the Liouville equation E2.14, and assuming that the phase density D vanishes at infinity, show that the collision integral of the Boltzmann equation 2.130 can be written as in Eq. 2.131. (*Hint:* Integrate the Liouville equation E2.14 over a pair of conjugated variables.)

2.12 **Collision invariants.** Consider the Boltzmann equation 2.130 with the collision integral 2.132. Prove that the collisions preserve number of particles, momentum (if lattice translation symmetry is preserved), and energy, namely

$$\int d\mathbf{p}\, \mathbf{I} = 0, \quad \text{particle-number conservation,}$$

$$\int d\mathbf{p}\, p_i\, \mathbf{I} = 0, \quad \text{momentum conservation,}$$

$$\int d\mathbf{p}\, p^2\, \mathbf{I} = 0, \quad \text{energy conservation,} \tag{E2.16}$$

where p_i ($i = x, y, z$) is a component of the particle momentum. The above relations are written for local collisions. If the collisions are not local an extra integration in space has to be performed. Show that the relaxation-time approximation 2.134 violates all these conservation laws.

2.13 **The H-theorem.** Consider the collision integral approximation 2.132 for the Boltzmann equation 2.130. Prove that the Boltzmann entropy

$$S_f(t) = -k_B \int d\mathbf{r}\, d\mathbf{p}\, f(\mathbf{r}, \mathbf{p}, t)\, \ln f(\mathbf{r}, \mathbf{p}, t), \tag{E2.17}$$

increases with time, namely

$$\frac{dS_f(t)}{dt} \geq 0. \tag{E2.18}$$

Show that, for any given time t, the equal sign holds only if f is a *local equilibrium distribution at every point in space*.

2.14 **Second law of thermodynamics.** Consider an arbitrary Hamiltonian $\hat{H}(\{\chi_\beta\}, t)$ which depends on some variables $\{\chi_\beta\}$ whose variation is related to the work W done on the system by external forces according to the linear relation

$$dW = \sum_\beta F_\beta\, d\chi_\beta, \tag{E2.19}$$

with

$$\hat{F}_\beta = \frac{\partial \hat{H}}{\partial \chi_\beta}. \tag{E2.20}$$

Show that the canonical entropy 2.146 is related to the heat dQ exchanged in an infinitesimal interval of time dt by the system with an external environment during a *quasi-static* transformation (namely, one in which the system finds itself always in equilibrium states) via

$$dS[\hat{\rho}_C^{eq}] = \frac{dQ}{\theta} \equiv \frac{\text{Tr}\{d\hat{\rho}\hat{H}\}}{\theta}, \quad \forall\, d\beta, d\chi_\beta, \tag{E2.21}$$

where
$$\hat{\rho}(t) = \frac{e^{-\beta \hat{H}(t)}}{\text{Tr}\{e^{-\beta \hat{H}(t)}\}}. \tag{E2.22}$$

Equation E2.21 is the second law of thermodynamics.

2.15 **Entropy production in the presence of stationary states.** Consider the Lindblad equation 1.73 with a time-independent Hamiltonian and Lindbladian. Assume that Eq. 1.73 admits a stationary solution $\hat{\rho}^{ss}$. Define the *entropy production rate*

$$\begin{aligned} r[\hat{\rho}(t)] &\equiv \frac{d(S[\hat{\rho}(t)|\hat{\rho}^{ss}])}{dt} \\ &\equiv -k_B \frac{d(\text{Tr}\{\hat{\rho}(t) \ln \hat{\rho}(t)\})}{dt} + k_B \frac{d(\text{Tr}\{\hat{\rho}(t) \ln \hat{\rho}^{ss}\})}{dt}. \end{aligned} \tag{E2.23}$$

Show that
$$r[\hat{\rho}(t)] \geq 0, \quad \forall t, \tag{E2.24}$$

and it is zero only if $\hat{\rho}(t) = \hat{\rho}^{ss}$. If the system dynamics follows the Lindblad equation, and there exists a stationary solution of such an equation, the entropy production rate is a convex functional in the Hilbert space of the system.

We can separate the entropy production rate in two parts
$$r[\hat{\rho}(t)] \equiv \frac{dS}{dt} + J, \tag{E2.25}$$

where S is the statistical entropy 2.142 and J may be identified with the *entropy flux*. Comparing Eq. E2.25 with Eq. E2.23 write the explicit form of this flux.

2.16 Consider two non-negative operators $\hat{A}$ and $\hat{B}$. Prove the following inequality
$$\text{Tr}\{\hat{A} \ln \hat{B}\} - \text{Tr}\{\hat{A} \ln \hat{A}\} \leq \text{Tr}\{\hat{B}\} - \text{Tr}\{\hat{A}\}. \tag{E2.26}$$

2.17 **Concavity of the statistical entropy.** Consider two statistical operators $\hat{\rho}_1$ and $\hat{\rho}_2$. Given a real number $0 < \alpha < 1$, use the relation E2.26 to prove that the statistical entropy 2.142 satisfies the following inequality
$$S[\alpha \hat{\rho}_1 + (1-\alpha) \hat{\rho}_2] \geq \alpha S[\hat{\rho}_1] + (1-\alpha) S[\hat{\rho}_2], \tag{E2.27}$$

with the equal sign only if $\hat{\rho}_1 = \hat{\rho}_2$.

Consider now a set of real positive numbers $\tilde{p}_j$, with $\sum_j \tilde{p}_j = 1$. Using the above inequality demonstrate that

$$S\left[\sum_j \tilde{p}_j \hat{\rho}_j(t)\right] \geq \sum_j \tilde{p}_j S[\hat{\rho}_j(t)], \qquad (E2.28)$$

which shows that the statistical mixture of states of the same system is more disordered than the sum of its parts.

3
Landauer approach

I am now ready to move on to the transport theories based on **Viewpoint 2**: the field is a consequence of carrier flow (see Sec. 1.6). Seeds of this approach are almost as old as quantum mechanics. For instance, one of the first applications of scattering theory to determine the electrical resistance at the contact between two conductors can be found in the work of J. Frenkel published in 1930 (Frenkel, 1930). Here, the calculation of the current is done by assuming that electrons on the two sides of the contact, at their own *local* equilibrium distribution, have a finite transmission probability to tunnel across the potential barrier induced by the junction. This result anticipated the concept that, under specific conditions, the conductance of a given system sandwiched between electrodes can be related to its transmission properties, namely to the probability for electrons to "cross" the system in going from one electrode to the other.

However, the idea that scatterers can induce, self-consistently, local fields which "act back" on the carrier dynamics was pioneered by Landauer (Landauer, 1957), and has contributed tremendously to our understanding of electron transport in mesoscopic and nanoscopic systems. It is a major conceptual departure from the theories I have discussed in the previous chapter.

Another fundamental result of the Landauer approach is that a finite resistance emerges even if the transmission probability of the sample is unity; a fact I have anticipated – though from a different perspective – in Sec. 2.3.4. This result requires a non-trivial understanding of the role of the sample versus all other elements of an electrical circuit (Imry, 1986).

Before proceeding, I want to stress once again – as done by Landauer himself in several of his publications – that the approach I am going to introduce is, first and foremost, a *viewpoint*, not a set of equations! This viewpoint rests on specific physical assumptions that may or may not be satisfied in all experimental realizations of transport in nanoscale systems.

I will review here these assumptions and derive the equations that are now known as Landauer's formulas. My introduction to this approach will rely first on physical arguments, followed by a formal derivation of these formulas using tools of scattering theory.

3.1 Formulation of the problem

In Chapter 1 I have discussed the different ways one can employ to initiate current flow. For instance, I have considered the discharge of a finite capacitor. In that case, I have argued that in the limit of infinite electrode size, while keeping the conductor in between, and its contact with the electrodes fixed, we expect the time-averaged current to realize the steady-state condition 1.13.

Alternatively, we can employ batteries attached to the electrodes, as represented in Fig. 1.3(c), to maintain such a stationarity condition. In both cases, the problem is still fundamentally time dependent and statistical: only the current averaged over time can be assumed constant, while its value at every given time still varies. In addition, we generally measure only the current, and not all quantum numbers necessary to specify the state of the system completely, so that we are forced to solve the equation of motion 1.60 for the statistical operator for the macro-state of the system.

In the case of the battery – where by "battery" I mean here any element of the circuit external to the nanojunction – we would have to solve for the total Hamiltonian $\hat{H}_{tot}$ that describes the dynamics of the electrons and ions of the nanoscale system we are interested in (let us call this Hamiltonian $\hat{H}_S$), plus the electrons and ions of the battery (call it $\hat{H}_{\text{battery}}$), and the interactions between these two systems, $\hat{H}_{\text{int}}$,[1]

$$\hat{H}_{tot} = \hat{H}_S + \hat{H}_{\text{battery}} + \hat{H}_{\text{int}}. \tag{3.1}$$

Approximation 1: *Open quantum systems*

However, the Hamiltonian of the battery is really too complicated to write down. We thus proceed as follows. As I have discussed in Sec. 1.2, the

---

[1] Equation 3.1 must be interpreted as

$$\hat{H}_{tot} = \hat{H}_S \otimes \hat{1}_{\text{battery}} + \hat{1}_S \otimes \hat{H}_{\text{battery}} + \hat{H}_{\text{int}},$$

where $\hat{1}_{\text{battery}}$ is the identity operator on the Hilbert space of the battery degrees of freedom, and $\hat{1}_S$ is the identity operator on the Hilbert space of the system.

3.1 Formulation of the problem

battery effectively acts as a reservoir of electrons, and we may thus replace it with the infinite capacitor, which, due to the infinite Poincaré recurrence time, also acts (at both sides of the junction) as a reservoir of electrons. The effective problem we are going to describe is thus represented in Fig. 3.1 with the nanoscale junction sandwiched between two large chunks of material, each *open* to a reservoir of electrons. Electrons come in and out of the two reservoirs, with the electrons coming into the reservoir completely uncorrelated from the electrons going out of it (see Sec. 1.2).

Since the reservoirs need to represent a battery, the energy required to extract an electron from one reservoir and bring it into the system must be different from the energy required to extract that electron from the other reservoir and bring it into system. We have discussed in Sec. 2.4 that such energy is the electrochemical potential. We thus assume that the electrochemical potential associated with one reservoir, call it the left reservoir, μ_L, differs from the electrochemical potential of the right reservoir, μ_R, by the bias (Fig. 3.1)

$$V = \frac{\mu_L - \mu_R}{e}, \tag{3.2}$$

which is precisely the relation 2.93 for a voltmeter we have derived in Sec. 2.4 via different arguments. We are thus assuming that *the bias is the electrochemical potential difference per unit charge between two electron reservoirs, as ideally measured by a voltmeter attached to them.*

Microscopically, the above procedure can be formally realized by *tracing out* all degrees of freedom associated with the battery from the statistical operator $\hat{\rho}_{tot}(t)$ of the total system, and maintaining only the degrees of freedom associated with the nanoscale system we are interested in[2]

$$\hat{\rho}_S(t) = \text{Tr}_{\text{battery}}\{\hat{\rho}_{tot}(t)\}. \tag{3.3}$$

However, I have shown in Sec. 1.4.5 that, in general, this does not lead to a closed equation of motion for the reduced statistical operator $\hat{\rho}_S(t)$. If we assume that the battery degrees of freedom are *dense* in energy, and all their correlation times are very short compared to the electron dynamics – namely, the degrees of freedom of the battery are much faster than the degrees of freedom of the system – we may apply the approximations we use to derive the Lindblad equation 1.73. In this case, we could argue that some closed form for the equation of motion of the reduced statistical operator of

---

[2] For the capacitor I can do the same: I can trace out the degrees of freedom belonging to a region of space far away from the junction.

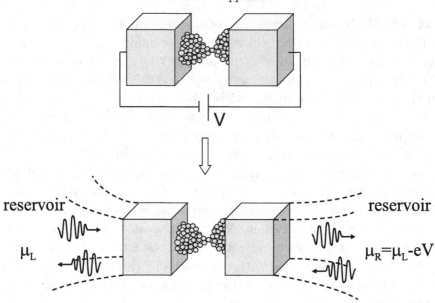

Fig. 3.1. First approximation towards a simplified theory of transport. The closed system, electron source plus electrode-junction-electrode structure, is replaced by the electrode-junction-electrode system *dynamically* coupled to two reservoirs at different electrochemical potentials.

the type

$$\frac{d\hat{\rho}_S(t)}{dt} = \mathcal{L}\hat{\rho}_S(t), \tag{3.4}$$

would be obtained, to be solved with some initial condition $\hat{\rho}_S(t = t_0)$, and $\mathcal{L}$ some super-operator.[3]

What we are now describing is the *dynamical* evolution of our system in interaction with the two reservoirs, each at its own electrochemical potential. This is still too complicated.

Approximation 2: *Ideal steady state*

As I have discussed in Sec. 1.4.5, Eq. 3.4 may or may not have a steady-state solution in the long-time limit (Eq. 1.74). If it does, this solution may not be unique. I *assume* here that Eq. 3.4 *does* admit a *unique* stationary solution (call it $\hat{\rho}_S^{ss}$), and instead of working with the time-dependent $\hat{\rho}_S(t)$,

---

[3] The form of this operator may be very complicated and may include the quantum "jumps" that the electrons experience when they enter or exit the left and right reservoirs. In this case, the statistical operator equation 3.4 contains terms representing states with different number of particles (see, e.g., Gurvitz, 1998).

I work with the *time-independent* $\hat{\rho}_S^{ss}$: we wait long enough for the system to reach a stationary solution (if it exists).[4] I thus assume that the steady-state condition on the current 1.13 is verified not just on average, but *at every instant of time*

$$\langle \hat{I} \rangle_t = \text{Tr}\{\hat{\rho}_S(t)\hat{I}\} \to \text{Tr}\{\hat{\rho}_S^{ss}\hat{I}\} = \langle \hat{I} \rangle = \text{constant}, \qquad (3.5)$$

namely the problem I want to describe is *ideally stationary*.

Note that the statistical operator $\hat{\rho}_S^{ss}$ is the long-time solution of the equation of motion of an *open* quantum system, not a *closed* one whose Hamiltonian is $\hat{H}_S$,[5] and is *not* a global equilibrium statistical operator. So, even if I am now considering an ideal steady state, the problem is still too complicated.[6] It would indeed be much easier if I could work with the Schrödinger equation 1.16 and not with an equation of the type 3.4.

Approximation 3: *"Openness" vs. boundary conditions*

We realize that once the problem has been rendered ideally stationary, the role of the reservoirs is simply to continually prepare electrons in the distant past, and far away from the nanoscale junction, into *wave-packets* which then move towards or away from the junction, without changing the current in time. We thus conveniently abandon the idea of solving the open quantum problem represented by an equation of the type 3.4, and replace the "openness" with appropriate boundary conditions.

These boundary conditions are precisely those I have just discussed: electrons are prepared in the distant past and far away from the junction into wave-packets. These wave-packets move towards the junction from regions of space I now call *leads* (which can be made of any material used to sandwich our junction), *scatter* on the junction potential, and subsequently move far away from it, without further scattering, so that they are (different) wave-

---

[4] I will discuss in Chapter 7, using the micro-canonical picture of transport, that a system may not reach a steady state for all possible initial conditions. Clearly, one can still *impose* an ideal steady state by an appropriate choice of $\hat{\rho}_S^{ss}$. For instance, one can work in a *coarse-grained* picture in which one defines a statistical operator over time scales larger than any possible *autocorrelation time* of the whole system – battery plus nanojunction – (see discussion in Sec. 5.1). Or we could think of this coarse-graining procedure being carried out by the measuring apparatus, which has a finite bandwidth. Either way, one loses information on the electron dynamics at time scales smaller than the autocorrelation time.

[5] Where the Hamiltonian $\hat{H}_S$ may be here an Hermitian operator different from the one appearing in Eq. 3.1, due to the *renormalization* of the energy states of the system in the presence of the interaction with the degrees of freedom of the battery.

[6] For one thing, it is not an easy task to determine the operator $\mathcal{L}$ from the interaction Hamiltonian $\hat{H}_{\text{int}}$.

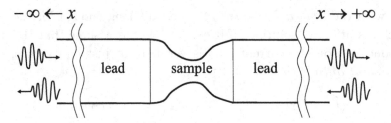

Fig. 3.2. Once the transport problem has been transformed into an ideally stationary one (*Approximation 2*), we can replace the dynamical coupling with the reservoirs with scattering boundary conditions at infinity. We are now working with a *closed* but *infinite* quantum system composed of our sample sandwiched between two leads (*Approximation 3*).

packets propagating in the leads in the distant future (Fig. 3.2). These are known as *scattering boundary conditions*.[7]

For conceptual convenience I now also allow the Hamiltonian $\hat{H}_S$ to describe an *infinite* system, so that when I discuss electrons far away from the junction I really mean *infinitely far* from it. As I will discuss later, this is also a necessary mathematical requirement for a scattering theory: if the Hamiltonian describes the dynamics of a large but finite system the convergence and existence, for $t \to \pm\infty$, of the scattering solutions 3.97 and 3.100 is not guaranteed (Newton, 1966).[8]

If the Hamiltonian $\hat{H}_S$ has eigensolutions

$$\hat{H}_S |\Psi_{E\alpha}\rangle = E |\Psi_{E\alpha}\rangle \tag{3.6}$$

in the continuum, and

$$\hat{H}_S |\Psi_{E_i\alpha}\rangle = E_i |\Psi_{E_i\alpha}\rangle \tag{3.7}$$

in the discrete part of the spectrum, the general stationary solution of the time-dependent Schrödinger equation 1.16 is

$$|\Psi(t)\rangle = \sum_{E_i\alpha} c_{E_i\alpha} |\Psi_{E_i\alpha}\rangle e^{-\frac{i}{\hbar}E_i t} + \sum_{\alpha} \int dE \, c_{E\alpha} |\Psi_{E\alpha}\rangle e^{-\frac{i}{\hbar}Et}, \tag{3.8}$$

where I have lumped in α any other possible quantum number the system may have, and the coefficients $c_{E_i\alpha}$ and $c_{E\alpha}$ are complex numbers.

---

[7] In certain literature these are also called "open" boundary conditions. I reserve the term "open" for systems in dynamical interaction with an environment, e.g., those evolving under the Lindblad equation 1.72. Scattering boundary conditions are for closed systems, namely those evolving under Hamiltonian dynamics 1.16.

[8] This is also the precise order of limits to obtain scattering solutions: first the system volume goes to infinity, and *then* the time can be taken to $\pm\infty$.

We are thus now working with a *closed* (but infinite) system in a *pure* state.

Loss of information in the Landauer approach

Note that unlike the Kubo formalism (Sec. 2.3), we have now lost information on the *history* of the system: in the present approach there is no way for us to determine whether the electron dynamics at a given time may affect the dynamics at later times. In other words, no dynamical information on the current density – which is definitely present in real systems – can be obtained after the static approximation 3.5 is made.[9]

In addition, *Approximation 3* is actually really strong: while in the case of the Kubo formalism I could determine (in principle) the amount of loss of information in going from the exact statistical operator to the approximate Kubo one (Eq. 2.154), I cannot do the same here. I have gone from the statistical operator of the full system – including the battery – to a reduced statistical operator describing the system of interest open to two reservoirs, back to a *different* closed system in a pure state. The amount of correlations I have lost in this process cannot be determined, not even in principle.

Nonetheless, the above three approximations are still not enough if we want to derive a closed form for the total current within this approach. We need to make another approximation.

Approximation 4: *Mean-field approximation*

Let us assume that the Hamiltonian $\hat{H}_S$ can be separated into at least two pieces[10]

$$\hat{H}_S = \hat{H}^{mf} + \hat{V}, \tag{3.9}$$

where $\hat{H}^{mf}$ is a Hamiltonian describing *independent* electrons – at most experiencing the mean field of other electrons – in the presence of the ions that do not belong to the nanoscale junction, and $\hat{V}$ is the interaction energy between electrons – beyond mean field – in the nanojunction, and the interaction energy of these electrons with the ions of the junction.[11]

Without this approximation no closed form for the electrical current can be obtained, and one should, in principle, determine the full many-body statistical operator, from which, via Eq. 1.86, the current may be computed.

---

[9] For instance, we cannot capture the non-linear dynamical effects I describe in Sec. 8.5.
[10] Here, as in most of the book, I will use the same symbol for potential and potential energy.
[11] In simpler words: $\hat{V}$ contains anything else that is *not* contained in the mean-field Hamiltonian $\hat{H}^{mf}$.

The reason is that, if the Hamiltonian $\hat{H}^{mf}$ were to describe truly interacting electrons – beyond mean field – away from the nanoscale junction – say their Hamiltonian contains terms of the type 2.96 – we would have to determine how two electrons evolve in time *in the presence* of all other electrons in the leads *before* they even experience the scattering at the nanostructure. This requires us to know the time evolution of three electrons in the presence of all other electrons, which in turn requires the calculation of the scattering properties of four electrons, and so forth. This generates the equivalent of the infinite BBGKY hierarchy of the Boltzmann approach (Sec. 2.7 and Appendix B). The BBGKY hierarchy is an *infinite number* of "nested" equations, which prevents us from expressing the total current in some "simple" analytical form. I will discuss these issues again in Chapter 4, where I introduce the non-equilibrium Green's function formalism.

The reader may also object that the partitioning 3.9 is arbitrary. Indeed, it is!

First of all, what portion of the total Hamiltonian represents the nanoscale system, and which one the remaining, is a choice that in part may be dictated by how tractable – either analytically or computationally – the resulting problem is, and in part by the physics of the specific problem at hand (see discussions in Secs. 3.5 and 4.4).

An additional complication arises if we consider true interacting electrons in the nanoscale junction: while partitioning the system in terms of ionic components is a well-defined procedure (e.g., we can easily say ion A belongs to the Hamiltonian $\hat{H}^{mf}$ while ion B to the potential $\hat{V}$), the same is not true for electronic interactions. Due to the long-range Coulomb interaction, we cannot, in principle, say that some interactions belong to one region of the system, and other interactions to another.[12]

Despite the above caveats I will assume the partition 3.9 can still be done. In addition, in this chapter I will assume *also* that the potential $\hat{V}$ describes (at most) mean-field interactions, and I will derive in Sec. 4.4 the current in the case in which interactions among electrons – beyond mean field – are included in $\hat{V}$.

Therefore, in the mean-field approximation we consider in this chapter for both $\hat{H}^{mf}$ and the scattering potential $\hat{V}$, we can work with just one *single-particle* Hamiltonian. This Hamiltonian could be for instance the Hartree

---

[12] We can rightfully assume that electrons deep in the electrodes screen Coulomb interactions quite effectively, but it is difficult to argue that these interactions are well screened at, or near, the nanojunction. The separation between a "perfectly" screened region and a partially screened one is thus fuzzy, and depends on the microscopic details of the nanojunction and the self-consistent charge distribution under current flow (see also discussion in Sec. 3.2).

3.1 Formulation of the problem

Hamiltonian

$$H_S = H_{\text{Hartree}}(\mathbf{r}) = -\frac{\hbar^2}{2m}\nabla^2 + V_H(\mathbf{r}) + V_{ext}(\mathbf{r}), \qquad (3.10)$$

where $V_H(\mathbf{r})$ is the Hartree potential energy

$$V_H(\mathbf{r}) = e^2 \int \frac{n(\mathbf{r}')}{|\mathbf{r} - \mathbf{r}'|} d\mathbf{r}', \qquad (3.11)$$

and $V_{ext}(\mathbf{r})$ an external static potential energy, such as the electron-ion interaction potential appearing in Eq. 2.94 and describing the electrode-nanojunction-electrode potential.

If we want to include some correlations among particles we may use the ground-state DFT Hamiltonian (Appendix D)

$$H_S = H_{KS}(\mathbf{r}) = -\frac{\hbar^2}{2m}\nabla^2 + V_H(\mathbf{r}) + V_{xc}(\mathbf{r}) + V_{ext}(\mathbf{r}), \qquad (3.12)$$

with $V_{xc}(\mathbf{r})$ a static exchange-correlation potential energy, which, once again, must be viewed in the present context as a mean-field approximation, even if we knew the exact functional (see also Sec. 3.11). This potential describes both the exchange interaction between electrons (due to the Pauli exclusion principle), and all possible static quantum correlations of the many-body electron system, which are generally unknown and hence represented by some approximate static functional of the density, see Appendix D. In addition, the potentials appearing in $\hat{H}_S$ may be *non-local* in space, if, e.g., a magnetic field is present (see Sec. 3.10).

Therefore, in what follows I will assume that our single-particle Hamiltonian can be generally written as (for a local potential)

$$H_S = -\frac{\hbar^2}{2m}\nabla^2 + V(\mathbf{r}), \qquad (3.13)$$

where, once again, the potential energy $V(\mathbf{r})$ may be the sum of the Hartree energy, exchange-correlation energy, the energy due to the electron-ion interaction, and any other possible external potential energy. It describes all these interactions for the complete electrode-nanojunction-electrode system.

Here, I have also implicitly assumed that electrons scatter with *static* ions. By doing so I am considering only *elastic* scattering where the energy is conserved, while single-particle momentum changes. As I have discussed in Sec. 2.2.1 this amounts to saying that the time it takes electrons to relax energy due to any inelastic effect is much longer than the time it takes electrons to traverse the junction. Similarly for the time it takes them to relax phase, as if all the energy relaxation and all dephasing effects occur in the reservoirs.[13]

This is clearly an idealization. In Chapter 6 I will describe interactions with ionic vibrations, and in Sec. 8.6 I will introduce the local electron heating effect due to inelastic electron-electron interactions.

In terms of the *magnitude* of the *total resistance*, however, in many cases, considering the scattering problem as *phase-coherent* turns out to be not a bad approximation. The reason is that the majority of the *resistance* is due to *elastic* scattering at the nanojunction, while the inelastic component generally gives a small contribution, both from electron-phonon and electron-electron interactions. Nonetheless, this does not mean these effects can be neglected, and, as I will show later, they indeed generate very interesting phenomena.

Approximation 5: *Independent channels and their energy filling*

After all these approximations I am now left with essentially a static – and deterministic – *single-particle* problem that is definitely easier to deal with than the true non-equilibrium statistical problem I have been referring to all along. To the attentive reader, however, the "deterministic" – or pure-state – aspect of all this may seem a bit far-fetched. In fact, we are assuming that the state vector can be developed as in Eq. 3.8, where in the present context we define

Channel: *A set of quantum numbers $\{E, \alpha\}$ that describes a scattering solution.*

How do we know that the source of electrons prepares them in a pure state – even if they interact at a mean-field level?

It is indeed very likely that electrons are initially prepared in a mixed state: we do not measure all their quantum numbers. And even if we could measure *all* quantum numbers necessary to specify the system completely at a given time (a practically impossible proposition), by focusing only on the degrees of freedom of single electrons we are losing information on all the correlations with the other electrons (similar to what happens in the Boltzmann approach where we focus on the single-particle distribution function – see discussion in Sec. 2.8.3).

Even if we work within a single-particle picture with mixed states, in order

---

[13] In Sec. 3.9.1 I will describe an idealized procedure (due to Büttiker, 1986) to generalize the Landauer approach to include dephasing without energy relaxation.

3.1 Formulation of the problem

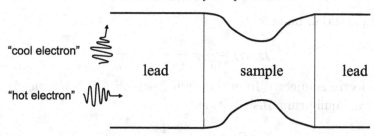

Fig. 3.3. Electrons injected from one side of the battery with a large component of momentum in the direction of the sample are generally "hotter" – farther from local equilibrium – at the boundary with the leads than those injected in a direction with large transverse momentum.

to neglect the correlations between different states we need to assume that the off-diagonal elements of the density matrix – which connect different channels, i.e., any two sets of quantum numbers $\{E, \alpha\}$ and $\{E, \alpha'\}$ – are *exactly* zero. From the discussion in Sec. 2.2.1, we conclude this is possible only if the system interacts with an environment, and does so in such a fast way that these off-diagonal elements are zero almost immediately, *before* we can treat the corresponding states as scattering states. In other words, we need to assume that the system has somehow evolved into a totally *incoherent* (independent) set of (single-particle) channels. Whether this is a good approximation for all physical systems considered experimentally, and under all experimental conditions, remains an outstanding issue.

Finally, even if the channels are independent – do not *mix* – and correspond to the dynamics of single particles, how do we *populate* them? In other words: what is their distribution function? This is a non-trivial point since we are faced with many physically plausible choices (Büttiker *et al.*, 1985). If you think I am being too picky, consider this example.

Take, for instance, electrons injected at a given time from one terminal of the battery. Suppose some of them are injected in a direction of motion that has a large component along the direction of global current flow, and others have a large component in the transverse direction (Fig. 3.3). These latter electrons will take longer to reach the junction, and will thus have more time to relax to whatever local equilibrium energy-momentum distribution they have at the leads. The former set will have instead a much smaller time to relax. Therefore, these electrons will be "hotter" – more precisely, farther from local equilibrium – than the other set.

We neglect these possible differences and *assume* that the electrons are injected from the left reservoir – we call them *right-moving* electrons – with a local equilibrium distribution appropriate to the corresponding electrochemical potential, namely

$$f_L(E) = \frac{1}{e^{(E-\mu_L)/k_B\theta} + 1}, \qquad (3.14)$$

and the electrons injected from the right reservoir – *left-moving* electrons – have a local equilibrium distribution

$$f_R(E) = \frac{1}{e^{(E-\mu_R)/k_B\theta} + 1}, \qquad (3.15)$$

which corresponds to the electrochemical potential of the right reservoir.[14]

This is one of the *many possible choices* for the distribution functions, and it is indeed reasonable. It amounts to saying that despite the presence of the current there are regions of space where the current density is *exactly* zero, and thus the system is truly at equilibrium. However, similar to the approximation of independent channels, this is a strong assumption which may not be satisfied in all systems of interest, and under all experimental conditions.

With the above choice of local distributions, if we call $|\Psi^L\rangle$ a general single-particle state for electrons injected from the left, and $|\Psi^R\rangle$, the corresponding state from the right, we can write the stationary single-particle statistical operator of this problem as the "incoherent" sum of two statistical operators, one describing left-moving electrons and one describing right-moving electrons:[15]

$$\hat{\rho}_S^{ss} = \sum_L |\Psi^L\rangle f_L \langle\Psi^L| + \sum_R |\Psi^R\rangle f_R \langle\Psi^R|. \qquad (3.16)$$

The above five approximations constitute the core of the Landauer approach. The physics represented by this approach is condensed into Fig. 3.4. Our nanojunction (a nanotube, a molecule, etc.) is connected to two chunks of conducting material we have called leads. Electrons are free to move in the leads without scattering. These leads are ideally "open" in the infinite far left and in the infinite far right to two reservoirs as indicated in the figure, even though, as I have discussed above, the analytical problem I am considering does not involve the dynamical coupling with the reservoirs.

---

[14] Note that by replacing the electron sources with reservoirs, we are conceptually using one of their main properties I have discussed in Sec. 1.2: *dilution* of degrees of freedom. The very fact that a reservoir has an infinite number of degrees of freedom allows us to state that electrons that are *emitted* from a reservoir are not the same ones that are *absorbed* by the same reservoir: the *left-moving* and *right-moving* electrons are *uncorrelated* (see also Sec. 3.8.2). In addition, we are also assuming that the current density is zero in a reservoir, so that the latter is in a well-defined *global* equilibrium state.

[15] This *ad-hoc* statistical operator can be thought of as representing a "mixture" of single-particle states in equilibrium with two reservoirs: left-moving channels in equilibrium with the right reservoir, the right-moving with the left reservoir.

3.2 Local resistivity dipoles and the "field response"

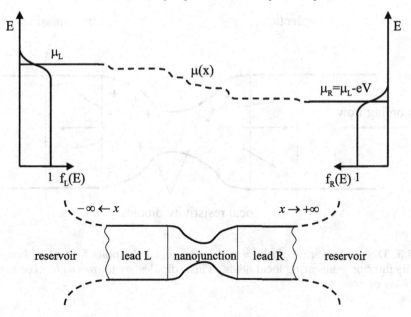

Fig. 3.4. Schematic of the system described within the Landauer approach. The nanojunction is sandwiched between two ideal (scattering-free) leads. Particles are injected at the infinite far left and infinite far right with two different local equilibrium distributions (*Approximation 5*). The infinite far regions may be conceptually thought of as two "reservoirs", even though these do not interact *dynamically* with the system. The (macroscopic-averaged) local electrochemical potential $\mu(x)$ varies along the whole structure.

Electrons are injected at the infinite far left at the local equilibrium distribution f_L, Eq. 3.14, and at the infinite far right, at the local distribution f_R, Eq. 3.15. Conceptually, we are also assuming that the reservoirs are connected to the leads *adiabatically*, namely electrons will be absorbed into the reservoirs without experiencing any further reflection (see also Sec. 2.3.4). The reservoirs are just there to define the left-moving and right-moving distributions 3.14 and 3.15 without otherwise affecting the scattering properties of the system.

3.2 Local resistivity dipoles and the "field response"

Why is the Landauer approach different from the other approaches I have discussed in the previous chapter? The conceptual difference is the following: even if we have a bias applied to the system represented by the electrochemical potential difference 3.2, we do not use that bias as a perturbation to some Hamiltonian. Instead, the bias here is a *boundary condition* on the system,

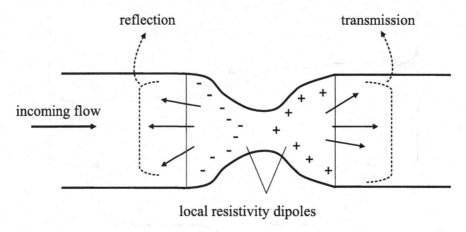

Fig. 3.5. Due to scattering of waves at the sample, local dipoles form at its boundary or in its interior, generating local fields. These dipoles are known as *local* or *residual resistivity dipoles*.

where wave-packets with given momenta carry the current across the nanojunction. Due to the scattering of these wave-packets with the potential $\hat{V}$ of the junction, each electron has a finite probability to be transmitted in any given direction (probability of changing momentum), and consequently a probability to be reflected (Fig. 3.5).

A finite probability of being reflected means that charge accumulates on one side of the scattering center, with consequent depletion on the other side of the obstacle, or if the scattering potential $\hat{V}$ extends over a large region of space, accumulation and depletion occur also in the interior of this region (Fig. 3.5). Since charge accumulates at the boundary of the scattering region or in its interior, it will create, *self-consistently*, local dipoles known as *local* or *residual resistivity dipoles* (Landauer, 1957; Landauer and Woo, 1974). These dipoles develop over length scales of the order of a *screening length*, which (loosely speaking) is the length over which the electric field of a local charge immersed in the electron liquid decays due to the response of the liquid to the local disturbance. In bulk materials, the screening length can be anywhere between 1 Å (in metals) to hundreds of angstroms (in semiconductors).

At nanojunctions (whether metallic or semiconducting), however, the screening length may be much larger than in the bulk due to the reduced dimensions of the junction, and consequent inability of the electrons to respond to local disturbances as effectively as in the bulk – an effect we may call *partial screening*. The evaluation of screening lengths in nanoscale systems thus strongly depends on the atomic details of the junction.[16]

The above shows that, unlike the assumption 2.10 in the Drude model, part of the momentum generated by the field that drives the current is "lost" in the creation of these dipoles. These dipoles create local microscopic electric fields which due to partial screening may be locally much larger than any other macroscopic electric field in the system. We may say the field is the "response" to an incident flux.

Local resistivity dipoles are very important in the interaction among electrons and ions, and they are the main source of current-induced forces (see Sec. 6.5). Note also that these dipoles are *not* a property of the ground state of the system, showing once more that electrical conduction is intrinsically an out-of-equilibrium problem.

3.3 Conduction from transmission

Let us now quantify what I have just discussed. Consider again the junction configuration represented in Fig. 3.4 with our nanojunction in between two ideal (scattering-free) leads "open" to two reservoirs whose only role is to define the left-moving 3.14 and right-moving 3.15 local distributions.

3.3.1 Scattering boundary conditions

The leads define a convenient region where our scattering states can be developed into an appropriate basis of the Hilbert space with consequent definition of channels. To simplify even more (although this step is not necessary) I will assume that the leads are identical (both as material properties and shape), and contain electrons free to travel in the x direction (translationally invariant in the x direction) but confined in the $y-z$ direction (Fig. 3.6).

The Hamiltonian $\hat{H}_S$ in Eq. 3.13 thus satisfies the asymptotic conditions

$$\lim_{x \to -\infty} H_S = -\frac{\hbar^2}{2m} \nabla^2 + V_L(\mathbf{r}_\perp) \equiv H_L, \quad (3.17)$$

and

$$\lim_{x \to +\infty} H_S = -\frac{\hbar^2}{2m} \nabla^2 + V_R(\mathbf{r}_\perp) \equiv H_R, \quad (3.18)$$

---

[16] In other words, such a calculation cannot be done as in the bulk, where one can introduce a well-defined dielectric constant (see, e.g., Ashcroft and Mermin, 1975).

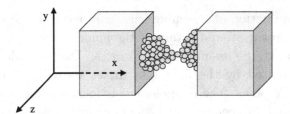

Fig. 3.6. The leads are assumed to confine electrons in the $y-z$ direction.

with $V_L(\mathbf{r}_\perp)$ a generic single-particle potential that confines electrons in the transverse $y-z$ plane in the L electrode, and similarly for the R electrode.

The eigenstates of the "asymptotic" Hamiltonians H_L and H_R can be easily found. For instance, the Schrödinger equation for the left Hamiltonian reads

$$\left[-\frac{\hbar^2}{2m}\nabla^2 + V_L(\mathbf{r}_\perp)\right]\psi_{\alpha k}(\mathbf{r}) = E_\alpha(k)\,\psi_{\alpha k}(\mathbf{r}), \tag{3.19}$$

which can be separated into a "longitudinal" equation in the x direction

$$-\frac{\hbar^2}{2m}\frac{\partial^2}{\partial x^2}e^{ikx} = \frac{\hbar^2 k^2}{2m}e^{ikx}, \tag{3.20}$$

and a "transverse" equation in the $y-z$ plane

$$\left[-\frac{\hbar^2}{2m}\nabla_\perp^2 + V_L(\mathbf{r}_\perp)\right]u_\alpha(\mathbf{r}_\perp) = \epsilon_\alpha\, u_\alpha(\mathbf{r}_\perp). \tag{3.21}$$

The full solution of Eq. 3.19 is thus

$$\psi_{\alpha k}(\mathbf{r}) = \sqrt{\frac{1}{L_x}}\,u_\alpha(\mathbf{r}_\perp)\,e^{ikx}, \qquad -\infty < k < +\infty, \tag{3.22}$$

with energies (which form *subbands*)

$$E_\alpha(k) = \epsilon_\alpha + \frac{\hbar^2 k^2}{2m} \equiv \epsilon_\alpha + \frac{1}{2}mv_\alpha^2(k), \tag{3.23}$$

where I have introduced a normalization length L_x, and defined the electron velocity along the x direction

$$v_\alpha(k) = \frac{\hbar k}{m}. \tag{3.24}$$

The "transverse energies" depend on the geometry of the confining potential. For instance, for leads whose confining potential has an infinite strength (an infinite quantum well) with a rectangular transverse shape with length L_y in the y direction, and L_z in the z direction, these transverse energies are simply

$$\epsilon_{\alpha_y,\alpha_z} = \frac{\pi^2 \hbar^2}{2m}\left[\frac{\alpha_y^2}{L_y^2} + \frac{\alpha_z^2}{L_z^2}\right], \qquad (3.25)$$

with α_y and α_z positive integers that define a set of quantum numbers for this problem.

In the following, I will not choose any specific transverse shape for the potential of the leads, and assume we have calculated the energies ϵ_α exactly, with α representing the whole set of possible quantum numbers necessary to specify the states completely.[17]

The states 3.22 satisfy the continuum orthonormality condition

$$\langle \psi_{\alpha k}|\psi_{\alpha' k'}\rangle = \frac{1}{L_x}\int dx\, e^{i(k-k')x}\langle u_\alpha|u_{\alpha'}\rangle = \frac{2\pi}{L_x}\delta(k'-k)\delta_{\alpha\alpha'}, \qquad (3.26)$$

where in the last equality I have used the orthonormality condition

$$\langle u_\alpha|u_{\alpha'}\rangle = \delta_{\alpha\alpha'}, \qquad (3.27)$$

satisfied by the eigenstates of the transverse Schrödinger equation 3.21, and the Fourier transform of the δ-function.

We thus see that the transverse solutions describe *transverse modes* of discrete energies ϵ_α "embedded" in the continuum defined by the longitudinal solutions – in this case simple plane waves. Since we are working in a coherent-transport picture, to each single-particle energy E may correspond a certain number of transverse modes (or *channels*), and this number is fixed for each energy (see Fig. 3.7).

Given an energy, E, the number of channels at that energy, $N_c(E)$, is simply provided by those modes whose energy ϵ_α is smaller than E. In mathematical terms

$$N_c(E) = \sum_\alpha \Theta(E-\epsilon_\alpha), \qquad (3.28)$$

with Θ the Heaviside step function.

The number of channels in the leads of cross section S can be estimated as follows. In each direction of length $\approx \sqrt{S}$ can only "fit" as many channels as the wavelength of the electrons at a given energy E will allow. In the following, we will be mostly interested in the number of channels at the Fermi energy E_F, to which corresponds a wave-vector k_F. The number of channels at the Fermi energy is thus of the order of $N_c(E_F) \approx k_F^2 S$. This

---

[17] These states may even have a continuum component that represents electrons free to move *away* from the leads. These are not the ones we want to use to build our scattering theory of the lead-nanojunction-lead system.

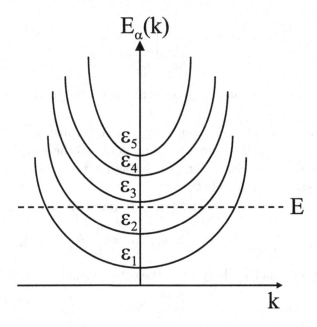

Fig. 3.7. Energy subbands as a function of the longitudinal wave-vector k. Each energy E corresponds to only a fixed number of channels.

argument can be extended to d dimensions with S the $(d-1)$-dimensional "cross section", so that

$$N_c(E_F) \approx k_F^{d-1} S. \quad (3.29)$$

This also shows that in a strictly 1D system the number of channels is 1.[18]

The states 3.22 represent *traveling* states, those whose probability density (their squared modulus) is finite everywhere in space. In addition, however, there exist solutions to Eq. 3.19 of the type

$$\psi_{\alpha k}(\mathbf{r}) = \sqrt{\frac{1}{L_x}}\, u_\alpha(\mathbf{r}_\perp)\, e^{-\tilde{k}x}, \qquad \tilde{k} > 0 \quad (3.30)$$

with energies

$$E_\alpha(k) = \epsilon_\alpha - \frac{\hbar^2 \tilde{k}^2}{2m}. \quad (3.31)$$

These are called *evanescent* states (or modes) because their probability density decays exponentially in space. While these states do not contribute to

---

[18] In quasi-1D systems, such as infinitely long armchair nanotubes, there may be other degeneracies, for instance, multiple transverse bands intersecting at the Fermi energy (for an extensive discussion of the properties of nanotubes see, e.g., Saito *et al.*, 1998).

the current directly, they need to be taken into account in a self-consistent theory of transport since they determine the correct total charge density in the system (Lang, 1992; Di Ventra and Lang, 2002). Similar considerations apply to the Hamiltonian H_R.

3.3.2 Transmission and reflection probabilities

We have determined the eigensolutions at the boundaries. We now need to determine the general solution of $\hat{H}_S$ (Eq. 3.13), which corresponds to the full lead-nanojunction-lead system. For a given energy E we thus need to solve

$$\left[-\frac{\hbar^2}{2m}\nabla^2 + V(\mathbf{r})\right]\Psi_{\alpha k}(\mathbf{r}) = E\Psi_{\alpha k}(\mathbf{r}). \qquad (3.32)$$

We can proceed in several equivalent ways. Let us start from the most intuitive one.

The solutions $\Psi_{\alpha k}(\mathbf{r})$ have to asymptotically merge with the eigenstates of the left and right Hamiltonians H_L and H_R, respectively. As discussed in *Approximation 5* of Sec. 3.1, we have two sets of traveling states, those moving from right to left, and those from left to right.[19] Let us consider the latter ones first. The results for the other set are obtained similarly.

Let us then consider an electron with energy E_i that at $x \to -\infty$ was in an initial eigenstate $\psi_{ik_i}(\mathbf{r})$ of the asymptotic Hamiltonian 3.17 at the *same* energy. This electron has thus a positive momentum $\hbar k_i$ such that

$$E_i(k_i) = \epsilon_i + \frac{\hbar^2 k_i^2}{2m}. \qquad (3.33)$$

At the nanojunction, this state may be very complicated, according to the form of the potential $V(\mathbf{r})$. Nonetheless, we expect that deep into the right electrode it is simply a linear combination of eigenstates of the asymptotic Hamiltonian H_R (with number of channels N_c^R at that energy[20]), or, equivalently, a linear combination of transmitted waves in the right lead (Fig. 3.5),

---

[19] From a mathematical point of view the states of the Hamiltonians 3.17, 3.18, and 3.13 that contribute to the current do not belong to the Hilbert space (they are not square-integrable). As discussed in Sec. 1.3.2 we should then work with eigendifferentials (see Eq. 1.32) or, more generally, with wave-packets. This requirement is even more important in the context of scattering theory since it has consequences on the convergence of the solutions of the Lippmann−Schwinger equation I will introduce in Sec. 3.4 (see Newton, 1966). To avoid complicating the notation, however, I will assume that the transformation to eigendifferentials or wave-packets is implicitly performed, and keep on using the notation of single eigenstates in the continuum.

[20] Under the assumption of identical leads, this is equal to N_c^L, the number of channels in the left lead at the same energy. However, for generality and clarity, in the following I maintain a distinct notation for the number of channels in the two leads.

namely

$$\Psi^+_{ik_i}(\mathbf{r}) \to \sum_{f=1}^{N_c^R} \mathcal{T}_{if} \psi_{fk_f}(\mathbf{r}), \qquad x \to +\infty, \qquad (3.34)$$

where $\mathcal{T}_{if}$ are complex numbers, and the symbol + for the wave-function is to remind us that $\Psi^+_{ik_i}(\mathbf{r})$ originates from a "past" wave-function $\psi_{ik_i}(\mathbf{r})$.[21] As discussed above, the total Hamiltonian $\hat{H}_S$ may also have evanescent modes. We assume that we are so deep into the right lead that these modes have zero contribution to 3.34.

Deep into the left lead, we do not expect $\Psi^+_{ik_i}(\mathbf{r})$ to be simply the incoming wave: electrons are scattered at the junction back to the left electrode (Fig. 3.5). We thus expect a solution of the type (again, with no contribution from evanescent modes)

$$\Psi^+_{ik_i}(\mathbf{r}) \to \psi_{ik_i}(\mathbf{r}) + \sum_{f=1}^{N_c^L} \mathcal{R}_{if} \psi_{fk_f}(\mathbf{r}), \qquad x \to -\infty, \qquad (3.35)$$

a linear combination of the incoming wave, and all possible reflected waves (their number is N_c^L) with momenta $\hbar k_f$ oriented in the negative x direction, which correspond to the *same* energy E_i of the incoming wave. The quantities $\mathcal{R}_{if}$ are complex numbers.

Let us now calculate the current across a given surface S – perpendicular to the direction x of global electron flow – carried by the wave $\Psi^+_{ik_i}(\mathbf{r})$. We can use the definition 1.27 of current operator (with only the paramagnetic current density component), and apply it to a single electron. We make the usual change $\mathbf{p} \to -i\hbar\partial/\partial\mathbf{r}$, and evaluate first the expectation value of the current density operator 1.26 over the state $|\Psi^+_{ik_i}\rangle$ (Exercise 3.1)

$$\begin{aligned}
\mathbf{j}(\mathbf{r}) &= \langle \Psi^+_{ik_i} | \hat{\mathbf{j}}(\mathbf{r}) | \Psi^+_{ik_i} \rangle \\
&= \frac{\hbar}{2im} \left[[\Psi^+_{ik_i}(\mathbf{r})]^* \frac{\partial \Psi^+_{ik_i}(\mathbf{r})}{\partial x} - \Psi^+_{ik_i}(\mathbf{r}) \frac{\partial [\Psi^+_{ik_i}(\mathbf{r})]^*}{\partial x} \right] \\
&= \frac{\hbar}{m} \mathrm{Im} \left\{ [\Psi^+_{ik_i}(\mathbf{r})]^* \frac{\partial \Psi^+_{ik_i}(\mathbf{r})}{\partial x} \right\}.
\end{aligned} \qquad (3.36)$$

We then integrate over a plane perpendicular to the x direction. This

---

[21] This terminology will become clearer in Sec. 3.4 where I will introduce the Lippmann–Schwinger equation.

gives the average current $I(E_i)$ carried by the state at energy E_i,

$$I(E_i) = e\langle \Psi^+_{ik_i} | \hat{I} | \Psi^+_{ik_i} \rangle$$
$$= \frac{e\hbar}{2im} \int_{-\infty}^{\infty} dy \int_{-\infty}^{\infty} dz \left[[\Psi^+_{ik_i}(\mathbf{r})]^* \frac{\partial \Psi^+_{ik_i}(\mathbf{r})}{\partial x} - \Psi^+_{ik_i}(\mathbf{r}) \frac{\partial [\Psi^+_{ik_i}(\mathbf{r})]^*}{\partial x} \right]. \tag{3.37}$$

We are working with a system in an ideal steady state. Therefore, this current cannot depend on the position of the surface at which we evaluate it.[22] We can thus evaluate it deep into the left lead, or, equivalently, deep into the right lead, and the resulting currents must be identical.

Deep into the left lead I thus replace 3.35 into 3.37, and obtain

$$I_L(E_i) = I_i(E_i) + \sum_{f=1}^{N_c^L} |\mathcal{R}_{if}|^2 I_f(E_i) = I_i(E_i) - \sum_{f=1}^{N_c^L} |\mathcal{R}_{if}|^2 |I_f(E_i)|$$
$$\equiv I_i(E_i) \left(1 - \sum_{f=1}^{N_c^L} R_{if}(E_i) \right), \tag{3.38}$$

where I have defined the following quantity (from Eq. 3.22)

$$I_i(E_i) = \frac{e\hbar}{2im} \int_{-\infty}^{\infty} dy \int_{-\infty}^{\infty} dz \left[\psi^*_{ik_i}(\mathbf{r}) \frac{\partial \psi_{ik_i}(\mathbf{r})}{\partial x} - \psi_{ik_i}(\mathbf{r}) \frac{\partial \psi^*_{ik_i}(\mathbf{r})}{\partial x} \right]$$
$$= \frac{\hbar k_i}{mL_x} = \frac{v_i(k_i)}{L_x}. \tag{3.39}$$

In a similar way, I_f, the currents reflected back into the left lead, are

$$I_f(E_i) = \frac{\hbar k_f}{mL_x} = \frac{v_f(k_f)}{L_x}, \tag{3.40}$$

which, due to the fact that in this case all wave-vectors k_f point in the negative x direction, they have the opposite sign than the incident current I_i – hence the negative sign in Eq. 3.38. The quantity

$$R_{if}(E_i) \equiv |\mathcal{R}_{if}|^2 \frac{|I_f(E_i)|}{|I_i(E_i)|} \tag{3.41}$$

is known as the *reflection probability* for a wave incident on the nanostructure with momentum $\hbar k_i$ to be scattered *back* into the left lead in a state with momentum $\hbar k_f$, while the energy is conserved.

---

[22] The density of the system is independent of time, so the current density is divergence-less (see Eq. 1.11).

We can do the same for the asymptotic solution 3.34 deep into the right lead. We replace 3.34 into 3.37, and find

$$I_R(E_i) = \sum_{f=1}^{N_c^R} |\mathscr{T}_{if}|^2 |I_f(E_i)|$$

$$\equiv I_i(E_i) \sum_{f=1}^{N_c^R} T_{if}(E_i), \qquad (3.42)$$

where

$$T_{if}(E_i) \equiv |\mathscr{T}_{if}|^2 \frac{|I_f(E_i)|}{|I_i(E_i)|} \qquad (3.43)$$

is the *transmission probability* that the wave with initial momentum $\hbar k_i$ is *transmitted* across the nanojunction into the right lead in a final state with momentum $\hbar k_f$, at the same energy.

As stated before, in an ideal steady state the two currents $I_L(E_i)$ and $I_R(E_i)$ have to be identical. Therefore, from Eqs. 3.38 and 3.42, we find the relation between the reflection and transmission probabilities for a wave incident from the left lead

$$\boxed{\sum_{f=1}^{N_c^R} T_{if}(E_i) + \sum_{f=1}^{N_c^L} R_{if}(E_i) = 1, \qquad \psi_{ik_i}(\mathbf{r}) \in L.} \qquad (3.44)$$

The whole procedure I have outlined so far can be repeated for a wave incident from the *right* lead. In this case the wave is transmitted into the left lead and reflected back into the right lead. Therefore, the above relation becomes

$$\boxed{\sum_{f=1}^{N_c^L} T_{if}(E_i) + \sum_{f=1}^{N_c^R} R_{if}(E_i) = 1, \qquad \psi_{ik_i}(\mathbf{r}) \in R.} \qquad (3.45)$$

Equations 3.44 and 3.45 simply state the conservation of particle flux. In other words, given a particle incoming from either direction of motion, it can be either reflected or transmitted; it cannot be "lost".

Finally, due to time-reversal invariance the above relations must hold even if we revert the velocities of initial and final states, by changing simultaneously the directionality of the scattering process. For instance, a process in which a wave scatters from the left with momentum $\hbar k_i$ into a right state with momentum $\hbar k_f$ is equivalent to a process in which a wave from the right with momentum $-\hbar k_f$ scatters into the state on the left with momentum

$-\hbar k_i$. The relations between the transmission and reflection probabilities of these reversed processes can be obtained directly from Eqs. 3.44 and 3.45 by simply exchanging i with f, and L and with R. We thus get

$$\sum_{i=1}^{N_c^L} T_{fi}(E_f) + \sum_{i=1}^{N_c^R} R_{fi}(E_f) = 1, \qquad \psi_{fk_f}(\mathbf{r}) \in R \qquad (3.46)$$

and

$$\sum_{i=1}^{N_c^R} T_{fi}(E_f) + \sum_{i=1}^{N_c^L} R_{fi}(E_f) = 1, \qquad \psi_{fk_f}(\mathbf{r}) \in L. \qquad (3.47)$$

3.3.3 Total current

Due to the independent-channel assumption (*Approximation 5* of Sec. 3.1), the total current in the system is the sum of *all* currents carried by all states (channels) at *all* energies. We thus need to integrate in energy, multiply by the *density of states* (number of channels per unit energy) for each momentum direction, and sum over all incident channels (both right-moving and left-moving).

Since each channel represents a one-dimensional problem, the density of states per spin for a momentum $\hbar k_i$ (of given direction) is simply

$$D_i(E_i)dE_i = \frac{L_x}{2\pi}dk_i \iff D_i(E_i) = \frac{L_x}{2\pi}\frac{dk_i}{dE_i} = \frac{L_x}{2\pi\hbar v_\alpha(k_i)}. \qquad (3.48)$$

Once again, we are dealing with an ideal steady state so that we can calculate this current anywhere in space. We choose an arbitrary point deep into the right lead. From Eq. 3.38 and Eq. 3.42, and their equivalent quantities for a wave incident from the right lead, we then get the total current (the factor 2 is for spin degeneracy, and the energy integration goes from $-\infty$ to $+\infty$)

$$
\begin{aligned}
I &= e \overset{\text{spin}}{\overbrace{2}} \int dE \Bigg\{ \underbrace{\sum_{i=1}^{N_c^L} \sum_{f=1}^{N_c^R} D_i(E_i) I_i(E_i) T_{if}(E_i)}_{\text{from L to R}} \\
&\qquad\qquad - \sum_{i=1}^{N_c^R} D_i(E_i) I_i(E_i) \Bigg[1 - \underbrace{\sum_{f=1}^{N_c^R} R_{if}(E_i)}_{\text{from R reflected back to R}} \Bigg] \Bigg\} \\
&= e\, 2 \int dE \Bigg\{ \underbrace{\sum_{i=1}^{N_c^L} \sum_{f=1}^{N_c^R} D_i(E_i) I_i(E_i) T_{if}(E_i)}_{\text{from L to R}} \\
&\qquad\qquad - \underbrace{\sum_{i=1}^{N_c^R} \sum_{f=1}^{N_c^L} D_i(E_i) I_i(E_i) T_{if}(E_i)}_{\text{from R to L}} \Bigg\} \\
&= e\, 2 \frac{1}{2\pi\hbar} \int dE \left\{ \mathcal{T}_{LR}(E) - \mathcal{T}_{RL}(E) \right\} \\
&= 0,
\end{aligned} \qquad (3.49)
$$

where, in the second equality, I have used the relation 3.45. I have also used Eqs. 3.39 and 3.40 for the current per channel, and Eq. 3.48 for the density of states per spin – for a given direction of momentum – to show that their product is simply the constant $1/2\pi\hbar$. In addition, in the third equality of Eq. 3.49, I have defined the total *transmission coefficient* at a given energy

$$
\mathcal{T}_{RL}(E) = \sum_{i=1}^{N_c^R} \sum_{f=1}^{N_c^L} T_{if}(E) = \sum_{i=1}^{N_c^R} \tilde{T}_i(E), \qquad \text{from R to L,} \qquad (3.50)
$$

and

$$
\mathcal{T}_{LR}(E) = \sum_{i=1}^{N_c^L} \sum_{f=1}^{N_c^R} T_{if}(E) = \sum_{i=1}^{N_c^L} \tilde{T}_i(E), \qquad \text{from L to R.} \qquad (3.51)
$$

The quantities $\tilde{T}_i(E)$ and $\tilde{T}_i(E)$ in Eqs. 3.50 and 3.51, respectively, are the total transmission probability

$$\tilde{T}_i(E) = \sum_{f=1}^{N_c^L} T_{if}(E), \qquad i \in R, \qquad (3.52)$$

for a wave incident from the right to be scattered into *any* of the left channels, and

$$T_i(E) = \sum_{f=1}^{N_c^R} T_{if}(E), \qquad i \in L, \qquad (3.53)$$

the total transmission probability that a wave incident from the left is scattered into *any* of the right channels.

Finally, since the particle flux must be conserved (Eqs. 3.44 and 3.45), the total transmission coefficient from left to right must be equal to the total transmission coefficient from right to left (Exercise 3.12)

$$\mathcal{T}_{RL}(E) = \mathcal{T}_{LR}(E) \equiv \mathcal{T}(E), \qquad \text{flux conservation}, \qquad (3.54)$$

a property I have used in the last equality of Eq. 3.49 to show that the total current is zero!

How come we have obtained zero current? The reason is that in the above calculation I have assumed that the two classes of channels, right-to-left moving, and left-to-right moving, are *equally populated*. If this is true, then the amount of current that flows from left to right is exactly the *same* amount that flows from right to left, and the net current is zero.

This is the equivalent result we obtain in a semiclassical theory of conduction in solids, where it is easy to show that electrons in *filled bands do not conduct*, and only electrons in *partially filled bands carry current* (Ashcroft and Mermin, 1975).

The situation here is analogous, and indeed with *Approximation 5* we have assumed that right-traveling states are populated up to the left electrochemical potential, μ_L, according to the local distribution function 3.14, while those moving from right to left (left-moving) are populated up to the right electrochemical potential, μ_R, with local distribution 3.15. The two electrochemical potentials differ, according to our *Approximation 1*, by $\mu_L - \mu_R = eV$. We thus need to modify the result in Eq. 3.49 to take this population imbalance into account.

This can be easily done by using the statistical operator 3.16. By repeating the above calculations, with now the left-moving channels populated differently than the right-moving ones, we get (spin included)

$$I = e\text{Tr}\{\rho_S^{ss}\hat{I}\} = \frac{e}{\pi\hbar}\int dE\,[f_L(E)\mathcal{T}_{LR}(E) - f_R(E)\mathcal{T}_{RL}(E)], \quad (3.55)$$

and using the flux conservation relation 3.54 we get

$$\boxed{I = \frac{e}{\pi\hbar}\int_{-\infty}^{+\infty} dE\,[f_L(E) - f_R(E)]\,\mathcal{T}(E)} \quad (3.56)$$

This current is not zero precisely because it is the difference between two opposite currents from states which are populated *differently*.

The zero-bias limit of the current

In the limit of zero bias, namely in the limit $\mu_L - \mu_R \to 0$, we can approximate the above equation even further. In this limit we may Taylor-expand the left local distribution function

$$f_L(E) = f_R(E) - \left.\frac{\partial f_R(E)}{\partial E}\right|_{\mu_R}(\mu_L - \mu_R) + \mathcal{O}\left[(\mu_L - \mu_R)^2\right] \quad (3.57)$$

and obtain from Eq. 3.56, to first order in $(\mu_L - \mu_R)$

$$I = \frac{2e}{h}(\mu_L - \mu_R)\int dE\left(-\left.\frac{\partial f_R(E)}{\partial E}\right|_{\mu_R}\right)\mathcal{T}(E), \quad (\mu_L - \mu_R) \to 0. \quad (3.58)$$

If we now set the temperature θ to zero, the local Fermi–Dirac distribution, $f_R(E)$, becomes a step function (see Sec. 1.4.6) and its energy derivative a δ-function centered at the right electrochemical potential μ_R. In the limit of zero temperature Eq. 3.58 then becomes

$$I = \frac{2e}{h}(\mu_L - \mu_R)\mathcal{T}(E = \mu_R)$$
$$= \frac{2e^2}{h}\mathcal{T}(E = \mu_R)V; \quad (\mu_L - \mu_R) \to 0, \ \theta \to 0. \quad (3.59)$$

Alternatively, I could have chosen to expand the *right* local Fermi–Dirac distribution

$$f_R(E) = f_L(E) - \left.\frac{\partial f_L(E)}{\partial E}\right|_{\mu_L}(\mu_R - \mu_L) + \mathcal{O}\left[(\mu_L - \mu_R)^2\right] \quad (3.60)$$

and replace this into 3.56, and obtain, to first order in $(\mu_L - \mu_R)$,

$$I = \frac{2e^2}{h}\mathcal{T}(E = \mu_L)V; \quad (\mu_L - \mu_R) \to 0, \ \theta \to 0, \quad (3.61)$$

with now the transmission coefficient evaluated at the *left* electrochemical potential. Which one is then correct, Eq. 3.59 or Eq. 3.61? They are both correct in the following sense.

Since we are really working in the limit in which μ_L and μ_R differ only slightly from each other, we can assume that they both differ negligibly from the Fermi energy, E_F, of the electron gas at equilibrium, with μ_L an infinitesimal energy ϵ *above* the Fermi energy

$$\mu_L \approx E_F + \epsilon, \qquad (3.62)$$

and μ_R an infinitesimal energy ϵ *below* the Fermi energy

$$\mu_R \approx E_F - \epsilon, \qquad (3.63)$$

or, equivalently, by eliminating ϵ from these two equations,

$$\frac{\mu_L + \mu_R}{2} \approx E_F. \qquad (3.64)$$

The transmission coefficient evaluated at the left or right electrochemical potential is thus

$$\mathcal{T}(E_F \pm \epsilon) = \mathcal{T}(E_F) \pm \left.\frac{\partial \mathcal{T}(E)}{\partial E}\right|_{E_F} \epsilon + \mathcal{O}(\epsilon^2). \qquad (3.65)$$

If the transmission coefficient varies slowly with energy at the Fermi level, we may neglect its energy derivative and, to zero order in ϵ, we can make the further approximation

$$\mathcal{T}(E_F) \approx \mathcal{T}(E = \mu_R) \approx \mathcal{T}(E = \mu_L); \qquad (\mu_L - \mu_R) \to 0, \ \theta \to 0. \qquad (3.66)$$

With this approximation we finally get

$$I = \frac{2e^2}{h}\mathcal{T}(E_F)V; \qquad (\mu_L - \mu_R) \to 0, \ \theta \to 0, \qquad (3.67)$$

where I have now eliminated the ambiguity over the energy at which the transmission coefficient is evaluated: when the two electrochemical potentials differ infinitesimally from each other, and the transmission coefficient does not vary appreciably with energy in the energy window $\mu_L - \mu_R$ (and the temperature is zero), we may evaluate the transmission coefficient at the Fermi energy of the system at equilibrium.[23]

Clearly, this does *not* mean that the current is an equilibrium property, it simply means that under the above conditions the transmission properties of our non-interacting system may be *approximated* with the transmission properties of those electrons at the Fermi level of the equilibrium system.

---

[23] The zero temperature limit may not be necessary if the transmission coefficient varies negligibly over the energy window $k_B \theta$.

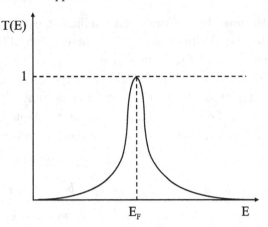

Fig. 3.8. An electronic resonance close to the Fermi energy is characterized by a transmission probability peaked around that energy.

A large energy variation at the Fermi level may occur if the system has an *electronic resonance* at that energy, namely a transmission coefficient that is peaked in a narrow region of energy around the Fermi energy (see Fig. 3.8). In that case, we cannot make the approximation 3.66, and we need to work with the full expression 3.56 for the current.

3.3.4 Two-probe conductance

At this point we should be content since we have calculated the most important quantity: the current. In many cases, however, it is of interest to know one of its *derived* quantities: the conductance (or its inverse, the resistance). Here we face another fundamental question:

How do we define the conductance?

This is another non-trivial issue.

Probes

In Sec. 2.1 I have defined the linear-response conductance as the ratio between the current I and the voltage drop V. However, this definition says nothing about the actual procedure to *measure* these two quantities. In fact, there are several possible answers to the above question according to the experimental way we measure – or *probe* – the current and the other relevant quantity, the bias, we use to define the conductance.

For example, one could couple somewhere along the leads *capacitative probes* which ideally measure local *electrostatic potentials* (Landauer, 1989). If this coupling is *not invasive*, one can define the conductance in terms of the electrostatic potential difference measured at these probes, as I will show in Chapter 7. Clearly, due to the microscopic features of the materials composing the leads, these potentials may vary fast in regions of space the size of a lattice constant. One can then define a macroscopic average of these potentials as in 2.99, and define the conductance accordingly.

In addition, as has also been shown numerically (Lang, 1995; Di Ventra and Lang, 2002), in most cases of interest, electrochemical potential differences – for instance, as ideally "measured" at the reservoirs – are very close to electrostatic potential differences deep into the leads – again, in a macroscopically averaged sense.

The fact that these two quantities are not very different on length scales much larger than the electron screening length is due to the electrons ability to respond to any local disturbance by rearranging themselves, in such a way that these disturbances are effectively screened.[24]

Therefore, employing electrostatic or electrochemical potential differences as biases in the definition of conductance may not lead to appreciable numerical differences.

To the above considerations, we need to add that in many instances the measurements of electrochemical potentials – and to some extent also of electrostatic potentials – are *invasive*. The measurement probes thus influence the transport properties of our system; a fact I will stress in Sec. 3.9 when discussing multi-probe measurements and their analogy with dephasing effects.

Two-probe conductance

In this section, I will assume that we ideally measure electrochemical potential differences at the "location" of the reservoirs I have introduced to replace the true electron source. The bias is then $V = (\mu_L - \mu_R)/e$, and I define the conductance as in Eq. 2.4. This definition, together with the current 3.67, gives us (Büttiker *et al.*, 1985)

---

[24] One then expects appreciable differences between the electrochemical potential and the electrostatic one only over regions of space shorter than the screening length.

$$\boxed{\begin{aligned} G_{\text{2-probe}} = \frac{I}{V} &= \frac{2e^2}{h} \mathcal{T}(E_F) \\ &= \frac{2e^2}{h} \sum_{i=1}^{N_c^L} \sum_{f=1}^{N_c^R} T_{if}(E_F) \\ &= \frac{2e^2}{h} \sum_{i=1}^{N_c^R} \sum_{f=1}^{N_c^L} T_{if}(E_F); \quad (\mu_L - \mu_R) \to 0, \ \theta \to 0. \end{aligned}} \quad (3.68)$$

This result finally shows what I anticipated at the beginning of this chapter: under specific conditions and approximations, the conductance of non-interacting electrons can be written in terms of their transmission properties. In the limits of zero bias and zero temperature, and away from electronic resonances, these transmission properties are those of the electrons at the Fermi level – a conclusion I reached in Sec. 2.3.4 using the Kubo formalism (see Eq. 2.86). From Eq. 3.67 it is also evident that, if the transmission coefficient is unity, the conductance assumes the *quantized* value of Eq. 2.89

$$G_0 = \frac{2e^2}{h}; \quad \mathcal{T}(E_F) = 1, \ (\mu_L - \mu_R) \to 0, \ \theta \to 0. \quad (3.69)$$

From this and Eq. 3.68, it is also clear that the quantized conductance does not always imply that the non-interacting system has only one conducting channel in both the left and right leads. Indeed, since $T_{if} \leq 1$, we could have many – possibly infinite – channels whose cumulative effect is to sum to a transmission coefficient of unit value, and thus provide a quantized conductance. I will return to this point in Sec. 3.7.4.

Finally, if we are not working in the limits for which Eq. 3.68 is valid, the transmission coefficient $\mathcal{T}$ may depend on both energy *and* bias, so that the total current 3.56 may depend non-linearly on bias: $I \equiv I(V)$. In this case, we may define the *differential conductance* (or simply conductance) at any given bias V_0 as the bias derivative of the total current evaluated at V_0,

$$\boxed{G_{\text{2-probe}}(V_0) = \left.\frac{dI(V)}{dV}\right|_{V_0}.} \quad (3.70)$$

To distinguish it from another conductance I will define in Sec. 3.8, the present one is also called *two-probe* conductance. Once more, both 3.70, or its zero-bias limit, 3.68, must be understood in terms of the bias as ideally measured by probes which "couple" with the reservoirs.

Experimental verification of quantized conductance

The first experiments that clearly showed quantization of conductance were done on quantum point contacts created from the 2D electron gas at the interface of GaAs-AlGaAs heterostructures (van Wees et al., 1988; Wharam et al., 1988). In those experiments, the width of the quantum point contact was controlled by a gate lithographically built in proximity to the heterostructure (see inset in the left panel of Fig. 3.9). By applying a negative voltage to the gate, electrons are depleted underneath it, and transport occurs only in the formed quantum-point-contact region.

The conductance was then found to decrease in steps of about $2e^2/h$ with increasing negative gate voltage – and thus with decreasing constriction effective width.

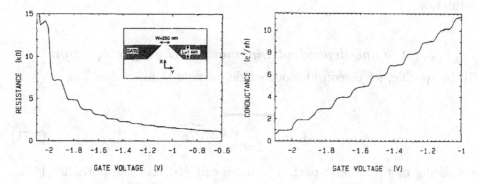

Fig. 3.9. Left panel: Resistance steps as a function of negative gate voltage observed in the transport properties of quantum point contacts. These are formed by depleting regions of space of a 2D electron gas with a gate fabricated in proximity to the 2D gas, as shown in the inset. Right panel: Conductance steps as a function of gate voltage. The conductance data have been corrected via the subtraction of a "series resistance". Reprinted with permission from van Wees et al., 1988.

This is shown in Fig. 3.9 where the actual conductance evaluated as the ratio between the current and the applied bias has been "corrected" with what has been called a "series resistance" (van Wees et al., 1988; Wharam et al., 1988). In reality, the experimental configuration used was closer to what we will define as a *four-probe* conductance measurement (Sec. 3.8), where the current is driven by two probes (which can be thought of as the two "reservoirs" of Fig. 3.4), and the voltage is measured by two other probes at different locations along the sample. The interpretation of these experiments cannot, therefore, be done by simple application of Eq. 3.68 (Landauer, 1989). I will come back to this point in Sec. 3.8.

3.4 The Lippmann–Schwinger equation

My goal now is to rewrite the single-particle current 3.56 in a different but *equivalent* form, which provides additional insight into the transport problem. To do this I need to derive an important equation of formal scattering theory: the *Lippmann–Schwinger equation*. I could simply state this equation and refer the reader to other textbooks for a complete account. However, for completeness, and because I believe the steps leading to it shed light on the basic physical and mathematical assumptions of scattering theory, I will spend few pages on its derivation and discussion.

Let us start from the time-*dependent* Lippmann–Schwinger equation. Its time-*independent* version can be easily derived from it using a Fourier transformation.

3.4.1 Time-dependent Lippmann–Schwinger equation

To be specific, let us consider our single-particle Hamiltonian 3.71

$$H_S = \overbrace{-\frac{\hbar^2}{2m}\nabla^2}^{\hat{H}_0} + \overbrace{V(\mathbf{r})}^{\hat{V}}, \qquad (3.71)$$

consisting of the kinetic part, which we call $\hat{H}_0$, and the potential $\hat{V}$ describing the scattering due to the nanoscale junction.[25] This potential is assumed to be time-independent.

We can solve two different time-dependent Schrödinger equations with given initial conditions. The first one

$$i\hbar \frac{d|\Psi(t)\rangle}{dt} = \hat{H}_S |\Psi(t)\rangle = (\hat{H}_0 + \hat{V})|\Psi(t)\rangle, \quad |\Psi(t_0)\rangle, \qquad (3.72)$$

for the full Hamiltonian, and the second one

$$i\hbar \frac{d|\Psi_0(t)\rangle}{dt} = \hat{H}_0 |\Psi_0(t)\rangle, \quad |\Psi_0(t_0)\rangle, \qquad (3.73)$$

for the Hamiltonian $\hat{H}_0$ in the absence of the scattering potential $\hat{V}$.

---

[25] In the following discussion, the Hamiltonian $\hat{H}_0$ need not contain only the kinetic energy of the electron: it may be some arbitrary single-particle Hamiltonian. As before, $\hat{V}$ contains anything else that is not contained in $\hat{H}_0$.

3.4.1.1 Green's functions

The above equations have the general form

$$\hat{L}\psi(t) = f(t) \tag{3.74}$$

with

$$\hat{L} = i\hbar\frac{d}{dt} - \hat{H} \tag{3.75}$$

or

$$\hat{L} = i\hbar\frac{d}{dt} - \hat{H}_0 \tag{3.76}$$

a linear differential operator, and $f(t) = 0$. From the general theory of differential equations we then know that there exists a quantity, called *Green's function* or *propagator*, $\hat{G}(t)$, satisfying the equation of motion

$$\hat{L}\hat{G}(t) = \hat{1}\delta(t). \tag{3.77}$$

In reality, to *each* of the Schrödinger equations 3.72 and 3.73 we can associate an equation of motion for *two* types of Green's functions. To Eq. 3.72 corresponds

$$\left(i\hbar\frac{\partial}{\partial t} - \hat{H}_S\right)\hat{G}^{\pm}(t) = \hat{1}\delta(t), \tag{3.78}$$

which represents two equations of motion for the Green's functions $\hat{G}^+$ and $\hat{G}^-$ with boundary conditions

$$\begin{aligned}\hat{G}^+(t) = 0 \quad & t < 0, \quad \text{retarded},\\ \hat{G}^-(t) = 0 \quad & t > 0, \quad \text{advanced}.\end{aligned} \tag{3.79}$$

A formal solution of 3.78 for $\hat{G}^+$ with these boundary conditions is

$$\hat{G}^+(t) = \begin{cases} -\dfrac{i}{\hbar}e^{-i\hat{H}_S t/\hbar} & t > 0, \\ 0 & t < 0, \end{cases} \tag{3.80}$$

which can be checked by direct substitution into 3.78.

We note that $\hat{G}^+(t)$ (for $t > 0$) is proportional to the time-evolution operator $U(t, t_0)$ of Eq. 1.21 when the Hamiltonian does not depend on time. In fact, using the formal solution 3.80 we can rewrite Eq. 1.17 – which relates the state vector $|\Psi(t_0)\rangle$ at time t_0 to the state vector $|\Psi(t)\rangle$ at time $t > t_0$ – as

$$|\Psi(t)\rangle = i\hbar\hat{G}^+(t - t_0)|\Psi(t_0)\rangle, \quad t > t_0, \tag{3.81}$$

which shows that the Green's function $\hat{G}^+$ *propagates* the state vector $|\Psi(t)\rangle$, and contains the whole *history* of its time evolution – similarly to the retarded response function $\chi_{AB}(t-t')$, Eq. 2.47, to which it may be related (Mahan, 1990). This is the reason $\hat{G}^+$ is called the *retarded* Green's function or propagator.

Similarly, we can now find the formal solution of 3.78 for $\hat{G}^-$. This is

$$\hat{G}^-(t) = \begin{cases} 0 & t > 0, \\ \dfrac{i}{\hbar} e^{-i\hat{H}_s t/\hbar} & t < 0, \end{cases} \tag{3.82}$$

and relates the state vector $|\Psi(t_0)\rangle$ at time t_0 to a *past* state vector $|\Psi(t)\rangle$ at time $t < t_0$ as

$$|\Psi(t)\rangle = -i\hbar \hat{G}^-(t-t_0)|\Psi(t_0)\rangle, \quad t < t_0, \tag{3.83}$$

which shows that the Green's function $\hat{G}^-$ "backtracks" a state vector from a "present" time, t_0, to a "past" time t. It thus requires knowledge in *advance* of the state vector $|\Psi(t)\rangle$ we want to determine, hence the name *advanced* Green's function.

From the formal solutions 3.80 and 3.82 we also see immediately that since the Hamiltonian is Hermitian the retarded and advanced Green's functions are related via

$$[G^+(t)]^\dagger = G^-(-t). \tag{3.84}$$

I can now repeat all the above steps for the Hamiltonian $\hat{H}_0$ and its time-dependent Schrödinger equation 3.73. I then define

$$\left(i\hbar \frac{\partial}{\partial t} - \hat{H}_0\right) \hat{G}_0^\pm(t) = \hat{1}\delta(t), \tag{3.85}$$

for the associated retarded $\hat{G}_0^+$ and advanced $\hat{G}_0^-$ Green's functions with boundary conditions

$$\hat{G}_0^+(t) = \begin{cases} -\dfrac{i}{\hbar} e^{-i\hat{H}_0 t/\hbar} & t > 0, \\ 0 & t < 0, \end{cases} \tag{3.86}$$

and

$$\hat{G}_0^-(t) = \begin{cases} 0 & t > 0, \\ \dfrac{i}{\hbar} e^{-i\hat{H}_0 t/\hbar} & t < 0, \end{cases} \tag{3.87}$$

respectively.

These Green's functions – which sometimes I shall call "free" because they do not include scattering via the potential $\hat{V}$ – share the same properties 3.81 and 3.83 of $\hat{G}^+$ and $\hat{G}^-$ (clearly with respect to the state vectors

3.4 The Lippmann–Schwinger equation

$|\Psi_0(t)\rangle\rangle$, and the property 3.84.

Relation between $\hat{G}^\pm$ and $\hat{G}_0^\pm$

Let us now relate all quantities associated with the Hamiltonian $\hat{H}_S$, with the corresponding quantities associated with the Hamiltonian $\hat{H}_0$. I can formally rewrite Eq. 3.85 as

$$\left(i\hbar\frac{\partial}{\partial t} - \hat{H}_0\right) = \hat{1}\delta(t)\left[\hat{G}_0^\pm(t)\right]^{-1}, \qquad (3.88)$$

which replaced in 3.78 gives

$$\hat{1}\delta(t)\left[\hat{G}_0^\pm(t)\right]^{-1}\hat{G}^\pm(t) - \hat{V}\hat{G}^\pm(t) = \hat{1}\delta(t), \qquad (3.89)$$

or

$$\begin{aligned}\hat{1}\delta(t)\hat{G}^\pm(t) &= \hat{1}\delta(t)\hat{G}_0^\pm(t) + \hat{G}_0^\pm(t)\hat{V}\hat{G}^\pm(t) \\ &= \hat{1}\delta(t)\hat{G}_0^\pm(t) + \hat{G}^\pm(t)\hat{V}\hat{G}_0^\pm(t),\end{aligned} \qquad (3.90)$$

where the last equality comes from the fact that I could have followed a different route by formally inverting Eq. 3.78, and replacing it into 3.85.

If we integrate the above equation for $\hat{G}^+$ from t_0 to $t > t_0$ we finally get the Lippmann–Schwinger equation which relates $\hat{G}^+$ to $\hat{G}_0^+$

$$\begin{aligned}\hat{G}^+(t - t_0) &= \hat{G}_0^+(t - t_0) + \int_{t_0}^t dt'\, \hat{G}_0^+(t - t')\hat{V}\hat{G}^+(t' - t_0) \\ &= \hat{G}_0^+(t - t_0) + \int_{t_0}^t dt'\, \hat{G}^+(t - t')\hat{V}\hat{G}_0^+(t' - t_0).\end{aligned} \qquad (3.91)$$

If we integrate Eq. 3.90 from $t < t_0$ to t_0 we obtain the corresponding equation that relates $\hat{G}^-$ to $\hat{G}_0^-$

$$\begin{aligned}\hat{G}^-(t - t_0) &= \hat{G}_0^-(t - t_0) + \int_t^{t_0} dt'\, \hat{G}_0^-(t - t')\hat{V}\hat{G}^-(t' - t_0) \\ &= \hat{G}_0^-(t - t_0) + \int_t^{t_0} dt'\, \hat{G}^-(t - t')\hat{V}\hat{G}_0^-(t' - t_0).\end{aligned} \qquad (3.92)$$

3.4.1.2 Dyson's equation and self-energy

I can rewrite Eqs. 3.91 and 3.92 in a different form that is the basis for a perturbation expansion of the full retarded and advanced Green's functions. Let us consider Eq. 3.91. To a first approximation (called the *Born approximation*) I can replace the full Green's function $\hat{G}^+$ in the integral of Eq. 3.91 with the "unperturbed" Green's function $\hat{G}_0^+$. I can then replace this "new" approximate full Green's function back into the right-hand side of Eq. 3.91, and so on. The Green's function $\hat{G}^+$ may be thus expanded (if the series expansion converges) as

$$\hat{G}^+(t-t_0) = \hat{G}_0^+(t-t_0) + \int_{t_0}^t dt' \, \hat{G}_0^+(t-t')\hat{V}\hat{G}_0^+(t'-t_0)$$
$$+ \int_{t_0}^t dt' \int_{t_0}^{t'} dt'' \, \hat{G}_0^+(t-t')\hat{V}\hat{G}_0^+(t'-t'')\hat{V}\hat{G}_0^+(t''-t_0) + \cdots \tag{3.93}$$

The above series expansion has an intuitive physical interpretation which can be visualized with the diagrams in Fig. 3.10. The full propagator (represented by the vertical double lines) is the sum of a succession of scattering events. The simplest one is a no-scattering event: the system propagates in time from t_0 to t via the unperturbed (free) propagator G_0^+ (represented by a single line). The first-order (in $\hat{V}$) process consists in the free propagation from t_0 to t', at which point a scattering event occurs, and then a subsequent free propagation from t' till time t. The second-order process is a free propagation from t_0 to t'', at which point a scattering event occurs, followed by a free propagation from t'' to t', a scattering event at t', and a subsequent free propagation from t' to t. And so on and so forth.

If the series expansion converges we can "lump" the effects of all these scattering events into a single quantity we call retarded *self-energy* and represent with the symbol $\hat{\Sigma}^+$ (see Fig. 3.10). We thus write the series expansion as

$$\begin{aligned}\hat{G}^+(t-t_0) &= G_0^+(t-t_0) \\ &+ \int_{t_0}^t dt' \int_{t_0}^{t'} dt'' \, \hat{G}_0^+(t-t')\hat{\Sigma}^+(t'-t'')\hat{G}^+(t''-t_0) \\ &= G_0^+(t-t_0) \\ &+ \int_{t_0}^t dt' \int_{t_0}^{t'} dt'' \, \hat{G}^+(t-t')\hat{\Sigma}^+(t'-t'')\hat{G}_0^+(t''-t_0), \end{aligned} \tag{3.94}$$

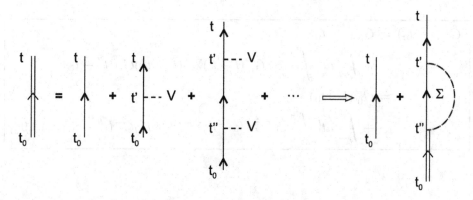

Fig. 3.10. Perturbation expansion of the total Green's function (represented by two vertical lines), from t_0 to t, as a sum of scattering processes, each with a fixed number of scattering events: free propagation (a vertical line), plus one scattering event at time t' due to potential V, etc. If the series expansion converges, we can lump the action of all scattering events into a *self-energy* Σ.

where, once again, I could interchange $\hat{G}^+$ with $\hat{G}_0^+$ in the last equality as in Eq. 3.91. Equation 3.94 is referred to as the *Dyson equation*.

By comparing 3.94 with 3.91 we see that, in the present case, the self-energy term is simply

$$\hat{\Sigma}(t' - t'') = \hat{V}\delta(t' - t''), \qquad \text{mean-field approximation.} \qquad (3.95)$$

Therefore, in this effective single-particle (mean-field) scattering problem I gain nothing by using the Dyson equation 3.94 compared to the Lippmann–Schwinger equation 3.91.[26] However, when interactions among particles are present, the self-energy does not have the form Eq. 3.95, and the Dyson equation is indeed more useful as it allows us to include in the self-energy the effects of these particle-particle interactions in a compact way. I will come back to these issues in Chapter 4.

We can follow the above procedure for the *advanced* Green's function $\hat{G}^-$ as well. In that case, the scattering events occur in the *future* and by defining an advanced self-energy $\hat{\Sigma}^-$, we obtain the following Dyson's equation ($t < t_0$)

---

[26] An Hermitian self-energy means an infinite state lifetime (see Sec. 4.2.4).

$$\hat{G}^-(t-t_0) = G_0^-(t-t_0)$$
$$+ \int_t^{t_0} dt' \int_{t'}^{t_0} dt''\, \hat{G}_0^-(t-t')\hat{\Sigma}^-(t'-t'')\hat{G}^-(t''-t_0)$$
$$= G_0^-(t-t_0)$$
$$+ \int_t^{t_0} dt' \int_{t'}^{t_0} dt''\, \hat{G}^-(t-t')\hat{\Sigma}^-(t'-t'')\hat{G}_0^-(t''-t_0). \quad (3.96)$$

Incoming and outgoing states

I will use the Dyson equation later in the book. For now, let us relate the states $|\Psi_0\rangle$ of the Hamiltonian $\hat{H}_0$ with the states $|\Psi\rangle$ of the full Hamiltonian $\hat{H}_S$. One of our assumptions is that in the infinite past the electrons did not experience the scattering potential $\hat{V}$, namely, their evolution was governed by the time-dependent Schrödinger equation 3.73. Let us call the state governed by this evolution an *incoming* state $|\Psi_0^{\text{in}}\rangle$. We have also assumed that in the infinite past this state coincided with the state $|\Psi\rangle$ of the *full* Hamiltonian $\hat{H}_S$. We can then make the following *ansatz*

$$|\Psi_0^{\text{in}}(t)\rangle \equiv i\hbar \lim_{t_0 \to -\infty} \hat{G}_0^+(t-t_0)|\Psi(t_0)\rangle, \quad (3.97)$$

namely the *incoming* free state vector $|\Psi_0^{\text{in}}\rangle$ is the result of the time evolution from the infinite past of the true state $|\Psi(t_0)\rangle$. This evolution, however, is carried by the free Hamiltonian $\hat{H}_0$. This is a strong assumption, which may or may not be verified for all physical systems under all experimental conditions; and it is precisely the condition for the very existence of a scattering theory.

From Eqs. 3.72 and 3.85 we get the following identity

$$\frac{\partial |\Psi_0^{\text{in}}(t)\rangle}{\partial t_0} = i\hbar \frac{\partial}{\partial t_0}[\hat{G}_0^+(t-t_0)|\Psi(t_0)\rangle]$$
$$= i\hbar \frac{\partial \hat{G}_0^+(t-t_0)}{\partial t_0}|\Psi(t_0)\rangle + i\hbar \hat{G}_0^+(t-t_0)\frac{\partial |\Psi(t_0)\rangle}{\partial t_0}$$
$$= \left[-\hat{1}\delta(t-t_0) - \hat{H}_0\hat{G}_0^+(t-t_0) + \hat{G}_0^+(t-t_0)\hat{H}_0 \right.$$
$$\left. + \hat{G}_0^+(t-t_0)\hat{V}\right]|\Psi(t_0)\rangle$$
$$= -\hat{1}\delta(t-t_0)|\Psi(t_0)\rangle + \hat{G}_0^+(t-t_0)\hat{V}|\Psi(t_0)\rangle, \quad (3.98)$$

3.4 The Lippmann–Schwinger equation

where, in the last equality, I have also used the fact that $\hat{G}_0$ and $\hat{H}_0$ commute (this is obvious from Eq. 3.86).

Let us now integrate the above equation from $t_0 \to -\infty$ to $t_0 \to +\infty$. Using the definition of incoming state 3.97 we finally get

$$|\Psi^+(t)\rangle = |\Psi_0^{in}(t)\rangle + \int_{-\infty}^{+\infty} dt'\, G_0^+(t-t')\hat{V}|\Psi^+(t')\rangle. \qquad (3.99)$$

This is also called the Lippmann–Schwinger equation. In the above, I have also used the notation $|\Psi^+(t)\rangle$ for the full state of the system to remind us that it is a state originating from an *incoming* free state.

I can redo the above procedure by defining an *outgoing* state

$$|\Psi_0^{out}(t)\rangle \equiv -i\hbar \lim_{t_0 \to +\infty} \hat{G}_0^-(t-t_0)|\Psi(t_0)\rangle, \qquad (3.100)$$

which represents a free state generated in the *infinite future* by the full state $|\Psi(t_0)\rangle$. Similarly to what we did for the incoming state 3.97, by calculating the time derivative of the outgoing state 3.100 we get the Lippmann–Schwinger equation

$$|\Psi^-(t)\rangle = |\Psi_0^{out}(t)\rangle + \int_{-\infty}^{+\infty} dt'\, G_0^-(t-t')\hat{V}|\Psi^+(t')\rangle. \qquad (3.101)$$

Having defined the incoming and outgoing states, I can now think about my scattering process differently. My state vector $|\Psi(t)\rangle$ originates in the distant past from the incoming state $|\Psi_0^{in}(t)\rangle$, which evolves according to the propagator $\hat{G}^+$

$$|\Psi^+(t)\rangle \equiv i\hbar \lim_{t_0 \to -\infty} \hat{G}^+(t-t_0)|\Psi_0^{in}(t_0)\rangle, \qquad (3.102)$$

or it originates in the distant future from the outgoing state $|\Psi_0^{out}(t)\rangle$ which evolves according to $\hat{G}^-$

$$|\Psi^-(t)\rangle \equiv -i\hbar \lim_{t_0 \to +\infty} \hat{G}^-(t-t_0)|\Psi_0^{out}(t_0)\rangle. \qquad (3.103)$$

By following the same procedure we used to derive Eqs. 3.99 and 3.101 we then obtain from these states the following Lippmann–Schwinger equations (Exercise 3.2)

$$|\Psi^+(t)\rangle = |\Psi_0^{in}(t)\rangle + \int_{-\infty}^{+\infty} dt'\, G^+(t-t')\hat{V}|\Psi_0^{in}(t')\rangle \qquad (3.104)$$

and

$$|\Psi^-(t)\rangle = |\Psi_0^{\text{out}}(t)\rangle + \int_{-\infty}^{+\infty} dt' \, G^-(t-t')\hat{V}|\Psi_0^{\text{out}}(t')\rangle, \qquad (3.105)$$

which differ from Eqs. 3.99 and 3.101 because they require information on the *full* Green's functions $\hat{G}^+$ and $\hat{G}^-$. The use of Eqs. 3.99 and 3.101 or their equivalent, Eqs. 3.104 and 3.105, thus really depends on what type of information one has on the system dynamics.

3.4.2 Time-independent Lippmann–Schwinger equation

Once we have derived the time-dependent Lippmann–Schwinger equations, the corresponding time-*independent* versions can be easily obtained via Fourier transforming the time-dependent Green's functions.[27] We thus define

$$\hat{G}^+(E) = \int_{-\infty}^{+\infty} dt \, e^{iEt/\hbar} e^{-\epsilon t/\hbar} \hat{G}^+(t) = \int_{0}^{+\infty} dt \, e^{iEt/\hbar} e^{-\epsilon t/\hbar} \hat{G}^+(t), \qquad (3.106)$$

where the last equality comes about because of the formal solution of the full propagator 3.80. Similarly, for all other Green's functions

$$\hat{G}^-(E) = \int_{-\infty}^{0} dt \, e^{iEt/\hbar} e^{+\epsilon t/\hbar} \hat{G}^-(t), \qquad (3.107)$$

for the full propagator, and

$$\hat{G}_0^+(E) = \int_{0}^{+\infty} dt \, e^{iEt/\hbar} e^{-\epsilon t/\hbar} \hat{G}_0^+(t), \qquad (3.108)$$

$$\hat{G}_0^-(E) = \int_{-\infty}^{0} dt \, e^{iEt/\hbar} e^{+\epsilon t/\hbar} \hat{G}_0^-(t), \qquad (3.109)$$

for the free propagators. In the above, I have introduced an infinitesimal number $\epsilon > 0$, and corresponding exponential terms, which guarantee that the integrals are convergent.[28] By replacing the Green's functions 3.80 and 3.82 in 3.106 and 3.107, respectively, we get

$$\hat{G}^+(E) = \frac{\hat{1}}{E + i\epsilon - \hat{H}_S} \qquad \text{retarded}, \qquad (3.110)$$

---

[27] Remember that all quantities appearing in the various Lippmann–Schwinger equations depend at most on time differences only, so that their Fourier transform leads to a single energy function.

[28] For instance, the integral for $\hat{G}^+$ runs from $t=0$ to $t \to +\infty$, hence the exponential factor $e^{-\epsilon t/\hbar}$ guarantees convergence for $t \to +\infty$.

and
$$\hat{G}^-(E) = \frac{\hat{1}}{E - i\epsilon - \hat{H}_S} \quad \text{advanced.} \quad (3.111)$$

It will sometimes be useful to write the above two equations in a compact form by defining the complex number $z = E + i\epsilon$ for the retarded Green's function, and $z = E - i\epsilon$ for the advanced Green's function. In this case, both of the above two equations can be written as

$$\hat{G}(z) = \frac{\hat{1}}{z - \hat{H}_S}, \quad (3.112)$$

which corresponds to the two different Eqs. 3.110 and 3.111.

The corresponding quantities for the free Hamiltonian $\hat{H}_0$ can be obtained by replacing 3.86 and 3.87 in 3.108 and 3.109, respectively

$$\hat{G}_0^+(E) = \frac{\hat{1}}{E + i\epsilon - \hat{H}_0} \equiv \frac{\hat{1}}{z - \hat{H}_0} \quad \text{retarded,} \quad (3.113)$$

and

$$\hat{G}_0^-(E) = \frac{\hat{1}}{E - i\epsilon - \hat{H}_0} \equiv \frac{\hat{1}}{z - \hat{H}_0} \quad \text{advanced.} \quad (3.114)$$

As evident from their analytical structure, all the Green's functions $\hat{G}^+(E)$, $\hat{G}^-(E)$, $\hat{G}_0^+(E)$ and $\hat{G}_0^-(E)$ have poles in correspondence to the eigenvalues of the respective Hamiltonians. For instance, $\hat{G}^+(E)$ has poles in correspondence to the eigenvalues of the Hamiltonian $\hat{H}_S$.

For the *bound states* spectrum of $\hat{H}_S$ (see Eq. 3.6 where here, however, the states are single-particle states), $\hat{G}^+(E)$ has simple poles. In the continuum (i.e., from $E = 0$ to $E \to +\infty$, Eq. 3.7) these poles merge to form a *branch cut* in the positive energy real axis, and we must calculate $\hat{G}^+(E)$ in the complex energy plane by approaching the branch cut from *above*, i.e., for $\text{Im}(z) \to 0^+$. Similar considerations apply to the other Green's functions, the advanced ones being calculated by approaching the branch cut from below in the complex energy plane: $\text{Im}(z) \to 0^-$.

Using the definitions 3.110 and 3.111 for the full retarded and advanced Green's functions, respectively, we see that they are related via

$$[G^+(E)]^\dagger = G^-(E), \quad (3.115)$$

which can also be derived from the Fourier transform of the time-dependent relation 3.84. Similarly for the Green's functions associated with the free Hamiltonian $\hat{H}_0$.

It is now an easy task to write the various Lippmann–Schwinger equations in their static version by Fourier transforming their time-dependent counterpart. We get from 3.91 and 3.92 the following equations

$$\hat{G}^{\pm}(E) = \hat{G}_0^{\pm}(E) + \hat{G}_0^{\pm}(E)\hat{V}\hat{G}^{\pm}(E)$$
$$= \hat{G}_0^{\pm}(E) + \hat{G}^{\pm}(E)\hat{V}\hat{G}_0^{\pm}(E), \quad (3.116)$$

for the Green's functions, and from the Eqs. 3.99, 3.101, 3.104 and 3.105

$$|\Psi^{+}(E)\rangle = |\Psi_0^{\text{in}}(E)\rangle + G_0^{+}(E)\hat{V}|\Psi^{+}(E)\rangle$$
$$= |\Psi_0^{\text{in}}(E)\rangle + G^{+}(E)\hat{V}|\Psi_0^{\text{in}}(E)\rangle \quad (3.117)$$

and

$$|\Psi^{-}(E)\rangle = |\Psi_0^{\text{out}}(E)\rangle + G_0^{-}(E)\hat{V}|\Psi^{-}(E)\rangle$$
$$= |\Psi_0^{\text{out}}(E)\rangle + G^{-}(E)\hat{V}|\Psi_0^{\text{out}}(E)\rangle, \quad (3.118)$$

for the scattering states, where I have defined the Fourier transform of a general state

$$|\Psi(E)\rangle = \int_{-\infty}^{+\infty} dt\, e^{iEt/\hbar}|\Psi(t)\rangle. \qquad (3.119)$$

Also, for later use, I write the Fourier transform of the Dyson equations 3.94 and 3.96 ($\hat{\Sigma}^{\pm}(E)$ is the Fourier transform of $\hat{\Sigma}^{\pm}(t-t')$)

$$\hat{G}^{\pm}(E) = \hat{G}_0^{\pm}(E) + \hat{G}_0^{\pm}(E)\hat{\Sigma}^{\pm}(E)\hat{G}^{\pm}(E)$$
$$= \hat{G}_0^{\pm}(E) + \hat{G}^{\pm}(E)\hat{\Sigma}^{\pm}(E)\hat{G}_0^{\pm}(E), \quad (3.120)$$

even though, in the present single-particle scattering problem, this form is equivalent to the Lippmann–Schwinger equations 3.116.

Equation 3.120 can also be written as

$$\hat{G}^{\pm}(E) = \frac{\hat{1}}{[\hat{G}_0^{\pm}(E)]^{-1} - \hat{\Sigma}^{\pm}(E)} = \frac{\hat{1}}{E \pm i\epsilon - \hat{H}_0 - \hat{\Sigma}^{\pm}(E)}, \qquad (3.121)$$

a form I will use later.

Green's function spectral representation

Let us consider for instance the eigenstates of $\hat{H}_S$ given by the single-particle analogue of Eqs. 3.6 and 3.7 (similar considerations apply to $\hat{H}_0$). These eigenstates satisfy the *resolution of the identity*

$$\hat{1} = \sum_{E_i \alpha} |\Psi_{E_i \alpha}\rangle\langle\Psi_{E_i \alpha}| + \sum_\alpha \int_0^{+\infty} dE\, |\Psi_{E\alpha}\rangle\langle\Psi_{E\alpha}|. \qquad (3.122)$$

By applying 3.122 to the operator $\hat{G}(z)$ (Eq. 3.112) and using Eqs. 3.6 and 3.7 we get the *spectral representation* of the Green's function

$$\hat{G}(z) = \sum_{E_i \alpha} \frac{|\Psi_{E_i \alpha}\rangle\langle\Psi_{E_i \alpha}|}{z - E_i} + \sum_\alpha \int_0^{+\infty} dE\, \frac{|\Psi_{E\alpha}\rangle\langle\Psi_{E\alpha}|}{z - E}, \qquad (3.123)$$

which again represents two equations, one for $\hat{G}^+(E)$ and one for $\hat{G}^-(E)$. Similarly for all other Green's functions.

Density of states operator

Using the Green's functions spectral representation 3.123, we can now calculate several quantities we will make use of later on. First of all, we realize that the operator

$$\hat{D}(E) = i \lim_{\epsilon \to 0} \frac{\hat{G}^+(E) - \hat{G}^-(E)}{2\pi} \qquad (3.124)$$

counts the number of states at a given energy, and is thus called *density of states* operator. Indeed, if we replace 3.123 into 3.124 we get

$$\hat{D}(E) = \frac{1}{\pi} \lim_{\epsilon \to 0} \sum_{E_i \alpha} |\Psi_{E_i \alpha}\rangle \frac{\epsilon}{(E - E_i)^2 + \epsilon^2} \langle\Psi_{E_i \alpha}|$$

$$+ \frac{1}{\pi} \lim_{\epsilon \to 0} \sum_\alpha \int_0^{+\infty} dE'\, |\Psi_{E'\alpha}\rangle \frac{\epsilon}{(E - E')^2 + \epsilon^2} \langle\Psi_{E'\alpha}|$$

$$= \sum_{E_i \alpha} |\Psi_{E_i \alpha}\rangle \delta(E - E_i) \langle\Psi_{E_i \alpha}|$$

$$+ \sum_\alpha \int_0^{+\infty} dE'\, |\Psi_{E'\alpha}\rangle \delta(E - E') \langle\Psi_{E'\alpha}|. \qquad (3.125)$$

In the last equality I have used the definition of the δ-function

$$\delta(x) = \frac{1}{\pi} \lim_{\epsilon \to 0} \frac{\epsilon}{x^2 + \epsilon^2}. \qquad (3.126)$$

The above density of states operator can be evaluated in any given basis. If we choose the position representation, we get the *local* density of states

$$D(\mathbf{r}, E) \equiv \langle \mathbf{r}|\hat{D}(E)|\mathbf{r}\rangle$$
$$= i \lim_{\epsilon \to 0} \frac{G^+(\mathbf{r},\mathbf{r},E) - G^-(\mathbf{r},\mathbf{r},E)}{2\pi}$$
$$= -\frac{1}{\pi}\text{Im}[G^+(\mathbf{r},\mathbf{r},E)], \qquad (3.127)$$

which counts the number of states at a given energy at a point in space $\mathbf{r}$. In the above I have also written the Green's functions $\hat{G}^\pm$ in the position basis

$$G^\pm(\mathbf{r},\mathbf{r}',E) \equiv \langle \mathbf{r}|\hat{G}^\pm(E)|\mathbf{r}'\rangle. \qquad (3.128)$$

3.4.2.1 Free-particle Green's functions in one dimension

As an example, and since we will make use of it later, let us consider the position representation of the Green's function associated with the Hamiltonian,

$$\hat{H}_0 = \frac{\hat{p}_x^2}{2m}, \qquad (3.129)$$

of a free particle in one dimension along the cartesian direction x. The associated Green's functions $\hat{G}_0^\pm$ in the position basis are

$$G_0^\pm(x,x',E) \equiv \langle x|\hat{G}_0^\pm(E)|x'\rangle, \qquad (3.130)$$

where the energy $E = \hbar^2 k^2/2m$. We need to calculate just one of these Green's functions, say $G^+(x,x',E)$. The other one is related to this one via 3.115.

The eigenstates of 3.129 are plane waves (see Eq. 3.20) which I denote with $\langle x|\psi_k\rangle = \psi_k(x) = e^{ikx}$. Using the spectral representation 3.123 I can thus write

$$G_0^+(x,x',E) = \frac{1}{2\pi} \lim_{\epsilon \to 0} \int_{-\infty}^{+\infty} dk' \frac{\langle x|\psi_{k'}\rangle\langle\psi_{k'}|x'\rangle}{E + i\epsilon - \hbar^2 k'^2/2m}$$
$$= \frac{1}{2\pi} \lim_{\epsilon \to 0} \int_{-\infty}^{+\infty} dk' \frac{e^{ik'(x-x')}}{E + i\epsilon - \hbar^2 k'^2/2m}$$
$$= \frac{m}{i\hbar^2} \frac{e^{ik|x-x'|}}{k} \quad k > 0, \qquad (3.131)$$

where in the last step I have integrated in the complex plane by closing the contour of integration in the complex upper-half plane for $x > x'$ – thus obtaining the residue of the pole with $\text{Im}(k) > 0$ in that region – and in the lower-half plane for $x < x'$ – which includes the other pole with $\text{Im}(k) < 0$ (Exercise 3.6).

Since $k > 0$, if $x > x'$ the above Green's function is

$$G_0^+(x > x', E) = \frac{m}{i\hbar^2}\frac{e^{ik(x-x')}}{k} \qquad (3.132)$$

which, if we fix x', is proportional to a plane wave of given energy "leaving" the point x', and moving from left to right (right-moving wave). On the other hand, for $x < x'$

$$G_0^+(x < x', E) = \frac{m}{i\hbar^2}\frac{e^{-ik(x-x')}}{k} \qquad (3.133)$$

is proportional to a plane wave leaving from the point x', and moving from right to left (left-moving wave). This may be interpreted as if the point x' is a "source" of right-moving and left-moving waves.

The corresponding advanced Green's function is then

$$G_0^-(x, x', E) = [G_0^+(x, x', E)]^\dagger = [G_0^+(x', x, E)]^* = -\frac{m}{i\hbar^2}\frac{e^{-ik|x-x'|}}{k} \qquad k > 0, \qquad (3.134)$$

for which the above considerations can be reversed: the point x' can be considered now as a "sink" of right-moving and left-moving waves.

In the above I have considered propagating states (plane waves). I have discussed in Sec. 3.3.1 that we also need to consider evanescent modes (Eqs. 3.30 and 3.31). These correspond to negative energies or imaginary wave-vector $k = i\tilde{k}$, with $\tilde{k} > 0$. If we replace this imaginary wave-vector in 3.131 we get

$$G_0^+(x, x', E) = -\frac{m}{\hbar^2}\frac{e^{-\tilde{k}|x-x'|}}{\tilde{k}} \qquad E = -\frac{\hbar^2\tilde{k}^2}{2m} < 0 \qquad \tilde{k} > 0, \qquad (3.135)$$

and

$$G_0^-(x, x', E) = [G_0^+(x, x', E)]^\dagger = -\frac{m}{\hbar^2}\frac{e^{-\tilde{k}|x-x'|}}{\tilde{k}} = G_0^+(x, x', E), \qquad (3.136)$$

which are simply proportional to the evanescent mode wave-functions (see Eq. 3.30).

We now have all the formal tools of single-particle scattering theory to recalculate the total current 3.56 in another equivalent way.

3.5 Green's functions and self-energy

For clarity, let us refer to Fig. 3.11. Regions L and R are two semi-infinite chunks of conductors that sandwich a central region C (the sample). The

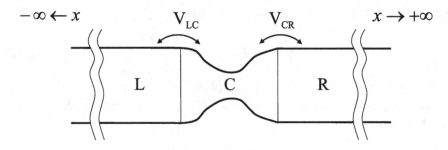

Fig. 3.11. A finite sample C coupled to two semi-infinite leads via some operators. The leads do not couple with each other directly.

latter could again be a molecule or any other nanoscale structure. The two chunks of conductors may or may not be identical in all their properties.

I now make a different partition of the Hamiltonian $\hat{H}_S$, Eq. 3.9 (see, e.g., Kurth et al., 2005). I assume that regions L and C are coupled by some potential $(\hat{V}_{LC} + \hat{V}_{LC}^{\dagger})$, while regions C and R, by a potential $(\hat{V}_{CR} + \hat{V}_{CR}^{\dagger})$ (see Fig. 3.11), with $\hat{V}_{LC}$ and $\hat{V}_{CR}$ some operators. We also *assume* that regions L and R are decoupled. Physically this means there is no direct tunneling between the two regions.

In certain nanoscale systems, for instance, a single molecule or, in general, a small cluster of atoms between two bulk electrodes (see Figs. 1.1 and 1.2), this approximation must be handled with care. If the two chunks of conductors contain all ions, except those of the molecule or cluster, the electrode surfaces of the bulk conductors may be separated by just a few angstroms. In this case, some finite electronic coupling (tunneling) between the two surfaces is present.

Most importantly, these surfaces are generally charged (as I have anticipated in Sec. 1.1, and I will discuss again in Chapter 7) so that direct Coulomb interactions may affect the electronic states of the two electrodes without the sample. We can, however, choose our "sample" region to extend several atomic layers inside the bulk electrodes where screening is essentially complete – possibly within microscopic Friedel-like oscillations. By doing so we may assume the above coupling negligible.

In addition, let us assume the potentials $\hat{V}_{LC}$ and $\hat{V}_{CR}$ to be "short range", in the sense that their spatial ranges do not overlap in the sample region.[29]

---

[29] Which is the same as saying that we can indeed define an "interfacial region" between the left electrode and the central region *distinct* from the "interfacial region" between the right electrode and the central region.

3.5 Green's functions and self-energy

Again, in nanoscale systems this approximation may not be quite valid due to partially screened Coulomb interactions.[30]

The *isolated* semi-infinite conductors are described by the Hamiltonians $\hat{H}_L$ and $\hat{H}_R$. To these Hamiltonians are associated the Green's functions $\hat{G}_L(z)$, $\hat{G}_R(z)$, respectively, according to the definition 3.112. The isolated sample is described by the Hamiltonian $\hat{H}_C$, with Green's function $\hat{G}_C(z)$. Here we assume to know these Green's functions exactly. Our goal is to determine the Green's function $\hat{G}(z)$, not of $\hat{H}_C$, but of the sample *in the presence* of the coupling with the electrodes. To do this let us proceed as follows.

With these partitions and definitions, the total Hamiltonian is

$$\hat{H}_S = \hat{H}_L + \hat{H}_R + \hat{H}_C + \hat{V}_{LC} + \hat{V}_{LC}^\dagger + \hat{V}_{CR} + \hat{V}_{CR}^\dagger. \tag{3.137}$$

The Schrödinger equation can be formally written in matrix form

$$\begin{bmatrix} \hat{H}_L & \hat{V}_{LC} & 0 \\ \hat{V}_{LC}^\dagger & \hat{H}_C & \hat{V}_{CR}^\dagger \\ 0 & \hat{V}_{CR} & \hat{H}_R \end{bmatrix} \begin{bmatrix} |\phi_L\rangle \\ |\phi_C\rangle \\ |\phi_R\rangle \end{bmatrix} = E \begin{bmatrix} |\phi_L\rangle \\ |\phi_C\rangle \\ |\phi_R\rangle \end{bmatrix}. \tag{3.138}$$

The elements of the vector

$$\begin{bmatrix} |\phi_L\rangle \\ |\phi_C\rangle \\ |\phi_R\rangle \end{bmatrix} \tag{3.139}$$

are the single-particle wave-functions associated with the Hamiltonians of the three regions. For instance, we denote with $|\phi_L\rangle$ the solutions to the equation $\hat{H}_L|\phi_L\rangle = E|\phi_L\rangle$. The matrix equation 3.138 can be easily solved and gives the three equations

$$\hat{H}_L|\phi_L\rangle + \hat{V}_{LC}|\phi_C\rangle = E|\phi_L\rangle, \tag{3.140}$$

$$\hat{V}_{LC}^\dagger|\phi_L\rangle + \hat{H}_C|\phi_C\rangle + \hat{V}_{CR}^\dagger|\phi_R\rangle = E|\phi_C\rangle, \tag{3.141}$$

and

$$\hat{V}_{CR}|\phi_C\rangle + \hat{H}_R|\phi_R\rangle = E|\phi_R\rangle. \tag{3.142}$$

We can rearrange 3.140 to get

$$(E - \hat{H}_L)|\phi_L\rangle = \hat{V}_{LC}|\phi_C\rangle. \tag{3.143}$$

---

[30] In practice, all parameters entering these potentials need to be determined self-consistently. This helps reduce, to some degree, the arbitrariness of their value, but clearly does not change the assumed *form* of the coupling potentials.

If we make the usual substitution $E \to E \pm i\epsilon \equiv z$ (see Eq. 3.112), to analytically continue the Green's functions in the complex plane, then the left-hand side of Eq. 3.143 is simply the inverse of the Green's function $\hat{G}_L(z)$, that is

$$|\phi_L\rangle = \hat{G}_L(z)\hat{V}_{LC}|\phi_C\rangle. \tag{3.144}$$

Similarly, from Eq. 3.142

$$|\phi_R\rangle = \hat{G}_R(z)\hat{V}_{CR}|\phi_C\rangle. \tag{3.145}$$

These last two equations directly relate the wave-function of the sample with the wave-function of the electrodes. By multiplying both sides of Eqs. 3.144 and 3.145 by $\langle\phi_C|$ we get

$$\langle\phi_C|\phi_L\rangle = \langle\phi_C|\hat{G}_L(z)\hat{V}_{LC}|\phi_C\rangle, \tag{3.146}$$

and

$$\langle\phi_C|\phi_R\rangle = \langle\phi_C|\hat{G}_R(z)\hat{V}_{CR}|\phi_C\rangle, \tag{3.147}$$

i.e., there is a finite probability amplitude of finding an electron of the central region C in both L and R regions. If we label ϵ_k the eigenenergies of the Hamiltonian $\hat{H}_C$ of the *isolated* central region, this is equivalent to saying that there is a finite probability for an electron in the sample (with one of the energies ϵ_k) to *tunnel* to the left and right electrodes. Since we are assuming that $\hat{H}_L$ and $\hat{H}_R$ describe semi-infinite electrodes, then their spectrum is continuous, and the discrete states of the isolated sample broaden into resonances (see Fig. 3.12).[31] The energy ϵ_k is therefore "shifted" (or *renormalized*) by the presence of the electrode states and, at the same time, the electron acquires a *lifetime* to scatter from the central region into the electrodes. Let us expand on this concept even more.

We replace $|\phi_L\rangle$ and $|\phi_R\rangle$ in 3.141 to get

$$\left[E - \hat{H}_C - \hat{V}_{LC}^\dagger \hat{G}_L(z)\hat{V}_{LC} - \hat{V}_{CR}^\dagger \hat{G}_R(z)\hat{V}_{CR}\right]|\phi_C\rangle = 0. \tag{3.148}$$

The operators

$$\hat{\Sigma}_L(z) \equiv \hat{V}_{LC}^\dagger \hat{G}_L(z)\hat{V}_{LC}, \tag{3.149}$$

and

$$\hat{\Sigma}_R(z) \equiv \hat{V}_{CR}^\dagger \hat{G}_R(z)\hat{V}_{CR}, \tag{3.150}$$

---

[31] Those with renormalized energy higher than the energy of the bottom of the continuum spectrum of both the Hamiltonians $\hat{H}_L$ and $\hat{H}_R$. Those that end up between the bottom of the continuum of $\hat{H}_L$ and the bottom of the continuum of $\hat{H}_R$ are exponentially decaying into the left electrode and standing waves into the right electrode – assuming the bottom of the continuum of H_L is higher in energy than the corresponding one for H_R (Lang, 1995; Di Ventra and Lang, 2002).

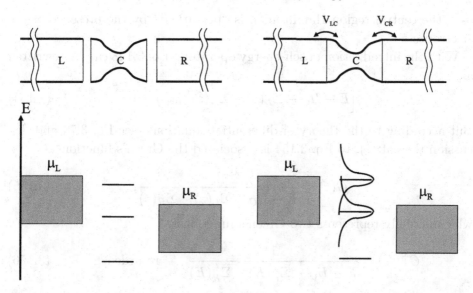

Fig. 3.12. When the central region is coupled to two semi-infinite electrodes, its energy states are "renormalized", and those whose renormalized energy is higher than the bottom of the continuum spectrum of the electrodes' Hamiltonians broaden into resonances.

are again called *self-energy* operators.

As stated before, $z = E \pm i\epsilon$ so that in reality $\hat{\Sigma}_L(z)$ defines two self-energy operators,

$$\hat{\Sigma}_L^+(E) \equiv \hat{\Sigma}_L(E + i\epsilon) \qquad \text{retarded,} \tag{3.151}$$

$$\hat{\Sigma}_L^-(E) \equiv \hat{\Sigma}_L(E - i\epsilon) \qquad \text{advanced,} \tag{3.152}$$

and similarly for $\hat{\Sigma}_R(z)$.

However, unlike the self-energy term 3.95 I have introduced in Sec. 3.4 when discussing the Lippmann–Schwinger equation, the self-energy operators here are non-Hermitian: $\hat{\Sigma}_{L,R}^\dagger(z) \neq \hat{\Sigma}_{L,R}(z)$. Indeed, from the general relation between the retarded and advanced Green's functions, Eq. 3.115, we get from Eqs. 3.149 and 3.150

$$(\hat{\Sigma}^+)_{L,R}^\dagger(E) = \hat{\Sigma}_{L,R}^-(E), \tag{3.153}$$

i.e., the advanced self-energy is the Hermitian conjugate of the retarded one, and vice versa. The fact that in the present case these operators are not Hermitian is a consequence of the partitioning 3.137. While in Sec. 3.4 the scattering potential $\hat{V}$ acted on the states of the free (electrodes) Hamiltonian, here the "interface" potentials act on the states of the central region, as if the central region Hamiltonian is "perturbed" by the presence of the electrodes.

With the introduction of self-energy operators, Eq. 3.148 can be rewritten as

$$\left[E - \hat{H}_C - \hat{\Sigma}_L(z) - \hat{\Sigma}_R(z)\right] |\phi_C\rangle = 0. \tag{3.154}$$

But according to the theory of differential equations (see Eq. 3.74 and discussion thereafter), to Eq. 3.154 is associated the Green's function

$$\hat{G}(z) = \frac{\hat{1}}{E - \hat{H}_C - \hat{\Sigma}_L(z) - \hat{\Sigma}_R(z)}, \tag{3.155}$$

which actually represents two Green's functions

$$\hat{G}^+(E) = \frac{\hat{1}}{E - \hat{H}_C - \hat{\Sigma}_L^+(E) - \hat{\Sigma}_R^+(E)} \quad \text{retarded}, \tag{3.156}$$

and

$$\hat{G}^-(E) = \frac{\hat{1}}{E - \hat{H}_C - \hat{\Sigma}_L^-(E) - \hat{\Sigma}_R^-(E)} \quad \text{advanced}. \tag{3.157}$$

Equation 3.155 (and each of the Eqs. 3.156 and 3.157) is of the Dyson form 3.121 for a single particle in the central region "interacting" with the L and R electrodes. It can be put in the equivalent form 3.120 by using the Green's function $\hat{G}_C(z)$ of the isolated central region, and defining the total self-energy $\hat{\Sigma}(z) = \hat{\Sigma}_L(z) + \hat{\Sigma}_R(z)$:

$$\hat{G}(z) = \hat{G}_C(z) + \hat{G}_C(z)\hat{\Sigma}(z)\hat{G}(z) = \hat{G}_C(z) + \hat{G}(z)\hat{\Sigma}(z)\hat{G}_C(z), \tag{3.158}$$

which can be verified by direct substitution into Eq. 3.155. Again, Eq. 3.158 represents two distinct equations, one for the retarded Green's function and one for the advanced Green's function.

Let us note that, by choosing the partition 3.137, we have transformed the original problem of the sample plus the electrodes into the problem of the central region "open" to the electrodes via the self-energies.[32] The self-energies take care of the scattering due to the presence of the electrodes. As I will show in a moment, however, this is *equivalent* to the single-particle scattering problem we have solved to derive the total current 3.56, and does not add any new physics to the solution of the Lippmann–Schwinger equations 3.117 and 3.118.

Also, the reader should note that even if the equation 3.158 has a formal

---

[32] Clearly, this is not an open quantum system in the sense that it is not dynamically coupled to reservoirs.

3.5 Green's functions and self-energy

structure similar to the many-body Dyson equation I will discuss in Chapter 4, I have derived it here without ever employing many-body perturbation theory. The steps leading to Eq. 3.158 – and the equation itself – should not be thus confused with the non-equilibrium Green's function formalism of Chapter 4.

Energy "renormalization" and state "lifetime"

Without the self-energy terms the Green's functions $\hat{G}^+(E)$ and $\hat{G}^-(E)$ (Eqs. 3.156 and 3.157) have poles in correspondence to the eigenenergies ϵ_k of $\hat{H}_C$. We can call these "zero-order poles" of the Green's functions. The self-energies therefore make $\hat{G}^+(E)$ and $\hat{G}^-(E)$ analytical in correspondence to these poles by "renormalizing" the energies of $\hat{H}_C$ (Fig. 3.12).

We can see this even better if we write

$$\hat{\Sigma}_L(z) = \text{Re}\{\hat{\Sigma}_L(z)\} + i\text{Im}\{\hat{\Sigma}_L(z)\}, \quad (3.159)$$

in terms of the real and imaginary parts (and similarly for the right electrode).[33] Equation 3.155 then becomes

$$\hat{G}(z) = \frac{\hat{1}}{E - \hat{H}_C - \text{Re}\{\hat{\Sigma}_L(z) + \hat{\Sigma}_R(z)\} - i\text{Im}\{\hat{\Sigma}_L(z) + \hat{\Sigma}_R(z)\}}, \quad (3.160)$$

which shows that the energies ϵ_k of $\hat{H}_C$ are "shifted" (or "renormalized").

To first order, we can estimate the amount of this renormalization by replacing in the argument of the self-energy the "unperturbed" energy (zero-order pole) ϵ_k, namely $\text{Re}\{\Sigma_L(\epsilon_k) + \hat{\Sigma}_R(\epsilon_k)\}$ is the approximate energy shift (see also discussion in Sec. 4.2.4).

Using the relations 3.156 and 3.157 it is also easy to show that

$$[\hat{G}^+(E)]^{-1} - [\hat{G}^-(E)]^{-1} = \left(\hat{\Sigma}_L^-(E) - \hat{\Sigma}_L^+(E)\right) + \left(\hat{\Sigma}_R^-(E) - \hat{\Sigma}_R^+(E)\right)$$

$$\equiv i\left[\hat{\Gamma}_L(E) + \hat{\Gamma}_R(E)\right] \equiv i\hat{\Gamma}(E), \quad (3.161)$$

which can be equivalently written as

$$i\left[\hat{G}^+(E) - \hat{G}^-(E)\right] = \hat{G}^+(E)\hat{\Gamma}\hat{G}^-(E). \quad (3.162)$$

In the above equations I have defined the quantities

$$\hat{\Gamma}_{L,R}(E) = i[\hat{\Sigma}_{L,R}^+(E) - \hat{\Sigma}_{L,R}^-(E)] = -2\text{Im}\{\hat{\Sigma}_{L,R}^+(E)\}. \quad (3.163)$$

Recalling the definition of density of states operator, Eq. 3.124, we see that

---

[33] These must be interpreted as the real and imaginary parts of the matrix elements of the corresponding operator in a given basis.

the right-hand side of Eq. 3.162 is simply proportional to the single-particle density of states of the *entire* system (central region plus electrodes).[34]

In the absence of the imaginary part of the self-energies, the eigensolutions corresponding to the full Green's function $\hat{G}(z)$ would only be shifted with respect to the energies ϵ_k, but would still belong to the real energy axis. They would thus be stationary states of some Hamiltonian.

The imaginary parts move these solutions *off* the real energy axis, inside the complex plane. An imaginary energy means a "decay" of the solution (which, in a dynamical sense, means a non-unitary evolution of the state of the system), so that the imaginary parts can be interpreted as the "rate" at which the electrons are "scattered out" of the states of the free Hamiltonian $\hat{H}_C$ (see also Sec. 4.2.4). We thus say that due to (elastic) scattering the eigenstates of $\hat{H}_C$ acquire an "elastic lifetime".[35]

To first order again, this lifetime can be estimated by replacing the energy argument of the imaginary part of the self-energy with the unperturbed energies ϵ_k, so that (from the retarded Green's function)

$$\tau^{el}(\epsilon_k) = -\frac{\hbar}{\operatorname{Im}\{\Sigma_L^+(\epsilon_k) + \Sigma_R^+(\epsilon_k)\}}. \tag{3.164}$$

I stress here that the "broadening" and "lifetime" of the single-particle states are not due to interactions among electrons, but are simply due to the fact that the central region is coupled to two electrodes.

Discrete (or tight-binding) space representation

Before making a connection with formal scattering theory, let us understand better the advantage of assuming the potentials $\hat{V}_{LC}$ and $\hat{V}_{CR}$ to be short range. From a numerical standpoint, even if the Hamiltonians $\hat{H}_L$ and $\hat{H}_R$ describe two semi-infinite electrodes, the short-range potentials $\hat{V}_{LC}$ and $\hat{V}_{CR}$ limit the size of the non-zero matrix elements we can construct from the self-energy operators in any given *finite* basis set. In a position representation,

---

[34] In Sec. 4.2.4, I will show that an equation similar to 3.162 can be defined in the many-body case as well, with the left-hand side of Eq. 3.162 corresponding to the spectral function (see Eq. 4.54).

[35] Both the state renormalization and lifetime are energy-dependent and, in general, not a simple function of the energy. This is true also for the many-body case (see discussion in Sec. 4.2.2).

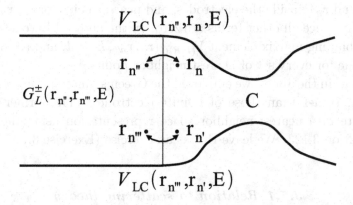

Fig. 3.13. Matrix elements of the operator $\hat{V}_{LC}$ entering the calculation of the self-energy $\Sigma_L^\pm(\mathbf{r}_n, \mathbf{r}_{n'}, E)$ in a nearest-neighbor discrete-space representation. The self-energy also contains the Green's function $G_L^\pm(\mathbf{r}_{n''}, \mathbf{r}_{n'''}, E)$ of the semi-infinite electrode.

by choosing two points $\mathbf{r}, \mathbf{r}'$ *inside* the central region (with obvious notation)

$$\begin{aligned}\Sigma_{L,R}^\pm(\mathbf{r}, \mathbf{r}', E) &\equiv \langle\mathbf{r}|\hat{\Sigma}_{L,R}^\pm(E)|\mathbf{r}'\rangle \\ &= \int d\mathbf{r}'' d\mathbf{r}''' \langle\mathbf{r}|\hat{V}_{(L,R)C}^\dagger|\mathbf{r}''\rangle\langle\mathbf{r}''|\hat{G}_{L,R}^\pm(E)|\mathbf{r}'''\rangle\langle\mathbf{r}'''|\hat{V}_{(L,R)C}|\mathbf{r}'\rangle \\ &= \int d\mathbf{r}'' d\mathbf{r}''' V_{(L,R)C}^\dagger(\mathbf{r}, \mathbf{r}'', E) G_{L,R}^\pm(\mathbf{r}'', \mathbf{r}''', E) V_{(L,R)C}(\mathbf{r}''', \mathbf{r}', E),\end{aligned}$$
(3.165)

with the integrals extending over all (semi-infinite) space but converging very fast, in view of the fact that $\langle\mathbf{r}|\hat{V}_{LC}|\mathbf{r}'\rangle$ and $\langle\mathbf{r}|\hat{V}_{CR}|\mathbf{r}'\rangle$ are short range.[36]

If a finite discrete representation of space is used,

$$\mathbf{r} \to \mathbf{r}_n, \quad \text{discrete-space representation,} \tag{3.166}$$

the only non-zero matrix elements contributing to $\hat{\Sigma}_{L,R}(\mathbf{r}_n, \mathbf{r}_{n'}, E)$ are those containing the (finite number of) hopping terms between sites one retains in the representation. All other matrix elements are zero.

In the simplest tight-binding model, which keeps only the nearest-neighbor hopping terms, from 3.165 we get for points $\{\mathbf{r}_n, \mathbf{r}_{n'}\}$ inside the central region (Fig. 3.13)

$$\begin{aligned}\Sigma_{L,R}^\pm(\mathbf{r}_n, \mathbf{r}_{n'}, E) &= V_{(L,R)C}^\dagger(\mathbf{r}_n, \mathbf{r}_{n''}, E) G_{L,R}^\pm(\mathbf{r}_{n''}, \mathbf{r}_{n'''}, E) V_{(L,R)C}(\mathbf{r}_{n'''}, \mathbf{r}_{n'}, E) \\ &= t^2\, G_{L,R}^\pm(\mathbf{r}_{n''}, \mathbf{r}_{n'''}, E),\end{aligned}$$
(3.167)

---

[36] For instance, for the L lead $\mathbf{r}''$ and $\mathbf{r}'''$ are inside the semi-infinite L lead space region.

with $\mathbf{r}_{n''}$ and $\mathbf{r}_{n'''}$ inside the electrodes, and nearest-neighbors of $\mathbf{r}_n$ and $\mathbf{r}_{n'}$, respectively, since all other terms are zero. In the above, I have also defined the tight-binding matrix element $\hat{V}_{(L,R)C}(\mathbf{r}_n, \mathbf{r}_{n''}, E) = t$, and assumed it to be the same for every set of nearest-neighbor points.

Note that, in the above, we employed the Green's functions of *semi-infinite* electrodes, rather than those of infinite electrodes. The former are easy to compute in a nearest-neighbor space representation using the spectral representation 3.123. We leave this as an exercise (Exercise 3.7).

3.5.1 Relation to scattering theory

Let us now relate these results to the single-particle (static) scattering theory of Sec. 3.4.2. This will allow us to quantify what we mean by "tunneling" between the central region and the two electrodes, and rewrite the total current 3.56 in another equivalent form.

We do this by employing a different partitioning. Let us assume that $\hat{H}_L$ and $\hat{H}_R$ represent two identical *infinite* electrodes.[37] Similar to what we did in Sec. 3.3.1 let us assume that they are described by the translationally invariant Hamiltonians (see Eqs. 3.17 and 3.18)

$$-\frac{\hbar^2}{2m}\nabla^2 + V_{L,R}(\mathbf{r}_\perp), \tag{3.168}$$

with $V_{L,R}(\mathbf{r}_\perp)$ a generic single-particle potential that confines electrons in the transverse $y-z$ plane in the L and R electrodes, respectively. We have calculated the eigenstates of these Hamiltonians, and they are given by Eq. 3.22 for traveling states, namely they are separable into a transverse wave-function with channel index α, and a longitudinal plane wave.

From Eq. 3.131 we know that for a free particle moving in the x direction the retarded Green's function is simply

$$G^+(x, x', E) = \frac{m}{i\hbar^2}\frac{e^{ik|x-x'|}}{k}, \quad E = \frac{\hbar^2 k^2}{2m} > 0, \ k > 0. \tag{3.169}$$

Since the transverse and longitudinal motions are decoupled, the spectral representations 3.123 of the retarded Green's functions associated with the Hamiltonians 3.168 are simply (replace 3.22 into 3.123 and use 3.169)

$$G^+_{L,R}(\mathbf{r}, \mathbf{r}', E) \equiv G^+_0(\mathbf{r}, \mathbf{r}', E) = \frac{1}{i\hbar}\sum_\alpha u^*_\alpha(\mathbf{r}_\perp)\frac{e^{ik|x-x'|}}{v_\alpha(k)}u_\alpha(\mathbf{r}'_\perp), \tag{3.170}$$

---

[37] The electrodes need not to be identical. Here, we assume this to simplify the discussion.

with (as in Eq. 3.23)

$$E \equiv E_\alpha(k) = \epsilon_\alpha + \frac{\hbar^2 k^2}{2m}, \qquad (3.171)$$

where ϵ_α and $v_\alpha(k)$ are defined in Eqs. 3.21 and 3.24, respectively.

The Hamiltonian $\hat{H}_C$ of the central region is instead of the form

$$-\frac{\hbar^2}{2m}\nabla^2 + V_C(\mathbf{r}), \qquad (3.172)$$

with $V_C(\mathbf{r})$ a generic (but of finite range) single-particle potential.

An electron with energy $E = E_i(k_i)$ in an incoming state $\psi_{ik_i}(\mathbf{r})$ in the L electrode, with positive momentum $\hbar k_i$ (right-moving), would be scattered (according to the partition 3.137 of the total Hamiltonian $\hat{H}_S$) by the potential

$$\hat{V} = \hat{V}_C + \hat{V}_{LC} + \hat{V}_{LC}^\dagger + \hat{V}_{CR} + \hat{V}_{CR}^\dagger, \qquad (3.173)$$

which we have all along assumed to be of finite range. To be precise, this potential contains everything else that is *not* included in the Hamiltonians of the electrodes.

The scattered wave-function $\Psi_{ik_i}^+(\mathbf{r})$ is then related to the incoming state $\psi_{ik_i}(\mathbf{r})$ according to the Lippmann–Schwinger equation 3.117 in the position representation[38]

$$\Psi_{ik_i}^+(\mathbf{r}) = \psi_{ik_i}(\mathbf{r}) + \int d\mathbf{r}'\, G_0^+(\mathbf{r},\mathbf{r}',E)\, V(\mathbf{r}')\, \Psi_{ik_i}^+(\mathbf{r}'), \qquad (3.174)$$

where $V(\mathbf{r}')$ are the positional elements of the operator $\hat{V}$, and the integration extends over all space. From the meaning of the Green's function $\hat{G}_0^+$ (see discussion following Eq. 3.131), the above equation shows once again what I have anticipated in Sec. 3.3.2: the wave-function 3.174 is the combination of an incident wave $\psi_{ik_i}(\mathbf{r})$ plus a linear combination of *outgoing* waves.

From the Lippmann–Schwinger equation 3.118 in the positional basis, I can also calculate the linearly independent solution

$$\Psi_{ik_i}^-(\mathbf{r}) = \psi_{ik_i}(\mathbf{r}) + \int d\mathbf{r}'\, G_0^-(\mathbf{r},\mathbf{r}',E)\, V(\mathbf{r}')\, \Psi_{ik_i}^-(\mathbf{r}'), \qquad (3.175)$$

which is the combination of an outgoing wave $\psi_{ik_i}(\mathbf{r})$ plus a linear combination of *incoming* waves.

---

[38] If the potential $\hat{V}$ is non-local, the single integral in Eq. 3.174 becomes a double integral, i.e.,

$$\Psi_{ik_i}^+(\mathbf{r}) = \psi_{ik_i}(\mathbf{r}) + \int d\mathbf{r}' \int d\mathbf{r}''\, G_0^+(\mathbf{r},\mathbf{r}',E)\, V(\mathbf{r}',\mathbf{r}'')\, \Psi_{ik_i}^+(\mathbf{r}'').$$

In the absence of a magnetic field, the above Green's functions cannot depend on the order of the position indexes

$$G_0^\pm(\mathbf{r},\mathbf{r}',E) = G_0^\pm(\mathbf{r}',\mathbf{r},E), \quad \text{in the absence of a magnetic field.} \quad (3.176)$$

By transposing Eq. 3.174, using the general relation 3.115, and comparing the result with 3.174 we easily see that

$$\Psi^-_{ik_i}(\mathbf{r}) = \left[\Psi^+_{i(-k_i)}(\mathbf{r})\right]^*. \quad (3.177)$$

With the use of Eq. 3.170, the Lippmann–Schwinger equation 3.174 can be written as

$$\Psi^+_{ik_i}(\mathbf{r}) = \psi_{ik_i}(\mathbf{r}) + \frac{1}{i\hbar}\sum_{f=1}^{N_c^L}\frac{1}{v_f(k_f)}\int d\mathbf{r}'\, u_f^*(\mathbf{r}'_\perp)e^{ik_f|x-x'|}u_f(\mathbf{r}'_\perp)V(\mathbf{r}')\Psi^+_{ik_i}(\mathbf{r}'). \quad (3.178)$$

Let us now look at the asymptotic form of the scattered wave-function. With x' fixed, for $x \to -\infty$, deep inside the L lead, $\Psi_{ik_i}(\mathbf{r})$ has the form

$$\lim_{x\to-\infty}\Psi^+_{ik_i}(\mathbf{r}) = \psi_{ik_i}(\mathbf{r})$$
$$+ \frac{1}{i\hbar}\sum_{f=1}^{N_c^L}\frac{e^{-ik_f x}u_f(\mathbf{r}_\perp)}{v_f(k_f)}\int d\mathbf{r}'\, e^{ik_f x'}u_f^*(\mathbf{r}'_\perp)V(\mathbf{r}')\Psi^+_{ik_i}(\mathbf{r}'), \quad (3.179)$$

which, with the use of Eq. 3.22, can be written as[39]

$$\lim_{x\to-\infty}\Psi^+_{ik_i}(\mathbf{r}) = \psi_{ik_i}(\mathbf{r})$$
$$+ \frac{L_x}{i\hbar}\sum_{f=1}^{N_c^L}\frac{\psi_{fk_f}(\mathbf{r})}{v_f(k_f)}\int d\mathbf{r}'\, \psi^*_{fk_f}(\mathbf{r}')V(\mathbf{r}')\Psi^+_{ik_i}(\mathbf{r}'). \quad (3.180)$$

But we know from Sec. 3.3.2 that the general form of the scattered wavefunction deep into the L lead where evanescent modes do not contribute is given by Eq. 3.35.

By comparing 3.180 with 3.35 we see that

$$\mathscr{R}_{if} \equiv \frac{L_x}{i\hbar v_f(k_f)}\int d\mathbf{r}'\, \psi^*_{fk_f}(\mathbf{r}')V(\mathbf{r}')\Psi^+_{ik_i}(\mathbf{r}'), \quad (3.181)$$

[39] Note that in our notation states of channels f belonging to the L electrode represent waves moving *away* from the junction in the direction of *negative* x axis, i.e., they correspond to a momentum $-\hbar k_f$, with $k_f > 0$.

which can be written in the more compact form

$$\mathscr{R}_{if} \equiv \frac{L_x}{i\hbar v_f(k_f)} \langle \psi_{fk_f} | \hat{V} | \Psi^+_{ik_i} \rangle. \qquad (3.182)$$

The T matrix

I have calculated the above quantity by using the Green's function of the leads only. But I have also shown in Sec. 3.4 that the Lippmann–Schwinger equation can also be written in terms of the lead-sample-lead Green's function $G^+(\mathbf{r},\mathbf{r}',E)$, which is precisely the positional matrix element of the operator 3.156,[40]

$$\Psi^+_{ik_i}(\mathbf{r}) = \psi_{ik_i}(\mathbf{r}) + \int d\mathbf{r}' \, G^+(\mathbf{r},\mathbf{r}',E) \, V(\mathbf{r}') \, \psi_{ik_i}(\mathbf{r}'), \qquad (3.183)$$

which we equivalently write as

$$|\Psi^+_{ik_i}\rangle = |\psi_{ik_i}\rangle + \hat{G}^+ \hat{V} |\psi_{ik_i}\rangle. \qquad (3.184)$$

By replacing 3.184 into 3.182 we get

$$\mathscr{R}_{if} = \frac{L_x}{i\hbar v_f(k_f)} \left[\langle \psi_{fk_f} | \hat{V} | \psi_{ik_i} \rangle + \langle \psi_{fk_f} | \hat{V} \hat{G}^+ \hat{V} | \psi_{ik_i} \rangle \right]. \qquad (3.185)$$

The quantity in the square bracket is precisely the matrix element t_{if}

$$t_{if}(E_i) = \langle \psi_{fk_f} | \hat{V} | \psi_{ik_i} \rangle + \langle \psi_{fk_f} | \hat{V} \hat{G}^+ \hat{V} | \psi_{ik_i} \rangle, \qquad (3.186)$$

of an operator known as the T *matrix*

$$\boxed{\hat{T} = \hat{V} + \hat{V}\hat{G}^+\hat{V},} \qquad T \text{ matrix.} \qquad (3.187)$$

Transmission and reflection amplitudes

We can then write[41]

$$\mathscr{R}_{if} = \frac{L_x}{i\hbar v_f(k_f)} t_{if}. \qquad (3.188)$$

---

[40] The assumption we have made so far, i.e., no direct tunneling between the electrodes, is not necessary for the connection we make with scattering theory. Relaxation of this assumption would only change the *form* of the scattering potential $\hat{V}$ and $G^+(\mathbf{r},\mathbf{r}',E)$ (Eq. 3.156) but not the discussion that follows.

[41] All these quantities are evaluated at the same energy E_i, so I will drop it from the notation.

Using this relation I can rewrite the reflection probability 3.41 for an incident wave coming from channel i in lead L being reflected back into the channel f in the *same* lead as

$$R_{if} = |\mathscr{R}_{if}|^2 \frac{|I_f|}{|I_i|} = |\mathscr{R}_{if}|^2 \frac{|v_f|}{|v_i|} = \frac{L_x^2}{\hbar^2} \frac{|t_{if}|^2}{|v_i||v_f|}, \qquad (3.189)$$

with all quantities defined as in Eq. 3.41.

From this expression it is natural to define a *reflection amplitude* for a wave coming from the L electrode, and reflected back into the same electrode

$$r_{if} = \mathscr{R}_{if} \sqrt{\frac{|v_f|}{|v_i|}} = \frac{L_x}{i\hbar\sqrt{|v_f||v_i|}} t_{if}, \qquad (3.190)$$

so that the asymptotic form of the wave-function deep into the L lead is

$$\Psi^+_{ik_i}(\mathbf{r}) \to \psi_{ik_i}(\mathbf{r}) + \sum_{f=1}^{N_c^L} r_{if} \sqrt{\frac{|v_i|}{|v_f|}} \psi_{fk_f}(\mathbf{r}), \qquad x \to -\infty. \qquad (3.191)$$

Starting from Eq. 3.178 and following the same arguments we can look at the asymptotic form of $\Psi^+_{ik_i}(\mathbf{r})$ at $x \to +\infty$[42]

$$\lim_{x \to +\infty} \Psi^+_{ik_i}(\mathbf{r}) = \psi_{ik_f}(\mathbf{r})$$

$$+ \frac{1}{i\hbar} \sum_{f=1}^{N_c^R} \frac{e^{ik_f x} u_f(\mathbf{r}_\perp)}{v_f(k_f)} \int d\mathbf{r}' e^{-ik_f x'} u_f^*(\mathbf{r}'_\perp) V(\mathbf{r}') \Psi^+_{ik}(\mathbf{r}').$$

(3.192)

Comparing this with expression 3.34 deep into the R electrode, where evanescent modes do not contribute, we obtain

$$\mathscr{T}_{if} = \delta_{if} + \frac{L_x}{i\hbar v_f(k_f)} \int d\mathbf{r}' \psi^*_{fk_f}(\mathbf{r}') V(\mathbf{r}') \Psi^+_{ik}(\mathbf{r}'). \qquad (3.193)$$

Introducing 3.184 into 3.193, and recalling the definition 3.186 of the T matrix elements, we get

$$\mathscr{T}_{if} = \delta_{if} + \frac{L_x}{i\hbar v_f(k_f)} t_{if}. \qquad (3.194)$$

The transmission probability T_{if}, from channel i in lead L, into a given channel f in lead R is thus

$$T_{if} = |\mathscr{T}_{if}|^2 \frac{|I_f|}{|I_i|} = \delta_{if} \left[1 + \frac{1}{i\hbar} \frac{t_{if} - t^*_{if}}{|I_i|} \right] + \frac{L_x^2}{\hbar^2} \frac{|t_{if}|^2}{|v_i||v_f|}. \qquad (3.195)$$

---

[42] In this case, states of channels f in the R lead represent waves moving outwards in the direction of *positive* x axis, i.e., they correspond to a momentum $+\hbar k_f$ with $k_f > 0$.

Equations 3.189 and 3.195 show the relation between Dyson's equation 3.158 and scattering theory.

From Eq. 3.195, I can now define the *transmission amplitude* for a wave coming from the L electrode and propagating into the R electrode

$$\tau_{if} = \delta_{if} + \frac{L_x}{i\hbar\sqrt{|v_f||v_i|}} t_{if}, \tag{3.196}$$

so that the form of the wave-function deep into the R electrode is

$$\Psi^+_{ik_i}(\mathbf{r}) \to \sum_{f=1}^{N_c^R} \tau_{if} \sqrt{\frac{|v_i|}{|v_f|}} \psi_{fk_f}(\mathbf{r}), \qquad x \to +\infty. \tag{3.197}$$

3.6 The S matrix

In addition to the above scattering processes there are also those corresponding to a wave coming from the R electrode with momentum $-\hbar k_i$ ($k_i > 0$) which is partly reflected back into the same lead (channel f in the R lead), and partly transmitted into the L lead (channel f in the L lead). For these processes we can, therefore, define the reflection amplitude $\tilde{r}_{if}$ so that the wave-function deep into the R electrode is

$$\Psi^+_{i(-k_i)}(\mathbf{r}) \to \psi_{i(-k_i)}(\mathbf{r}) + \sum_{f=1}^{N_c^R} \tilde{r}_{if} \sqrt{\frac{v_i}{v_f}} \psi_{fk_f}(\mathbf{r}), \qquad x \to +\infty. \tag{3.198}$$

We can also define the transmission amplitude $\tilde{\tau}_{if}$, so that the wave-function deep into the L electrode is

$$\Psi^+_{i(-k_i)}(\mathbf{r}) \to \sum_{f=1}^{N_c^L} \tilde{\tau}_{if} \sqrt{\frac{v_i}{v_f}} \psi_{fk_f}(\mathbf{r}), \qquad x \to -\infty. \tag{3.199}$$

Since the wave-function described by the asymptotic solutions 3.198 and 3.199 is linearly independent from the one described by 3.191 and 3.197, we conclude that, for each energy, the general solution to the Schrödinger equation is a linear combination of the two. Let us write this general solution.

In order to simplify the discussion let us first assume that there is only one initial channel i, and one final channel f. No summation then appears in Eqs. 3.191, 3.197, 3.198 and 3.199. The general solution to the Schrödinger equation has then the form

$$\Psi(\mathbf{r}, E) = a_L \Psi^+_{ik_i}(\mathbf{r}) + a_R \Psi^+_{i(-k_i)}(\mathbf{r}), \tag{3.200}$$

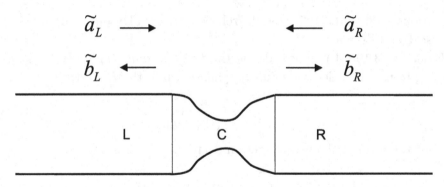

Fig. 3.14. Incoming $\tilde{a}$ and outgoing $\tilde{b}$ flux amplitudes are related via the $\mathcal{S}$ matrix of the sample.

where E is the energy defined in Eq. 3.23.

Using Eqs. 3.191, 3.197, 3.198 and 3.199, the asymptotic form of $\Psi(\mathbf{r}, E)$ is thus

$$\Psi(\mathbf{r}, E) \rightarrow \begin{cases} a_L \psi_{ik_i}(\mathbf{r}) + \left(a_L\, r_{if}\sqrt{\frac{v_i}{v_f}} + a_R\, \tilde{\tau}_{if}\sqrt{\frac{v_i}{v_f}}\right)\psi_{fk_f}(\mathbf{r}) & x \to -\infty, \\ a_R\, \psi_{i(-k_i)}(\mathbf{r}) + \left(a_R \tilde{r}_{if}\sqrt{\frac{v_i}{v_f}} + a_L\, \tau_{if}\sqrt{\frac{v_i}{v_f}}\right)\psi_{fk_f}(\mathbf{r}) & x \to +\infty. \end{cases} \quad (3.201)$$

Let us define

$$\tilde{b}_L \equiv \sqrt{|v_f|}\, b_L = a_L\, r_{if}\sqrt{|v_i|} + a_R\, \tilde{\tau}_{if}\sqrt{|v_i|} \equiv \tilde{a}_L\, r_{if} + \tilde{a}_R\, \tilde{\tau}_{if}, \quad (3.202)$$

and

$$\tilde{b}_R \equiv \sqrt{|v_f|}\, b_R = a_R\, \tilde{r}_{if}\sqrt{|v_i|} + a_L\, \tau_{if}\sqrt{|v_i|} \equiv \tilde{a}_R\, \tilde{r}_{if} + \tilde{a}_L\, \tau_{if}, \quad (3.203)$$

where I have defined the *flux amplitudes*, $\tilde{b}_{L,R} \equiv \sqrt{|v_f|}\, b_{L,R}$ and $\tilde{a}_{L,R} \equiv \sqrt{|v_i|}\, a_{L,R}$.

By noting that, irrespective of the sign of the momentum $\hbar k_f$, $\psi_{fk_f}(\mathbf{r})$ represents a wave moving *outward* from the nanojunction, while $\psi_{ik_i}(\mathbf{r})$ is a wave moving *inward* towards the nanojunction, deep in both L and R leads, Eq. 3.200 defines a linear relation between the *flux amplitudes* of outgoing and incoming waves (see Fig 3.14). This relation can be written in the following matrix form

$$\begin{pmatrix} \tilde{b}_L \\ \tilde{b}_R \end{pmatrix} = \begin{bmatrix} r_{if} & \tilde{\tau}_{if} \\ \tau_{if} & \tilde{r}_{if}, \end{bmatrix} \begin{pmatrix} \tilde{a}_L \\ \tilde{a}_R \end{pmatrix}. \quad (3.204)$$

The matrix

$$S = \begin{bmatrix} r_{if} & \tilde{\tau}_{if} \\ \tau_{if} & \tilde{r}_{if} \end{bmatrix}, \qquad (3.205)$$

is called the S *matrix*. It relates the outgoing flux amplitudes $\tilde{b}_{L,R}$ to the incoming flux amplitudes $\tilde{a}_{L,R}$. With this matrix, Eq. 3.204 can be concisely written as

$$\begin{pmatrix} \tilde{b}_L \\ \tilde{b}_R \end{pmatrix} = S \begin{pmatrix} \tilde{a}_L \\ \tilde{a}_R \end{pmatrix}. \qquad (3.206)$$

We can now easily generalize these results to N channels in the L electrode and N channels in the R electrode. The S matrix is a $2N \times 2N$ matrix similarly defined. If we label these channels by $i, f = 1, 2, \ldots, N$ then Eq. 3.206 has the general form

$$\begin{pmatrix} \tilde{b}_L^1 \\ \tilde{b}_L^2 \\ \vdots \\ \tilde{b}_L^N \\ \tilde{b}_R^1 \\ \tilde{b}_R^2 \\ \vdots \\ \tilde{b}_R^N \end{pmatrix} = \left[\begin{array}{cccc|cccc} r_{11} & r_{12} & \cdots & r_{1N} & \tilde{\tau}_{11} & \tilde{\tau}_{12} & \cdots & \tilde{\tau}_{1N} \\ r_{21} & r_{22} & & \vdots & \tilde{\tau}_{21} & \tilde{\tau}_{22} & & \vdots \\ \vdots & & \ddots & & \vdots & & \ddots & \\ r_{N1} & \cdots & & r_{NN} & \tilde{\tau}_{N1} & \cdots & & \tilde{\tau}_{NN} \\ \hline \tau_{11} & \tau_{12} & \cdots & \tau_{1N} & \tilde{r}_{11} & \tilde{r}_{12} & \cdots & \tilde{r}_{1N} \\ \tau_{21} & \tau_{22} & & \vdots & \tilde{r}_{21} & \tilde{r}_{22} & & \vdots \\ \vdots & & \ddots & & \vdots & & \ddots & \\ \tau_{N1} & \cdots & & \tau_{NN} & \tilde{r}_{N1} & \cdots & & \tilde{r}_{NN} \end{array} \right] \begin{pmatrix} \tilde{a}_L^1 \\ \tilde{a}_L^2 \\ \vdots \\ \tilde{a}_L^N \\ \tilde{a}_R^1 \\ \tilde{a}_R^2 \\ \vdots \\ \tilde{a}_R^N \end{pmatrix}. \qquad (3.207)$$

The matrix in Eq. 3.207 is thus block-diagonal, with the left-half of the matrix corresponding to reflection (top block) and transmission (bottom block) amplitudes of waves incident from the L lead, and the right-half of the matrix corresponding to reflection (bottom block) and transmission (top block) amplitudes of waves incident from the R lead.

However, the number of channels in the L lead, N_c^L, does not need to be the same as the number of channels in the R lead, N_c^R. In general, the S matrix is a $(N_c^L + N_c^R) \times (N_c^L + N_c^R)$ matrix which can be written in the compact form

$$S = \begin{pmatrix} r_{N_c^L \times N_c^L} & \tilde{\tau}_{N_c^L \times N_c^R} \\ \tau_{N_c^R \times N_c^L} & \tilde{r}_{N_c^R \times N_c^R} \end{pmatrix}, \qquad (3.208)$$

with the blocks r, τ, $\tilde{r}$ and $\tilde{\tau}$ matrices of dimensions indicated as subscripts (e.g., $r_{N_c^L \times N_c^L}$ is a matrix with $N_c^L \times N_c^L$ elements). Since the number of channels depends on the energy E (Eq. 3.28) the size of this matrix (not just its elements) depends on the energy E.

Recalling the relations 3.44 and 3.45 between the transmission and reflection probabilities, it is easy to show that the $\mathcal{S}$ matrix is unitary, i.e., (Exercise 3.11)

$$\hat{\mathcal{S}}\hat{\mathcal{S}}^\dagger = \hat{\mathcal{S}}^\dagger \hat{\mathcal{S}} = \hat{1}, \qquad (3.209)$$

which is another way of expressing particle flux conservation.[43] One can also proceed backward: define the $\mathcal{S}$ matrix and its unitary property 3.209, and from there obtain the relations 3.44 and 3.45 (Exercise 3.12).

Finally, we can express the total transmission coefficient 3.51 in terms of the elements of the $\mathcal{S}$ matrix. From Eq. 3.51 and the definition 3.196 of transmission amplitude, the total transmission coefficient from left to right at a given energy E is

$$\mathcal{T}_{LR}(E) = \sum_{i=1}^{N_c^L} \sum_{f=1}^{N_c^R} T_{if} = \sum_{i=1}^{N_c^L} \sum_{f=1}^{N_c^R} \tau_{if}\tau_{if}^* = \text{Tr}\{\tau\tau^\dagger\} = \text{Tr}\{\tau^\dagger\tau\}, \qquad (3.210)$$

which is equal to the total transmission coefficient $\mathcal{T}_{RL}(E)$ from right to left (Eq. 3.54).

Since the above expression is now written in terms of a trace, from the properties of the latter, one can evaluate that expression in any basis. It is sometimes convenient to use a basis, known as the basis of *eigenchannels*, in which the matrix $\tau^\dagger\tau$ is diagonal with eigenvalues T_n ($1 \leq n \leq N_c^L$), and the matrix $r^\dagger r$ diagonal with eigenvalues $1 - T_n$. In the eigenchannel basis 3.210 reduces to

$$\begin{aligned}\mathcal{T}_{LR}(E) &= \mathcal{T}_{RL}(E) \\ &= \text{Tr}\{\tau^\dagger\tau\} = \text{Tr}\{\tau\tau^\dagger\} \\ &= \sum_{n=1}^{N_c^L} T_n(E), \qquad \text{eigenchannels basis.}\end{aligned} \qquad (3.211)$$

3.6.1 Relation between the total Green's function and the $\mathcal{S}$ matrix

It is also possible to relate the lead-sample-lead Green's function $G^+(\mathbf{r}, \mathbf{r}', E)$ directly to the $\mathcal{S}$ matrix (Fisher and Lee, 1981). This relation is actually very convenient as it will allow us to write the total transmission coefficient Eq. 3.210 in a compact form containing the total Green's function. This will

---

[43] Unitarity of the $\mathcal{S}$ matrix is also valid for potentials that are not invariant under time-reversal symmetry.

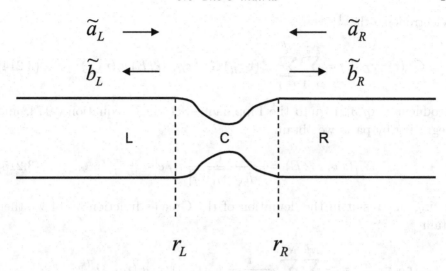

Fig. 3.15. The points $\mathbf{r}_L$ and $\mathbf{r}_R$ are in the two leads where the total Green's function is related to the S matrix.

complete our connection between the scattering approach and the approach in which we have introduced the self-energies representing the leads.

Let us consider Eq. 3.110 and rewrite it in the equivalent form (in the limit of $\epsilon \to 0$)

$$\hat{G}^+(E)\left(E - \frac{\hat{\mathbf{p}}^2}{2m} - \hat{V}\right) = \hat{1}. \qquad (3.212)$$

Multiply both sides of this equation on the right by the vector $|\mathbf{r}'\rangle$, and on the left by $\langle\mathbf{r}|$. Recalling that $\langle\mathbf{r}|\mathbf{r}'\rangle = \delta(\mathbf{r}-\mathbf{r}')$, we obtain the differential equation satisfied by the position representation of the Green's function, $G^+(\mathbf{r},\mathbf{r}',E)$,

$$G^+(\mathbf{r},\mathbf{r}',E)V(\mathbf{r}') = \left(E + \frac{\hbar^2}{2m}\nabla'^2\right)G^+(\mathbf{r},\mathbf{r}',E) - \delta(\mathbf{r}-\mathbf{r}'), \qquad (3.213)$$

where the symbol ∇' means differentiation with respect to $\mathbf{r}'$.

Take now a point $\mathbf{r}_L$ inside the L electrode where the amplitudes of the S matrix are evaluated, and an equivalent point $\mathbf{r}_R$ in the R electrode (see Fig. 3.15). Using the transverse wave-functions $u_\alpha(\mathbf{r}_\perp)$ of the L and R regions, we then define the Green's functions $G^+_{if}(x_R, x_L, E)$ associated with channel i at point x_L in the L electrode, and channel f at point x_R in the

R electrode so that[44]

$$G^+(\mathbf{r}_R, \mathbf{r}_L, E) = \sum_{i=1}^{N_c^L} \sum_{f=1}^{N_c^R} u_f^*(\mathbf{r}_{\perp R}) G_{if}^+(x_R, x_L, E) u_i(\mathbf{r}_{\perp L}). \quad (3.214)$$

Introducing Eq. 3.213 into the Lippmann–Schwinger equation 3.174, and integrating by parts we obtain

$$G_{if}^+(x_R, x_L, E) = -\frac{i}{\hbar\sqrt{|v_i||v_f|}} \tau_{if}\, e^{i(k_i x_L - k_f x_R)}. \quad (3.215)$$

Putting this result in the definition of the Green's function 3.214 we then obtain

$$G^+(\mathbf{r}_R, \mathbf{r}_L, E) = -\sum_{i=1}^{N_c^L}\sum_{f=1}^{N_c^R} \frac{i}{\hbar\sqrt{v_i v_f}}\, u_f^*(\mathbf{r}_{\perp R})\, \tau_{if}\, u_i(\mathbf{r}_{\perp L})\, e^{i(k_i x_L - k_f x_R)}. \quad (3.216)$$

Following the same procedure for two points $\mathbf{r}_L' < \mathbf{r}_L$ in the left electrode, a similar expression can be derived that relates the total Green's function to the reflection amplitudes (Exercise 3.14):

$$G^+(\mathbf{r}_L, \mathbf{r}_L', E) = -\sum_{i=1}^{N_c^L}\sum_{f=1}^{N_c^L} \frac{i}{\hbar v_i}\, u_f^*(\mathbf{r}_{\perp L}') \left[\delta_{if}\, e^{i k_i (x_L' - x_L)} \right.$$
$$\left. + r_{if}\sqrt{\frac{v_i}{v_f}}\, e^{-i(k_f x_L' + k_i x_L)} \right] u_i(\mathbf{r}_{\perp L}). \quad (3.217)$$

Equations 3.215 and 3.217 are the desired expressions relating the Green's function to the elements of the S matrix. We can now express the total transmission coefficient 3.210 in terms of the Green's function by inverting Eq. 3.216: multiply it by $u_f(\mathbf{r}_{\perp R})\, \tau_{if}\, u_i^*(\mathbf{r}_{\perp L})\, e^{-i(k_i x_L - k_f x_R)}$, integrate over the coordinates $\mathbf{r}_L$ and $\mathbf{r}_R$, and use the orthonormality condition 3.27 between the transverse wave-functions. The transmission amplitudes are thus

$$\tau_{if} = i\hbar\sqrt{v_i v_f} \int d\mathbf{r}_{\perp L} \int d\mathbf{r}_{\perp R}\, u_f^*(\mathbf{r}_{\perp R})\, G^+(\mathbf{r}_{\perp R}, \mathbf{r}_{\perp L}, E)\, u_i(\mathbf{r}_{\perp L}), \quad (3.218)$$

where I have defined

$$G^+(\mathbf{r}_{\perp R}, \mathbf{r}_{\perp L}, E) \equiv G^+(\mathbf{r}_{\perp R}, x_R = 0; \mathbf{r}_{\perp L}, x_L = 0; E). \quad (3.219)$$

---

[44] This separation can be done within the assumption that our problem is separable into a longitudinal and a transverse component deep into the two leads.

Inserting 3.218 in 3.210 we get

$$\mathcal{T}_{LR} = \sum_{i=1}^{N_c^L} \sum_{f=1}^{N_c^R} \tau_{if} \tau_{if}^*$$

$$= \hbar^2 \sum_{i=1}^{N_c^L} \sum_{f=1}^{N_c^R} v_i v_f \int\int\int\int u_f^*(\mathbf{r}_{\perp R})\, G^+(\mathbf{r}_{\perp R}, \mathbf{r}_{\perp L}, E)\, u_i(\mathbf{r}_{\perp L})$$
$$\times u_f(\mathbf{r}'_{\perp R}) \left[G^+(\mathbf{r}'_{\perp R}, \mathbf{r}'_{\perp L}, E)\right]^* u_i^*(\mathbf{r}'_{\perp L})\, d\mathbf{r}_{\perp L}\, d\mathbf{r}_{\perp R}\, d\mathbf{r}'_{\perp L}\, d\mathbf{r}'_{\perp R}$$
$$= \int\int\int\int \Gamma_R(\mathbf{r}'_{\perp R}, \mathbf{r}_{\perp R})\, G^+(\mathbf{r}_{\perp R}, \mathbf{r}_{\perp L}, E)\, \Gamma_L(\mathbf{r}_{\perp L}, \mathbf{r}'_{\perp L})$$
$$\times G^-(\mathbf{r}'_{\perp L}, \mathbf{r}'_{\perp R}, E)\, d\mathbf{r}_{\perp L}\, d\mathbf{r}_{\perp R}\, d\mathbf{r}'_{\perp L}\, d\mathbf{r}'_{\perp R}, \qquad (3.220)$$

where in the last equality I have made use of Eq. 3.115, and I have defined

$$\Gamma_R(\mathbf{r}'_{\perp R}, \mathbf{r}_{\perp R}) = \sum_{f=1}^{N_c^R} u_f(\mathbf{r}'_{\perp R})\, \hbar v_f\, u_f^*(\mathbf{r}_{\perp R}), \qquad (3.221)$$

and similarly for $\Gamma_L(\mathbf{r}_{\perp L}, \mathbf{r}'_{\perp L})$:

$$\Gamma_L(\mathbf{r}_{\perp L}, \mathbf{r}'_{\perp L}) = \sum_{i=1}^{N_c^L} u_i(\mathbf{r}_{\perp L})\, \hbar v_i\, u_i^*(\mathbf{r}'_{\perp L}). \qquad (3.222)$$

If we use again a discrete-space representation (Eq. 3.166), then Eq. 3.220 can be compactly written in operator notation as

$$\boxed{\mathcal{T}_{LR} = \mathrm{Tr}\{\hat{\Gamma}_R\, \hat{G}^+\, \hat{\Gamma}_L\, \hat{G}^-\},} \qquad (3.223)$$

which, due to flux conservation, is equivalent to the reversed relation

$$\boxed{\mathcal{T}_{RL} = \mathrm{Tr}\{\hat{\Gamma}_L\, \hat{G}^+\, \hat{\Gamma}_R\, \hat{G}^-\}.} \qquad (3.224)$$

The operators $\hat{\Gamma}_R$ and $\hat{\Gamma}_L$ are related to the self-energies $\hat{\Sigma}_R$ and $\hat{\Sigma}_L$ we have defined in Eqs. 3.149 and 3.150, via

$$\hat{\Gamma}_R = \mathrm{i}[\hat{\Sigma}_R^+ - \hat{\Sigma}_R^-] = -2\mathrm{Im}\{\hat{\Sigma}_R^+\}, \qquad (3.225)$$

and similarly for $\hat{\Gamma}_L$. These are precisely the relations 3.163. This is easy to prove if we choose as discrete-space representation the simplest orthogonal tight-binding model with only nearest-neighbor interactions (see discussion after Eq. 3.166 and Exercise 3.8).

Recalling the relation between the retarded and advanced self-energies

3.153, and the definition of "elastic lifetime" τ_k^{el}, Eq. 3.164, we can formally write

$$\hat{\Gamma}_R(E) = \frac{2\hbar}{\tau_R^{el}(E)}, \qquad (3.226)$$

and

$$\hat{\Gamma}_L(E) = \frac{2\hbar}{\tau_L^{el}(E)}, \qquad (3.227)$$

where I have defined

$$\tau_{R(L)}^{el}(E) \equiv -\frac{\hbar}{\text{Im}\{\hat{\Sigma}_{R(L)}^+(E)\}}. \qquad (3.228)$$

Therefore, $\hat{\Gamma}_{R(L)}$ can be interpreted as quantities proportional to the "rates" for electrons to scatter (tunnel) *elastically* from the central region into the electrodes (and vice versa). When either one of $\hat{\Gamma}_{R(L)}$ is small, electrons have a low probability to tunnel through the nanojunction, and from Eq. 3.224 we see that, for $\hat{\Gamma}_{R(L)} \to 0$, $\mathcal{T}_{LR}$ tends to zero.

Finally, if we assume that the local distribution functions of the two reservoirs are centered at different electrochemical potentials, μ_L and μ_R for the L and R leads, respectively, so that the bias is $V = (\mu_L - \mu_R)/e$, the total current flowing across the central region is (spin included)[45]

$$\boxed{I = \frac{e}{\pi\hbar}\int_{-\infty}^{+\infty} dE\,[f_L(E) - f_R(E)]\,\text{Tr}\{\hat{\Gamma}_R\,\hat{G}^+\,\hat{\Gamma}_L\,\hat{G}^-\}} \qquad (3.229)$$

with all quantities in Eq. 3.224 dependent on energy and external bias.[46]

I stress once more that Eq. 3.229 is valid only for single electrons experiencing at most mean-field interactions, both in the leads and in the sample region. I will show in Chapter 4 that under the *same* approximations of the Landauer approach (see Sec. 3.1), it can be derived using non-equilibrium Green's functions. However, Eq. 3.229 should not be confused with such formalism: I have derived it here without ever employing many-body perturbation theory.[47] Equation 3.229 is simply a single-particle equation for the current derived from single-particle scattering theory.

---

[45] Replace $\mathcal{T}(E) \equiv \mathcal{T}_{LR}(E) = \mathcal{T}_{RL}(E)$ in Eq. 3.56, with $\mathcal{T}_{LR}(E)$ given in 3.224.

[46] As noted before, a practical calculation of $\mathcal{T}_{LR}(E)$ would require a self-consistent determination of the different quantities entering Eq. 3.229 in the presence of the bias.

[47] It is the same as saying that one can derive the Hartree–Fock propagator as a mean-field solution of the interacting version of the Dyson equation 4.33 (a mathematical exercise equivalent to shooting a fly with a rocket!). But we never call the Hartree–Fock equations Dyson equations.

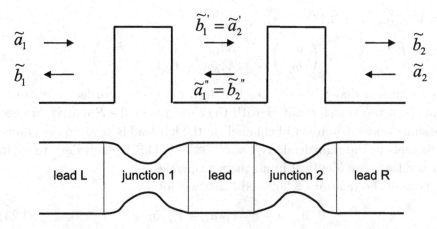

Fig. 3.16. Lower panel: Schematic of two junctions connected in series by a common lead. Upper panel: Schematic of the energy profile along the structure. At every given energy, waves transmitted inside the central lead from junction 1 become the incoming waves for the scattering into junction 2. Similarly, waves reflected from junction 2 become left-moving waves incident on junction 1. These waves interfere with each other.

3.7 The transfer matrix

Now that I have calculated the total current, I want to discuss how this current changes if instead of a single junction we consider many junctions. For instance, if the system is composed of two junctions in series, and we follow the dynamics of one electron from left to right, we expect this electron to scatter first on the first junction, being partly reflected and partly transmitted towards the second junction, where it scatters again, with partial transmission and partial reflection. This reflected wave can then *interfere* with the wave that was transmitted in between the two junctions (Fig 3.16). This interference occurs if the reflected and transmitted waves are fully coherent. If total loss of coherence occurs in between the two junctions we then expect the two scattering processes at the two different junctions to be *uncorrelated* from each other, and we can then treat them independently (Fig 3.16).

In order to deal with this problem we introduce a quantity slightly different than the S matrix. As discussed in Sec. 3.6 the S matrix acts on the *incoming* flux amplitudes to give the *outgoing* flux amplitudes.

Let us define, instead, the *transfer matrix* (or $\mathcal{M}$ matrix) as the one which relates the flux amplitudes in the *left* lead with the flux amplitude in the *right* lead (see Fig. 3.14)

$$\begin{pmatrix} \tilde{b}_R \\ \tilde{a}_R \end{pmatrix} = \mathcal{M} \begin{pmatrix} \tilde{a}_L \\ \tilde{b}_L \end{pmatrix} = \begin{bmatrix} \mathcal{M}_{LL} & \mathcal{M}_{LR} \\ \mathcal{M}_{RL} & \mathcal{M}_{RR} \end{bmatrix} \begin{pmatrix} \tilde{a}_L \\ \tilde{b}_L \end{pmatrix}. \qquad (3.230)$$

For a reason that will become clear in a moment, in order to relate the elements of the transfer matrix with the elements of the $\mathcal{S}$ matrix, we need to assume that the number of channels in the left lead is *equal* to the number of channels in the right lead: $N_c^L = N_c^R \equiv N_c$. This is equivalent to saying that the leads are identical in all their properties.

Inverting the relations 3.202 and 3.203 we obtain

$$\tilde{a}_R = -\tilde{\tau}_{if}^{-1} r_{if} \tilde{a}_L + \tilde{\tau}_{if}^{-1} \tilde{b}_L, \qquad (3.231)$$

and

$$\tilde{b}_R = (\tau_{if} - \tilde{r}_{if} \tilde{\tau}_{if}^{-1} r_{if}) \tilde{a}_L + \tilde{r}_{if} \tilde{\tau}_{if}^{-1} \tilde{b}_L. \qquad (3.232)$$

By comparing with the definition 3.230 of transfer matrix, and generalizing to N_c channels, we get

$$\mathcal{M}_{LL} = \tau^{\dagger -1}, \qquad \mathcal{M}_{LR} = \tilde{r} \tilde{\tau}^{-1}, \qquad (3.233)$$
$$\mathcal{M}_{RL} = -\tilde{\tau}^{-1} r, \qquad \mathcal{M}_{RR} = \tilde{\tau}^{-1}, \qquad (3.234)$$

with r, τ, $\tilde{r}$ and $\tilde{\tau}$ matrices of dimensions $N_c \times N_c$.[48] It is now evident why I had to assume that the number of channels must be equal in both leads: I had to invert the transmission submatrices.

The great physical and analytical advantage of the $\mathcal{M}$ matrix is that it is *multiplicative*, namely, if I can divide my scattering region into, say, N regions of space, each described by an $\mathcal{M}_i$ matrix, then the *total* transfer matrix is simply the product of these single matrices

$$\mathcal{M} = \mathcal{M}_N \mathcal{M}_{N-1} \cdots \mathcal{M}_1. \qquad (3.235)$$

This property is trivial to prove by recalling that each $\mathcal{M}_i$ matrix relates flux amplitude on one side of the scattering region to the other side. So, the flux amplitudes of the left-most scattering region are transferred to the region of space in between the first and the second scattering regions (Fig. 3.16). These transferred amplitudes are now incoming amplitudes for this second region, which are then transferred to the region of space between the third and the second scatterer, and so on and so forth.

---

[48] In 3.234 I have used the identity

$$\tau - \tilde{r} \tilde{\tau}^{-1} r = \tilde{\tau}^{\dagger -1}.$$

3.7.1 Coherent scattering of two resistors in series

As an example of the above property let us consider the single-particle scattering due to two junctions placed in series as shown schematically in Fig. 3.16, separated by a region of space where electrons do not experience any scattering. We call the first junction, resistor 1, the second, resistor 2. We assume that the region of space that separates the two resistors has the same physical properties as the two leads at the opposite end of the scattering region. To simplify the derivation let us also assume that there is only one incoming and one outgoing channel, namely $N_c = 1$, so that I can drop the indexes i, f in all transmission and reflection quantities.

From the definition 3.234, the transfer matrix of each resistor is (I have also used the relations E3.13)

$$\mathcal{M} = \begin{pmatrix} 1/\tau^* & -r^*/\tau^* \\ -r/\tilde{\tau} & 1/\tilde{\tau} \end{pmatrix}. \tag{3.236}$$

Using the multiplicative property of the $\mathcal{M}$ matrix 3.235, we get

$$\mathcal{M} = \begin{pmatrix} 1/\tau^* & -r^*/\tau^* \\ -r/\tilde{\tau} & 1/\tilde{\tau} \end{pmatrix} = \begin{pmatrix} 1/\tau_1^* & -r_1^*/\tau_1^* \\ -r_1/\tilde{\tau}_1 & 1/\tilde{\tau}_1 \end{pmatrix} \begin{pmatrix} 1/\tau_2^* & -r_2^*/\tau_2^* \\ -r_2/\tilde{\tau}_2 & 1/\tilde{\tau}_2 \end{pmatrix}. \tag{3.237}$$

The total transmission amplitude is then

$$\tau = \frac{t_1 t_2}{1 - \tilde{r}_1 r_2}. \tag{3.238}$$

The total transmission probability from the far left lead to the far right lead is thus (all quantities depend on the same energy E)

$$T = \tau \tau^* = \frac{T_1 T_2}{1 + R_1 R_2 - 2\sqrt{R_1 R_2} \cos \delta}, \quad \text{coherent scattering}, \tag{3.239}$$

where $\delta = \arg(\tilde{r}_1 r_2) = \arg(\tilde{r}_1) + \arg(r_2)$ is the global phase acquired by the single particle in scattering from resistor 1 to resistor 2, back to resistor 1, back again across resistor 2. It is the phase acquired due to the *coherent superposition* of waves *inside* the region separating the two resistors.

3.7.1.1 Resonant tunneling

Equation 3.239 simplifies considerably if in the neighborhood of a given energy $E = E_R$ the transmission probability T is a very sharp function of energy (see Fig. 3.17).[49] A sharp function of energy means that there are several electron reflections at each resistor, so that, in a dynamical sense, an electron coming, say, from resistor 1 "spends" a lot of time inside the

---

[49] A similar derivation of the resonant tunneling transmission can be found in Datta, 1995.

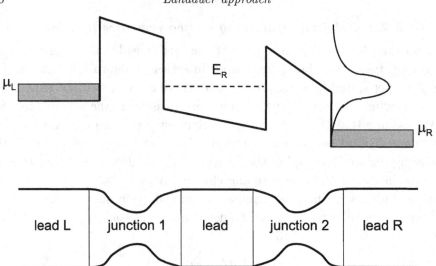

Fig. 3.17. Lower panel: Schematic of two junctions connected in series by a common lead that may realize a resonant-tunneling configuration. Upper panel: Schematic of the energy profile along the structure for the biased structure. At a certain energy E_R the transmission probability is a very sharp function of energy, and resonant tunneling occurs in the neighborhood of that energy.

region between resistor 1 and 2 before "escaping" beyond resistor 2. We can therefore assume the reflection probabilities $R_1(E) \approx R_2(E) \approx 1$ at the resonance energy E_R. By rewriting Eq. 3.239 in a different form

$$T = \frac{T_1 T_2}{(1 - \sqrt{R_1 R_2})^2 + 2\sqrt{R_1 R_2}\,(1 - \cos\delta)}, \quad (3.240)$$

the assumption $R_1(E) \approx R_2(E) \approx 1$ ($T_1(E) \approx T_2(E) \ll 1$) can be used to simplify 3.240 (to first order in T_1 and T_2)

$$T \approx \frac{T_1 T_2}{\left(\dfrac{T_1 + T_2}{2}\right)^2 + 2\,(1 - \cos\delta)}. \quad (3.241)$$

At resonance, the phase shift $\delta(E)$ is also maximum (i.e., it is an integer multiple of 2π, see, e.g., Newton, 1966) so that we can Taylor-expand the term $1 - \cos\delta$ in the above equation

$$1 - \cos\delta(E) \approx \frac{1}{2}\left(\frac{d\delta(E)}{dE}\right)^2 (E - E_R)^2. \quad (3.242)$$

If we now define the quantities

$$\Gamma_{1,2} = T_{1,2} \left[\frac{d\delta(E)}{dE} \right]^{-1}, \qquad (3.243)$$

associated with the left and right transmission probabilities, we finally get

$$T(E) \approx \frac{\Gamma_1 \Gamma_2}{\left(\frac{\Gamma_1 + \Gamma_2}{2}\right)^2 + (E - E_R)^2} = \left(\frac{\Gamma_1 \Gamma_2}{\Gamma_1 + \Gamma_2}\right) \frac{\Gamma}{(\Gamma/2)^2 + (E - E_R)^2}, \qquad (3.244)$$

where in the last equality I have defined $\Gamma = \Gamma_1 + \Gamma_2$. Expression 3.244 has the form of a *Lorentzian* function, with an effective broadening given by Γ, a weighted broadening introduced by the "coupling" of the resonant state E_R with the continuum of states in the leads. In Sec. 4.66 I will show that

$$A(E) = \frac{\Gamma}{(\Gamma/2)^2 + (E - E_R)^2} \qquad (3.245)$$

has the same form of the spectral function (or local density of states) for non-interacting electrons, with Γ the "rate" for a particle in the state E_R to scatter out of that state. I then anticipate that Γ_1 and Γ_2 may be interpreted as the "rates" at which a particle with initial energy E_R "escapes" into the left or right lead, similar to the rates 3.226 and 3.227 for a particle to scatter elastically into the left and right lead.

Under similar conditions, we can generalize the above result to multiple resonances, provided one assumes that they are sufficiently separated in energy so that their contribution is additive. With the obtained transmission function, the total current can then be calculated as in 3.56.

3.7.2 Incoherent scattering of two resistors in series

If the single electron scatters on the first resistor, and immediately after this it "loses memory" of this scattering event, it approaches the second resistor as if coming from a "reservoir" (see discussion in Sec. 3.1). This way, when the electron scatters on the second resistor, the phase it has acquired from the first scattering event is completely *uncorrelated* from the phase it will acquire from the second scattering process with resistor 2, so that the transmission out of resistor 1 is uncorrelated with the incident and reflected waves of resistor 2 (Fig. 3.18). The scattering off these two resistors can be thus thought of as *sequential* (hence sometimes the name *sequential tunneling*). In order for this condition to be satisfied, we need to have strong dephasing processes inside the region separating the two resistors.

However, we still assume here that these dephasing processes do *not* change single-particle energy (unlike, e.g., electron-phonon scattering). How do we calculate then the total transmission probability in this case? Let us proceed as follows.

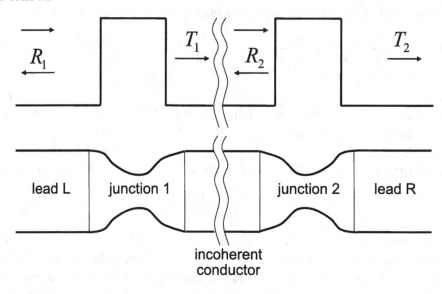

Fig. 3.18. Same as in Fig. 3.16 but now the conductor between the two electrodes creates incoherence between the transmission out of junction 1 and the incident and reflected waves of junction 2.

Since the phases of the scattering events at the two resistors are assumed uncorrelated, I need to determine the *intrinsic* resistance of these two resistors, and then sum them up in series. In Sec. 2.3.4 I have shown that the quantized conductance can be thought of as that associated with an ideal wire connected to two reservoirs, one at each end: the quantized conductance originates from the scattering off the "contact" between these reservoirs and the leads (Imry, 1986).

This is what I need here: the problem at hand resembles that of *three* reservoirs, two at the far end of the total structure formed by the two resistors, and one in between the two resistors (Fig. 3.18). Let us call G_i the *intrinsic* conductance of one of the two resistors (for simplicity, we still work with one channel, in the limit of zero bias, and zero temperature). The total resistance of that resistor is then the sum of its intrinsic resistance $R_i = G_i^{-1}$ and the "contact" resistance with the reservoirs:

$$G^{-1} = \frac{1}{T}\frac{h}{2e^2} = G_i^{-1} + \frac{h}{2e^2}, \qquad (3.246)$$

3.7 The transfer matrix

from which we get the intrinsic conductance (spin degeneracy included)

$$\frac{1}{R_i} = G_i = \frac{2e^2}{h}\frac{T}{1-T} = \frac{2e^2}{h}\frac{T}{R}, \tag{3.247}$$

which shows an important fact: if the transmission probability $T \to 1$, the intrinsic conductance diverges, $G_i \to \infty$ (the intrinsic resistance tends to zero), as one would expect when electrons move from the left lead to the right lead and no scattering occurs in the wire.

The above result must be true for both resistors, and for the whole structure as well. If no energy is lost in between the two resistors, using simple circuit theory we can now sum the intrinsic resistances of the two resistors to obtain the total intrinsic resistance. We get

$$G_i^{-1} = \frac{h}{2e^2}\frac{R}{T} = G_1^{-1} + G_2^{-1} = \frac{h}{2e^2}\frac{R_1}{T_1} + \frac{h}{2e^2}\frac{R_2}{T_2}, \tag{3.248}$$

or if we extract the total transmission probability T

$$T = \frac{T_1 T_2}{1 - R_1 R_2}, \quad \text{incoherent scattering}, \tag{3.249}$$

which is quite different from the result we have found for coherent transport between the two resistors, Eq. 3.239.

We can now generalize the above result to N resistors incoherently coupled in series, to obtain

$$\frac{1-T(N)}{T(N)} = \frac{1-T_1}{T_1} + \frac{1-T_2}{T_2} + \cdots + \frac{1-T_N}{T_N}, \tag{3.250}$$

which simplifies if we assume all resistors equal, with transmission probability T

$$T(N) = \frac{T}{N(1-T)+T}. \tag{3.251}$$

I will come back later to the example above. It has indeed more profound implications than I have discussed above, as it allows us to interpret conductance measurements when multiple *non-invasive* probes are employed. For now, I want to derive the relation between the conductance and the $\mathcal{M}$ matrix, which will allow us to distinguish among different transport regimes.

3.7.3 Relation between the conductance and the transfer matrix

The transfer matrix is *pseudo-unitary*, namely it satisfies

$$\mathcal{M}^\dagger \Sigma \mathcal{M} = \Sigma \tag{3.252}$$

where

$$\Sigma = \begin{pmatrix} \hat{1}_{N_c \times N_c} & \hat{0}_{N_c \times N_c} \\ \hat{0}_{N_c \times N_c} & -\hat{1}_{N_c \times N_c} \end{pmatrix}. \tag{3.253}$$

This property derives directly from the unitarity of the $\mathcal{S}$ matrix 3.209 (Exercise 3.15).

Now, let us invert the transfer matrix to get

$$\mathcal{M}^{-1} = \Sigma \mathcal{M}^\dagger \Sigma = \begin{pmatrix} \mathcal{M}_{LL}^\dagger & -\mathcal{M}_{RL}^\dagger \\ -\mathcal{M}_{LR}^\dagger & \mathcal{M}_{RR}^\dagger \end{pmatrix}, \tag{3.254}$$

which is easy to show from the pseudo-unitarity of the $\mathcal{M}$ matrix 3.252. With this result we then find the following identity (this is quite a tedious calculation and I leave it as an exercise, Exercise 3.16)

$$\left[\mathcal{M}^\dagger \mathcal{M} + (\mathcal{M}^\dagger \mathcal{M})^{-1} + 2 \cdot \hat{1}\right]^{-1} = \frac{1}{4} \begin{pmatrix} \tau^\dagger \tau & \hat{0} \\ \hat{0} & \tilde{\tau} \tilde{\tau}^\dagger \end{pmatrix}. \tag{3.255}$$

I have previously shown that the total transmission coefficient can be written in terms of the matrix τ as in Eq. 3.210. From this and Eq. 3.255 we thus conclude that (since $\text{Tr}\{\tau^\dagger \tau\} = \text{Tr}\{\tilde{\tau} \tilde{\tau}^\dagger\}$)

$$\mathcal{T}_{LR}(E) = \mathcal{T}_{RL}(E) = 2\text{Tr}\left\{\left[\mathcal{M}^\dagger \mathcal{M} + (\mathcal{M}^\dagger \mathcal{M})^{-1} + 2 \cdot \hat{1}\right]^{-1}\right\}. \tag{3.256}$$

The conductance can then be expressed in terms of the transfer matrix. For instance, in the limit of zero bias and zero temperature, from Eq. 3.68 we get (spin degeneracy is included)

$$G = \frac{4e^2}{h} \text{Tr}\left\{\left[\mathcal{M}^\dagger \mathcal{M} + (\mathcal{M}^\dagger \mathcal{M})^{-1} + 2 \cdot \hat{1}\right]^{-1}\right\}, \quad (\mu_L - \mu_R) \to 0, \ \theta \to 0. \tag{3.257}$$

This is often referred to as the *Pichard formula* (Pichard, 1990).

3.7.4 Localization, ohmic and ballistic regimes

The relation 3.257 between the transfer matrix and the conductance allows us to discuss the different transport regimes I have anticipated in Sec. 2.3.4. Since the conductance 3.257 is now written in terms of the matrix $\mathcal{M}^\dagger \mathcal{M}$, it is easier if we work in the basis that diagonalizes this matrix (note that $\mathcal{M}^\dagger \mathcal{M}$ is Hermitian). The N_c eigenvalues of this matrix can be found from its characteristic polynomial

$$p(t) = \det\left(t\hat{1} - \mathcal{M}^\dagger \mathcal{M}\right), \tag{3.258}$$

where $\hat{1}$ is the $2N_c \times 2N_c$ identity matrix.

By using the pseudo-unitary property of the transfer matrix 3.252, we obtain that $\mathcal{M}^\dagger = \Sigma \mathcal{M}^{-1} \Sigma$ and $\mathcal{M} = \Sigma \mathcal{M}^{\dagger -1} \Sigma$. From 3.258 we then get

$$\begin{aligned} p(t) &= \det\left(t\,\hat{1} - \Sigma\mathcal{M}^{-1}\Sigma\Sigma\mathcal{M}^{\dagger-1}\Sigma\right) \\ &= \det\left(t\,\hat{1} - \Sigma\mathcal{M}^{-1}\mathcal{M}^{\dagger-1}\Sigma\right) \\ &= \det\left(t\,\hat{1} - (\mathcal{M}^\dagger\mathcal{M})^{-1}\right) \\ &= t^{2N_c}\det\frac{(t^{-1}\hat{1} - \mathcal{M}^\dagger\mathcal{M})}{\det(\mathcal{M}^\dagger\mathcal{M})}, \end{aligned} \qquad (3.259)$$

where, from the first to the second equality, I have used the fact that $\Sigma\Sigma = \hat{1}$ (from Eq. 3.253), from the third to the forth, the property that the determinant of a multiple of the unit matrix minus a *similarity transformation* of a matrix[50] is equal to the characteristic polynomial of that matrix, and finally I have used the distributive property of determinants (for two square matrices A and B, $\det(AB) = \det(A)\det(B)$) and the fact that the dimensions of $\mathcal{M}$ are $2N_c \times 2N_c$, so that $2N_c$ is an even number.

We thus see from the above that whenever we have a solution of $p(t) = 0$ we also have a solution of $p(t^{-1}) = 0$. This implies that the eigenvalues of $\mathcal{M}^\dagger\mathcal{M}$ are the set of pairs $\{t_i, t_i^{-1}\}$, with $i = 1, \ldots, N_c$. Using these eigenvalues we can then rewrite the conductance 3.257 as

$$G = \frac{2e^2}{h}\sum_{i=1}^{N_c}\frac{4}{t_i + t_i^{-1} + 2}. \qquad (3.260)$$

The eigenvalues of $\mathcal{M}^\dagger\mathcal{M}$ are non-negative, and since they come in pairs $\{t_i, t_i^{-1}\}$, we can always assume that each $t_i \geq 1$.

If the length of the *total* scattering region is L (this is the length over which we define the scattering properties of our sample, that may be separated into sub-scattering regions) we can then define a *localization length* ζ_i for *each* of these eigenvalues via

$$t_i \equiv e^{2L/\zeta_i} \qquad (3.261)$$

or, equivalently,

$$\zeta_i = \frac{2L}{\ln t_i}, \qquad (3.262)$$

---

[50] Given two square matrices A and B, a similarity transformation of A with respect to B is BAB^{-1}. In our case, $A = \mathcal{M}^{-1}\mathcal{M}^{\dagger-1}$, and $B = \Sigma$.

where we can order $\lambda_1 < \lambda_2 < \cdots < \lambda_{N_c}$, so that $\zeta_1 > \zeta_2 > \cdots > \zeta_{N_c}$ (all these quantities depend on the energy of the electrons).

By replacing these definitions in 3.260 we rewrite the two-probe conductance 3.257 as[51]

$$G = \frac{2e^2}{h} \sum_{i=1}^{N_c} \frac{2}{1 + \cosh(2L/\zeta_i)}, \quad (\mu_L - \mu_R) \to 0, \quad \theta \to 0. \qquad (3.263)$$

We can now discuss three different regimes:

Localization regime: $L > \zeta_1$. The length of the scattering region is larger than the largest localization length. If we keep the number of channels fixed, but increase the length of the scattering region L, then one can prove that the localization lengths ζ_i tend to well-defined values (Pichard, 1990). The physical reason for this is that, if in L we have a number N of scattering regions, each described by an $\mathcal{M}_i$ matrix, by letting the total scattering region go to the thermodynamic limit (increase L to infinity but also the total number N, so that their ratio is constant), then the ensemble of these matrices has a *probability distribution* (hence the name *random matrices*), and we can treat their effect on an electron transferring from one side of the sample to the other as a *random* process. In the above thermodynamic limit, one can then resort to a theorem known as *Oseledec's theorem* – or the *multiplicative ergodic theorem* (Oseledec, 1968) – which states that the inverse of each localization length ζ_i^{-1} converges to a definite *Lyapunov* exponent as $L \to \infty$.[52]

If this holds, then from Eq. 3.263 we see that the conductance is dominated by the largest of the localization lengths. According to our ordering this is ζ_1, so that in the (thermodynamic) limit

$$G(L) \to \frac{2e^2}{h} 4 e^{-L/\zeta_1}, \quad (\mu_L - \mu_R) \to 0, \quad \theta \to 0, \quad L \to \infty, \qquad (3.264)$$

and in the limit of $(\mu_L - \mu_R) \to 0$, the localization length ζ_1 is that of electrons at the Fermi level.

This shows that the conductance tends to zero with the length of the system, so that the system becomes an *insulator*. Note that unlike the result

---

[51] I used the definition
$$\cosh(x) = \frac{e^x + e^{-x}}{2}.$$

[52] Loosely speaking a *Lyapunov* exponent provides a measure of how fast two solutions of the same dynamical equation of motion, that originate from two different, but infinitesimally close initial conditions, depart from each other during evolution.

I discussed in Sec. 2.3.4, here the insulating character of the system is due to quantum-mechanical localization of the wave-functions, and not to the semiclassical scattering of an infinite 1D system.

Ohmic regime: $\zeta_1 > L > \zeta_{N_c}$. The length of the scattering region is finite, with a value in between the largest and the smallest localization lengths. The channels that have a localization length $\zeta_{\text{closed}} \ll 2L$ are such that $\cosh(2L/\zeta_{\text{closed}}) \sim \frac{1}{2}\exp(2L/\zeta_{\text{closed}}) \gg 1$, so that their contribution to the conductance 3.263 is negligible. We call these channels *closed*.

On the other hand, for those channels with a localization length $\zeta_{\text{open}} \gg 2L$, $\cosh(2L/\zeta_{\text{open}}) \sim 1$ so that, from Eq. 3.263, they contribute a conductance $2e^2/h$ (spin included) to the total conductance, and we call them *open*. Following this definition we can thus write[53]

$$N_c = N_c^{\text{closed}} + N_c^{\text{open}}, \qquad (3.265)$$

with

$$N_c^{\text{open}} = \sum_{i=1}^{N_c} \Theta(\zeta_i - 2L). \qquad (3.266)$$

The conductance 3.263 can then be written as

$$G(L) \simeq \frac{2e^2}{h} N_c^{\text{open}}. \qquad (3.267)$$

However, I have argued in Sec. 3.3.1 that the number of channels at the Fermi level is of the order of $N_c(E_F) \approx k_F^{d-1} S$, where S is the $(d-1)$-dimensional cross section of the leads. If N_c^{open} is only a fraction of these channels, with this fraction determined by the ratio between the elastic scattering length λ I have defined in Sec. 2.3.4, and the length L of the scattering region, namely $N_c^{\text{open}} = N_c \lambda/L$, we say the system is in a semiclassical regime, or in the *ohmic regime*. The conductance is then

$$G(L) \simeq \frac{2e^2}{h} N_c \frac{\lambda}{L} \simeq \frac{2e^2}{h} k_F^{d-1} \lambda \frac{S}{L}, \qquad (3.268)$$

which depends on the length of the scattering region. Apart from the geometric factor $\Omega_d/d(2\pi)^{d-1}$, this is the same result we have obtained in Sec. 2.3.4 using the Kubo formalism within the semiclassical Drude approximation for independent electrons.

From the above we also see the following. If the number of channels tends to infinity with $L \to \infty$, then we cannot apply Oseledec's theorem, since we

---

[53] Note that this is a somewhat arbitrary separation, since a channel is neither open nor closed if $\zeta_i \sim 2L$.

are now dealing with infinite-dimensional transfer matrices. Nonetheless, if the largest localization length remains finite in the thermodynamic limit, we see from 3.263 that the conductance goes to zero, and the system is an insulator. On the other hand, if ζ_1 diverges in such a way that the ratio $2L/\zeta_1$ is constant, from the same equation we get a finite conductance: the system is a *conductor*.

Ballistic regime: $L < \zeta_{N_c}$. From the above discussion, this means *all* channels are open ($N_c^{\text{open}} = N_c$) so that the conductance is

$$G \simeq \frac{2e^2}{h} N_c^{\text{open}} \simeq \frac{2e^2}{h} N_c \simeq \frac{2e^2}{h} k_F^{d-1} S, \qquad (3.269)$$

which is independent of the length of the scattering region (see also discussion in Sec. 2.3.4).

The above categories are clearly an idealization, and mesoscopic and nanoscopic systems do not follow just one of the above classifications. In the following chapters, when describing transport within a single-particle picture of electrons interacting at a mean-field level, we will either consider the ballistic regime, or in between the localization and ballistic regimes. We will call this the *quasi-ballistic* regime.

3.8 Four-probe conductance in the non-invasive limit

So far, I have only considered two terminals, one on the left of our sample and one on the right. In order to define the corresponding two-probe conductance 3.70 I have also assumed that the bias $V = (\mu_L - \mu_R)/e$ is the one ideally "measured" at the reservoirs positions.

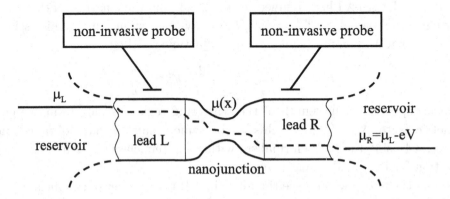

Fig. 3.19. Ideal probes coupled non-invasively with the leads, measuring the local electrochemical potential at the leads.

3.8 Four-probe conductance in the non-invasive limit

What if I introduce two measurement *probes* closer to the junction, in the region of the leads, so that these probes can ideally measure the *local* electrochemical potential in the region where they are connected without, however, perturbing the current appreciably (see Fig. 3.19). The experimental situation I am now considering is the following: I measure the current as that created by the scattering at the sample by carriers originating from the electron sources, but I measure the bias *along* the leads.

Due to electron scattering at the boundaries of the sample, and consequent charge accumulation and depletion on the opposite sides of the sample, there is clearly no reason to believe that these two *non-invasive* probes would measure the same bias as the one measured by the probes "connected" to the reservoirs.

3.8.1 Single-channel case

In order to calculate the bias *in* the leads in this non-invasive limit I need to calculate the charge density that accumulates in that region of space due to scattering at the sample. For simplicity, consider first the case of a single channel, $N_c = 1$.

In the left lead, we have the charge of all the waves, with different energy E, traveling from left to right, plus the charge of all the waves reflected back to the left lead with, say, probability $R(E)$. These are populated according to the local distribution $f_L(E)$. In addition, always in the left lead, there is the charge of the waves moving from right to left with, say, probability $\tilde{T}(E)$, and populated according to the local distribution $f_R(E)$. At each energy E there corresponds a well-defined density of states $D(E)$ per spin and per direction of motion (Eq. 3.48), so that the total number density (per spin) is

$$n_L = \int dE\, D(E) \left\{ [\;\overbrace{1}^{\text{from L to R}} + \overbrace{R(E)}^{\text{from L to L}}\;] f_L(E) + \overbrace{\tilde{T}(E)}^{\text{from R to L}} f_R(E) \right\}. \tag{3.270}$$

Similar considerations apply for the number density in the right lead (with all the quantities "reversed" in directionality)

$$n_R = \int dE\, D(E) \left\{ [\;\overbrace{1}^{\text{from R to L}} + \overbrace{\tilde{R}(E)}^{\text{from R to R}}\;] f_R(E) + \overbrace{T(E)}^{\text{from L to R}} f_L(E) \right\}. \tag{3.271}$$

I now *assume* that with the above number densities are associated two *local* electrochemical potentials $\bar{\mu}_L$ and $\bar{\mu}_R$ (not to be confused with the *chemical* potential 2.91) *in the leads*. This is clearly an idealization for two reasons: (1) as discussed in Sec. 2.4, the local electrochemical potential is the one ideally measured by a local probe in equilibrium with the system, and which does not perturb the system's properties appreciably. (2) Microscopically, due to the atomic geometry of the leads, all local properties may vary appreciably within a lattice distance. The local electrochemical potential I am talking about here is then the macroscopically averaged one I discussed in Sec. 2.4 (see Eq. 2.99).

With these approximations, I then define

$$n_L \equiv 2 \int dE\, D(E)\, f(E - \bar{\mu}_L), \qquad (3.272)$$

and

$$n_R \equiv 2 \int dE\, D(E)\, f(E - \bar{\mu}_R), \qquad (3.273)$$

where $f(E - \bar{\mu}_L)$ and $f(E - \bar{\mu}_R)$ are Fermi–Dirac distributions, like 3.14 or 3.15, with electrochemical potentials, $\bar{\mu}_L$ and $\bar{\mu}_R$, respectively. In the above two equations, the factor of 2 is *not* due to spin degeneracy, but due to the two possible directions of motion – left to right, and right to left. Note that, as in the case of left and right reservoirs, the very fact that these local equilibrium distributions are Fermi–Dirac distributions is a very strong assumption – see discussion of *Approximation 5* in Sec. 3.1. Electrons may be "hot" or "cool" according to their direction of motion with respect to global current flow.

Let us now Taylor-expand *all* the local distribution functions as in Eq. 3.57, as well as *all* the transmission and reflection probabilities as in Eq. 3.65, and work in the limit of zero temperature. In the limit of $(\mu_L - \mu_R) \to 0$ (and hence $(\bar{\mu}_L - \bar{\mu}_R) \to 0$) to first order in the expansion of the local distribution functions, and zero order in the expansion of the transmission and reflection probabilities, by equating 3.272 with 3.270, and 3.273 with 3.271 we get

$$2(\mu_L - \bar{\mu}_L) = (\mu_L - \mu_R)\tilde{T}(E_F) \implies \bar{\mu}_L = \mu_L - \frac{1}{2}\tilde{T}(E_F)(\mu_L - \mu_R) \qquad (3.274)$$

$$2(\mu_R - \bar{\mu}_R) = (\mu_R - \mu_L)T(E_F) \implies \bar{\mu}_R = \mu_R + \frac{1}{2}T(E_F)(\mu_L - \mu_R), \qquad (3.275)$$

3.8 Four-probe conductance in the non-invasive limit

from which I get

$$(\bar{\mu}_L - \bar{\mu}_R) = \left(1 - \frac{1}{2}T(E_F) - \frac{1}{2}\tilde{T}(E_F)\right)(\mu_L - \mu_R)$$
$$= (1 - T(E_F))(\mu_L - \mu_R), \qquad (3.276)$$

where the last equality comes from the conservation of particle flux 3.54.

Equation 3.276 defines the bias in the leads

$$V_{\text{leads}} = \frac{\bar{\mu}_L - \bar{\mu}_R}{e}. \qquad (3.277)$$

The current instead is still given by Eq. 3.67, which in the present single-channel case is (spin is now included)

$$I = \frac{2e^2}{h}T(E_F)\frac{(\mu_L - \mu_R)}{e}. \qquad (3.278)$$

With the bias in the leads 3.277, and the "reservoir current" 3.278 we can now define the conductance

$$\boxed{G_{\text{4-probe}} = \frac{I}{V_{\text{leads}}} = \frac{2e^2}{h}\frac{T(E_F)}{1 - T(E_F)}; \quad (\mu_L - \mu_R) \to 0, \; \theta \to 0}$$
$$(3.279)$$

which is known as *four-probe* conductance, precisely because two probes measure the current, while two others measure *non-invasively* the bias, with the latter being measured in the lead region. It reduces to the *two-probe* conductance 3.68 – which corresponds to the bias ideally measured at the reservoir positions – only in the limit of small transmission probability: $T(E_F) \ll 1$. In general, the four-probe conductance is *larger* than the two-probe conductance.

In the opposite limit of perfect transmission, $T(E_F) = 1$, we get

$$G_{\text{4-probe}} \to \infty; \qquad T(E_F) \to 1, \qquad (3.280)$$

similar to what we found when deriving the intrinsic resistance of the system, in the discussion of scattering at two resistors coupled incoherently (Sec. 3.7.2). This shows that the four-probe conductance is nothing other than the intrinsic conductance 3.247 of the scattering region, without accounting for the "contact scattering" between the leads and the reservoirs.

3.8.2 Geometrical "dilution"

If the leads have a cross section W and the sample represents a narrow constriction of cross section w (as in the experiments that show conductance steps – see Sec. 3.3.4), the transmission probability can be assumed to be of the order of $T \sim w/W$. From Eq. 3.279 we then get[54]

$$G_{4\text{-probe}} \sim \frac{2e^2}{h} \frac{w/W}{1 - w/W}, \qquad (3.281)$$

while the corresponding two-probe conductance, with the same approximations would be

$$G_{2\text{-probe}} \sim \frac{2e^2}{h} \frac{w}{W}. \qquad (3.282)$$

The above shows that when the ratio $w/W \to 0$, namely when the leads cross section tends to infinity (while keeping the nanojunction cross section fixed), the four-probe conductance tends to the two-probe conductance

$$G_{4\text{-probe}} \to G_{2\text{-probe}}; \qquad \text{leads with infinite cross section}, \qquad (3.283)$$

which shows that the leads "behave" like reservoirs when their cross section tends to infinity, in line with what I have discussed in Sec. 1.2: an infinite system behaves like a reservoir.

It also shows that the main property of a reservoir (or bath) is the *geometrical dilution* of degrees of freedom. There is *no need* to introduce inelastic scattering for a reservoir to behave as such. When the size of a system is infinite, any particle wave-packet "entering" this system is "diluted" into an infinite number of degrees of freedom, without ever returning to its initial state (infinite Poincaré recurrence time, Eq. 1.15)[55].

3.8.3 Multi-channel case

For the multi-channel case, it is now a simple matter of adding more channels in the left lead and in the right lead. As in Eq. 3.53 let us define

$$T_i = \sum_{f=1}^{N_c^R} T_{if}, \qquad i \in L \qquad (3.284)$$

the total transmission probability for an incoming wave i from the left lead to be transmitted into *any* of the channels f in the right lead. Similarly

$$R_i = \sum_{f=1}^{N_c^L} R_{if}, \qquad i \in L \qquad (3.285)$$

[54] For a much more thorough discussion of this case see Landauer, 1989.
[55] Note that this is true also for a 1D system connected to 1D infinite leads: there is no need for a 1D system to "open up" into a wider region so long as the electrochemical potential difference is "measured" at the leads which due to their infinite size play the role of reservoirs (Bushong et al., 2005).

3.8 Four-probe conductance in the non-invasive limit

the total reflection probability for an incoming wave i from the left lead to be *reflected* back into *any* of the channels f in the left lead.

The corresponding quantities for a wave incident from the right will be labeled as $\tilde{T}_i$ and $\tilde{R}_i$, with i now a wave from the right lead.

By repeating the same arguments of the single-channel case, I can calculate the number density (per spin) in the left lead by summing over all the initial channels

$$n_L = \sum_{i=1}^{N_c^L} \int dE\, D_i(E) \left\{ [\ \overbrace{1}^{\text{from L to R}} + \overbrace{R_i(E)}^{\text{from L to L}}\ ]f_L(E) + \overbrace{\tilde{T}_i(E)}^{\text{from R to L}} f_R(E) \right\}$$

$$= 2 \sum_{i=1}^{N_c^L} \int dE\, D_i(E)\, f(E - \bar{\mu}_L), \qquad (3.286)$$

where here $D_i(E)$ is the density of states per spin, per direction of motion, of each incoming channel i (see Eq. 3.48). The number density per spin in the right lead is

$$n_R = \sum_{i=1}^{N_c^R} \int dE\, D_i(E) \left\{ [\ \overbrace{1}^{\text{from R to L}} + \overbrace{\tilde{R}_i(E)}^{\text{from R to R}}\ ]f_R(E) + \overbrace{T_i(E)}^{\text{from L to R}} f_L(E) \right\}$$

$$= 2 \sum_{i=1}^{N_c^R} \int dE\, D_i(E)\, f(E - \bar{\mu}_R). \qquad (3.287)$$

By Taylor-expanding the distribution functions and the transmission and reflection probabilities, in the zero-temperature limit, to lowest order in these expansions I finally get (use also Eq. 3.48)

$$\bar{\mu}_L = \mu_L - \frac{1}{2} \langle \tilde{T}(E_F) \rangle (\mu_L - \mu_R), \qquad (3.288)$$

$$\bar{\mu}_R = \mu_R + \frac{1}{2} \langle T(E_F) \rangle (\mu_L - \mu_R), \qquad (3.289)$$

with

$$\langle \tilde{T}(E_F) \rangle = \frac{\sum_{i=1}^{N_c^L} \tilde{T}_i\, |v_i|^{-1}}{\sum_{i=1}^{N_c^L} |v_i|^{-1}} \qquad (3.290)$$

the average of the transmission probabilities weighted over the inverse channel velocities.[56] Similarly

$$\langle T(E_F) \rangle = \frac{\sum_{i=1}^{N_c^R} T_i |v_i|^{-1}}{\sum_{i=1}^{N_c^R} |v_i|^{-1}}. \qquad (3.291)$$

Once more the electrochemical potential difference in the leads is

$$(\bar{\mu}_L - \bar{\mu}_R) = \left(1 - \frac{1}{2}\langle T(E_F) \rangle - \frac{1}{2}\langle \tilde{T}(E_F) \rangle\right)(\mu_L - \mu_R), \qquad (3.292)$$

where now I cannot use the relation 3.54 for these averaged transmission functions.

By defining the four-probe conductance as in 3.279, I get its multi-channel version with spin included, and in the limits $(\mu_L - \mu_R) \to 0$, $\theta \to 0$ (see also Exercise 3.17)

$$G_{4\text{-probe}} = \frac{2e^2}{h} \frac{\sum_{i=1}^{N_c^R} T_i}{1 - \frac{1}{2}\frac{\sum_{i=1}^{N_c^L} \tilde{T}_i |v_i|^{-1}}{\sum_{i=1}^{N_c^L} |v_i|^{-1}} - \frac{1}{2}\frac{\sum_{i=1}^{N_c^R} T_i |v_i|^{-1}}{\sum_{i=1}^{N_c^R} |v_i|^{-1}}}. \qquad (3.293)$$

Since the transmission probabilities are real numbers between zero and one, from Eq. 3.293 we see that, as in the single-channel case, the four-probe conductance is larger than the two-probe one. The latter is reached when *all* the transmission probabilities $T_i \ll 1$ and $\tilde{T}_i \ll 1$. In the opposite limit of all transmission probabilities equal to 1, the four-probe conductance tends to infinity as in the single-channel case.

I stress here that Eq. 3.293 (or its single-channel version 3.279) has been derived in the limit of zero "reservoir bias" and zero temperature. If either of these two limits is not valid one has to solve the integral equations 3.286 and 3.287 for the number density, from which one can then derive the current and conductance. However, in most cases, this procedure can only be done numerically.

If we repeat the same arguments on the geometrical dilution of wider leads attached to a narrow constriction (Eqs. 3.281 and 3.282), and assume *all* transmission probabilities equal to w/W, we get in this multi-channel case the four-probe conductance (assume also $N_c^L = N_c^R = N_c^W$ the number

---

[56] Note that due to the presence of terms of the type $|v_i|^{-1}$, these averages may contain divergences corresponding to states with zero velocity (e.g., the velocities at some high-symmetry points in the Brillouin zone of the material of the leads (Ashcroft and Mermin, 1975)). In real materials, due to the presence of inelastic scattering, or simply imperfections of the material itself, these singularities are likely to be smeared out (Landauer, 1989).

of channels in the wider leads)

$$G_{4\text{-probe}} \sim \frac{2e^2}{h} N_c^W \frac{w/W}{1 - w/W}, \tag{3.294}$$

to be compared with the corresponding two-probe conductance

$$G_{2\text{-probe}} \sim \frac{2e^2}{h} N_c^W \frac{w}{W}, \tag{3.295}$$

which shows once more that by taking the limit of infinite cross section the two conductances approach each other.[57]

We can expand on this result even more by realizing that the ratio $w/W \sim N_c^N/N_c^W$, i.e., the ratio of the width of the narrow constriction with respect to the width of the wider leads is approximately equal to the ratio of the number of channels in the narrow constriction (as if it were infinite along the x axis) and in the wider leads. By replacing this ratio in Eq. 3.294 we get (Landauer, 1989)

$$G_{4\text{-probe}} \sim \frac{2e^2}{h} N_c^N \left(\frac{N_c^W}{N_c^W - N_c^N} \right). \tag{3.296}$$

This result shows that if the number of channels in the large leads is much larger than the number of channels in the narrow constriction, $N_c^W \gg N_c^N$, the four-probe conductance reduces to

$$G_{4\text{-probe}} \to \frac{2e^2}{h} N_c^N, \quad N_c^W \gg N_c^N, \tag{3.297}$$

a conductance that is proportional to the number of channels in the narrow constriction, as if the latter were an infinitely long narrow wire. By changing the width w of this wire one then changes the number of channels in a discrete way (Eq. 3.28), and hence the conductance changes in steps. The values of these steps, however, are not simply multiples of $2e^2/h$, but contain a correction of the order of $N_c^W/(N_c^W - N_c^N)$ (from Eq. 3.296), which is not a "series resistance" but a term arising from the mismatch of number of channels between the narrow and the wide region. This type of result is precisely what has been found in the experiments on the 2D electron gas I have discussed in Sec. 3.3.4 (see Fig. 3.9).

3.9 Multi-probe conductance in the invasive limit

I can now discuss the case in which our voltage probes are *invasive*, namely they affect the current appreciably, but only elastic scattering occurs at the

---

[57] In the limit of $W \to \infty$ also the number of channels tends to infinity. The ratio N_c^W/W, however, stays constant.

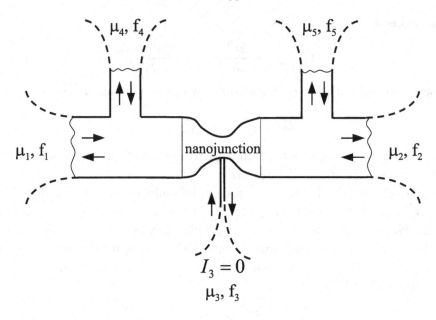

Fig. 3.20. Multi-probe configuration with invasive probes. The ones in which we impose the current to be zero are called *floating probes*.

sample. This generalization was done by Büttiker (Büttiker, 1988). An advantage of this generalization is that it allows us to introduce dephasing processes – while still conserving energy – in a phenomenological way. A schematic of the system and probes I have in mind is represented in Fig. 3.20. Two probes are "connected" to our ideal reservoirs which "drive" the current in the system, and for consistency in notation I now call them probe 1 (corresponding to the L reservoir) and probe 2 (the R reservoir).

I assume that all other probes measure a local electrochemical potential μ_α, and they are connected to the sample via some ideal leads (like the probes 1 and 2 are connected to the sample via the L lead and the R lead). We thus expect them to perturb the current set by reservoirs 1 and 2. We also assume that electrons in the regions to which these probes are connected have local Fermi–Dirac distribution functions $f_\alpha(E)$. As a further simplification, we impose that current cannot tunnel *directly* between any pairs of leads. The only current that we allow to flow is the one that goes across the sample.

The choice of which probes drive the current and which ones are used to measure the bias is clearly arbitrary. I may, for instance, assume that the current is driven by, say, probes 4 and 5 of Fig. 3.20, and the electrochemical potential difference is measured, say, between probes 1 and 2.

Let us now define the total transmission function of waves incident from a lead α from channels i, and transmitted into lead β into channels f,

$$\mathcal{T}_{\alpha\beta} = \sum_{i=1}^{N_c^\alpha} \sum_{f=1}^{N_c^\beta} T_{if}, \qquad \alpha \neq \beta. \tag{3.298}$$

Due to particle flux conservation 3.54, this transmission function must satisfy the symmetry

$$\mathcal{T}_{\alpha\beta} = \mathcal{T}_{\beta\alpha}, \qquad \text{flux conservation}, \tag{3.299}$$

and more generally

$$\sum_{\alpha \neq \beta} \mathcal{T}_{\alpha\beta} = \sum_{\alpha \neq \beta} \mathcal{T}_{\beta\alpha}, \qquad \text{flux conservation.} \tag{3.300}$$

It is now easy to generalize the current 3.55 to this multi-probe case: the current in probe α is the difference between the current flowing *into* that probe from all other probes, and the current that flows *out* of probe α into all other probes:

$$\boxed{I_\alpha = \sum_{\beta \neq \alpha} \frac{e}{\pi\hbar} \int dE \left[f_\alpha(E) \mathcal{T}_{\alpha\beta}(E) - f_\beta(E) \mathcal{T}_{\beta\alpha}(E) \right]} \tag{3.301}$$

which (together with Eq. 3.299) shows that if all electrochemical potentials are equal, $f_\alpha(E) = f_\beta(E)$, $\forall \alpha, \beta$, the current 3.301 vanishes, as required.

To simplify the discussion and obtain an analytical expression for the electrochemical potential of the probes, I will assume from now on that any two biases $V_{\alpha\beta} = (\mu_\alpha - \mu_\beta)/e$ are negligibly small. In addition, I will assume that all probes have a temperature close to zero. In this limit (by Taylor-expanding all quantities and retaining only the lowest-order terms), Eq. 3.301 becomes

$$I_\alpha = \sum_{\beta \neq \alpha} \frac{e}{\pi\hbar} \left[\mu_\alpha \mathcal{T}_{\alpha\beta}(E_F) - \mu_\beta \mathcal{T}_{\beta\alpha}(E_F) \right]. \tag{3.302}$$

3.9.1 Floating probes and dephasing

Equation 3.302 is most useful when one wants to introduce dephasing phenomenologically, while conserving energy. To simplify the discussion, consider first three probes (for instance, probes 1, 2 and 3 in Fig. 3.20). The current is set by the probes 1 and 2, while we let the electrochemical potential of probe 3 vary ("float") so that the current in probe 3 is zero: $I_3 = 0$. These are called *floating probes* (sometimes also called *voltage probes*) because one adjusts their electrochemical potential so that the current flowing in them is zero (Engquist and Anderson, 1981).

From Eq. 3.302 we then find

$$I_3 = 0 = \mu_3 (T_{13} + T_{23}) - \mu_1 T_{13} - \mu_2 T_{23}, \quad (3.303)$$

from which we get

$$\mu_3 = \frac{\mu_1 T_{13} + \mu_2 T_{23}}{T_{13} + T_{23}}. \quad (3.304)$$

The sum of all currents in all probes must be zero, which is easy to see from Eq. 3.302 (and using the symmetry 3.299), namely

$$\sum_\alpha I_\alpha = 0. \quad (3.305)$$

In the present case $I_3 = 0$, so that

$$I_1 + I_2 + I_3 = I_1 + I_2 = 0 \quad \Longrightarrow \quad I_1 = -I_2. \quad (3.306)$$

We can then just compute the current, say, in probe 1. From Eq. 3.302, and using the electrochemical potential μ_3, Eq. 3.304, we finally obtain

$$I_1 = \frac{e}{\pi \hbar} \left(T_{12} + \frac{T_{13} T_{32}}{T_{13} + T_{23}} \right) (\mu_1 - \mu_2). \quad (3.307)$$

This result is very instructive and can be interpreted as follows. The current in the absence of probe 3 is simply given by the transmission function T_{12}. Probe 3 *measures* the electrochemical potential μ_3, and it does so by perturbing the system. This perturbation (or coupling) is proportional to the transmission functions T_{13} for waves coming from lead 1 and entering lead 3, and T_{23} for waves from lead 2 entering lead 3 (and the reversed processes as well). These waves are *uncorrelated*: those electrons from lead 1 that enter probe 3 are not the same ones getting out of probe 3 into probe 2. This is one way of thinking about dephasing processes, which change the single-particle phase without changing the energy of the system (see discussion in Sec. 2.2.1).

Clearly, if *either* one of the probe connections is lost (1 with 3, or 2 with 3), the respective transmission functions are zero, and the current is the same as in the absence of these dephasing effects.

The above arguments can be extended to several floating probes with electrochemical potentials $\mu_3, \mu_4, \ldots \mu_F$, such that $I_3 = I_4 = \cdots = I_F = 0$ (still assuming the current is carried from probes 1 and 2). If the transmission function between the leads of *any* two floating probes is zero, it is then easy to show that the current I_1 is

$$I_1 = \frac{e}{\pi\hbar}\left(\mathcal{T}_{12} + \sum_{\beta=3}^{F} \frac{\mathcal{T}_{1\beta}\mathcal{T}_{\beta 2}}{\mathcal{T}_{1\beta} + \mathcal{T}_{2\beta}}\right)(\mu_1 - \mu_2). \qquad (3.308)$$

This expression may be, for instance, used to simulate local dephasing processes whose strength is defined by the transmission functions $\mathcal{T}_{1\beta}$ and $\mathcal{T}_{2\beta}$. The latter can then be used as parameters that control the strength of the dephasing effects.

Effect of dephasing on the current

From Eq. 3.308 it seems that the presence of floating probes always increases the current with respect to its value in the absence of dephasing. This is not always true as one can see by starting from the most general expression 3.301. For instance, for three probes we get from Eq. 3.301 (flux conservation implies $\mathcal{T}_{12} = \mathcal{T}_{21}$ and $\mathcal{T}_{13} = \mathcal{T}_{31}$)

$$I_1 = \frac{e}{\pi\hbar}\int dE\,[f_1(E) - f_2(E)]\,\mathcal{T}_{12}(E) + \frac{e}{\pi\hbar}\int dE\,[f_1(E) - f_3(E)]\,\mathcal{T}_{31}(E). \qquad (3.309)$$

For a floating probe to behave as such we only need to impose that the current at probe 3 be zero:

$$I_3 = \frac{e}{\pi\hbar}\int dE\,[f_3(E) - f_2(E)]\,\mathcal{T}_{32}(E) + \frac{e}{\pi\hbar}\int dE\,[f_3(E) - f_1(E)]\,\mathcal{T}_{31}(E) = 0, \qquad (3.310)$$

which (given the temperatures in all three probes) provides us with the electrochemical potential μ_3 of probe 3 as a function of μ_1, μ_3, $\mathcal{T}_{31}$ and $\mathcal{T}_{32}$ (which generalizes Eq. 3.304 to finite bias and temperature). This, however, does not make the second term on the right-hand side of Eq. 3.309 zero. Indeed, according to the value of the electrochemical potential μ_3 with respect to μ_1 (and also the relative temperature differences in the different probes) one obtains different results. For instance, if probes 1 and 3 have the same temperature, the condition $\mu_3 > \mu_1$ implies that the current I_1 is lower than the current in the absence of dephasing. The opposite occurs when $\mu_3 < \mu_1$, and dephasing actually "helps" electron motion. This scenario may be even more complicated if the floating probes are at a different temperature than probes 1 and 2.

3.10 Generalization to spin-dependent transport

In all the calculations we have done so far the spin has entered only as a degeneracy factor of 2 both in the current 3.56, and in the conductance 3.68.

In this section I want to generalize those expressions to take into account the spin explicitly, as an independent degree of freedom. This is necessary, for instance, when we consider the interaction with an external magnetic field, or the lead material is ferromagnetic.

As I have discussed in Sec. 2.5.2, in the presence of spin interactions, such as spin-orbit coupling, or coupling of the electrons with nuclear spins, the total electronic spin $\hat{S}$ is not a good quantum number, and one cannot, in principle, separate the current into spin components. Following the same arguments as in that section, I will neglect such effects on the basis that the spin relaxation is generally "slow" compared to any other time scale in the system.[58]

In the spirit of the single-particle Landauer picture we have developed, I will still assume that electrons may only interact at a mean-field level both in the leads and in the scattering region. I am then only concerned with the effect of a static magnetic field on the spin transport of these "mean-field electrons". At the end of this section, I will briefly discuss two possibly important many-body effects, which may modify the spin conductance in nanoscale junctions, even in the absence of a magnetic field.

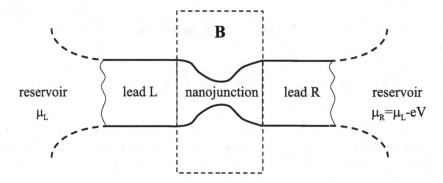

Fig. 3.21. In order to develop a scattering theory of conduction in the presence of a magnetic field **B**, we assume that the field acts only in the region of the sample, and does not perturb the states in the leads.

If a static (not necessarily uniform) magnetic field $\mathbf{B}(\mathbf{r})$ is present, by neglecting any relativistic effect, I need to modify the Hamiltonian 3.13 as

---

[58] This approximation may not be valid for all nanoscale systems, and thus needs to be handled with care case by case (see also Sec. 3.10.3).

follows (see also Sec. 2.3)[59]

$$H_S(\mathbf{B}) = \frac{1}{2m}\left[-i\hbar\nabla - \frac{e}{c}\mathbf{A}(\mathbf{r})\right]^2 + V(\mathbf{r}) + \mu_B\, g_0\, \frac{1}{2}\hat{\boldsymbol{\sigma}}\cdot\mathbf{B}(\mathbf{r}), \qquad (3.311)$$

where $\mathbf{A}(\mathbf{r})$ is the vector potential associated with the field,

$$\mathbf{B}(\mathbf{r}) = \nabla\times\mathbf{A}(\mathbf{r}), \qquad (3.312)$$

$\hat{\boldsymbol{\sigma}} \equiv (\hat{\sigma}_x, \hat{\sigma}_y, \hat{\sigma}_z)$ (with $\hat{\mathbf{S}} = \hbar\hat{\boldsymbol{\sigma}}/2$) are Pauli matrices (see discussion in Sec. 2.5.2), $\mu_B = |e|\hbar/2mc = 0.579\times 10^{-8}$ eV/G, is the Bohr magneton, and g_0 is known as the g-factor. Its value is close to 2.

Hamiltonian 3.313 is the Pauli limit of the relativistic Dirac equation when one neglects terms of order v^2/c^2, with v the electron velocity and c the speed of light (Messiah, 1961). Due to the two possible electron spin components in any given spatial direction, Hamiltonian 3.313 acts on *spinors* (two-component wave-functions) not simple scalar wave-functions. However, we can simplify further if we assume the spin is spatially oriented with the *local* external magnetic field. The scalar product in 3.313 then just separates into the two projections of the spin on the local orientation of the field.[60]

With this approximation the Hamiltonian H_S separates into two distinct Hamiltonians of the type

$$H_S^s(\mathbf{B}) = \frac{1}{2m}\left[-i\hbar\nabla - \frac{e}{c}\mathbf{A}(\mathbf{r})\right]^2 + V(\mathbf{r}) + \mu_B\, g_0\, s\, B(\mathbf{r}), \qquad (3.313)$$

with $s = \pm 1/2$. The eigenstates of this Hamiltonian will then acquire a quantum number s, in addition to the other quantum numbers

$$H_S^s(\mathbf{B})|\Psi_{ik_i,s}\rangle = E|\Psi_{ik_i,s}\rangle. \qquad (3.314)$$

In order to develop a scattering theory in the presence of a magnetic field, I also have to assume that the field only extends over the region of the sample I am interested in, namely (see Eqs. 3.17 and 3.18)

$$\lim_{x\to\pm\infty} H_S^s = \frac{1}{2m}\left[-i\hbar\nabla - \frac{e}{c}\mathbf{A}(\mathbf{r})\right]^2 + V_{L,R}(\mathbf{r}_\perp), \qquad (3.315)$$

so that the influence of the field does not carry over deep into the leads (see Fig. 3.21). However, we know that even if we impose the field to be zero in a given region of space, its influence on the electron dynamics still persists due to the presence of the vector potential in the kinetic part of the Hamiltonian. Indeed, imposing $\mathbf{B} = 0$ in the leads does not mean that the

---

[59] Here we are also neglecting any magnetic screening effect, and magnetic currents (see discussion in Sec. 1.2).
[60] This approximation is valid, for instance, if the spatial variation of the magnetic field is slow.

vector potential **A** is also zero in that region (that is why I had to keep it in 3.315[61]).

At first sight, this seems quite an annoyance, since we cannot carry over all the theoretical apparatus of transverse modes we have introduced in Sec. 3.3.1, namely we cannot write the eigenstates of 3.315 into the product of a transverse and a longitudinal component (see Eq. 3.22).[62] Here, *gauge invariance* comes to the rescue. This invariance tells us that a given vector potential **A** can be transformed into another vector potential **A**′ via

$$\mathbf{A}'(\mathbf{r}) = \mathbf{A}(\mathbf{r}) + \nabla \lambda(\mathbf{r}), \qquad (3.316)$$

with λ an arbitrary scalar function of position.[63]

Thanks to this freedom, we can now choose the vector potential oriented in the longitudinal direction only, even if we have many leads (Baranger and Stone, 1989). For instance, we can choose the vector potential *in each lead* as

$$(A_x, A_y, A_z) = (-B\,y, 0, 0), \qquad (3.317)$$

which is equivalent to a constant magnetic field in the z direction, and a divergence-less vector potential. With this choice of gauge, the eigenstates of 3.315 can now be written in the form 3.22.

In the following, we will assume the above gauge transformation has been done for each lead, so that we can define transverse modes in the lead regions. We can now solve the scattering problem of a single particle scattered by the sample potential in the presence of a magnetic field acting only in the region of the sample. For the last step we just need to determine the scattering potential in the presence of the field, and plug this potential into the Lippmann–Schwinger equation 3.117. This is easily done by writing explicitly all terms in the Hamiltonian 3.315 that contain the vector potential.

Equation 3.314 thus becomes

$$\left[-\frac{\hbar^2}{2m}\nabla^2 + V(\mathbf{r})\right]\Psi_{ik_i,s}(\mathbf{r}) + \int d\mathbf{r}'\, V_{\mathbf{B},s}(\mathbf{r},\mathbf{r}')\Psi_{ik_i,s}(\mathbf{r}') = E\,\Psi_{ik_i,s}(\mathbf{r}), \qquad (3.318)$$

---

[61] A typical example of this is the Aharonov–Bohm effect, where the quantum dynamics of a charged particle is affected even in regions of space of zero magnetic field (see, e.g., Sakurai, 1967).

[62] I call this simply an annoyance because I can still diagonalize 3.315 and use those solutions as my initial scattering states in the leads. Their physical interpretation in terms of left and right moving waves, however, is a bit trickier.

[63] In the present time-independent problem the scalar potential is identically invariant, because λ is independent of time.

where

$$V_{\mathbf{B},s}(\mathbf{r},\mathbf{r}') = \frac{ie\hbar}{2m}[\mathbf{A}(\mathbf{r}) + \mathbf{A}(\mathbf{r}')] \cdot \nabla \delta(\mathbf{r}-\mathbf{r}') + \frac{e^2}{2m} A^2(\mathbf{r}) \delta(\mathbf{r}-\mathbf{r}')$$
$$+ \mu_B\, g_0\, s\, B(\mathbf{r})\, \delta(\mathbf{r}-\mathbf{r}'), \qquad (3.319)$$

as can be shown via direct substitution into 3.318. The potential 3.319 is *non-local*, so that the Lippmann–Schwinger equation 3.117 needs to be modified into

$$\Psi^+_{ik_i,s}(\mathbf{r}) = \psi_{ik_i}(\mathbf{r}) + \int d\mathbf{r}' \int d\mathbf{r}''\, G^+_{\mathbf{B},s}(\mathbf{r},\mathbf{r}',E)\, V'_{\mathbf{B},s}(\mathbf{r}',\mathbf{r}'')\, \psi_{ik_i}(\mathbf{r}'') \quad (3.320)$$

with

$$V'_{\mathbf{B},s}(\mathbf{r},\mathbf{r}') = V(\mathbf{r}) + V_{\mathbf{B},s}(\mathbf{r},\mathbf{r}') \qquad (3.321)$$

the total scattering potential, sum of the local potential $V(\mathbf{r})$ and the non-local one 3.319.[64] $G^+_{\mathbf{B},s}(\mathbf{r},\mathbf{r}',E)$ are the positional matrix elements of the retarded Green's function

$$\hat{G}^+_{\mathbf{B},s}(E) = \frac{\hat{1}}{E + i\epsilon - \hat{H}^s_S(\mathbf{B})} \qquad (3.322)$$

associated with the full Hamiltonian 3.313. Note that I have assumed the "unperturbed" scattering states $\psi_{ik_i}(\mathbf{r})$ to be independent of the particular spin orientation, in line with my choice of no magnetic field effect in the leads (the electrons are *spin unpolarized* in the leads). If electrons are spin polarized in the leads (for instance, in the case of ferromagnetic leads) the states $\psi_{ik_i}(\mathbf{r})$ carry a spin quantum number as well.[65]

The potential $V'_{\mathbf{B},s}(\mathbf{r},\mathbf{r}')$ satisfies the following symmetry

$$V'_{\mathbf{B},s}(\mathbf{r},\mathbf{r}') = V'_{-\mathbf{B},-s}(\mathbf{r}',\mathbf{r}), \qquad (3.323)$$

---

[64] V' is non-local but Hermitian (Exercise 3.18).
[65] For instance, for a nanojunction sandwiched between two ferromagnetic leads, in the absence of a magnetic field, the transmission coefficient depends on the relative magnetization of the two ferromagnets. For parallel magnetization, the total transmission coefficient is (Zwolak and Di Ventra, 2002)

$$\mathcal{T} = \mathcal{T}_{\uparrow\uparrow} + \mathcal{T}_{\downarrow\downarrow},$$

with $\mathcal{T}_{\uparrow\uparrow}$ the transmission coefficient of majority-spin electrons in one lead (say the left lead) scattered into majority-spin states in the other lead (and similarly for the minority-spin electrons transmission coefficient $\mathcal{T}_{\downarrow\downarrow}$). When the magnetization of the two ferromagnets is anti-parallel, majority-spin electrons in the left lead scatter into minority-spin states in the opposite lead (now oriented as the majority-spin states of left lead), with transmission coefficient $\mathcal{T}_{\uparrow\downarrow}$. Similarly for minority-spin electrons in the left lead. For identical leads $\mathcal{T}_{\uparrow\downarrow} = \mathcal{T}_{\downarrow\uparrow}$ so that the total transmission coefficient for the anti-parallel magnetization is

$$\mathcal{T} = 2\mathcal{T}_{\uparrow\downarrow}.$$

This transmission coefficient is usually smaller than for the parallel magnetization case, so that the two resistances ($R_{\uparrow\uparrow}$ for parallel magnetization, and $R_{\uparrow\downarrow}$ for anti-parallel magnetization) are different. By changing relative magnetization, one thus realizes a *spin valve* characterized which is easy to demonstrate by recalling the property of the derivative of the δ-function, $\delta'(-x) = -\delta'(x)$, and that by changing the sign of **B**, we change the sign of **A** as well. Given this property, and the definition of the retarded Green's function 3.322, we find that the latter (and corresponding advanced Green's function $\hat{G}^-_{\mathbf{B},s}(E)$) does not satisfy the property 3.176 any longer. Instead, one gets

$$G^{\pm}_{\mathbf{B},s}(\mathbf{r},\mathbf{r}',E) = G^{\pm}_{-\mathbf{B},-s}(\mathbf{r}',\mathbf{r},E), \quad \text{in the presence of a magnetic field.} \tag{3.324}$$

3.10.1 Spin-dependent transmission functions

Within all the above approximations and assumptions, it is now straightforward to generalize all the results I have developed for spin-unpolarized electrons to the present case. Each quantity, whether a transmission or reflection probability, or a transmission function, depends now on a spin index, and is a function of a magnetic field **B**, in addition to its dependence on bias. I can then redo all calculations from Sec. 3.3.1 on, with just the caveat that the total current now contains a diamagnetic contribution (see Eq. 1.25). Equation 3.37 for the state 3.320 then reads

$$\begin{aligned} I(E_i) &= e\langle \Psi^+_{ik_i,s} | \hat{I} | \Psi^+_{ik_i,s} \rangle \\ &= \frac{e\hbar}{2im} \int_{-\infty}^{\infty} dy \int_{-\infty}^{\infty} dz \left[[\Psi^+_{ik_i,s}(\mathbf{r})]^* \frac{\partial \Psi^+_{ik_i,s}(\mathbf{r})}{\partial x} - \Psi^+_{ik_i,s}(\mathbf{r}) \frac{\partial [\Psi^+_{ik_i,s}(\mathbf{r})]^*}{\partial x} \right] \\ &\quad - \frac{e^2}{mc} \int_{-\infty}^{\infty} dy \int_{-\infty}^{\infty} dz \, [\Psi^+_{ik_i,s}(\mathbf{r})]^* A_x(\mathbf{r}) \Psi^+_{ik_i,s}(\mathbf{r}). \end{aligned} \tag{3.325}$$

For instance, we can define the transmission function *per spin* for waves moving from right to left

$$T^s_{RL}(E,\mathbf{B}) = \sum_{i=1}^{N_c^R} \sum_{f=1}^{N_c^L} T^s_{if}(E,\mathbf{B}) = \sum_{i=1}^{N_c^R} \tilde{T}^s_i(E,\mathbf{B}) \quad \text{from R to L,} \tag{3.326}$$

with $T^s_{if}(E)$ the transmission probability of an incoming wave in channel i in the left lead to be transmitted into an outgoing wave in channel f in the right lead.

by a *magnetoresistance*

$$R = \frac{R_{\uparrow\uparrow} - R_{\uparrow\downarrow}}{R_{\uparrow\uparrow} + R_{\uparrow\downarrow}}.$$

3.10 Generalization to spin-dependent transport

The same quantity can be defined for waves moving from left to right

$$\mathcal{T}_{LR}^s(E, \mathbf{B}) = \sum_{i=1}^{N_c^L} \sum_{f=1}^{N_c^R} T_{if}^s(E, \mathbf{B}) = \sum_{i=1}^{N_c^L} T_i^s(E, \mathbf{B}) \qquad \text{from L to R.} \quad (3.327)$$

If we also *assume* that the local Fermi distributions 3.14 and 3.15 do *not* depend on the spin polarization,[66] the total current is, from Eq. 3.56,

$$\boxed{I = \frac{e}{h} \int dE \, [f_L(E) - f_R(E)] \mathcal{T}(E, \mathbf{B})} \qquad (3.328)$$

with

$$\mathcal{T}(E, \mathbf{B}) = \mathcal{T}_{LR}(E, \mathbf{B}) = \mathcal{T}_{LR}^{1/2}(E, \mathbf{B}) + \mathcal{T}_{LR}^{-1/2}(E, \mathbf{B}) = \mathcal{T}_{RL}(E, \mathbf{B}), \quad (3.329)$$

the total transmission coefficient of both spin components. The last equality comes from particle flux conservation (Exercise 3.19).

From Eq. 3.328 and definition 3.68 of two-probe conductance in the limit of zero temperature and zero bias we get, in the presence of a magnetic field

$$\boxed{G = \frac{I}{V} = \frac{e^2}{h} \mathcal{T}(E_F, \mathbf{B}); \qquad (\mu_L - \mu_R) \to 0, \ \theta \to 0,} \qquad (3.330)$$

while the general definition for finite bias and temperature remains the same as in 3.70 with the only difference that now the current is given by Eq. 3.328.

3.10.2 Multi-probe conductance in the presence of a magnetic field

Similarly, we can extend the discussion of multi-probe conductance in the presence of a magnetic field located in the region of the sample, when invasive probes are present. Given any two probes α and β, we define the total transmission function per spin

$$\mathcal{T}_{\alpha\beta}^s(E, \mathbf{B}) = \sum_{i=1}^{N_c^\alpha} \sum_{f=1}^{N_c^\beta} T_{if}^s(E, \mathbf{B}), \qquad (3.331)$$

which, due to current conservation, must satisfy the relation

$$\mathcal{T}_{\alpha\beta}^s(E, \mathbf{B}) = \mathcal{T}_{\beta\alpha}^{-s}(E, -\mathbf{B}), \qquad (3.332)$$

---

[66] This is quite a strong approximation: we need to assume that *both* the charge *and* the spin relax to an unpolarized local equilibrium distribution.

and by extension of Eq. 3.300 and 3.302 to the magnetic case, we also have

$$\sum_{\alpha \neq \beta} \mathcal{T}_{\alpha\beta}(\mathbf{B}) = \sum_{\alpha \neq \beta} \mathcal{T}_{\beta\alpha}(\mathbf{B}) \qquad \text{flux conservation,} \qquad (3.333)$$

and

$$I_\alpha = \sum_{\beta \neq \alpha} \frac{e}{\pi\hbar} [\mu_\alpha \mathcal{T}_{\alpha\beta}(E_F, \mathbf{B}) - \mu_\beta \mathcal{T}_{\beta\alpha}(E_F, \mathbf{B})]. \qquad (3.334)$$

An interesting result of all the above is that, in linear response, if one reverses the magnetic field at the same time as the current and the voltage probes one employs to define the resistance, the latter remains unchanged (this is one form of *Onsager's relations* of systems close to equilibrium (Balian, 1991)).

To illustrate this, let us consider, as in Sec. 3.9.1, several probes. For simplicity consider only four probes, two of which (say numbers 3 and 4) are floating, i.e., $I_3 = I_4 = 0$, and measure the resistance

$$R_{12}^{34}(\mathbf{B}) = \frac{(\mu_3 - \mu_4)}{eI_1} \qquad (3.335)$$

with $I_1 = -I_2$ the obvious generalization of 3.308 in the presence of a magnetic field for the current measured in lead 1, driven by the electrochemical potential difference between μ_1 and μ_2, but in the presence of probes 3 and 4.

We can also make the probes 1 and 2 float ($I_1 = I_2 = 0$), and define the resistance in terms of their electrochemical potentials, while measuring the current $I_3 = -I_4$ at probe 3

$$R_{34}^{12}(\mathbf{B}) = \frac{(\mu_1 - \mu_2)}{eI_3}. \qquad (3.336)$$

The Onsager relation then reads (Exercise 3.20)

$$R_{12}^{34}(\mathbf{B}) = R_{34}^{12}(-\mathbf{B}), \qquad (3.337)$$

and it has been verified experimentally (Benoit *et al.*, 1986).

As in the absence of a magnetic field, the generalization of these relations to finite biases and temperatures is not obvious, which shows once again the difficulty in deriving general properties of systems out of equilibrium.

3.10.3 *Local resistivity spin dipoles and dynamical effects*

In the above, I have completely neglected any electron interaction which cannot be treated as contribution to the single-particle transmission probabilities. As anticipated in Chapter 1, nanoscale systems are characterized by large current densities so that electron interactions play an important role. Here, I need to distinguish two effects.

Local resistivity spin dipoles

We have seen in Sec. 3.2 that, due to the finite probability of electrons being reflected at the nanoscale junction, the self-consistent charge distribution forms local resistivity dipoles, i.e., dipoles of local charge at the boundaries of the junction, and for an extended junction, in its interior.

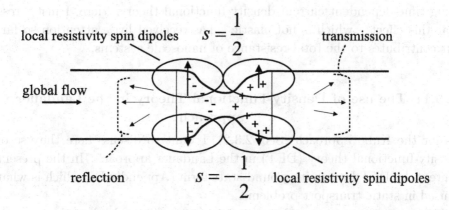

Fig. 3.22. In the case of spin-dependent transport, charge accumulation at the sample boundaries or in its interior is accompanied by spin accumulation, creating *local resistivity spin dipoles*.

In the case of spin-dependent transport, the same dipole is now both a charge dipole and a *local resistivity spin dipole* (Mal'shukov and Chu, 2006), namely one has an accumulation of spins of one polarity on one side of the junction, and a depletion of spins of the same polarity on the other side of the junction (see Fig. 3.22). This means that if one considers two different spin currents – say one moving in one direction and the other moving in the opposite direction – these two currents interact with each other via Coulomb interaction. If this interaction is strong, under certain conditions we may also expect one of the two spin currents to be *blockaded*, since minority spin electrons may impede majority spin electrons to cross the nanojunction (see also Sec. 4.5).[67] This effect has not received much attention yet.

---

[67] A similar phenomenon occurs when one *extracts* electrons from a semiconductor to a ferromagnet with 100% spin polarization: the minority spins accumulate at the interface between the semiconductor and the ferromagnet to such an extent that they may block the extraction of majority spins (Pershin and Di Ventra, 2007).

Dynamical spin resistance

A second effect is related to the true electron *dynamics* in nanojunctions, and the electron *viscosity*, a many-body effect beyond the single-particle mean-field theory of this chapter.

As I will show in Chapter 8 the electron liquid behaves to some degree like a classical liquid, with an effective viscosity due to Coulomb interactions. This interaction creates a *friction* between the two spin components which contributes to the total resistance of the system. In Sec 8.4 I will estimate this resistance for the spin-unpolarized case (in the linear-response regime) using time-dependent current-density-functional theory. Here, I just stress that this effect – which is not elastic – has received little attention so far, but contributes to the total resistance of nanoscale systems.

3.11 The use of Density-Functional Theory in the Landauer approach

As for the Kubo approach (Sec. 2.3.2), I want to discuss here the use of density-functional theory (DFT) in the Landauer approach. In the present context, by DFT I mean its ground-state form (Appendix D), which is what is used in static transport problems.

The present literature on the subject is vast and dates back to the first calculations of steady-state transport in atomic and molecular junctions carried out using the scattering approach I have described in this chapter (Lang, 1995; Di Ventra *at al.*, 2000; Xue *et al.*, 2001; Damle *et al.* 2001; Palacios *at al.*, 2002). It is not my intention to review the literature, only to point out the general procedure used, and the fundamental limitations of employing ground-state DFT within this approach.

As I have discussed in Sec. 3.1, *Approximation 4* of the Landauer approach requires the use of mean-field Hamiltonians. As an *approximation* one can then use Hamiltonian 3.12 from *ground-state* DFT (see also Appendix D), which I rewrite here for convenience

$$H_S = H_{KS}(\mathbf{r}) = -\frac{\hbar^2}{2m}\nabla^2 + V_H(\mathbf{r}) + V_{xc}(\mathbf{r}) + V_{ext}(\mathbf{r}), \qquad (3.338)$$

with V_H the Hartree term 3.11, and $V_{xc}(\mathbf{r})$ a static exchange-correlation potential (which may even assume to know *exactly*). V_{ext} is the effective potential describing the confinement of the electrons in the lead-junction-lead system, as represented by static electron-ion potentials (see Eq. 6.4 in Sec. 6.1.1), plus the non-local potential 3.319 if one includes a magnetic field.

One can then solve the Lippmann–Schwinger equation 3.104 directly, with the scattering potential represented by the sum of the above terms. The total current is then calculated via Eq. 3.37 integrated in energy according to Eq. 3.56. Equivalently, one can solve for the Green's function 3.156 with an appropriate "central" region, and evaluate the self-energy 3.151 for the "contact" between the left lead and the central region, and similarly for the right lead. The total current may then be calculated using Eq. 3.229.

As I have shown in this chapter, the above two procedures are *exactly equivalent* in the present effective single-particle problem. Both are equivalent ways of calculating the current within a single-particle scattering theory of conduction.

Self-consistency

Whatever the preferred route chosen, due to the presence of the Hartree potential, one needs to solve the problem *self-consistently* via the coupling with the Poisson equation

$$\nabla^2 \phi(\mathbf{r}) = -4\pi e\, n(\mathbf{r}), \quad (3.339)$$

with $\phi(\mathbf{r})$ the electrostatic potential 2.94, and n the total number density.

Since the ground-state density can be determined, in principle, exactly using static DFT, if one knows the exact functional (see Appendix D), it is actually more convenient to calculate the *extra* number density $\delta n(\mathbf{r}) = n(\mathbf{r})[V] - n(\mathbf{r})[V = 0]$ induced by the presence of a bias V, and the corresponding electrostatic potential difference $\delta\phi(\mathbf{r})$, via

$$\nabla^2 \delta\phi(\mathbf{r}) = -4\pi e\, \delta n(\mathbf{r}). \quad (3.340)$$

Starting with a guess of this potential, one can then iterate the calculation of the Lippmann–Schwinger equation (or Eq. 3.156), by computing the new potential (including the new exchange-correlation potential V_{xc}), from which a new density can be determined, and hence a new potential. And so on and so forth, until *self-consistency* in the density, or potential, is reached.

Non-variational properties of the current

To the above considerations, I add one more that has implications in practical calculations of the electrical conductance: the average current is generally *not a variational quantity* (Yang et al., 2002). This means the following.

Consider a Hamiltonian $\hat{H}$, and its *ground-state* wave-function $|\psi_{GS}\rangle$. Denote with $|\psi\rangle$ an arbitrary wave-function in the Hilbert space of the system, and construct the functional

$$E[|\psi\rangle] = \frac{\langle\psi|\hat{H}|\psi\rangle}{\langle\psi|\psi\rangle}. \qquad (3.341)$$

One can then easily prove the following *variational theorem* of quantum mechanics (Messiah, 1961): the ground-state energy of the system $E_{GS} = \langle\psi_{GS}|\hat{H}|\psi_{GS}\rangle$ is always smaller than $E[|\psi\rangle]$

$$E[|\psi\rangle] \geq E_{GS}, \qquad (3.342)$$

with the equal sign only when the wave-function $|\psi\rangle$ coincides (up to a phase) with the ground-state wave-function $|\psi_{GS}\rangle$.

This is a very powerful tool in practical calculations of the ground-state properties of a given system, because one can use a *finite* set of *trial* wave-functions, which may depend on some parameters (for instance, one can choose the set of plane waves $\langle\mathbf{r}|\psi_\mathbf{k}\rangle = e^{i\mathbf{k}\cdot\mathbf{r}}$, with $\{\mathbf{k}\}$ a finite set of wave-vectors), and vary these wave-functions in such a way that the energy $E[|\psi\rangle]$ evaluated from these trial wave-functions approaches the ground-state energy E_{GS} (in the case of plane waves, by increasing the range of $\mathbf{k}$ vector values).

With the above theorem, the "quality" of a given set of trial wave-functions to represent the ground-state energy, with respect to another, is easy to check. If a new set of trial wave-functions provides a larger value of the energy with respect to the previous one, it is a "worse choice" than the previous set.

A variational theorem does not generally hold for the electrical current, even at steady state,[68] and this represents a serious limitation in practical calculations.

For instance, if one chooses to work within the Landauer approach, and, in practical computations, represents the scattering wave-functions $|\Psi_{ik_i}^+\rangle$ as a linear combination of a (necessarily) finite basis set, the approach to the asymptotic value of the current $I(E_i) = e\langle\Psi_{ik_i}^+|\hat{I}|\Psi_{ik_i}^+\rangle$ with increasing number of elements in the basis set, does not occur variationally – the current may "fluctuate" above and below the asymptotic value $I(E_i)$ with increasing number of elements in the set (Yang *et al.*, 2002). This means that the "quality" of a given set of trial wave-functions cannot be easily determined as in the ground state. For instance, a set of trial wave-functions which

---

[68] A minimum principle still applies to the scattering amplitudes for a spherically symmetric potential at zero energy for the very special case when (i) a single s-wave channel contributes to scattering at that energy, and (ii) the potential does not support bound states or, if it does, the radial part of the Hamiltonian has negative expectation value on the bound states trial wave-functions (Schwinger, 1947).

gives a reasonable account of the ground-state energy E_{GS}, may not be an equally good choice for the calculation of the current.

Fundamental limitations of ground-state DFT in transport

The above procedures still do not answer the fundamental question:

How accurate is a ground-state DFT calculation of conductance?

I am asking here the question with respect to the true *many-body* current evaluated for a *well-defined* nanojunction. I am not comparing directly with experiments, because the currents there contain many effects – which I will discuss in the next chapters – not just the elastic ones discussed here, and may actually relate to microscopic structures of which we do not know all the atomic details. This is a very important point, not always appreciated in the literature.

While in mesoscopic systems a few-atom difference in atomic geometry may not affect the current considerably, in nanoscopic systems a single-atom change in the microscopic configuration of the junction, or even the presence of unintentional dopants in proximity to the junction, may lead to large differences in the total current (see, e.g., Yang *et al.*, 2003); a fact that has also been shown experimentally (see, e.g., Cui *et al.*, 2003). The main reason for this behavior is the partial electronic screening in nanoscale systems, with consequent large variations of the electrostatic potential at the nanostructure, and the sensitivity of this potential to microscopic (atomic) details.

So, given a well-defined nanostructure, how different is the current evaluated using ground-state DFT (even if we know the *exact* ground-state functional) within the Landauer approach, from the current the true many-body system develops in time (namely the current 1.86)? The answer is: we do not know!

The reason has nothing to do with whether we know the exact ground-state functional or not. Even if we assume that *Approximations* 1 – 5 of the Landauer approach provide an acceptable description of the dynamics of the true many-body system in an ideal steady state, by employing *ground-state* DFT (even with the exact functional) we are effectively *using a ground-state theory for a non-equilibrium problem*. For instance, as I will discuss in Sec. 8.4, ground-state DFT cannot capture dynamical properties of the electron liquid, such as its viscous dynamics.

This is reflected in the fact that, unlike the generalizations of DFT to dynamical problems (see Chapter 7 and Appendices E, F and G), there is *no theorem* that guarantees that the total current calculated with ground-state DFT is the same as the true many-body current. Not even in the limit of zero bias.

Indeed, as I have discussed at length, and emphasized once more in this chapter, even in the limit of zero bias, the current is *not* a ground-state property (see, e.g., the discussion in Sec. 2.3 regarding the Kubo approach). Therefore, the use of ground-state DFT in combination with the Landauer approach must be understood as a type of mean-field approximation, even if we know the exact ground-state functional.[69]

Nonetheless, in specific cases, such as transport properties of metallic quantum point contacts (see Sec. 1.1), the current obtained within this approach is very close to the experimental values (Di Ventra and Lang, 2002). While this may be taken as a demonstration of the validity of ground-state DFT in transport problems, it is fair to say that the same results may be obtained *without* DFT, and in many cases, even without the Hartree interaction at all, namely within a true non-interacting electron picture.

I will show in Chapter 7 that the above issues do not appear in the *dynamical* formulations of DFT (whether for closed or open quantum systems): the total current obtained using the dynamical versions of DFT is *exactly* equal to the total current of the true many-body system, if one knows the exact *dynamical* exchange-correlation potential. This fundamental result can be used to develop new approximate functionals (as always, the exact functional is unknown), and represents a solid starting point for developing exact theories of electrical conduction.

Summary and open questions

In this chapter I have introduced a single-particle scattering approach to steady-state transport, known as the Landauer approach. This approximates the true non-equilibrium problem according to the following assumptions:

- The junction is connected to a pair of defect-free metallic leads, each of which is "connected" to its own distant infinite heat-particle reservoir. The pair of reservoirs – that are not, however, dynamically coupled to the system – represents the battery.
- Each reservoir defines the electrochemical potential appropriate to the

---

[69] In the same vein, there is currently much discussion on improving *ground-state* DFT functionals to capture phenomena, such as the Coulomb blockade (see Sec. 4.5). Once again, these "improvements" miss the very fundamental (non-equilibrium) nature of the transport problem, and cannot thus answer the posed question satisfactorily.

bulk of that reservoir and a local Fermi distribution according to which single-particle states are populated.

- Each "injected" electron is then assumed to travel undisturbed down the respective lead to the junction, where it is scattered and is transmitted, with a finite probability, into the other lead.

- From there it flows, without further disturbance, into the other reservoir.

The reservoirs are conceptual constructs that allow us to map the transport problem onto a truly stationary scattering one, in which the time derivative of the average total current, *and of all other local physical properties of the system*, is zero. By doing so, however, we arbitrarily enforce a specific steady state whose microscopic nature is not, in reality, known *a priori*.

The above construct can only capture mean-field properties of the electron dynamics. It is thus very plausible in the case of non-interacting electrons. In the case of interacting electrons, however, it is not at all obvious that the steady state in the Landauer picture is the same as that which would be established *dynamically* by the electrons originating from the battery and flowing across the junction. It is also not obvious whether the same steady state can be reached with different initial conditions or if it can be realized with microscopically different many-body states (Di Ventra and Todorov, 2004).

Finally, I have pointed out that the use of ground-state DFT in the Landauer approach must be interpreted as a type of mean-field approximation, even if the exact ground-state exchange-correlation functional is known. Ground-state DFT – and in fact the Landauer approach itself – misses some dynamical properties of the electron liquid; properties that are specific to the true non-equilibrium many-body nature of the transport problem. I will come back to all these issues in Chapters 7 and 8.

Exercises

3.1 Show that from the current density operator 1.22 one gets the form 3.37 for the total current.

3.2 Prove Eqs. 3.104 and 3.105.

3.3 **Additivity of the Lippmann–Schwinger equation.** Consider two scattering potentials $\hat{V}_1$ and $\hat{V}_2$ and construct the potential $\hat{V} = \hat{V}_1 + \hat{V}_2$. Prove that the Lippmann–Schwinger equations 3.117

and 3.118 satisfy the additivity property

$$|\Psi^{+,-}(E)\rangle = |\Psi_0^{\text{in,out}}(E)\rangle + G^{+,-}(E)\hat{V}|\Psi_0^{\text{in,out}}(E)\rangle$$
$$= |\Psi_0^{\text{in,out}}(E)\rangle + G^{+,-}(E)\hat{V}_1|\Psi_0^{\text{in,out}}(E)\rangle$$
$$+ G^{+,-}(E)\hat{V}_2|\Psi_0^{\text{in,out}}(E)\rangle, \quad (\text{E3.1})$$

namely, the actions of these two potentials can be added separately.

3.4 **Principal-value Green's function.** Using the identity

$$\frac{1}{x+i\epsilon} = \mathcal{P}\frac{1}{x} - i\pi\delta(x), \quad (\text{E3.2})$$

with $\mathcal{P}$ the Cauchy principal value

$$\mathcal{P}\frac{1}{x} = \mathcal{P}\int_{-a}^{+a} dx \frac{1}{x} = \lim_{\epsilon \to 0} \left\{ \int_{-a}^{-\epsilon} dx \frac{1}{x} + \int_{+\epsilon}^{+a} dx \frac{1}{x} \right\}, \quad (\text{E3.3})$$

show that the time-independent retarded and advanced Green's functions may be written as

$$\hat{G}^{\pm}(E) = \mathcal{P}\frac{\hat{1}}{E - \hat{H}} \mp i\pi\delta(E - \hat{H}) \equiv G^{\mathcal{P}}(E) \mp i\pi\delta(E - \hat{H}), \quad (\text{E3.4})$$

where

$$\delta(E - \hat{H})|\Psi(E')\rangle = \delta(E - E')|\Psi(E')\rangle. \quad (\text{E3.5})$$

The quantity $G^{\mathcal{P}}(E)$ is known as the *principal-value Green's function*. With this relation, show that the density of states operator can be written as in Eq. 3.124.

Using the resolution of the identity 3.122 show that for any arbitrary state $|\Psi\rangle$

$$\int_{-\infty}^{+\infty} dE \langle \Psi|\hat{D}(E)|\Psi\rangle = 1. \quad (\text{E3.6})$$

3.5 **Bound states.** Consider the Lippmann–Schwinger equations 3.117 and 3.118. If we set the incoming and outgoing solutions to zero, no propagating states exist. However, one can still obtain *bound-state* solutions. Determine the condition for the existence of bound states.

3.6 **Free-particle Green's functions.** Prove Eq. 3.131 and extend the result to a free particle in three dimensions. Show that in this case the positional matrix elements of the retarded and advanced Green's functions are (with $E = \hbar^2 k^2/2m$ and $k > 0$)

$$G_0^{\pm}(\mathbf{r}, \mathbf{r}', E) = -\frac{m}{2\pi\hbar^2} \frac{e^{\pm ik|\mathbf{r}-\mathbf{r}'|}}{|\mathbf{r}-\mathbf{r}'|}. \quad (\text{E3.7})$$

3.7 **Green's functions of semi-infinite leads.** Consider the simplest nearest-neighbor tight-binding representation with a constant hopping integral t (Sec. 3.5). Use the spectral representation 3.123 to show that the retarded and advanced Green's functions of semi-infinite leads are

$$G^{\pm}(\mathbf{r}_n, \mathbf{r}_{n'}, E) = -\frac{1}{t} \sum_{\alpha} u_\alpha(\mathbf{r}_{\perp n}) e^{\pm i k_\alpha d} u_\alpha^*(\mathbf{r}_{\perp n'}), \qquad (E3.8)$$

where $|u_\alpha\rangle$ are the transverse solutions of Eq. 3.21, d is the longitudinal lattice distance, and k_α is the longitudinal wave-vector, solution of Eq. 3.24. (*Hint:* Realize that at the bounded edge of the leads the wave-functions have to vanish.)

3.8 Write Eqs. 3.221 and 3.222 in the simplest nearest-neighbor tight-binding representation with constant hopping integral t. Referring to the self-energies 3.167 show that relation 3.225 is satisfied.

3.9 **1D versus 3D free-electron density of states.** Use the 1D free-electron Green's functions 3.131 and 3.134 to calculate the local density of states

$$\langle \mathbf{r} | \hat{D}(E) | \mathbf{r} \rangle = \frac{m}{\pi \hbar} \frac{1}{\sqrt{2mE}}, \qquad E > 0, \qquad (E3.9)$$

which is independent of space. Redo the calculation for a free particle in three dimensions and show that

$$\langle \mathbf{r} | \hat{D}(E) | \mathbf{r} \rangle = \frac{m}{2\pi^2 \hbar^3} \sqrt{2mE}, \qquad E > 0. \qquad (E3.10)$$

3.10 Using the scattering solutions 3.174 and 3.175 show that the $\mathcal{S}$ matrix can be equivalently written as

$$\langle \Psi_f^-(E') | \Psi_i^+(E) \rangle = S_{if}(E) \delta(E' - E). \qquad (E3.11)$$

3.11 **Unitarity of the $\mathcal{S}$ matrix.** Use the relations 3.44, 3.45, 3.46 and 3.47 to prove that

$$\hat{\mathcal{S}} \hat{\mathcal{S}}^\dagger = \hat{\mathcal{S}}^\dagger \hat{\mathcal{S}} = \hat{1}. \qquad (E3.12)$$

3.12 Using the unitarity of the $\mathcal{S}$ matrix prove that

$$\begin{aligned}
r r^\dagger + \tilde{r} \tilde{r}^\dagger &= r^\dagger r + \tau^\dagger \tau = \hat{1}_{N_c^L \times N_c^L}, \\
\tau \tau^\dagger + \tilde{r} \tilde{r}^\dagger &= \tilde{\tau}^\dagger \tilde{\tau} + \tilde{r}^\dagger \tilde{r} = \hat{1}_{N_c^R \times N_c^R}, \\
r \tau^\dagger + \tilde{\tau} \tilde{r}^\dagger &= r^\dagger \tilde{\tau} + \tau^\dagger \tilde{r} = \hat{0}_{N_c^L \times N_c^R}, \\
\tau r^\dagger + \tilde{r} \tilde{\tau}^\dagger &= \tilde{\tau}^\dagger r + \tilde{r}^\dagger \tau = \hat{0}_{N_c^R \times N_c^L},
\end{aligned} \qquad (E3.13)$$

where $\hat{1}$ and $\hat{0}$ are the unit matrix and a matrix with all elements zero, respectively, with dimensions given by the subscripts. Derive from these the current conservation relations 3.44 and 3.45. Using also Eqs. 3.46 and 3.47 derive Eq. 3.54.

3.13 **Optical theorem.** Show that the $\mathcal{S}$ matrix and the T matrix are related via

$$\hat{\mathcal{S}} = \hat{1} - 2\pi i \hat{T}. \tag{E3.14}$$

Using the unitarity property of the $\mathcal{S}$ matrix prove the *optical theorem*

$$\frac{1}{2}i(\hat{T}^\dagger - \hat{T}) = -\pi \hat{T} \hat{T}^\dagger = -\pi \hat{T}^\dagger \hat{T}. \tag{E3.15}$$

3.14 Derive Eq. 3.217.

3.15 **Pseudo-unitarity of the transfer matrix.** Using the unitarity of the $\mathcal{S}$ matrix 3.209 prove that the $\mathcal{M}$ matrix is *pseudo-unitary*, Eq. 3.252. Show that pseudo-unitarity of the transfer matrix implies the current conservation relations 3.44 and 3.45.

3.16 Prove relation 3.255.

3.17 Derive Eq. 3.293.

3.18 Prove that the non-local potential 3.319 is Hermitian.

3.19 Show that particle flux conservation implies Eq. 3.329.

3.20 **Onsager relation and the $\mathcal{S}$ matrix.** Prove Eq. 3.337. Show also that due to relations 3.332 the $\mathcal{S}$ matrix satisfies

$$\mathcal{S}(-\mathbf{B}) = [\mathcal{S}(\mathbf{B})]^T, \tag{E3.16}$$

with the symbol T indicating the transpose of the matrix. Equation E3.16 is equivalent to the Onsager's relation 3.337.

3.21 **The transfer Hamiltonian approach.** (Bardeen, 1961) Refer to Fig. 3.11 and consider a Hamiltonian of the type

$$\hat{H} = \hat{H}_L + \hat{H}_R + \hat{H}_T, \tag{E3.17}$$

where $\hat{H}_{L,R}$ describe the Hamiltonians of the semi-infinite left and right electrodes, and $\hat{H}_T$ is called the *tunneling Hamiltonian* and connects the states of the two electrodes. The electrodes are biased, with V the bias. Take for simplicity the electrodes to be translationally invariant in the y, z direction (i.e., in the direction perpendicular to current flow). Call x_L the position of the contact between the left electrode and the central region and x_R the position between the central region and the right electrode. Assume that the tunneling Hamiltonian describes a potential barrier $U(x)$ of width d. Work within a single-particle picture, and assume that the wave-function inside the leads is

$$\Psi_i^{\text{leads}} = \sqrt{\frac{2}{L}} e^{i(k_y y + k_z z)} \sin(k_x x), \quad \text{(E3.18)}$$

with L some normalization length. Inside the barrier consider the WKB approximation (see, e.g., Messiah, 1961)

$$\Psi_i^{\text{barrier}} = \frac{1}{2}\sqrt{\frac{2}{L}} e^{i(k_y y + k_z z)} e^{-\int_{x_a}^{x} |k_x| dx}, \quad x_L < x < x_R, \quad \text{(E3.19)}$$

where $k_x = \sqrt{2m(U(x) - k_y^2 - k_z^2)}/\hbar$.

The left and right electrodes are independent of each other so that the above are good solutions for the left electrode alone when $x < x_R$, and for the right electrode alone when $x > x_L$.

The transition probability for an electron to transfer from one initial state $|i\rangle$ of one electrode to a state $|f\rangle$ of the other electrode can be estimated using perturbation theory via (see also Sec. 6.2.1 of Chapter 6)

$$P_{i \to f} = \frac{2\pi}{\hbar} |M_{if}|^2 \delta(E_f - E_i), \quad \text{(E3.20)}$$

with E_i and E_f the energies of those states, and M_{if} the transfer matrix element. Show that this matrix element can be written as

$$M_{if} = \frac{\hbar^2}{2m} \int\int dy\, dz \left(\Psi_i^* \frac{\partial \Psi_f}{\partial x} - \Psi_i \frac{\partial \Psi_f^*}{\partial x} \right)\bigg|_{x=a}, \quad \text{(E3.21)}$$

where a is any point inside the barrier. Show also that this matrix element decays exponentially with the barrier width d. Finally, prove that the current flowing from one electrode to the other can be expressed as

$$I = \frac{4\pi e}{\hbar} \sum_{k_i} \sum_{k_f} |M_{if}|^2 \left[f(E_i) - f(E_f - eV) \right]$$
$$\times D_i(E_i) D_f(E_f - eV) \delta(E_i - E_f), \quad \text{(E3.22)}$$

with $f(E)$ the Fermi–Dirac distribution function, and $D(E)$ the density of states.

3.22 **Non-variational properties of the current.** Consider the following functional (Yang et al., 2002)

$$A[|\chi\rangle] = \frac{|\langle \Psi_0 | \hat{V} | \chi \rangle|^2}{\langle \chi | \hat{V} - \hat{V} \hat{G}_0 \hat{V} | \chi \rangle}, \quad \text{(E3.23)}$$

where the Green's function $\hat{G}_0$ and potential $\hat{V}$ are those appearing in the Lippmann–Schwinger equations 3.117 and 3.118, with $|\Psi_0\rangle$ either an incoming or an outgoing state. Call $|\Psi\rangle$ a general solution of these equations. Show first that if $|\chi\rangle = c|\Psi\rangle$, with c a proportionality constant, then $A[|\chi\rangle] = \langle\Psi_0|V|\Psi\rangle$. In this case, the functional A is proportional to the scattering amplitude in the limit of high energies and/or weak potentials (the T matrix, Eq. 3.187, is $\hat{T} \approx \hat{V}$, in those limits). Now, demonstrate that for any arbitrary variation of $|\chi\rangle$ of the type

$$|\chi\rangle = c|\Psi\rangle + \delta|\chi\rangle \tag{E3.24}$$

the functional E3.23 is stationary. Stationarity does not imply a minimum or a maximum of the functional. This thus shows that the current (which is proportional to the modulus square of the scattering amplitudes, Eq. 3.186) is not necessarily a variational quantity.

4
Non-equilibrium Green's function formalism

Within the Landauer approach to conduction discussed in the previous chapter, electron interactions have been included only at the mean-field level. This is quite a strong approximation, especially in nanojunctions, where large current densities are common. I therefore want to go beyond this level of description.

I can follow two different routes. (1) I can develop a *functional* theory of quantum correlations via *time-dependent* effective single-particle equations, as I will do in Chapter 7 where I will use *dynamical* density-functional theories within the micro-canonical approach to conduction. (2) I can try to solve the time-dependent Schrödinger equation 1.16 – or its mixed-state version, the Liouville–von Neumann equation 1.60 – directly.

Written this way, this last proposition seems hopeless. In reality, we can employ a many-body technique, known as the *non-equilibrium Green's function formalism* (NEGF), also referred to as the *Keldysh formalism* (Keldysh, 1964; Kadanoff and Baym, 1962), which allows us, at least in principle, to do just that: solve the time-dependent Schrödinger equation for an interacting many-body system exactly, from which one can, in principle, calculate the time-dependent current. This is done by solving equations of motion for specific time-dependent single-particle Green's functions, from which the physical properties of interest, such as the charge and current densities, can be obtained.

I have stressed the term "time-dependent" several times, because, as I will show in a moment, the NEGF is "exact" only when one solves the *time-dependent* Schrödinger equation for a *closed* quantum system, subject to deterministic perturbations: the system is closed but not necessarily isolated. These perturbations may drive the system far away from its initial state of thermodynamic equilibrium (Keldysh, 1964).

Specific cases, such as steady-state transport, are usually described within the NEGF by making almost all the *assumptions* and *approximations* we have employed to develop the Landauer approach. Therefore, apart from the – surely non-trivial – addition of interactions in the sample beyond mean field, which represents the main application of the NEGF to transport problems, we cannot say that, in these cases, this approach is more "exact" than the Landauer one with respect to the dynamics of the true many-body system (see discussion in Sec. 3.1 regarding the loss of information in the Landauer approach).[1]

So, how do we proceed? We follow steps similar to those we have used in the Kubo formalism of Sec. 2.3 to determine the response of a closed quantum system to an external time-dependent perturbation. The major difference with the Kubo formalism is that we do not limit ourselves to weak perturbations (linear response), but we calculate the full response *to all orders of perturbation theory.*

Here, the term "perturbation" refers to *many-body* perturbation, so that the NEGF can be strictly applied only to those physical systems and processes where many-body perturbation theory holds. Examples of problems beyond standard many-body perturbation techniques are the Kondo effect, i.e., the entanglement of a localized spin with the spins of nearby conduction electrons (Kondo, 1964), or the Luttinger liquid, a form of correlated state of the electron liquid in strictly 1D systems (Luttinger, 1963). While these phenomena are interesting and may occur, under certain conditions, in nanoscale systems as well (see e.g., Yu and Natelson, 2004; Park *et al.*, 2000), their theoretical description is beyond the scope of this book.

Before proceeding, I want to stress that the NEGF has found many applications, including the study of superconductivity and superfluidity. For examples of its applicability other than those presented here, I refer the reader to the review by Rammer and Smith (1986), or the extensive book by Haug and Jauho (1996) fully dedicated to the theory and application of this approach.

My goal for this book is to outline the general assumptions underlying the NEGF, the approximations behind its application to steady-state transport, and its analogy with the classical Boltzmann equation 2.7. I will then apply it to the case in which electrons in the nanojuction (but not in the leads, otherwise no closed form for the current can be obtained) experience interactions beyond mean field. For this application, the general idea of a nanostructure connected, via scattering-free non-interacting leads, to two "reservoirs" is still borrowed from Landauer's original formulation. I will fi-

---

[1] As I will also discuss later, additional approximations are necessary when one wants to treat electron-electron or electron-phonon interactions explicitly.

nally show that within the *same* Landauer approximations, we recover, as we should, the current formulas derived from the single-particle non-interacting scattering theory of the previous chapter.[2]

A final note: I will guide the reader through an explicit example of many-body perturbation theory in Sec. 6.2.2, when discussing electron-phonon interaction in a nanojunction. However, it is beyond the scope of this book to develop the full formalism of many-body perturbation theory, with all its diagrammatic formulation and analytical properties. I refer the reader to existing textbooks, e.g., Mattuck (1976) or Mahan (1990) for a thorough description of this technique.

In this chapter, I will also make use of second quantization. A primer of this formalism is in Appendix A. The reader may find extended accounts of second quantization in any advanced textbook on quantum mechanics (Mattuck's, in particular, has a wonderful pedagogical introduction).

Since I will use the NEGF only in a few other sections of this book, the reader who is not interested in this many-body technique may safely avoid this chapter at first reading, without much detrimental effect on the comprehension of most of the following chapters.

4.1 Formulation of the problem

Before the derivation of the NEGF equations let us spell out the physical systems to which these equations apply.

Assumption 1: *Closed quantum systems*

The NEGF is a formalism to describe the dynamics of *closed* quantum systems. We thus want to solve the time-dependent Schrödinger equation 1.16, or the Liouville–von Neumann equation 1.60 if the system is in a mixed state. The NEGF does *not* apply to *open* quantum systems (those interacting *dynamically* with external environments), for the simple reason that there is no many-body perturbation technique for systems evolving beyond Hamiltonian dynamics, e.g., those obeying the Lindblad equation 1.73.[3]

---

[2] This is probably the reason why in certain literature the term NEGF is often, and mistakenly, used to indicate the non-interacting current formula 3.229 which, as I have shown, can be derived from scattering theory *without* ever employing the many-body perturbation techniques I will discuss in this chapter.

[3] This does not mean that one cannot use the Keldysh formalism to derive *approximate* equations of motion (like the Lindblad equation 1.73) for some reduced statistical operator of an open system. An example of this is the application of functional methods of field theory to coupling of matter with an electromagnetic field in *open* quantum electrodynamics. Due to the linear (and local) coupling between current densities and radiation fields, one can still derive an exact

As in the Kubo approach (Sec. 2.3) I thus assume that the system is described, before we switch on a time-dependent perturbation, by a many-body *time-independent* Hamiltonian $\hat{\mathbf{H}}_0$. In this chapter, in order to distinguish the operators that refer to *interacting* particles from those referring to *non-interacting* particles, I will use the notation that letters in boldface indicate interacting many-particle operators.

The Hamiltonian $\hat{\mathbf{H}}_0$, may be for instance the one in the Kubo formalism, Eq. 2.29, describing the interaction of many electrons among themselves and with static ions. That Hamiltonian can be separated into a sum of single-particle Hamiltonians, describing the kinetic energy of each electron and its interaction with the static ions, and the Coulomb interaction part.

This is just one possible choice of many-body Hamiltonian. For instance, in Chapter 5 I will discuss the case of electron-phonon interactions, and the Hamiltonian in that case also contains the interaction with the phonons.

Without loss of generality, I will then assume that $\hat{\mathbf{H}}_0$ can be written as the sum of one-particle Hamiltonians $\hat{h}_0^i$ and a Hamiltonian $\hat{\mathbf{W}}$ describing interactions among particles,

$$\mathbf{H}_0 = \sum_i \hat{h}_0^i + \hat{\mathbf{W}}. \qquad (4.1)$$

Each single-particle Hamiltonian $\hat{h}_0^i$ has eigenstates $|\phi_k\rangle$ with energies ϵ_k (we drop the superscript i to simplify the notation)

$$h_0 |\phi_k\rangle = \epsilon_k |\phi_k\rangle. \qquad (4.2)$$

Assumption 2: *Initial global equilibrium*

The system is at the initial time[4] $t = t_0$ in *global* canonical equilibrium with a bath at temperature θ, namely it is described by the canonical statistical operator (see Eq. 1.78)[5]

$$\hat{\rho}(t \leq t_0) = \frac{e^{-\beta \hat{\mathbf{H}}_0}}{\mathcal{Z}_C} \equiv \hat{\rho}_C^{eq}, \qquad (4.3)$$

form for the reduced statistical operator for the matter degrees of freedom only, provided the initial state of the radiation field has a Gaussian distribution (Breuer and Petruccione, 2002).

[4] Where we may take $t_0 \to -\infty$. If we want to describe steady-state transport this limit must be taken *after* the limit of system size to infinity.

[5] As in the Kubo formalism, we may choose some arbitrary initial statistical operator, but this may lead to serious mathematical difficulties in the calculation of the expectation values of a given operator (see discussion after Eq.. 4.11).

with $\beta = 1/k_B\theta$, and θ the temperature of the bath.

The above means that, for $t < t_0$, we have allowed for some (very weak) interaction with an external environment so that equilibrium could be reached.

Assumption 3: *Adiabatic approximation*

At time t_0, we apply a time-dependent perturbation $\hat{H}'(t)$ so that the total Hamiltonian is for $t > t_0$

$$\hat{\mathbf{H}}(t) = \hat{\mathbf{H}}_0 + \hat{H}'(t), \qquad (4.4)$$

and we want to determine the effect of this perturbation on the system dynamics.

The perturbation $\hat{H}'(t)$ may not necessarily be an external field. It could be, as I will show in a moment, some form of internal coupling of the system.

However, we want to work with a closed quantum system and not an open one. We thus do not want the environment – which has generated the initial equilibrium statistical operator 4.3 – to induce transitions between the many-body states of the unperturbed Hamiltonian $\hat{\mathbf{H}}_0$ – which would destroy unitary evolution of our quantum states.

This means that we need to switch on the perturbation $\hat{H}'(t)$ *adiabatically* as we have assumed in the Kubo formalism (see discussion in Sec. 2.3). As previously anticipated, this approximation becomes less and less satisfied in the limit of zero frequency of the external perturbation (d.c. limit): waiting a longer time allows the environment to induce transitions between the states of $\hat{\mathbf{H}}_0$. These transitions would force the system to some new state, with a statistical operator different from the equilibrium one in Eq. 4.3.

Partitioning vs. partition-free approach

In the context of steady-state transport both the above initial state assumption and adiabatic approximation can be implemented in different ways, which may indeed *not* lead to the same physical result.

For instance, we may refer to the partition I have employed in Sec. 3.5, where the sample is sandwiched between two semi-infinite electrodes (see Fig. 4.1). We can then assume that for $t < t_0$ the two electrodes are separated from each other and from the sample (the coupling potentials $\hat{V}_{LC}$, and $\hat{V}_{CR}$ are zero for $t < t_0$, and there is no direct tunneling between the electrodes).

We can, however, assume the electrodes to be *in the presence of the bias*, in the sense that current is not allowed to flow but the electrodes are *each* in

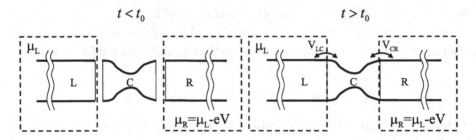

Fig. 4.1. One way to implement the switching-on of the external potential in the NEGF for transport. At $t < t_0$ two semi-infinite electrodes are decoupled from the sample and are at equilibrium with their respective reservoirs at different electrochemical potentials. At $t > t_0$ we switch on *adiabatically* the coupling potentials between the electrodes and the central region.

equilibrium with their own reservoirs at their own electrochemical potential (with the left electrochemical potential different than the right electrochemical potential), or, equivalently, there is an electron imbalance between the left electrode and the right electrode (Caroli et al., 1971).

The total statistical operator of this decoupled system at $t = t_0$ is then

$$\hat{\rho}(t_0) = \hat{\rho}_L(t_0) \otimes \hat{\rho}_C(t_0) \otimes \hat{\rho}_R(t_0), \qquad (4.5)$$

the tensor product of the equilibrium statistical operators of the electrodes and the central region.

At $t = t_0$ the couplings $\hat{V}_{LC}$, and $\hat{V}_{CR}$ between the electrodes and the sample are switched on *adiabatically* and current is allowed to flow (Fig. 4.1). Clearly, this procedure does not correspond to an actual experimental realization where the reverse is true: the physical couplings between the electrodes and the junction are there even *before* the bias is switched on. It is therefore not clear that the switch-on of the couplings *after* the bias leads to the *same* physical state as the correct order (if it leads to a steady state at all).

Alternatively, one could use a "partition-free" approach (Cini, 1980; Stefanucci and Almbladh, 2004) in which no separation of the physical system is done before the perturbation is switched on, with the global statistical operator given by the equilibrium operator 4.3. One then switches on, adiabatically, an external potential which in the long-time limit may reach, deep into the two electrodes, different values, in order to simulate the electrochemical potential difference as in Eq. 3.2 (if one is interested in the steady-state limit).

Also in this case, however, this procedure may not correspond to the correct physical scenario, since in an actual experiment the bias is not necessarily switched on adiabatically. In addition, one runs into the fundamental question of whether a steady state develops at all, and under which conditions. I will come back to these issues in Chapter 7.

4.1.1 Contour ordering

Within the above assumptions, our goal is now to calculate the expectation value of a given operator $\hat{A}$ (Eq. 1.52)

$$\langle \hat{A}_{\mathbf{H}}(t) \rangle = \text{Tr}\{\hat{\rho}_C^{eq} \hat{A}_{\mathbf{H}}(t)\}, \tag{4.6}$$

where

$$\hat{A}_{\mathbf{H}}(t) = \hat{U}(t_0, t) \hat{A} \hat{U}(t, t_0) \tag{4.7}$$

is the operator $\hat{A}$ written in the Heisenberg picture, with $U(t, t_0)$ the time-evolution operator of the *full* Hamiltonian 4.4 with the perturbation included. As in Eq. 1.19 the time-evolution operator satisfies the equation of motion

$$i\hbar \frac{\partial \hat{U}(t, t_0)}{\partial t} = \hat{\mathbf{H}}(t) \hat{U}(t, t_0), \tag{4.8}$$

with boundary condition $U(t_0, t_0) = 1$. I have written the formal solution of Eq. 4.8 in Eq. 1.18, which I re-write here for convenience

$$\hat{U}(t, t_0) = \text{T} \left\{ \exp\left[-\frac{i}{\hbar} \int_{t_0}^{t} dt' \hat{\mathbf{H}}(t') \right] \right\}, \tag{4.9}$$

with the symbol T indicating the time-ordering operator that orders products so that "past" times appear to the right.

By looking at the structure of the canonical statistical operator 4.3 and the definition of the time-evolution operator for a time-independent Hamiltonian 1.21, we realize that we can interpret the operator $e^{-\beta \hat{\mathbf{H}}_0}$ as the "evolution" of the system from the time t_0 to the complex time $t_0 - i\hbar\beta$[6]

$$\hat{U}(t_0 - i\hbar\beta, t_0) = \hat{U}(-i\hbar\beta) = e^{-\beta \hat{\mathbf{H}}_0}. \tag{4.10}$$

Using this result I can re-write the expectation value 4.6 (using the definition 4.3 of the canonical operator) as

$$\langle \hat{A}_{\mathbf{H}}(t) \rangle = \frac{\text{Tr}\{\hat{U}(t_0 - i\hbar\beta, t_0) \hat{U}(t_0, t) \hat{A} \hat{U}(t, t_0)\}}{\text{Tr}\{\hat{U}(t_0 - i\hbar\beta, t_0)\}}. \tag{4.11}$$

---

[6] I used a similar "trick" when discussing the fluctuation-dissipation theorem within the Kubo approach, see Eq. 2.65 in Sec. 2.3.3.

If we set the normalization factor in the denominator aside for a moment, the operator in the numerator of the expectation value 4.11 has the following structure with respect to the time evolution of a given state (think of this operator applied to a state on the right, so you read its time evolution from right to left)

$$\hat{U}(t, t_0) \to \hat{A} \to \hat{U}(t_0, t) \to \hat{U}(t_0 - i\hbar\beta, t_0), \qquad (4.12)$$

namely, the state evolves from the initial time t_0 till a time t, at which point the operator $\hat{A}$ acts. The subsequent evolution is then *backward* from time t to time t_0, after which the system "evolves" in the complex time from t_0 to $t_0 - i\hbar\beta$. This last step is also what is embodied in the denominator of Eq. 4.6, and represents the quantum correlations of the initial state of the system, which we have assumed to be in canonical equilibrium at t_0.

The expectation value of a given observable can then be represented as the result of the system evolution on a *contour*, as represented in Fig. 4.2. Note that the above re-interpretation of the equilibrium statistical operator on the complex time contour may not hold for an arbitrary initial statistical operator, making the calculation of the expectation value 4.6 more difficult (if not impossible) to carry out.

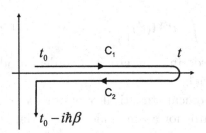

Fig. 4.2. Time contour used to evaluate the expectation value of a given operator at time t. The system evolves from an initial time t_0 to the time t, after which it returns to the initial time again, with further "propagation" in the complex time till $t_0 - i\hbar\beta$ (with $\beta = 1/k_B\theta$). The contour has thus two branches we label C_1 and C_2.

Since we have written all quantities in terms of time-evolution propagators on different parts of a contour, we can apply the same many-body perturbation techniques we apply to the equilibrium case, by just defining operations on a contour.

For instance, Eq. 4.7 can be re-written as (Rammer and Smith, 1986)

$$\hat{A}_{\mathbf{H}}(t) = \hat{U}_{\mathbf{H}_0}(t_0, t) \hat{A}_{\mathbf{H}_0}(t) \hat{U}_{\mathbf{H}_0}(t, t_0), \qquad (4.13)$$

where
$$\hat{U}_{\mathbf{H_0}}(t,t_0) = \mathrm{T}\left\{\exp\left[-\frac{\mathrm{i}}{\hbar}\int_{t_0}^{t}dt'\hat{H}'_{\mathbf{H_0}}(t')\right]\right\}, \qquad (4.14)$$

with
$$\hat{H}'_{\mathbf{H_0}}(t) = e^{\mathrm{i}\mathbf{H_0}(t-t_0)/\hbar}\,\hat{H}'(t)\,e^{-\mathrm{i}\mathbf{H_0}(t-t_0)}, \qquad (4.15)$$

the interaction representation of $H'(t)$.

The physical meaning of 4.13 is (reading that expression from right to left): propagation from t_0 to t, at which point the operator $\hat{A}$ acts, and subsequent propagation from t to t_0.

The operator $\hat{A}_{\mathbf{H}}(t)$ can then be written as the *contour time-ordering* (Rammer and Smith, 1986)

$$\hat{A}_{\mathbf{H}}(t) = \mathrm{T}_C\left\{\exp\left[-\frac{\mathrm{i}}{\hbar}\int_C dt'\hat{H}'_{\mathbf{H_0}}(t')\right]\hat{A}_{\mathbf{H_0}}(t)\right\}, \qquad (4.16)$$

where here C is the piece of contour in Fig. 4.2 that goes from time t_0 to time t, and back to time t_0, and the time-ordering on the contour must be interpreted as ordering products so that times occurring earlier on the contour appear to the right.

Since the above expression is now written in terms of the equilibrium Hamiltonian $\hat{H}_0$, perturbation techniques similar to the equilibrium ones can be applied. In fact, we can expand the time-evolution operator $\hat{U}_{\mathbf{H_0}}(t,t_0)$ in series (Mahan, 1990)

$$\hat{U}_{\mathbf{H_0}}(t,t_0) = \sum_{n=0}^{\infty}\frac{(-\mathrm{i})^n}{n!}\int_{t_0}^{t}dt_1\cdots\int_{t_0}^{t}dt_n\,\mathrm{T}\left\{\hat{H}'_{\mathbf{H_0}}(t_1)\cdots\hat{H}'_{\mathbf{H_0}}(t_n)\right\}, \qquad (4.17)$$

which represents the basis for a many-body perturbation expansion. In particular, since the single-particle *contour-ordered Green's function* I will define below can be written in terms of this time-evolution operator, a similar perturbation expansion for these Green's functions can be carried out.

At the cost of working on a time contour – hence with two time branches – we now have knowledge of the non-equilibrium state of our system at time t, simply on the basis of its state at the initial time. This procedure represents the core of the Keldysh formalism.

4.2 Equilibrium Green's functions

Before defining the single-particle non-equilibrium Green's functions let us step back for a moment and define single-particle Green's functions for an interacting system *at equilibrium*. This will allow me to define various important quantities (such as the self-energy and spectral function) that I will need later on. From now on second quantization is absolutely necessary (Appendix A).

In Sec. 3.4 I have defined advanced and retarded Green's functions for non-interacting single-particles in the context of scattering theory. Here I want to define similar Green's functions (or propagators) in the presence of interactions among particles. The physical meaning of these Green's functions will become clear as we go along.

As I will show in a moment, these propagators are not Green's functions in the usual mathematical sense, i.e., they are not solutions of a linear equation of the type 3.74. Their equation of motion instead generates the well-known BBGKY hierarchy of infinite equations (Appendix B).

4.2.1 Time-ordered Green's functions

Let us then consider a many-particle system described by the many-body Hamiltonian $\hat{\mathbf{H}}_0$. We define the field operator (see Appendix A)

$$\psi(\mathbf{r}) = \sum_k \langle \mathbf{r} | \phi_k \rangle \hat{a}_k, \quad \text{destroys particle at position } \mathbf{r}, \quad (4.18)$$

with $|\phi_k\rangle$ the single-particle states, solutions of Eq. 4.2, and $\hat{a}_k$ destroys a particle in state $|\phi_k\rangle$. Its Hermitian conjugate, $\psi^\dagger(\mathbf{r})$, creates a particle at position $\mathbf{r}$.

From this field operator we get the number density operator 1.24 in second-quantization form

$$\hat{n}(\mathbf{r}) \equiv \psi^\dagger(\mathbf{r})\psi(\mathbf{r}) = \sum_{k,k'} \langle \phi_{k'} | \mathbf{r} \rangle \langle \mathbf{r} | \phi_k \rangle \hat{a}^\dagger_{k'} \hat{a}_k. \quad (4.19)$$

If we take the expectation value of this operator over the canonical statistical operator 4.3 we obtain the number density of the system at equilibrium

$$n(\mathbf{r}) = \langle \hat{n}(\mathbf{r}) \rangle = \text{Tr}\{\hat{\rho}^{eq}_C \psi^\dagger(\mathbf{r})\psi(\mathbf{r})\}. \quad (4.20)$$

If the above average is taken over the grand-canonical statistical operator 1.76 with chemical potential $\bar{\mu}$ and the electrons are non-interacting, i.e. (see also Appendix A),

$$\hat{\mathbf{H}}_0 = \sum_i \hat{h}^i_0 = \sum_k \epsilon_k \hat{a}^\dagger_k \hat{a}_k, \quad (4.21)$$

one easily finds (Exercise 4.1)

$$n(\mathbf{r}) = \langle \hat{n}(\mathbf{r}) \rangle = \text{Tr}\{\hat{\rho}_G^{eq} \psi^\dagger(\mathbf{r})\psi(\mathbf{r})\} = \sum_k f_k |\langle \mathbf{r}|\phi_k \rangle|^2, \qquad (4.22)$$

where f_k is the average value of the number of particles in the state with energy ϵ_k when the non-interacting system is at equilibrium (cf. Eq. 1.80),

$$f_k = \frac{1}{e^{(\epsilon_k - \bar{\mu})/k_B \theta} + 1}, \qquad (4.23)$$

and I have used the fact that at thermal equilibrium each single-particle state $|\phi_k\rangle$ is populated according to (see Sec. 1.4.6)

$$\langle \hat{n}_k \rangle = \langle \hat{a}_k^\dagger \hat{a}_k \rangle = \text{Tr}\{\hat{\rho}_G^{eq} \hat{a}_k^\dagger \hat{a}_k\} = f_k. \qquad (4.24)$$

Let us now write the field operator 4.25 in the Heisenberg picture with respect to the Hamiltonian $\hat{\mathbf{H}}_0$ (whether for interacting or non-interacting particles)

$$\psi_{\mathbf{H}_0}(\mathbf{r}, t) = e^{i\hat{\mathbf{H}}_0 t/\hbar} \psi(\mathbf{r}) e^{-i\hat{\mathbf{H}}_0 t/\hbar}. \qquad (4.25)$$

We then define the following single-particle (or *two-times*) *time-ordered Green's function*[7]

$$\mathbf{G}(\mathbf{r}, t; \mathbf{r}', t') = -\frac{i}{\hbar} \langle \text{T}[\psi_{\mathbf{H}_0}(\mathbf{r}, t) \psi_{\mathbf{H}_0}^\dagger(\mathbf{r}', t')] \rangle$$

$$\equiv -\frac{i}{\hbar} \text{Tr} \left\{ \hat{\rho}_C^{eq} \text{T}[\psi_{\mathbf{H}_0}(\mathbf{r}, t) \psi_{\mathbf{H}_0}^\dagger(\mathbf{r}', t')] \right\}. \qquad (4.26)$$

Here $\text{T}[\ldots]$ is again the operation that time-orders the field operators, i.e., moves the operator with the earlier time variable to the right. For fermions it is[8]

$$\text{T}[\psi_{\mathbf{H}_0}(\mathbf{r}, t) \psi_{\mathbf{H}_0}^\dagger(\mathbf{r}', t')] = \Theta(t - t') \psi_{\mathbf{H}_0}(\mathbf{r}, t) \psi_{\mathbf{H}_0}^\dagger(\mathbf{r}', t')$$

$$- \Theta(t' - t) \psi_{\mathbf{H}_0}^\dagger(\mathbf{r}', t') \psi_{\mathbf{H}_0}(\mathbf{r}, t). \qquad (4.27)$$

The step function $\Theta(t - t')$ equals 1 for $t > t'$, and zero otherwise.

---

[7] For finite-temperature perturbation theory, we introduce *imaginary-time Green's functions* defined by replacing in Eq. 4.26 $\hat{\mathbf{H}}_0 \to \hat{\mathbf{H}}_0 - \bar{\mu}\hat{\mathbf{N}}$, with $\bar{\mu}$ the chemical potential and $\hat{\mathbf{N}}$ the total number operator 1.77. In addition, in order to have a perturbation theory formally equivalent to the zero-temperature case, we introduce the complex time $it \to \tau$, with t a real number. The canonical statistical operator appearing in Eq. 4.26 is also replaced by the grand-canonical statistical operator 1.76. The formal theory then follows similarly to what I discuss here (Mattuck, 1976). In the following, whenever we discuss finite-temperature Green's functions we will implicitly assume that their perturbation expansion has to be carried out by "transforming" them into the corresponding imaginary-time Green's functions.

[8] Note that the time-ordering operator for bosons has a positive sign and reads $\text{T}[\psi(\mathbf{r}, t)\psi^\dagger(\mathbf{r}', t')] = \Theta(t - t')\psi(\mathbf{r}, t)\psi^\dagger(\mathbf{r}', t') + \Theta(t' - t)\psi^\dagger(\mathbf{r}', t')\psi(\mathbf{r}, t)$.

By recalling the physical meaning of the field operator 4.25 and its Hermitian conjugate, as destroying or creating a particle, respectively, at position $\mathbf{r}$, we see that for $t > t'$ the Green's function 4.26 is essentially the probability amplitude that if at time t' a particle is created at position $\mathbf{r}'$ in the system in its equilibrium state, then the system at time t will be in the equilibrium state but with a particle destroyed in position $\mathbf{r}$. If we revert the times, i.e., for $t < t'$, $\mathbf{G}(\mathbf{r}, t; \mathbf{r}', t')$ is proportional to the probability amplitude that if at time t a *hole* is created at position $\mathbf{r}$, then the system at time t' will be in the same state but with the hole destroyed in position $\mathbf{r}'$.

4.2.1.1 Equation of motion for the equilibrium Green's function

The equation of motion for $\mathbf{G}(\mathbf{r}, t; \mathbf{r}', t')$ is easily derived by differentiating both sides of Eq. 4.26 with respect to time and taking into account the commutation relations of the field operators (Eqs. A.21 and A.22). This derivation is reported in Appendix B and I re-write the final result here for convenience:

$$\left(i\hbar\frac{\partial}{\partial t_1} + \frac{\hbar^2}{2m}\nabla_1^2\right)\mathbf{G}(\mathbf{r}_1, t_1; \mathbf{r}_1', t_1') = \delta(\mathbf{r}_1 - \mathbf{r}_1')\delta(t_1 - t_1')$$
$$- i\int d\mathbf{r}\, w(\mathbf{r}_1 - \mathbf{r})\, \mathbf{G}_2(\mathbf{r}, t_1, \mathbf{r}_1, t_1; \mathbf{r}, t_1^+, \mathbf{r}_1', t_1'), \quad (4.28)$$

where $\mathbf{G}_2$ is the *two-particle* (or *four-time*) Green's function (Eq. B.8), and t_1^+ means $\lim_{\epsilon \to 0+}(t_1 + \epsilon)$.[9]

For later use, and to avoid carrying too many indexes, it is convenient to denote with numbers any set of position and time variables, e.g., $1 \equiv (\mathbf{r}_1, t_1)$. Similarly, I will define $\delta(1, 1') = \delta(\mathbf{r}_1 - \mathbf{r}_1')\delta(t_1 - t_1')$. Using the properties of the δ-function I can then re-write the above equation as

$$\left(i\hbar\frac{\partial}{\partial t_1} + \frac{\hbar^2}{2m}\nabla_1^2\right)\mathbf{G}(1; 1') = \delta(1, 1') - i\int d2\, w(1, 2)\, \mathbf{G}_2(1, 2; 2^+, 1'), \quad (4.29)$$

with $w(1, 2) = \delta(t_1, t_2)w(\mathbf{r}_1 - \mathbf{r}_2)$.

The above is however a formal result since we need now the equation of motion for $\mathbf{G}_2$, which requires the equation of motion for the three-particle Green's function $\mathbf{G}_3$, and so on. This is the infinite BBGKY hierarchy which I have written explicitly in Appendix B.

---

[9] In the above, I have also assumed for simplicity that the single particles do not interact with any other single-particle potential, otherwise we need to subtract that potential in the parentheses of the left-hand side of Eq. 4.28. See Eq. 4.82.

4.2.2 Dyson's equation for interacting particles

Instead of solving the BBGKY hierarchy, we assume that all possible interaction effects of *all* particles on the *single-particle* dynamics can be lumped into a quantity known as the *irreducible self-energy* (if it exists), which we represent with the symbol $\mathbf{\Sigma}[\mathbf{G}]$, and which is a functional of the single-particle Green's function $\mathbf{G}$.

We have introduced a similar quantity in the scattering theory of conduction for non-interacting electrons (see Secs. 3.4.1.2 and 3.5). In Sec. 3.4.1.2, when discussing the Lippmann–Schwinger equation, the self-energy was not a useful concept, because in that non-interacting case it had the same form as the scattering potential of the full electrode-sample-electrode system (Eq. 3.95). In Sec. 3.5, when I partitioned the system into a central region and two electrodes, it represented the "interaction" of the central region with the electrodes, or better, the scattering of a single particle in the central region into the electrodes.

In this section, instead, $\mathbf{\Sigma}$ represents all interactions of a single particle with *all* other particles in the system, and its elastic scattering with the single-particle potential representing the static ions of the whole electrode-sample-electrode system (Fig. 4.1). Unfortunately, most of the time these two types of scattering cannot be separated from each other (we cannot say, e.g., the particle scatters with the potential of the nanojunction, and *independently* scatters with the other electrons or phonons). For all practical purposes, however, we will have to make this separation in order to obtain tractable equations.

For now, let us proceed by replacing

$$-\mathrm{i}\, w(1,2)\, \mathbf{G}_2(1,2;2^+,1') \to \mathbf{\Sigma}(1,2)\, \mathbf{G}(2;1') \qquad (4.30)$$

into Eq. 4.29 to get

$$\left(\mathrm{i}\hbar\frac{\partial}{\partial t_1} + \frac{\hbar^2}{2m}\nabla_1^2\right) \mathbf{G}(1;1') = \delta(1,1') + \int d2\, \mathbf{\Sigma}(1,2)\, \mathbf{G}(2;1'). \qquad (4.31)$$

By comparing the quantity in parentheses on the left-hand side of Eq. 4.31 with the definition of Green's functions in the true mathematical sense, Eq. 3.78, we recognize at once that with that term is associated the Green's function for a free particle, namely

$$\left(i\hbar\frac{\partial}{\partial t_1} + \frac{\hbar^2}{2m}\nabla_1^2\right)G_0(1;1') = \delta(1,1')$$

$$\Downarrow$$

$$\left(i\hbar\frac{\partial}{\partial t_1} + \frac{\hbar^2}{2m}\nabla_1^2\right) = \delta(1,1')G_0^{-1}(1;1'). \quad (4.32)$$

By replacing this result in Eq. 4.31 and integrating over one of the times to eliminate the δ-functions, we get (I re-instate the times and coordinates explicitly)

$$\begin{aligned}\mathbf{G}(\mathbf{r},t;\mathbf{r}',t') &= G_0(\mathbf{r},t;\mathbf{r}',t') \\ &+ \int dr_1 \int dr_2 \int dt_1 \int dt_2\, \mathbf{G}(\mathbf{r},t;\mathbf{r_1},t_1) \\ &\times \mathbf{\Sigma}(\mathbf{r_1},t_1;\mathbf{r_2},t_2)\, G_0(\mathbf{r_2},t_2;\mathbf{r}',t') \\ &= G_0(\mathbf{r},t;\mathbf{r}',t') \\ &+ \int dr_1 \int dr_2 \int dt_1 \int dt_2\, G_0(\mathbf{r},t;\mathbf{r_1},t_1) \\ &\times \mathbf{\Sigma}(\mathbf{r_1},t_1;\mathbf{r_2},t_2)\, \mathbf{G}(\mathbf{r_2},t_2;\mathbf{r}',t'). \quad (4.33)\end{aligned}$$

This is the many-body version of the Dyson equation 3.94.

At equilibrium, the above Green's functions and self-energy can depend on time differences only. If, in addition, the system is homogeneous, these quantities can only depend on differences in the position arguments. I can then simplify the above equation by Fourier transforming the Green's functions and self-energy. For instance, I can define the Fourier transform of the single-particle Green's function

$$\mathbf{G}(\mathbf{r}-\mathbf{r}';t-t') = \frac{1}{\hbar}\int \frac{dE}{2\pi}\int \frac{d\mathbf{k}}{(2\pi)^3} e^{-iE(t-t')/\hbar}e^{i\mathbf{k}(\mathbf{r}-\mathbf{r}')}\mathbf{G}(\mathbf{k};E), \quad (4.34)$$

and similarly for all other quantities. Equation 4.33 then becomes

$$\begin{aligned}\mathbf{G}(\mathbf{k};E) &= G_0(\mathbf{k};E) + \mathbf{G}(\mathbf{k};E)\,\mathbf{\Sigma}(\mathbf{k};E)\,G_0(\mathbf{k};E) \\ &= G_0(\mathbf{k};E) + G_0(\mathbf{k};E)\,\mathbf{\Sigma}(\mathbf{k};E)\,\mathbf{G}(\mathbf{k};E), \quad (4.35)\end{aligned}$$

which generalizes Eq. 3.120 to the interacting-particle case.

The single-particle non-interacting case

We easily see that the time-ordered Green's function 4.26 is a generalization to the interacting case of the Green's function 3.78, and its Fourier transform generalizes Eq. 3.112. Indeed, if $\hat{h}_0$ is the Hamiltonian of a single-particle, from Eq. 4.28 we get (the interaction term $w(\mathbf{r}-\mathbf{r}') = 0$)

$$\left(i\hbar\frac{\partial}{\partial t} - \hat{h}_0\right) G_0(\mathbf{r},t;\mathbf{r}',t') = \delta(\mathbf{r}-\mathbf{r}')\delta(t-t'). \tag{4.36}$$

If the system is in equilibrium, then the above Green's function can depend only on the difference $t - t'$,

$$\left(i\hbar\frac{\partial}{\partial t} - \hat{h}_0\right) G_0(\mathbf{r},\mathbf{r}';t-t') = \delta(\mathbf{r}-\mathbf{r}')\delta(t-t'). \tag{4.37}$$

Fourier transforming the above equation, and using the Fourier transform of the δ-function, $\delta(t) = \frac{1}{2\pi}\int d\omega\, e^{-i\omega t}$, we obtain

$$\left(E - \hat{h}_0\right) G_0(\mathbf{r},\mathbf{r}';E) = \delta(\mathbf{r}-\mathbf{r}'). \tag{4.38}$$

From the theory of differential equations we then know that with Eq. 4.38 is associated the linear-operator equation (see Eq. 3.74)

$$\left(E - \hat{h}_0\right) \phi_k(\mathbf{r}) = 0, \tag{4.39}$$

which is the stationary Schrödinger equation with eigenstates $\langle \mathbf{r}|\phi_k\rangle$. Therefore, $G_0(\mathbf{r},\mathbf{r}';E)$ is the positional matrix element of the Green's function 3.112.

4.2.3 More Green's functions

Analogous to what we have done in the non-interacting case we can also define retarded and advanced single-particle Green's functions for fermions

$$\mathbf{G}^+(\mathbf{r},t;\mathbf{r}',t') = -\frac{i}{\hbar}\Theta(t-t')\langle\{\psi(\mathbf{r},t),\psi^\dagger(\mathbf{r}',t')\}\rangle, \quad \text{retarded}, \tag{4.40}$$

$$\mathbf{G}^-(\mathbf{r},t;\mathbf{r}',t') = \frac{i}{\hbar}\Theta(t'-t)\langle\{\psi(\mathbf{r},t),\psi^\dagger(\mathbf{r}',t')\}\rangle, \quad \text{advanced}. \tag{4.41}$$

They generalize to the interacting case the corresponding non-interacting Green's functions Eqs. 3.80 and 3.82, respectively.[10]

---

[10] The corresponding functions for bosons are

$$\mathbf{G}^+(\mathbf{r},t;\mathbf{r}',t') = -\frac{i}{\hbar}\Theta(t-t')\langle[\psi(\mathbf{r},t),\psi^\dagger(\mathbf{r}',t')]\rangle, \quad \text{retarded}, \tag{4.42}$$

$$\mathbf{G}^-(\mathbf{r},t;\mathbf{r}',t') = \frac{i}{\hbar}\Theta(t'-t)\langle[\psi(\mathbf{r},t),\psi^\dagger(\mathbf{r}',t')]\rangle, \quad \text{advanced}. \tag{4.43}$$

We also define the propagators

$$\mathbf{G}^<(\mathbf{r},t;\mathbf{r}',t') = \frac{i}{\hbar}\langle \psi^\dagger(\mathbf{r}',t')\psi(\mathbf{r},t)\rangle, \quad \text{lesser than}, \quad (4.44)$$

and

$$\mathbf{G}^>(\mathbf{r},t;\mathbf{r}',t') = -\frac{i}{\hbar}\langle \psi(\mathbf{r},t)\psi^\dagger(\mathbf{r}',t')\rangle, \quad \text{greater than}, \quad (4.45)$$

which are called the *lesser* and *greater* Green's functions (or correlation functions), respectively, and are equal to $\mathbf{G}(\mathbf{r},t;\mathbf{r}',t')$ for $t<t'$ and $t>t'$, respectively.

In all the above definitions the symbol $\langle\cdots\rangle$ means trace over the initial statistical operator (whether canonical or grand-canonical), namely $\langle\cdots\rangle = \text{Tr}\{\hat{\rho}(t=t_0)\cdots\}$.

Density and current density from $\mathbf{G}^<$

The lesser and greater Green's functions are central quantities in the non-equilibrium problem since they are directly related to the number and current densities of the system. For instance, by comparing Eq. 4.20 and Eq. 4.44, we can write

$$n(\mathbf{r},t) = -i\hbar \lim_{\epsilon\to 0} \mathbf{G}^<(\mathbf{r},t;\mathbf{r},t+\epsilon), \quad (4.46)$$

with ϵ a positive number. Similarly, we can calculate the single-particle density matrix in the positional basis as

$$\rho(\mathbf{r},\mathbf{r}',t) = -i\hbar \lim_{\epsilon\to 0} \mathbf{G}^<(\mathbf{r},t;\mathbf{r}',t+\epsilon). \quad (4.47)$$

In Sec. 3.3.2 we have written the current density (without any diamagnetic term) in terms of the single-particle wave-functions (see Eq. 3.36), starting from the definition of current density operator (Eq. 1.22). We can re-write that expression for a general wave-function $\Psi(\mathbf{r},t)$ as

$$\mathbf{j}(\mathbf{r},t) = \frac{\hbar}{2im} \lim_{\mathbf{r}\to\mathbf{r}'} (\nabla_\mathbf{r} - \nabla_{\mathbf{r}'})\Psi^*(\mathbf{r}',t)\Psi(\mathbf{r},t). \quad (4.48)$$

The second-quantization form of this expression is (Appendix A)

$$\begin{aligned}\mathbf{j}(\mathbf{r},t) &= \frac{\hbar}{2im} \lim_{\mathbf{r}\to\mathbf{r}'}(\nabla_\mathbf{r}-\nabla_{\mathbf{r}'})\langle \psi^\dagger(\mathbf{r}',t)\psi(\mathbf{r},t)\rangle \\ &= \frac{\hbar}{2im}\lim_{\mathbf{r}\to\mathbf{r}'}(\nabla_\mathbf{r}-\nabla_{\mathbf{r}'})\text{Tr}\{\rho(t=t_0)\,\psi^\dagger(\mathbf{r}',t)\,\psi(\mathbf{r},t)\}.\end{aligned} \quad (4.49)$$

Comparing this with the definition of lesser Green's function 4.44, we immediately see that the current density and the lesser Green's function are related via

$$\mathbf{j}(\mathbf{r},t) = \frac{\hbar^2}{2m} \lim_{\mathbf{r}'\to\mathbf{r}} (\nabla_{\mathbf{r}'} - \nabla_{\mathbf{r}}) \lim_{\epsilon\to 0} \mathbf{G}^<(\mathbf{r},t;\mathbf{r}',t+\epsilon). \quad (4.50)$$

The above density and current density must be related via the continuity equation $\partial_t n(\mathbf{r},t) = -\nabla \cdot \mathbf{j}(\mathbf{r},t)$, so that any approximation that one makes to actually calculate the lesser Green's function $\hat{G}^<$ must satisfy this fundamental relation. These approximations are known as *conserving approximations*.

Since, the lesser Green's function is related to the current density, it also provides the *fluctuations* of the system, and hence it must be related to the dissipation occurring in the system's dynamics (see Eq. 2.73, where I have related the spontaneous current fluctuations to the conductivity of the system). I will discuss this fluctuation-dissipation relation in a moment.

I finally note that, following their definition, the four Green's functions 4.40, 4.41, 4.44 and 4.45 are related to each other via the condition (Haug and Jauho, 1996)

$$\hat{G}^+ - \hat{G}^- = \hat{G}^> - \hat{G}^<, \quad (4.51)$$

so that only three of them are linearly independent.

4.2.4 The spectral function

Let us now discuss what type of information we can extract from the retarded 4.40 and advanced 4.41 Green's functions. In the non-interacting scattering case (Sec. 3.4), the retarded and advanced Green's functions contained information on the *spectral* properties of the system (such as eigenvalues of the Hamiltonian and density of states, see Eq. 3.124). The same type of information is contained in the interacting version of these propagators.

To see this let us assume the many-body system is homogeneous so that the functions 4.40 and 4.41 depend only on the differences of space variables.[11] We can thus work with the Fourier representation in energy and momentum space of the retarded and advanced Green's functions 4.34.

We then define the *spectral function*

$$A(\mathbf{k};E) = i \frac{G^+(\mathbf{k};E) - G^-(\mathbf{k};E)}{2\pi} = i \frac{G^>(\mathbf{k};E) - G^<(\mathbf{k};E)}{2\pi}, \quad (4.52)$$

---

[11] The assumption of homogeneity can be relaxed. This would require working with two spatial coordinates. I choose not to do this here to simplify the notation.

where the last equality comes from Eq. 4.51. As in the non-interacting case, the retarded and advanced Green's functions satisfy the relation 3.115, $[\mathbf{G}^+(\mathbf{k};E)]^\dagger = \mathbf{G}^-(\mathbf{k};E)$, which can be easily demonstrated from the Fourier transforms of their respective definitions. The spectral function can then be written as (see also Eq. 3.127)

$$A(\mathbf{k};E) = -\frac{1}{\pi}\mathrm{Im}\left[\mathbf{G}^+(\mathbf{k};E)\right], \qquad (4.53)$$

or, equivalently, using the Dyson equation 4.35 for the retarded and advanced Green's functions

$$2\pi A(\mathbf{k};E) = i\, \mathbf{G}^+(\mathbf{k};E)\left[\mathbf{\Sigma}^+(\mathbf{k};E) - \mathbf{\Sigma}^-(\mathbf{k};E)\right]\mathbf{G}^-(\mathbf{k};E)$$
$$\equiv \mathbf{G}^+(\mathbf{k};E)\,\mathbf{\Gamma}(\mathbf{k};E)\,\mathbf{G}^-(\mathbf{k};E), \qquad (4.54)$$

where

$$\mathbf{\Gamma}(\mathbf{k};E) = i\left[\mathbf{\Sigma}^+(\mathbf{k};E) - \mathbf{\Sigma}^-(\mathbf{k};E)\right] = -2\mathrm{Im}\{\mathbf{\Sigma}^+(\mathbf{k};E)\}. \qquad (4.55)$$

By comparing the above two equations with the single-particle noninteracting analogues, Eqs. 3.162 and 3.163, it is clear that the spectral function generalizes to the interacting many-body case the concept of density of states 3.124 we have introduced for non-interacting systems. Let us expand a bit on this point.

First, using the definitions of retarded and advanced Green's functions 4.40 and 4.41 it is easy to show (see Exercise 4.2) that $A(\mathbf{k};E)$ is positive definite and, for fermions, satisfies

$$\int_{-\infty}^{+\infty} dE\, A(\mathbf{k};E) = 1, \qquad (4.56)$$

which immediately hints at a possible statistical interpretation of the spectral function.

The density of states can be evaluated as the integral of the spectral function over all possible $\mathbf{k}$ vectors

$$D(E) = \int \frac{d\mathbf{k}}{(2\pi)^3}\, A(\mathbf{k};E). \qquad (4.57)$$

The fluctuation-dissipation theorem

Another important relation, which we will make use of several times in the following, is between the spectral function and the lesser Green's function $\mathbf{G}^<$ (or the spectral function and the greater Green's function $\mathbf{G}^>$). This is also known as the *fluctuation-dissipation theorem*, like the one we have discussed in Sec. 2.3.3.

If we assume that the *exact* eigenstates of the many-body Hamiltonian $\hat{H}_0$ are known,[12] and the system is in grand-canonical equilibrium with a reservoir at a chemical potential $\bar{\mu}$ and temperature θ (Eq. 1.76), it is not difficult to show that $\mathbf{G}^<$ and $\mathbf{G}^>$ are related via (Exercise 4.3)

$$\mathbf{G}^<(\mathbf{k}; E) = -e^{\beta(E-\bar{\mu})}\mathbf{G}^>(\mathbf{k}; E), \qquad (4.58)$$

with $\beta = 1/k_B\theta$.

From the definition of spectral function 4.52, we then get

$$\mathbf{G}^<(\mathbf{k}; E) = 2\pi\, i f(E) A(\mathbf{k}; E), \qquad (4.59)$$

and

$$\mathbf{G}^>(\mathbf{k}; E) = -2\pi\, i\, [1 - f(E)]\, A(\mathbf{k}; E), \qquad (4.60)$$

where $f(E)$ is the Fermi–Dirac distribution 4.23.

The above relations show that the lesser and greater Green's functions contain information on the *occupations* of the states (due to the presence of the distribution $f(E)$). Also, since the spectral function is related to the imaginary part of the retarded Green's function (Eq. 4.53), the former is related to the *dissipation* of the system, in the sense that it contains information on how a particle in a given single-particle state scatters out of that state (see also discussion below). Equation 4.59 then relates the fluctuations at equilibrium (represented by $\mathbf{G}^<$) with the dissipations in the system as quantified by $A(\mathbf{k}; E)$. Relation 4.60 is similar but can be interpreted as referring to *holes* (due to the presence of the term $1-f(E)$), and not electrons.

The non-interacting limit

From the above discussion, it is now easy to obtain the form of the spectral function when interactions are switched off. By comparing Eq. 4.52 with Eq. 3.124 we see that for non-interacting systems the spectral function has the form

$$A(\mathbf{k}; E) = \delta(E - \epsilon_k), \qquad (4.61)$$

where ϵ_k are solutions of a general non-interacting single-particle Hamiltonian (Eq. 4.2).

---

[12] This is clearly a formal statement as, in general, the exact eigenstates of $\hat{H}_0$ are not known.

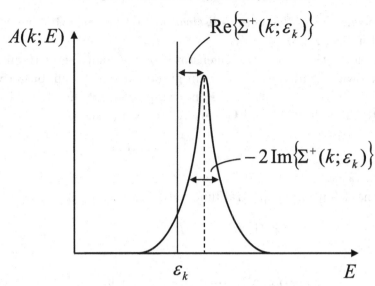

Fig. 4.3. Schematic of the spectral function for quasi-particles. In the absence of interactions, the spectral function is a δ-function centered at the energy of the non-interacting particle energies ϵ_k. Interactions shift the energy of the quasi-particles with respect to the unperturbed energies by an amount given by the real part of the self-energy Σ^+, and broaden the δ-function into a Lorentzian-type of function, with the width proportional to the imaginary part of the self-energy.

The density of states $D(E)$, Eq. 4.57, is then (the Tr here is over all single-particle states with wave-vector **k**)

$$D(E) = \text{Tr}\{A(\mathbf{k}; E)\} = \sum_{\epsilon_k} \delta(E - \epsilon_k), \quad (4.62)$$

which shows peaks in correspondence to the eigenenergies of the noninteracting Hamiltonian (Fig. 4.3).

From all this discussion, and property 4.56, we immediately see that the spectral function is nothing other than a *probability density*. Precisely, it is the probability density for increasing or decreasing the energy of the system by an infinitesimal amount dE via the addition (electron) or subtraction (hole) of a particle in the single-particle state with energy ϵ_k.

For a general many-body system $A(\mathbf{k}; E)$ does not have the simple form 4.61 and, especially for strongly correlated electron systems, most of the time it is not even known. However, for the electron systems – often called weakly interacting systems or Fermi liquids – we consider in this book we expect the *form* of the spectral function to be not so different from the non-interacting case.[13]

Energy "renormalization" and state "lifetime"

Let us now try to give a physical meaning to the self-energy. From the Dyson equation 4.35 the retarded Green's function is

$$\mathbf{G}^+(\mathbf{k}; E) = \frac{1}{E - \epsilon_k - \text{Re}\{\mathbf{\Sigma}^+(\mathbf{k}; E)\} - i\text{Im}\{\mathbf{\Sigma}^+(\mathbf{k}; E)\}} \quad (4.63)$$

with ϵ_k the single-particle energies of the (unperturbed) non-interacting Hamiltonian 4.2. This Green's function has poles in correspondence to the solutions of the equation[14]

$$E - \epsilon_k - \text{Re}\{\mathbf{\Sigma}^+(\mathbf{k}; E)\} - i\text{Im}\{\mathbf{\Sigma}^+(\mathbf{k}; E)\} = 0. \quad (4.64)$$

Quasi-particle approximation

The self-energy $\mathbf{\Sigma}^+(\mathbf{k}; E)$ contains information on *all* possible interactions of the single particle in state $|\phi_k\rangle$, with all other particles in the system. Therefore, without interactions $\text{Re}\{\mathbf{\Sigma}^+(\mathbf{k}; E)\} = \text{Im}\{\mathbf{\Sigma}^+(\mathbf{k}; E)\} = 0$, and the poles of the retarded Green's function occur precisely at the energy $E = \epsilon_k$. To first order we can thus replace this energy into Eq. 4.64 so that the poles occur now at the complex energies

$$E = \epsilon_k + \text{Re}\{\mathbf{\Sigma}^+(\mathbf{k}; \epsilon_k)\} + i\text{Im}\{\mathbf{\Sigma}^+(\mathbf{k}; \epsilon_k)\}, \quad \text{quasi-particle poles.} \quad (4.65)$$

Using the relation 4.53 between the spectral function and the retarded Green's function, together with Eq. 4.63, we then get, to first order,

$$A(\mathbf{k}; E) = -\frac{1}{\pi} \frac{\text{Im}\{\mathbf{\Sigma}^+(\mathbf{k}; \epsilon_k)\}}{[E - \epsilon_k - \text{Re}\{\mathbf{\Sigma}^+(\mathbf{k}; \epsilon_k)\}]^2 + [\text{Im}\{\mathbf{\Sigma}^+(\mathbf{k}; \epsilon_k)\}]^2}. \quad (4.66)$$

The spectral function, which in the case of non-interacting particles is a simple δ-function (Eq. 4.61), in the presence of interactions "broadens" into a Lorentzian function with a corresponding "broadening" of the density of states 4.62, and associated "shift" of the single-particle energies ϵ_k (Fig. 4.3).

---

[13] The energies ϵ_k may instead be quite different from the energies of the corresponding non-interacting system.

[14] It can be proven that, in general, the retarded Green's function $\hat{G}^+$ is analytical in the upper-half of the complex plane while the advanced one, $\hat{G}^-$, in the lower-half complex plane. They may have poles in the respective opposite half planes (Mattuck, 1976).

A measure of this broadening is the quantity (which is positive)

$$\Gamma(\mathbf{k}; \epsilon_k) = -2\mathrm{Im}\{\mathbf{\Sigma}^+(\mathbf{k}; \epsilon_k)\} = \mathrm{i}\left[\mathbf{\Sigma}^+(\mathbf{k}; \epsilon_k) - \mathbf{\Sigma}^-(\mathbf{k}; \epsilon_k)\right], \qquad (4.67)$$

which generalizes to the interacting case the "broadening" of the single-particle non-interacting states due to the presence of the electrodes (Eq. 3.163), as discussed in Sec. 3.5.

The corresponding states are called *quasi-particle* (or *dressed*) states with energies *renormalized* by the interactions

$$\epsilon_k \rightarrow \epsilon_k + \mathrm{Re}\{\mathbf{\Sigma}^+(\mathbf{k}; \epsilon_k)\}, \qquad \text{energy renormalization}, \qquad (4.68)$$

and the unperturbed states acquire a *lifetime*[15]

$$\tau_k = -\frac{\hbar}{\mathrm{Im}\{\mathbf{\Sigma}^+(\mathbf{k}; \epsilon_k)\}} = \frac{2\hbar}{\Gamma(\mathbf{k}; \epsilon_k)}, \qquad \text{state lifetime}. \qquad (4.69)$$

It can be shown (see, e.g., Mattuck, 1976) that for a homogeneous Fermi system at equilibrium the lifetime of a state close to the Fermi level E_F is $\tau_k(E) \sim (\epsilon_k - E_F)^{-2}$. At finite temperature θ_e, electrons within $k_B \theta_e$ from the Fermi energy have thus a relaxation time that scales with temperature as $\tau_k \sim 1/\theta_e^2$.[16]

When the energy of the state is too low or too large compared to E_F, we are not justified in making approximation 4.65 (the lifetime is very short), and the concept of quasi-particle is no longer valid.[17] However, as we have discussed at length, for the current (at least in the linear-response regime), only states close to the Fermi energy are important, so that the quasi-particle picture is meaningful.

Mean-field approximation

I am now in position to better clarify the concept of *mean-field approximation* that I have used several times in this book, without ever discussing its formal definition.

---

[15] In reality, the quasi-particle acquires also a weight $0 < Z_k \leq 1$ which mainly renormalizes the spectral weight at the Fermi energy. Due to this weight, the numerators of Eqs. 4.63 and 4.66 contain such numerical factors. In addition, to satisfy the sum rule 4.56 an extra function $\mathbf{F}(\mathbf{k}; E)$ needs to be added to Eqs 4.63 and 4.66 (see, e.g., Mattuck, 1976).

[16] At room temperature this relaxation time is of the order of 10^{-10} s, while the electron-phonon one is of the order of 10^{-14} s (see, e.g., Ashcroft and Mermin, 1976). By lowering the temperature, $\tau_k(E_F)$ increases due to a reduced scattering between quasi-particles at the Fermi level. In Appendix K I argue that the presence of the nanojunction puts an effective cutoff to this zero-temperature divergence.

[17] It does not make any sense to talk about a particle "dressed" by the interactions, if such particle has a very short "lifetime" to begin with.

A mean-field approximation (e.g., the Hartree approximation to the many-body Coulomb interaction, see Sec. 2.4) corresponds to quasi-particles that have *infinite lifetime*, i.e.,

$$\text{Im}\{\hat{\Sigma}^+(\mathbf{k};\epsilon_k)\} = 0 \implies \tau_k \to \infty, \quad \text{mean-field approximation}, \quad (4.70)$$

and for which the real part of the self-energy is independent of the energy E, and may depend only on the "unperturbed" state ϕ_k (call this quantity $\Delta_k \equiv \text{Re}\{\hat{\Sigma}^+(\mathbf{k};\epsilon_k)\}$). The quasi-particles in a mean-field approximation thus have an energy

$$\tilde{\epsilon}_k = \epsilon_k + \Delta_k, \quad \text{mean-field approximation}, \quad (4.71)$$

with $\tilde{\epsilon}_k$ eigensolutions of an effective single-particle Hamiltonian (e.g., the Hartree Hamiltonian 3.10).

Since their lifetime is infinite (zero imaginary part of the self-energy), the spectral function of these quasi-particles is (taking the limit $\text{Im}\{\hat{\Sigma}_k^+(\mathbf{k};\epsilon_k)\} \to 0^+$ in Eq. 4.66, and using the definition of δ-function, Eq. 3.126),

$$A(\mathbf{k}; E) = \delta(E - \tilde{\epsilon}_k), \quad \text{mean-field approximation}, \quad (4.72)$$

i.e., it is of the same form as the corresponding quantity for non-interacting particles, Eq. 4.61. Therefore, as I have already anticipated, there is *no formal difference* between truly non-interacting particles, and a single particle interacting with the mean field generated by the other particles (the difference being only in their energy spectrum). The form 4.72 also shows that in the mean-field approximation there is a well-defined (one-to-one) correspondence between energy and momentum (as for a classical particle). This correspondence is no longer true when interactions among particles are considered beyond mean field.

4.3 Contour-ordered Green's functions

I am now ready to move on to the non-equilibrium formalism. The non-equilibrium theory is formally equivalent to the equilibrium one, with the only difference that now the Green's functions (and corresponding equations of motion) are defined on a time contour where many-body perturbation theory can be applied.

From *Assumption 3*, at time $t = t_0$ we switch on a time-dependent perturbation adiabatically. We then define the following *contour-ordered Green's function*

$$\mathbf{G}(\mathbf{r},t;\mathbf{r}',t') = -\frac{i}{\hbar}\langle T_C[\psi_{\mathbf{H}}(\mathbf{r},t)\psi_{\mathbf{H}}^{\dagger}(\mathbf{r}',t')]\rangle$$

$$= -\frac{i}{\hbar}\mathrm{Tr}\{\hat{\rho}(t=t_0)\,T_C[\psi_{\mathbf{H}}(\mathbf{r},t)\psi_{\mathbf{H}}^{\dagger}(\mathbf{r}',t')]\}. \quad (4.73)$$

Here the field operators are written in the Heisenberg picture (Eq. 4.25) but with respect to the full Hamiltonian $\hat{\mathbf{H}}(t)$, not $\hat{\mathbf{H}}_0$, while the ensemble average is performed with the equilibrium statistical operator 4.3. The symbol T_C indicates the operation that time-orders the field operators on the contour C that goes from time t_0 and back to time t_0 of Fig. 4.2.

When one writes such a Green's function in terms of the time-evolution operators of the *unperturbed* Hamiltonian $\hat{\mathbf{H}}_0$, as done for the expectation value of a given operator (Eq. 4.11), the time variable may expand into the complex plane when the contour goes from t_0 to $t_0 - i\hbar\beta$ to take into account the initial correlations (see Fig. 4.2). Since now the different times lie on the two different sections of the contour C_1 and C_2 (see Fig. 4.2), the contour-ordered Green's function 4.73 represents four Green's functions

$$\mathbf{G}(\mathbf{r},t;\mathbf{r}',t') = \begin{cases} \mathbf{G}^C(\mathbf{r},t;\mathbf{r}',t'), & t,t' \in C_1, \\ \mathbf{G}^{>}(\mathbf{r},t;\mathbf{r}',t'), & t \in C_2, t' \in C_1, \\ \mathbf{G}^{<}(\mathbf{r},t;\mathbf{r}',t'), & t \in C_1, t' \in C_2, \\ \mathbf{G}^{\tilde{C}}(\mathbf{r},t;\mathbf{r}',t'), & t,t' \in C_2, \end{cases} \quad (4.74)$$

which it is convenient to write in matrix form as

$$\hat{\mathbf{G}} = \begin{bmatrix} \hat{\mathbf{G}}^C & -\hat{\mathbf{G}}^{<} \\ \hat{\mathbf{G}}^{>} & -\hat{\mathbf{G}}^{\tilde{C}} \end{bmatrix}. \quad (4.75)$$

In 4.74 $\hat{\mathbf{G}}^C$ is the time-ordered Green's function we have similarly defined in Eq. 4.26 for the equilibrium case, $\hat{\mathbf{G}}^{<}$ and $\hat{\mathbf{G}}^{>}$ the lesser and greater Green's functions which in equilibrium are 4.44 and 4.45, respectively, and $\hat{\mathbf{G}}^{\tilde{C}}$ is the *anti-time-ordered* Green's function, i.e., the Green's function $\hat{\mathbf{G}}^C$ where the time-ordering operation has the times reversed.

Following their definition, the Green's functions $\hat{\mathbf{G}}^C$ and $\hat{\mathbf{G}}^{\tilde{C}}$ can be written as

$$\mathbf{G}^C(\mathbf{r},t;\mathbf{r}',t') = \Theta(t-t')\mathbf{G}^{>}(\mathbf{r},t;\mathbf{r}',t') + \Theta(t'-t)\mathbf{G}^{<}(\mathbf{r},t;\mathbf{r}',t'), \quad (4.76)$$

$$\mathbf{G}^{\tilde{C}}(\mathbf{r},t;\mathbf{r}',t') = \Theta(t-t')\mathbf{G}^{<}(\mathbf{r},t;\mathbf{r}',t') + \Theta(t'-t)\mathbf{G}^{>}(\mathbf{r},t;\mathbf{r}',t'), \quad (4.77)$$

and thus satisfy the relation $\hat{\mathbf{G}}^C + \hat{\mathbf{G}}^{\tilde{C}} = \hat{\mathbf{G}}^{<} + \hat{\mathbf{G}}^{>}$, so that only three of them are linearly independent.

4.3 Contour-ordered Green's functions

Using the lesser and greater Green's functions we can also define non-equilibrium retarded and advanced Green's functions:

$$\mathbf{G}^+(\mathbf{r},t;\mathbf{r}',t') = \Theta(t-t')[\mathbf{G}^>(\mathbf{r},t;\mathbf{r}',t') - \mathbf{G}^<(\mathbf{r},t;\mathbf{r}',t')], \quad \text{retarded}, \quad (4.78)$$

$$\mathbf{G}^-(\mathbf{r},t;\mathbf{r}',t') = \Theta(t'-t)[\mathbf{G}^<(\mathbf{r},t;\mathbf{r}',t') - \mathbf{G}^>(\mathbf{r},t;\mathbf{r}',t')], \quad \text{advanced}, \quad (4.79)$$

which, as in the equilibrium case (see Eq. 4.51), satisfy the relation $\hat{G}^+ - \hat{G}^- = \hat{G}^> - \hat{G}^<$.

4.3.1 Equations of motion for non-equilibrium Green's functions

From the above definitions we see that the *form* of the contour-ordered Green's function 4.74 is similar to the equilibrium one. Therefore, a Dyson's equation similar to the equilibrium Eq. 4.33 must exist for the non-equilibrium Green's functions as well. Let us write it down.

In most practical cases the time-dependent perturbation $\hat{H}'(t)$ is written as a one-body potential $\hat{H}'(\mathbf{r})$, i.e., in second quantization (see Eq. A.26)

$$\hat{H}'(\mathbf{r},t) = \int d\mathbf{r}\, \psi^\dagger(\mathbf{r},t)\, H'(\mathbf{r})\, \psi(\mathbf{r},t). \tag{4.80}$$

With this potential and the free-particle Green's function $\hat{G}_0$ (Eq. 4.32), the Dyson equation for the non-equilibrium problem is thus (using again the abbreviated notation $(i) \equiv (\mathbf{r}_i, t_i)$, where the times may be complex)

$$\begin{aligned}
\mathbf{G}(1;1') &= G_0(1;1') + \int_{C'} d2\, G_0(1;2) H'(2) \mathbf{G}(2;1') \\
&\quad + \int_C d2 \int_C d3\, G_0(1;2) \mathbf{\Sigma}(2;3) \mathbf{G}(3;1') \\
&= G_0(1;1') + \int_{C'} d2\, \mathbf{G}(1;2) H'(2) G_0(2;1') \\
&\quad + \int_C d2 \int_C d3\, \mathbf{G}(1;2) \mathbf{\Sigma}(2;3) G_0(3;1'), \tag{4.81}
\end{aligned}$$

where the contour C is the complete contour of Fig. 4.2, that goes from t_0 to t'_1, back to t_0 and into the complex time plane till $t_0 - i\hbar\beta$, while the contour C' only goes from t_0 to t'_1, and back to t_0 along the real-time axis.[18]

---

[18] If the time t_0 is let to go to $-\infty$, and the interaction potential is still switched on adiabatically at $t_0 \to -\infty$, then the contribution to the time integration in Eq. 4.81 from the path that goes from t_0 to $t_0 - i\hbar\beta$ (see Fig. 4.2) can be neglected, and the two contours, C and C', coincide. This approximation neglects any transient phenomena, and the initial correlations of the system.

With the use of Eq. 4.32 the above equation can be equivalently written as

$$\left\{i\hbar\frac{\partial}{\partial t} + \frac{\hbar^2}{2m}\nabla_\mathbf{r}^2 - H'(\mathbf{r},t)\right\}\mathbf{G}(\mathbf{r},t;\mathbf{r}',t') = \delta(\mathbf{r}-\mathbf{r}')\delta(t-t')$$
$$+ \int d\mathbf{r}'' \int_C dt''\, \mathbf{\Sigma}(\mathbf{r},t;\mathbf{r}'',t'')\, \mathbf{G}(\mathbf{r}'',t'';\mathbf{r}',t') \qquad (4.82)$$

and

$$\left\{-i\hbar\frac{\partial}{\partial t'} + \frac{\hbar^2}{2m}\nabla_{\mathbf{r}'}^2 - H'(\mathbf{r}',t')\right\}\mathbf{G}(\mathbf{r},t;\mathbf{r}',t') = \delta(\mathbf{r}-\mathbf{r}')\delta(t-t')$$
$$+ \int d\mathbf{r}'' \int_C dt''\, \mathbf{G}(\mathbf{r},t;\mathbf{r}'',t'')\, \mathbf{\Sigma}(\mathbf{r}'',t'';\mathbf{r}',t'). \qquad (4.83)$$

Analogously to Eq. 4.75, the self-energy can be written in matrix form as

$$\hat{\mathbf{\Sigma}} = \begin{bmatrix} \hat{\mathbf{\Sigma}}^C & -\hat{\mathbf{\Sigma}}^< \\ \hat{\mathbf{\Sigma}}^> & -\hat{\mathbf{\Sigma}}^{\bar{C}} \end{bmatrix}, \qquad (4.84)$$

where

$$\mathbf{\Sigma}^C(\mathbf{r},t;\mathbf{r}',t') = \Theta(t-t')\mathbf{\Sigma}^>(\mathbf{r},t;\mathbf{r}',t') + \Theta(t'-t)\mathbf{\Sigma}^<(\mathbf{r},t;\mathbf{r}',t'), \quad (4.85)$$

$$\mathbf{\Sigma}^{\bar{C}}(\mathbf{r},t;\mathbf{r}',t') = \Theta(t-t')\mathbf{\Sigma}^<(\mathbf{r},t;\mathbf{r}',t') + \Theta(t'-t)\mathbf{\Sigma}^>(\mathbf{r},t;\mathbf{r}',t'). \quad (4.86)$$

The functions $\hat{\mathbf{\Sigma}}^<$ and $\hat{\mathbf{\Sigma}}^>$ are called *lesser* and *greater* self-energy, respectively, and are related to the irreducible self-energy $\hat{\mathbf{\Sigma}}$ appearing in Eq. 4.81 via

$$\mathbf{\Sigma}^<(\mathbf{r_1},t_1;\mathbf{r_2},t_2) = \mathbf{\Sigma}(\mathbf{r_1},t_1;\mathbf{r_2},t_2), \qquad t_1 < t_2, \qquad (4.87)$$

and

$$\mathbf{\Sigma}^>(\mathbf{r_1},t_1;\mathbf{r_2},t_2) = \mathbf{\Sigma}(\mathbf{r_1},t_1;\mathbf{r_2},t_2), \qquad t_1 > t_2. \qquad (4.88)$$

At *equilibrium*, these self-energies are related to the retarded $\hat{\mathbf{\Sigma}}^+$ and advanced $\hat{\mathbf{\Sigma}}^-$ self-energies, via the fluctuation-dissipation theorem (Eqs. 4.59 and 4.60), which reads (assuming the system is also homogeneous)

$$\mathbf{\Sigma}^<(\mathbf{k};E) = -2i\, f(E)\, \mathrm{Im}\{\mathbf{\Sigma}^+(\mathbf{k};E)\},$$
$$\mathbf{\Sigma}^>(\mathbf{k};E) = 2i\, [1-f(E)]\, \mathrm{Im}\{\mathbf{\Sigma}^+(\mathbf{k};E)\}. \qquad (4.89)$$

From the above equations we see that

$$i\left[\mathbf{\Sigma}^>(\mathbf{k};E) - \mathbf{\Sigma}^<(\mathbf{k};E)\right] = i\left[\mathbf{\Sigma}^+(\mathbf{k};E) - \mathbf{\Sigma}^-(\mathbf{k};E)\right]$$
$$= -2\mathrm{Im}\{\mathbf{\Sigma}^+(\mathbf{k};E)\} \equiv \mathbf{\Gamma}(\mathbf{k};E). \qquad (4.90)$$

4.3 Contour-ordered Green's functions

(I will write the explicit form of $\hat{\Sigma}^<$ for non-interacting particles in steady-state transport in Sec. 4.4.) Recalling the definition 4.69 of state lifetime in the quasi-particle approximation, we also see from Eq. 4.89 that $\Sigma^</\hbar$ and $\Sigma^>/\hbar$ must define some sort of "rate" for particles to scatter in and out of single particle states. In Sec. 4.6 I will provide a more precise meaning for these scattering rates when discussing the quantum Boltzmann equation.

Equation of motion for $\hat{G}^<$

As I have discussed in Sec. 4.2.3, the lesser Green's function is related to the density and current density of the system. Its equations of motion are thus central in transport theories. Here I write these equations, which I will use later on.

Using the matrix notation for Green's functions and self-energies as in Eqs. 4.75 and 4.84, and the equations of motion for the contour-ordered Green's function $\mathbf{G}$ (Eqs. 4.82 and 4.83), it is straightforward to find (Exercise 4.4)

$$\left(i\hbar\frac{\partial}{\partial t_1} + \frac{\hbar^2}{2m}\nabla^2_{\mathbf{r}_1} - H'(\mathbf{r}_1, t_1)\right)\mathbf{G}^<(\mathbf{r}_1, t_1; \mathbf{r}_2, t_2)$$
$$= \{\mathbf{\Sigma}^+\mathbf{G}^< + \mathbf{\Sigma}^<\mathbf{G}^+\}(\mathbf{r}_1, t_1; \mathbf{r}_2, t_2), \qquad (4.91)$$

and

$$\left(-i\hbar\frac{\partial}{\partial t_2} + \frac{\hbar^2}{2m}\nabla^2_{\mathbf{r}_2} - H'(\mathbf{r}_2, t_2)\right)\mathbf{G}^<(\mathbf{r}_1, t_1; \mathbf{r}_2, t_2)$$
$$= \{\mathbf{G}^+\mathbf{\Sigma}^< + \mathbf{G}^<\mathbf{\Sigma}^-\}(\mathbf{r}_1, t_1; \mathbf{r}_2, t_2), \qquad (4.92)$$

where, for two general functions $\mathbf{A}(\mathbf{r}_1, t_1; \mathbf{r}_2, t_2)$ and $\mathbf{B}(\mathbf{r}_1, t_1; \mathbf{r}_2, t_2)$, the symbol $\{\cdots\}$ indicates the convolution

$$\{\mathbf{AB}\}(\mathbf{r}_1, t_1; \mathbf{r}_2, t_2) = \int dt \int d\mathbf{r}\, \mathbf{A}(\mathbf{r}_1, t_1; \mathbf{r}, t)\mathbf{B}(\mathbf{r}, t; \mathbf{r}_2, t_2). \qquad (4.93)$$

Finally, if we "absorb" into the unperturbed Green's function $\hat{G}_0$ also the action of the external potential, namely we define it as

$$\left(i\hbar\frac{d}{dt} + \frac{\hbar^2}{2m}\nabla^2_{\mathbf{r}} - H'(\mathbf{r}, t)\right)G_0(\mathbf{r}, t; \mathbf{r}', t') = \delta(\mathbf{r} - \mathbf{r}')\delta(t - t'), \qquad (4.94)$$

from 4.81, one can write the equation of motion for the lesser Green's function as (Mahan, 1990)

$$\hat{\mathbf{G}}^< = (\hat{1} + \hat{\mathbf{G}}^+ \hat{\mathbf{\Sigma}}^+) \hat{G}_0^< (\hat{1} + \hat{\mathbf{\Sigma}}^- \hat{\mathbf{G}}^-) + \hat{\mathbf{G}}^+ \hat{\mathbf{\Sigma}}^< \hat{\mathbf{G}}^-, \qquad (4.95)$$

where the product of two functions actually means the integral of their mute variables as in Eq. 4.81, and $\hat{G}_0^<$ is equal to $G_0(\mathbf{r}, t; \mathbf{r}', t')$ for $t < t'$. The above equation is known as the *Keldysh equation* (Keldysh, 1964), and an equivalent one also holds for $\hat{\mathbf{G}}^>$.

We now have all the tools to study transport in nanoscale systems using the NEGF. In principle, we have the tools to study the dynamics of any closed many-body quantum system subject to an arbitrary deterministic perturbation, provided that the perturbation is switched on adiabatically, and the non-equilibrium problem is such that many-body perturbation theory is applicable. All many-body interactions are embedded in self-energies, and once these are known, all physical properties of interest can be calculated from appropriate Green's functions. In particular, the number density, current density, and spectral function can be calculated via Eqs. 4.46, 4.50 and 4.52, respectively.

I want to stress here that, if we knew *exactly* the lesser Green's function $\hat{\mathbf{G}}^<$ of a closed quantum system subject to an external perturbation, namely we knew it to all orders of many-body perturbation theory, the current density, and hence the total current across any give surface, would be *exact*. Clearly, this is just a formal statement, and indeed we need specific approximations to deal with particular cases. One of these is the following.

4.4 Application to steady-state transport

Let us apply the NEGF to the transport problem. We limit ourselves to the case of steady-state transport. Some examples of applications of this formalism to time-dependent phenomena can be found in Haug and Jahuo (1996). For steady-state transport we employ most of the approximations of the Landauer approach (Sec. 3.1), and few more. In particular, we consider

Approximations 1 to 3 of the Landauer approach

We assume that the electron sources are first replaced by the dynamical interaction with reservoirs (*Approximation 1: open quantum systems*). We then assume that we work in an ideal steady state (*Approximation 2: ideal steady state*), namely the statistical operator admits, in the long-time limit, a stationary solution as in Eq. 3.5. We subsequently replace the open quantum problem with a *closed* (and infinite) one with appropriate scattering-boundary conditions (*Approximation 3: "openness" vs. boundary conditions*). The system is thus composed of two semi-infinite leads, which connect to a central sample (our nanojunction). The leads are such that at their infinite boundary they "accept" electrons with different local distribution functions (Eqs. 3.14 and 3.15), with the electrochemical potential difference given by the bias, $V = (\mu_L - \mu_R)/e$. At this point, we have lost, *irreversibly*, information on the dynamics of the true system, electron sources plus nanostructures, and as in the Landauer approach, the amount of loss of information cannot be determined, not even in principle (see discussion in Sec. 3.1).[19]

Approximation 4: *Non-interacting electrons in the leads*

We make an additional mean-field approximation to handle electron interactions in the system (*Approximation 4: mean-field approximation* of the Landauer approach). Unlike the Landauer approach, however, we limit the mean-field approximation to the leads, and allow interactions, beyond mean field, inside the sample region. As I have anticipated, no closed form for the single-particle Green's functions (and hence for the current) can be obtained if interactions are also taken into account inside the leads.

These approximations are the starting point for a steady-state theory of transport within the NEGF. If no particular partition is made on the system (Cini, 1980), one may assume that the above approximations occurred at $t_0 \rightarrow -\infty$, with the additional assumption that at that time the external perturbation that creates the electrochemical potential difference $\mu_L - \mu_R$ deep into the two electrodes has been switched on adiabatically, and that this adiabatic switch-on leads to a steady state *independent* of the initial correlations.

Instead of the above partition-free procedure, I will employ the partition approach (Caroli *et al.*, 1971) represented in Fig. 4.1.[20] In addition to the above approximations, we then need to assume that at $t_0 \rightarrow -\infty$ the two leads are separated from each other and from the central region (see Fig. 3.11).

---

[19] Note that even if we did not make these approximations, and could determine the *exact* single-particle Green's functions, we would have information only on the single-particle degrees of freedom. As I have discussed in Sec. 2.8.3 this implies a loss of information with respect to the dynamics of the many-body system, unless one recovers, using, e.g., dynamical density-functional theories, the *exact* many-body wave-function.

[20] The derivation of the electrical current follows closely the one in Meir and Wingreen (1992); see also Haug and Jauho (1996).

In the infinite past, electrons in the leads are at equilibrium at their own electrochemical potential (the central region is also at its own equilibrium), but current is not allowed to flow between them. As I have already discussed, this is *not* what happens in an experimental realization of this problem, but it allows us to write down a closed equation for the current.

The left and right lead Hamiltonians have thus the following (mean-field) form

$$\hat{H}_{LR} = \sum_{k;b \in L,R} E_{kb} \, \hat{a}^{\dagger}_{kb} \hat{a}_{kb}, \qquad (4.96)$$

with $\hat{a}^{\dagger}_{kb}$, $\hat{a}_{kb}$ the operators that create and destroy, respectively, an electron in the state (channel) with energy E_{kb} in $b = L$ or $b = R$ electrode. To simplify the notation, I have also assumed that the wave-vector k takes only discrete values, but the results can be easily generalized to a continuum representation.

The Hamiltonian of the central region describes interacting electrons, and its form thus varies according to the model chosen to describe these interactions. If the electrons interact only with each other (no interaction with other excitations, e.g., phonons) then quite generally we can assume that this Hamiltonian is a function of a complete set of single-particle creation $\{\hat{c}^{\dagger}_n\}$ and annihilation $\{\hat{c}_n\}$ operators in the interacting region.[21] We indicate this by just writing

$$\hat{\mathbf{H}}_C = \hat{\mathbf{H}}_C(\{\hat{c}^{\dagger}_n\}; \{\hat{c}_n\}). \qquad (4.97)$$

The coupling between the central region and the electrodes is defined as

$$\hat{\mathbf{V}}_{bC} = \sum_{n,k} V_{n,kb} \, \hat{a}^{\dagger}_{kb} \, \hat{c}_n, \qquad (4.98)$$

with $b \in L, R$ and $V_{n,kb}$ some complex numbers.[22]

The above choice does not allow for direct tunneling between the L and R electrodes. The total Hamiltonian is thus (note the similar partition we

---

[21] As we will see in a moment this is enough to apply the NEGF technique and obtain a general *form* for the current. Its actual value clearly depends on the details of the interaction Hamiltonian. Also, as I will discuss in Sec. 6.2.2, the derivation can be similarly carried out in the case in which the electrons interact with other excitations, like phonons.

[22] Note that all the parameters appearing in this model should be determined self-consistently as they depend on the detailed electronic distribution under current flow.

have made in Eq. 3.137 in the Landauer approach)

$$\begin{aligned}\hat{H} &= \hat{H}_L + \hat{H}_R + \hat{H}_C + \hat{V}_{LC} + \hat{V}_{RC} + \hat{V}^\dagger_{LC} + \hat{V}^\dagger_{RC} \\ &= \sum_{k,b \in L,R} E_{kb}\, \hat{a}^\dagger_{kb} \hat{a}_{kb} + \mathbf{H}_C(\{\hat{c}^\dagger_n\}; \{\hat{c}_n\}) + \sum_{nk; b \in L,R} (V_{n,kb}\hat{a}^\dagger_{kb}\hat{c}_n + V^*_{kb,n}\hat{c}^\dagger_n \hat{a}_{kb}). \end{aligned} \qquad (4.99)$$

Since we are working in an ideal steady-state, all Green's functions may depend on time differences only, and we Fourier transform them in the energy domain. With this in mind, the Green's functions of the two isolated non-interacting leads, $\hat{g}_{kb}$, can be simply calculated using the equilibrium theory of Sec. 4.2. For instance, the lesser Green's function can be obtained from Eqs. 4.59 and 4.61

$$g^<_{kb}(E) = 2\pi \mathrm{i} f_b(E) \delta(E - E_{kb}), \qquad (4.100)$$

where $f_b(E)$ is the local Fermi distribution 4.23 in each of the two leads.

In the central region we define the single-particle time-ordered Green's function (from Eq. 4.26)

$$\mathbf{G}_{nm}(t - t') = -\frac{\mathrm{i}}{\hbar} \langle \mathrm{T}[\hat{c}^\dagger_m(t')\hat{c}_n(t)] \rangle, \qquad (4.101)$$

where n and m are the quantum numbers defining two arbitrary states of the central interacting region. From this we can define the corresponding retarded $\hat{\mathbf{G}}^+_{nm}$, and lesser $\hat{\mathbf{G}}^<_{nm}$ Green's functions according to Eqs. 4.40 and 4.44, respectively.

In Eq. 4.50 we have shown that the current density (and hence the total current) can be expressed in terms of the lesser Green's function. That expression, however, is less convenient in the present case, and instead we adopt a different but equivalent definition of current (see Footnote [22] in Sec. 1.3.2).

The current that flows from the L electrode into the central region is simply given by the expectation value of the rate of change of the total number operator (see Eq. A.13)

$$\hat{N}_L = \sum_{k,b \in L} \hat{a}^\dagger_{kb} \hat{a}_{kb} \qquad (4.102)$$

that counts electrons in the L electrode[23]

$$I_L = -e\left\langle \frac{d\hat{N}_L}{dt} \right\rangle = -\frac{ie}{\hbar}\langle[\hat{H},\hat{N}_L]\rangle = -\frac{ie}{\hbar}\langle[(\hat{V}_{LC}+\hat{V}_{LC}^\dagger),\hat{N}_L]\rangle, \quad (4.103)$$

where the second equality is from the Heisenberg equation of motion for the operators in that representation (Eq. B.3), and the last one is due to the fact that $\hat{H}_C$, $\hat{H}_L$, $\hat{H}_R$, and $\hat{V}_{RC}$ all commute with $\hat{N}_L$. The average is performed, as usual, over the equilibrium statistical operator. Note that in the above expression for the current I have not included the spin degeneracy. I will do that at the end of the calculations.

By replacing 4.98 into 4.103, and using the commutation relations of the creation and annihilation operators (Eqs. A.17 and A.18) we obtain

$$I_L = \frac{ie}{\hbar}\sum_{nk;b\in L}(V_{n,kb}\langle \hat{a}_{kb}^\dagger \hat{c}_n\rangle - V_{kb,n}^*\langle \hat{c}_n^\dagger \hat{a}_{kb}\rangle). \quad (4.104)$$

By comparing Eq. 4.44 with the averaged quantities in 4.104 we see that the latter are a type of lesser Green's functions. We then define

$$\mathbf{G}^<{}_{n,kb}(t-t') = \frac{i}{\hbar}\langle \hat{a}_{kb}^\dagger(t')\hat{c}_n(t)\rangle, \quad (4.105)$$

and

$$\mathbf{G}^<{}_{kb,n}(t-t') = \frac{i}{\hbar}\langle \hat{c}_n^\dagger(t')\hat{a}_{kb}(t)\rangle, \quad (4.106)$$

which satisfy the property $\mathbf{G}^<_{kb,n}(t=t') = -[\mathbf{G}^<_{n,kb}(t=t')]^*$, or in Fourier space $\mathbf{G}^<_{kb,n}(E) = -[\mathbf{G}^<_{n,kb}(E)]^*$. Using this property, and working in Fourier space, the current 4.104 is simply

$$I_L = \frac{2e}{\hbar}\int \frac{dE}{2\pi}\mathrm{Re}\left\{\sum_{n;k,b\in L} V_{n,kb}\mathbf{G}^<{}_{n;kb}(E)\right\}. \quad (4.107)$$

We are therefore left to determine $\hat{\mathbf{G}}^<_{n,kb}$. This can be obtained from the equation of motion 4.28 for the time-ordered Green's function $\mathbf{G}_{n,kb}(t-t') = \frac{i}{\hbar}\langle T[a_{kb}^\dagger(t')c_n(t)]\rangle$. In this case no infinite BBGKY hierarchy is generated (Appendix B) because we have assumed the leads non-interacting. The

---

[23] The definition of the current 4.103 provides a finite value only if there is no direct tunneling between the leads or, if that is the case, the electrodes have (i) finite cross section or, (ii) a cross section that, by moving towards the interior of the electrodes, increases slower than the exponential decay of the direct tunneling current across the electrodes. Otherwise the current is infinite.

equation of motion for $\mathbf{G}_{n,kb}(t-t')$ is (from Eq. 4.28)

$$-i\hbar\frac{\partial}{\partial t'}\mathbf{G}_{n,kb}(t-t') = \delta_{n,kb}\delta(t-t') + E_{kb}\mathbf{G}_{n,kb}(t-t') + \sum_{m}\mathbf{G}_{n,m}(t-t')V^*_{kb,m}, \quad (4.108)$$

which shows that $\hat{\mathbf{G}}_{n,kb}$ is written in terms of the central region Green's function, and the coupling between the central region and the leads.

From here, one needs to calculate the lesser Green's function $\hat{\mathbf{G}}^<_{n,kb}$, by first defining the contour-ordered version of $\hat{\mathbf{G}}_{n,kb}$ (see Eq. 4.73), and then obtaining the lesser Green's function from the correct segments of the contour (Eq. 4.74). Here I just give the final result and refer the reader to Haug and Jauho (1996) for the full derivation. One finds

$$\mathbf{G}^<_{n;k,b}(E) = \sum_m V^*_{kb,m}\left[\mathbf{G}^+_{nm}(E)g^<_{kb}(E) + \mathbf{G}^<_{nm}(E)g^-_{kb}(E)\right], \quad (4.109)$$

where $\hat{g}^-_{kb}$ is the advanced Green's function of the leads.

Introducing this result into Eq. 4.107 we obtain

$$I_L = \frac{2e}{\hbar}\int\frac{dE}{2\pi}\text{Re}\left\{\sum_{n,m;k,b\in L}V_{n,kb}V^*_{m,kb}\left[\mathbf{G}^+_{mn}(E)g^<_{kb}(E) + \mathbf{G}^<_{nm}(E)g^-_{kb}(E)\right]\right\}. \quad (4.110)$$

We can write this last equation in a compact form. Using the relation 3.48 for the density of states per channel, we can transform the momentum integration into an energy integration. We thus introduce the matrix

$$\mathbf{\Gamma}_L \equiv \{\Gamma_L(E_k)\}_{mn} = 2\pi\sum_{b\in L}D_b(E_k)V_{n,b}(E_k)V^*_{b,m}(E_k), \quad (4.111)$$

and similarly for the R electrode. Using Eq. 4.100, and the properties 3.84 and 4.51, the current in the L lead can be written as (Exercise 4.5)

$$I_L = \frac{ie}{\hbar}\int\frac{dE}{2\pi}\text{Tr}\left\{\mathbf{\Gamma}_L(E)\left[\mathbf{G}^<(E) + f_L(E)(\mathbf{G}^+(E) - \mathbf{G}^-(E))\right]\right\}, \quad (4.112)$$

with $\mathbf{G}^<(E)$, $\mathbf{G}^+(E)$ and $\mathbf{G}^-(E)$ indicating the lesser, retarded and advanced single-particle Green's functions of the central region *in the presence of the coupling with the leads*.

Following the same procedure we can calculate the current flowing from the R electrode into the central region, which in an ideal steady state is $I_R = -I_L$. The total current can then be written in the symmetrized version

$I = (I_L - I_R)/2$, or (I now include spin degeneracy)

$$I = 2\frac{ie}{2\hbar} \int \frac{dE}{2\pi} \text{Tr}\left\{[\mathbf{\Gamma}_L(E) - \mathbf{\Gamma}_R(E)]\mathbf{G}^<(E) \right. $$
$$\left. + [f_L(E)\mathbf{\Gamma}_L(E) - f_R(E)\mathbf{\Gamma}_R(E)][\mathbf{G}^+(E) - \mathbf{G}^-(E)]\right\}.$$
(4.113)

This equation will be our starting point to develop approximate theories of steady-state transport in the presence of interactions in the central region. Let us make some remarks about this result.

First of all, I stress again the approximate nature of Eq. 4.113, that we could derive only with non-interacting electrons in the leads. I also notice that, in the presence of interactions in the central region, the *form* of the current is very different than in the non-interacting case (cf. Eqs. 4.113 and 3.229). In particular, the concept of transmission coefficient we have defined in the non-interacting case is lost (we cannot simply say that the argument in the integral of Eq. 4.113 is a transmission coefficient like the one we have introduced in Chapter 3). For instance, one cannot relate the elements t_{fi} of the T matrix (see Sec. 3.5.1) to single-particle wavefunctions as we do in the non-interacting case. This shows that, even within the simplifications that led us to Eq. 4.113, in the presence of interactions there are contributions to the single-particle Landauer current 3.229 that cannot be included in the single-particle transmission coefficient (see also Sec. 8.4.2).

Finally, expression 4.113 contains the single-particle Green's functions $\mathbf{G}^<(E)$, $\mathbf{G}^+(E)$ and $\mathbf{G}^-(E)$ of the interacting region in the presence of current flow (i.e., in the steady state but still out of equilibrium). These Green's functions are generally not known, except for few specific cases – some of which will be discussed in Chapter 6.

Proportional coupling

A simplified version of Eq. 4.113 can be obtained if we assume $\mathbf{\Gamma}_L(E) = c\mathbf{\Gamma}_R(E)$ at all energies E, with c an arbitrary constant.[24] Replacing this in Eq. 4.113 we get (spin included)

$$I = \frac{2ie}{\hbar} \int \frac{dE}{2\pi} [f_L(E) - f_R(E)] \text{Tr}\left\{\frac{\mathbf{\Gamma}_L(E)\mathbf{\Gamma}_R(E)}{\mathbf{\Gamma}_L(E) + \mathbf{\Gamma}_R(E)}(\mathbf{G}^+(E) - \mathbf{G}^-(E))\right\}$$
$$= \frac{2e}{\hbar} \int dE\, [f_L(E) - f_R(E)] \text{Tr}\left\{\frac{\mathbf{\Gamma}_L(E)\mathbf{\Gamma}_R(E)}{\mathbf{\Gamma}_L(E) + \mathbf{\Gamma}_R(E)}\mathbf{A}(E)\right\}, \quad (4.114)$$

---

[24] This is a rare occurrence in physical systems. For one thing, it would require no self-consistent charge accumulation at the boundary between the central region and the electrodes.

where in the last equation I have used the definition of spectral function 4.52. Using the general property 4.54 we can write

$$2\pi A(E) = -2\mathbf{G}^+(E)\,\text{Im}\{\mathbf{\Sigma}^+(E)\}\,\mathbf{G}^+(E) \qquad (4.115)$$

and by defining

$$\mathbf{\Gamma}_{LR}(E) \equiv \frac{\mathbf{\Gamma}_L(E)\,\mathbf{\Gamma}_R(E)}{\mathbf{\Gamma}_L(E)+\mathbf{\Gamma}_R(E)}, \qquad (4.116)$$

we can write Eq. 4.114 as

$$I = \frac{e}{\pi\hbar}\int dE\,[f_L(E)-f_R(E)]\,\mathcal{T}_{LR}(E), \qquad (4.117)$$

where the quantity

$$\mathcal{T}_{LR}(E) = -2\,\text{Tr}\left\{\mathbf{\Gamma}_{LR}(E)\,\mathbf{G}^+(E)\,\text{Im}[\mathbf{\Sigma}^+(E)]\,\mathbf{G}^-(E)\right\}, \qquad (4.118)$$

seems similar to the transmission coefficient I have defined in Eq. 3.224. It is thus tempting to *interpret* it as a transmission coefficient. However, as I have discussed above this interpretation is not correct in the presence of interactions, and the similarity between Eq. 4.114 and Eq. 3.229 can be, at best, considered a formal analogy.

The non-interacting case

Finally, let us show that if the electrons in the central region are also non-interacting (or interacting at a mean-field level), we recover the non-interacting scattering expression 3.229 for the current. This must be the case, because in order to treat the steady-state transport problem within the NEGF we have made the *same* approximations used in the Landauer approach, except for the type of interactions in the central region.

In the non-interacting case the central Hamiltonian is of the type

$$\hat{H}_C = H_C(\{\hat{c}_n^\dagger\};\{\hat{c}_n\}) = \sum_n \epsilon_n\,\hat{c}_n^\dagger \hat{c}_n\,. \qquad (4.119)$$

Since we are now working with a true non-interacting problem at steady state, we can calculate the lesser Green's function of the central region directly from the Keldysh equation 4.95. Employing the definition 4.100 of the lesser Green's function for the non-interacting leads and the Dyson equation 4.35, the first term on the right-hand side of Eq. 4.95 vanishes

$$(\hat{1} + \hat{G}^+\hat{\Sigma}^+)\hat{G}_0^< (\hat{1} + \hat{\Sigma}^-\hat{G}^-) \propto \left(\hat{1} + \frac{\hat{\Sigma}^+}{E - E_{kb} - \hat{\Sigma}^+}\right)$$
$$\times \delta(E - E_{kb})\left(\hat{1} + \frac{\hat{\Sigma}^-}{E - E_{kb} - \hat{\Sigma}^-}\right)$$
$$= 0, \tag{4.120}$$

so that Eq. 4.35 is simply

$$G^<(E) = G^+(E)\,\Sigma^<(E)\,G^-(E). \tag{4.121}$$

The lesser self-energy is easily calculated from the equilibrium relation 4.89

$$\Sigma^<(E) = \Sigma_L^<(E) + \Sigma_R^<(E)$$
$$= i[f_L(E)\Gamma_L(E) + f_R(E)\Gamma_R(E)], \tag{4.122}$$

where $\Gamma_L(E)$ and $\Gamma_R(E)$ are the level widths defined in Eq. 3.163.

Introducing Eq. 4.122 into Eq. 4.121 we get

$$G^<(E) = if_L(E)G^+(E)\Gamma_L(E)G^-(E) + if_R(E)G^+(E)\Gamma_R(E)G^-(E). \tag{4.123}$$

To obtain the current, we then replace 4.123 into Eq. 4.113. Using the non-interacting version of relation 4.54, namely Eq. 3.162, which here reads $i(G^+ - G^-) = G^+(\Gamma_L + \Gamma_R)G^-$, and the cyclic property of the trace (Eq. E1.3 in Exercise 1.1), after some manipulation we recover, as expected, the expression 3.229 for the current in the non-interacting case (Exercise 4.6),

$$I = \frac{e}{\pi\hbar}\int_{-\infty}^{+\infty} dE\,[f_L(E) - f_R(E)]\,\mathrm{Tr}\{\hat{\Gamma}_R\hat{G}^+\hat{\Gamma}_L\hat{G}^-\}, \tag{4.124}$$

which we have obtained using single-particle scattering theory, without ever employing the NEGF.

4.5 Coulomb blockade

Let us now discuss an explicit application of the NEGF to an important case. Coulomb interactions among electrons contribute to an interesting phenomenon, known as *Coulomb blockade* (or its complementary *single-electron tunneling*). This phenomenon appears in structures similar to the ones I have discussed in Sec. 3.7.1.1, when describing resonant tunneling. The main difference between resonant tunneling and single-electron tunneling is that the first is an *independent-electron* effect, due to the geometrical confinement of single particles within a double-barrier structure like the one represented in Fig. 3.17. This geometrical confinement leads to an enhanced single-particle transmission probability across the whole structure at specific energies (resonant energies).

On the other hand, the Coulomb blockade effect is a *many-body* non-equilibrium phenomenon, and is related to charge quantization. It is generally realized in double-barrier structures like the one in Fig. 3.17, when there are large tunnel barriers[25] between the leads and the central region, which is equivalent to saying the central region is weakly coupled to the leads. These structures are also called *quantum dots*. Experiments in single-molecule structures have clearly shown Coulomb blockade features as discussed in this section (see, e.g., Yu and Natelson, 2004; and Park *et al.*, 2000).

"Orthodox" picture of Coulomb blockade

To explain this phenomenon in simple terms, consider the configuration in Fig. 4.4, where a central region ("island") is weakly coupled via tunnel barriers to two leads (see also Ferry and Goodnick, 1997). Now let us switch on a bias V between the leads so that current flows.

The central region plus the tunnel barriers have a finite capacitance C, and an associated electrostatic energy $E_Q = Q^2/2C$, where Q is the charge on each "plate" of the ideal capacitor that represents this region.[26]

If I now assume that an electron tunnels from one of the leads into the island, the electrostatic energy of the central region changes

$$E_Q = \frac{Q^2}{2C} \xrightarrow{\text{addition of 1 e}} \frac{(|Q|-|e|)^2}{2C}. \qquad (4.125)$$

We are also assuming that the above condition is *maintained* in time, namely it occurs at steady state (hence the charge inside the island does not equilibrate with the rest of the circuit). This requires a *continuous* supply of electrons from the leads to the central region, so that Coulomb blockade is a purely *non-equilibrium* effect. From this it is clear that ground-state theories, such *ground-state* density-functional theory, *cannot* capture this phenomenon at all (even if we had the exact functional, see also discussion in Sec. 3.11).

From an energetic point of view, the above addition of an electron is

---

[25] Either in their energy "height" or spatial width.

[26] From a microscopic point of view this can occur via the formation of a charge Q inside the island, and compensating image charges in the contact regions in proximity to the junction. I am also assuming for simplicity that the capacitances of the two barriers are identical, and no extra capacitance is involved. It is easy to generalize the discussion to two barriers of different capacitance, and the case in which a gate (and its capacitance) are added to the circuit.

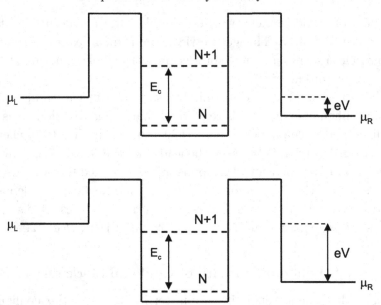

Fig. 4.4. Schematic of the energy diagram of a central region weakly coupled to two electrodes via tunnel junctions. $E_c = e^2/2C$ is the extra electrostatic energy necessary to add one electron to the island. The dotted lines in the central region indicate the energy of the many-body states of N (lower in energy) and $N+1$ (higher in energy) electrons. In the top panel, the bias is not enough to overcome the charging energy and Coulomb blockade occurs. In the bottom panel, the bias is large enough to overcome the charging energy and single-electron tunneling occurs.

favorable whenever the "new" electrostatic energy with the added electron is smaller than the corresponding energy in the absence of that electron

$$\frac{(|Q|-|e|)^2}{2C} \leq \frac{Q^2}{2C} \quad \Longrightarrow \quad |Q| \geq \frac{|e|}{2}. \tag{4.126}$$

Since we can relate the capacitance to the bias via $V = Q/C$,[27] we need a bias $|V| \geq |e|/2C$ for current to flow. Hence, electron transport is unfavorable when

$$|V| < \frac{|e|}{2C}, \quad \text{Coulomb blockade}, \tag{4.127}$$

and we say the system is in the *Coulomb blockade regime*, whereby the current is "blocked" despite the presence of a finite bias.

When the bias is

$$|V| \geq \frac{|e|}{2C}, \quad \text{single-electron tunneling}, \tag{4.128}$$

[27] Here I am assuming all the potential drop occurs at the central region and tunnel barriers.

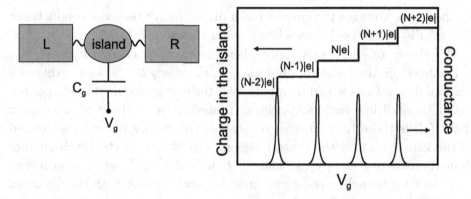

Fig. 4.5. Left panel: schematic of a central region locally coupled to a gate electrode and weakly coupled to two leads. The gate electrode controls the charge density in the central region. Right panel: by changing the gate voltage V_g (at a fixed leads voltage), the charge in the central region increases in steps, and the conductance $G = dI/dV_g$ shows peaks, at gate voltages $|V_g| = |e|(N+\frac{1}{2})/C_g$, with N the number of electrons in the central region and C_g the gate capacitance.

a single electron can tunnel through the structure.

The above arguments constitute the basis of the "orthodox" theory of Coulomb blockade (Averin and Likharev, 1986).

We can generalize the above discussion to extra electrons added to the central region, and we thus expect steps (and hence conductance spikes) in correspondence to the biases that overcome charging energies with increasing number of electrons. This is easier to obtain if the bias between the leads (sometimes referred to as "source-drain" bias) is kept fixed (and small), and a gate voltage V_g is locally applied to the central region with associated capacitance C_g (see Fig. 4.5). The role of the gate is to vary the electron density locally in the central region.

In this case, in the presence of N electrons in the central region, the electrostatic energy of the latter is

$$E_N = \frac{(Ne)^2}{2C_g} - V_g N |e|. \qquad (4.129)$$

The addition of an extra electron is favorable when the above energy is minimal with respect to the number N of electrons, namely when E_N coincides with E_{N+1}, which provides the gate voltage at which this condition is satisfied

$$E_{N+1} - E_N = 0 \quad \Longrightarrow \quad |V_g| = |e|\left(N + \frac{1}{2}\right)/C_g. \qquad (4.130)$$

At these gate voltages the current has a discontinuity (and the conductance $G = dI/dV_g$ shows peaks) as schematically shown in Fig. 4.5.

The above considerations apply when the charging energy, $E_c = e^2/2C$, is much larger than the thermal energy, $E_c = e^2/2C \gg k_B\theta$, otherwise thermal fluctuations would dominate over the energy required to charge the island. In addition, for single-electron tunneling to be observed, we expect that, if R is the total tunneling resistance, the energy uncertainty related to the capacitance of the central region, $\sim \hbar/RC$, must be much smaller than the charging energy E_c, namely $\hbar/RC \ll e^2/2C$, which also means that the total tunneling resistance must be much smaller than the quantum resistance (per spin), $R \ll h/e^2$.

Finally, we expect resonant tunneling features (if present) to overlap with single-electron tunneling effects. A clear difference between the two is that single-electron tunneling occurs at evenly spaced biases (Eq. 4.130), while the energy difference between resonances is not constant (it generally decreases with increasing energy).

NEGF approach to Coulomb blockade

Let us now discuss how the physics of Coulomb blockade can be obtained from the NEGF. We work within an ideal steady state, and employ the partition approach (with all its physical assumptions as discussed in Sec. 4.4), which has led us to Eq. 4.113 for the current.

The total Hamiltonian is still given by Eq. 4.99, with only two differences. We now adopt the following model Hamiltonian for the central region (Meir and Wingreen, 1992)

$$\hat{\mathbf{H}}_C = \hat{\mathbf{H}}_C(\{\hat{c}_n^\dagger\}; \{\hat{c}_n\}) = \sum_s \epsilon_s \, \hat{c}_s^\dagger \hat{c}_s + U \, \hat{n}_s \, \hat{n}_{-s}, \tag{4.131}$$

which describes two single-particle levels[28] with spin s and energy ϵ_s, and $\hat{n}_s = \hat{c}_s^\dagger \hat{c}_s$ is the occupation number of the state with spin s. The "on-site" energy U describes the Coulomb repulsion of electrons, and can be thus taken to be proportional to the charging energy of the central region, $U \sim e^2/C$.[29]

We also assume that the coupling of the central region with the electrodes

---

[28] These levels may or may not be degenerate.

[29] Once again, the value of this parameter should be determined self-consistently under current flow, and is not necessarily identical to the classical charging energy. I am not aware of any calculation where U is determined self-consistently in the presence of current.

is *spin-independent* so that the coupling terms 4.98 are of the form

$$\hat{\mathbf{V}}_{bC} = \sum_{n,k,s} V_{kb}\, \hat{a}^\dagger_{kbs}\, \hat{c}_s, \qquad (4.132)$$

where I have introduced a spin variable in the operators of the leads as well. The total Hamiltonian is then

$$\hat{\mathbf{H}} = \sum_{k,s,b\in L,R} E_{kb}\, \hat{a}^\dagger_{kbs}\hat{a}_{kbs} + \sum_s \epsilon_s\, \hat{c}^\dagger_s\hat{c}_s + U\,\hat{n}_s\,\hat{n}_{-s}$$

$$+ \sum_{nks;b\in L,R} (V_{kb}\hat{a}^\dagger_{kbs}\hat{c}_s + V^*_{kb}\hat{c}^\dagger_s\hat{a}_{kbs}), \qquad (4.133)$$

which has been used in literature to study a wealth of phenomena, not just the transport problem discussed here.

We now need to determine the contour-ordered Green's function of the central region, which I call $\hat{G}_{ss}$, from which all the Green's functions appearing in the equation for the current, Eq. 4.113, can be determined. Unfortunately, this Green's function cannot be determined in a closed form *in the presence* of the coupling with the leads (one has to solve the BBGKY hierarchy to all orders). Instead, one can proceed in a "semi-empirical" way as follows. Determine first the Green's function of the *isolated* central region. Call this equilibrium time-ordered Green's function $\hat{G}^{\text{isolated}}_{ss} = -\frac{i}{\hbar}\langle T\{\hat{c}_s(t)\,\hat{c}^\dagger_s(t')\}\rangle$. This Green's function can be determined from the equation of motion 4.28. The derivation is not difficult, but lengthy, and I write here only the final solution (see, e.g., Haug and Jauho, 1996)

$$\hat{G}^{\text{isolated}}_{ss}(z) = \frac{\langle \hat{n}_{-s}\rangle}{z - \epsilon_s - U} + \frac{1 - \langle \hat{n}_{-s}\rangle}{z - \epsilon_s}, \qquad (4.134)$$

with $z = E \pm i\delta$, and δ a small positive number. This Green's function has two poles corresponding to the energies $E_1 = \epsilon_s$ and $E_2 = \epsilon_s + U$; the first with an amplitude $(1 - \langle \hat{n}_{-s}\rangle)$, the second with amplitude $\langle \hat{n}_{-s}\rangle$.

One can then add the "perturbation" induced by the leads as if it is *independent* from the scattering induced by the electron-electron interactions. Intuitively we expect this "perturbation" to simply broaden the states of the central region. The complete Green's function $\hat{G}_{ss}$ with this perturbation included can then be obtained from the Dyson equation 4.35 with an appropriate self-energy term.

We choose the simplest one, namely the self-energy corresponding to *elastic* scattering of *non-interacting* electrons into the leads. This is the self-energy I have used in the Landauer approach, the sum of the left, Eq. 3.149, and right, Eq. 3.150, self-energies, which in the present case is simply

$$\Sigma_{\text{leads}}(z) = \sum_{kb} |V_{kb}|^2 g_{kb}(z), \tag{4.135}$$

where $\hat{g}$ is the equilibrium Green's function of the semi-infinite leads.

Using the Dyson equation 4.35 with the above self-energy, and $\hat{\mathbf{G}}_{ss}^{\text{isolated}}$ the "unperturbed" Green's function, we get for the retarded Green's function (and similarly for the advanced one)

$$\hat{\mathbf{G}}_{ss}^+(E) = \frac{E - \epsilon_s - (1 - \langle \hat{n}_{-s} \rangle) U}{(E - \epsilon_s - U)(E - \epsilon_s) - \Sigma_{\text{leads}}^+(E)(E - \epsilon_s - (1 - \langle \hat{n}_{-s} \rangle) U)}, \tag{4.136}$$

which, to first order in the self-energy term can be approximated as

$$\hat{\mathbf{G}}_{ss}^+(E) \simeq \frac{\langle \hat{n}_{-s} \rangle}{E - \epsilon_s - U - \Sigma_{\text{leads}}^+(E)} + \frac{1 - \langle \hat{n}_{-s} \rangle}{E - \epsilon_s - \Sigma_{\text{leads}}^+(E)}. \tag{4.137}$$

By comparing this last result with the Green's function of the isolated central region, Eq. 4.134, we realize that our initial goal has been accomplished: we have broadened the single-particle levels ϵ_s of the central region introducing an effective *elastic* lifetime due to the "interaction" with the leads.

Finally, the lesser Green's function we are missing to evaluate the total current 4.113, can be obtained directly from the Keldysh equation 4.95, which, again using the non-interacting self-energy, reads (cf. Eq. 4.121)

$$\mathbf{G}_{ss}^<(E) = \hat{\mathbf{G}}_{ss}^+(E) \, \Sigma^<(E) \, \hat{\mathbf{G}}_{ss}^-(E), \tag{4.138}$$

with

$$\Sigma_{\text{leads}}^<(E) = \sum_{kb} |V_{kb}|^2 g_{kb}^<(E), \tag{4.139}$$

where the lesser Green's functions $\hat{g}^<$ of the leads are given in equation 4.100.

By using all these Green's functions one can calculate the total current from Eq. 4.113. With specific choices of the parameters, such as the single-particle energy ϵ_s, on-site Coulomb repulsion U, etc., one finds reasonably good agreement between theory and experiment for single-electron tunneling (see, e.g., Meir et al., 1991), recovering the physics I have discussed within the "orthodox" theory of Coulomb blockade.

4.6 Quantum kinetic equations

I conclude this chapter with the application of the Keldysh formalism to the derivation of *quantum kinetic equations*, i.e., the quantum analogue of

4.6 Quantum kinetic equations

the Boltzmann equation (Sec. 2.7). This derivation will clarify further the meaning of lesser and greater self-energies.

In Sec. 2.7 I have shown that in a semiclassical theory of transport one can derive an equation of motion, known as the Boltzmann equation 2.130, for the *non-equilibrium* distribution function $f(\mathbf{r}, \mathbf{p}, t)$.

I want to show that from the equations of motion 4.82 and 4.83 of the non-equilibrium Green's functions, with suitable assumptions and approximations, we can derive an equivalent equation in the quantum case.

Due to the uncertainty principle, the position $\mathbf{r}$ and momentum $\mathbf{p}$ of a particle cannot be measured simultaneously with arbitrary accuracy. Therefore, the quantum version of the classical Boltzmann equation will be of some meaning when the position $\mathbf{r}$ of a particle can be specified on a scale larger than the typical smallest length scale of the system, but smaller than the wavelength of any external field. In Fermi systems the smallest length scale is the Fermi wavelength, which in metallic structures can be as small as the effective "size" of an atom (i.e., of the order of 1 Å), so that the quantum Boltzmann equation may have a range of applicability down to the nanoscale level.

Let us then outline the steps necessary to derive the quantum Boltzmann equation. The reader interested in a much more complete account of this derivation is urged to look into Kadanoff and Baym (1962).

We proceed by introducing the following *Wigner coordinates*[30]

$$\mathbf{R} = \frac{\mathbf{r}_1 + \mathbf{r}_2}{2}, \qquad T = \frac{t_1 + t_2}{2}, \qquad (4.140)$$

and

$$\mathbf{r} = \mathbf{r}_1 - \mathbf{r}_2, \qquad t = t_1 - t_2. \qquad (4.141)$$

In terms of these coordinates a general Green's function $\mathbf{G}(\mathbf{r}_1, t_1; \mathbf{r}_2, t_2)$ is

$$\mathbf{G}(\mathbf{r}, t; \mathbf{R}, t) = \mathbf{G}\left(\mathbf{R} + \frac{\mathbf{r}}{2}, T + \frac{t}{2}; \mathbf{R} - \frac{\mathbf{r}}{2}, T - \frac{t}{2}\right). \qquad (4.142)$$

---

[30] The first set of coordinates is called "slow" and the second "fast" because when a slowly varying (in $\mathbf{R}$ and T) external perturbation is applied, the lesser and greater Green's functions vary slowly with respect to $\mathbf{R}, T$, while they have sharp variations in the neighborhood of $\mathbf{r} = 0, t = 0$.

Let us then subtract Eq. 4.92 from 4.91 to get

$$\left(i\hbar\frac{\partial}{\partial t_1} + i\hbar\frac{\partial}{\partial t_2} + \frac{\hbar^2}{2m}\nabla^2_{\mathbf{r}_1} - \frac{\hbar^2}{2m}\nabla^2_{\mathbf{r}_2}\right.$$
$$\left. - H'(\mathbf{r}_1, t_1) + H'(\mathbf{r}_2, t_2)\right) \mathbf{G}^<(\mathbf{r}_1, t_1; \mathbf{r}_2, t_2)$$
$$= \{\mathbf{\Sigma}^+ \mathbf{G}^< + \mathbf{\Sigma}^< \mathbf{G}^+ - \mathbf{G}^+ \mathbf{\Sigma}^< - \mathbf{G}^< \mathbf{\Sigma}^-\}(\mathbf{r}_1, t_1; \mathbf{r}_2, t_2). \quad (4.143)$$

As in Eq. 4.93, the symbol $\{\cdots\}$ appearing in the above equation means convolution of the argument functions. We now express all quantities in 4.143 in terms of the Wigner coordinates 4.140 and 4.141, and use the relations

$$\left(\frac{\partial}{\partial t_1} + \frac{\partial}{\partial t_2}\right) \mathbf{G}^<(\mathbf{r}_1, t_1; \mathbf{r}_2, t_2) = \frac{\partial}{\partial T} \mathbf{G}^<(\mathbf{r}, t; \mathbf{R}, T) \quad (4.144)$$

and

$$\nabla^2_{\mathbf{r}_1} - \nabla^2_{\mathbf{r}_2} = 2\nabla_{\mathbf{R}} \cdot \nabla_{\mathbf{r}}. \quad (4.145)$$

Equation 4.143 then becomes[31]

$$\left[i\hbar\frac{\partial}{\partial T} + \frac{\hbar^2}{m}\nabla_{\mathbf{R}} \cdot \nabla_{\mathbf{r}} - H'\left(\mathbf{R} + \frac{\mathbf{r}}{2}, T + \frac{t}{2}\right)\right.$$
$$\left. + H'\left(\mathbf{R} - \frac{\mathbf{r}}{2}, T - \frac{t}{2}\right)\right] \mathbf{G}^<(\mathbf{r}, t; \mathbf{R}, T)$$
$$= \{\mathbf{\Sigma}^+ \mathbf{G}^< + \mathbf{\Sigma}^< \mathbf{G}^+ - \mathbf{G}^+ \mathbf{\Sigma}^< - \mathbf{G}^< \mathbf{\Sigma}^-\}(\mathbf{r}, t; \mathbf{R}, T). \quad (4.146)$$

Equation 4.146 is the desired *kinetic equation* for the lesser Green's function $\mathbf{G}^<$, from which one could, for instance, determine the current density via Eq. 4.50. In this form, however, it is not of much use and one needs approximations to obtain an equation that is computationally tractable. We will discuss only the approximations that allow us to derive the quantum version of the Boltzmann equation.[32]

We now assume that the potential H' varies slowly with the coordinates $\mathbf{R}$ and T so that we can make the following gradient expansion

$$H'\left(\mathbf{R} - \frac{\mathbf{r}}{2}, T - \frac{t}{2}\right) - H'\left(\mathbf{R} + \frac{\mathbf{r}}{2}, T + \frac{t}{2}\right) \simeq -\left(\mathbf{r} \cdot \nabla_{\mathbf{R}} + t\frac{\partial}{\partial T}\right) H'(\mathbf{R}, T). \quad (4.147)$$

---

[31] The convolution on the right-hand side of Eq. 4.146 is the same as in 4.93 except that the functions now depend on the coordinates $\mathbf{R}$, $\mathbf{r}$, T and t.

[32] Other approximations may be possible, which lead to different physical properties not contained in the Boltzmann equation (see, e.g., Kadanoff and Baym, 1962).

Equation 4.146 can then be approximated as

$$\left(i\hbar\frac{\partial}{\partial T} + \frac{\hbar^2}{m}\nabla_{\mathbf{R}}\cdot\nabla_{\mathbf{r}} - \left(\mathbf{r}\cdot\nabla_{\mathbf{R}} + t\frac{\partial}{\partial T}\right)H'(\mathbf{R},T)\right)\mathbf{G}^<(\mathbf{r},t;\mathbf{R},T)$$
$$= \{\mathbf{\Sigma}^+\mathbf{G}^< + \mathbf{\Sigma}^<\mathbf{G}^+ - \mathbf{G}^+\mathbf{\Sigma}^< - \mathbf{G}^<\mathbf{\Sigma}^-\}(\mathbf{r},t;\mathbf{R},T). \qquad (4.148)$$

Let us now define the following transformed function

$$\mathbf{G}^<(\mathbf{p},\omega;\mathbf{R},T) = \int d\mathbf{r}\int dt\, e^{-i\frac{\mathbf{p}}{\hbar}\cdot\mathbf{r}+i\omega t}\mathbf{G}^<(\mathbf{r},t;\mathbf{R},T). \qquad (4.149)$$

If we then Fourier transform Eq. 4.148 with respect to $\mathbf{r}$ and t we get

$$i\hbar\left(\frac{\partial}{\partial T} + \frac{\mathbf{P}}{m}\cdot\nabla_{\mathbf{R}} - \nabla_{\mathbf{R}}H'(\mathbf{R},T)\cdot\nabla_{\mathbf{p}} + \frac{\partial}{\partial T}H'(\mathbf{R},T)\frac{\partial}{\partial\hbar\omega}\right)\mathbf{G}^<(\mathbf{p},\omega;\mathbf{R},T)$$
$$\simeq \left[\mathbf{\Sigma}^+\mathbf{G}^< + \mathbf{\Sigma}^<\mathbf{G}^+ - \mathbf{G}^+\mathbf{\Sigma}^< - \mathbf{G}^<\mathbf{\Sigma}^-\right](\mathbf{p},\omega;\mathbf{R},T)$$
$$= \left[(\mathbf{\Sigma}^+ - \mathbf{\Sigma}^-)\mathbf{G}^< - (\mathbf{G}^+ - \mathbf{G}^-)\mathbf{\Sigma}^<\right](\mathbf{p},\omega;\mathbf{R},T)$$
$$= \left[\mathbf{\Sigma}^>\mathbf{G}^< - \mathbf{G}^>\mathbf{\Sigma}^<\right](\mathbf{p},\omega;\mathbf{R},T). \qquad (4.150)$$

The right-hand side of Eq. 4.150 has been approximated by retaining the lowest-order terms in the gradient expansion with respect to $\mathbf{R}$ and T, and in the last equality I have made use of the identities $\mathbf{\Sigma}^+ - \mathbf{\Sigma}^- = \mathbf{\Sigma}^> - \mathbf{\Sigma}^<$ (Eq. 4.90) and $\mathbf{G}^+ - \mathbf{G}^- = \mathbf{G}^> - \mathbf{G}^<$ (Eq. 4.51).

We already start to see that apart from the term $\frac{\partial}{\partial T}H'(\mathbf{R},T)\frac{\partial}{\partial\hbar\omega}$, the left-hand side of Eq. 4.150 has a structure similar to the classical Boltzmann equation 2.130, where $i\hbar\mathbf{G}^<$ plays the role of the non-equilibrium distribution function f, and the right-hand side replaces the collision integral $\mathbf{I}[f]$.

The term $\frac{\partial}{\partial T}H'(\mathbf{R},T)\frac{\partial}{\partial\hbar\omega}$ cannot appear in the classical case because the energy (or frequency) is not a variable independent of the momentum. For the same reason, it cannot appear in a mean-field approximation where there is a one-to-one correspondence between energy and momentum (see Eq. 4.72). Let us then proceed as follows.

At equilibrium, the Green's function $\hat{\mathbf{G}}^<$ is related to the equilibrium Fermi–Dirac distribution 4.23, and the spectral function A via the relation 4.59. It thus seems natural to make the following *local equilibrium ansatz* in the non-equilibrium case as well

$$\mathbf{G}^<(\mathbf{p},\omega,\mathbf{R},T) = 2\pi i f(\mathbf{p},\mathbf{R},T)\, A(\mathbf{p},\omega,\mathbf{R},T),$$
$$\mathbf{G}^>(\mathbf{p},\omega,\mathbf{R},T) = -2\pi i[1 - f(\mathbf{p},\mathbf{R},T)]\, A(\mathbf{p},\omega;\mathbf{R},T), \qquad (4.151)$$

with f some non-equilibrium local distribution function.

In the above ansatz we have also implicitly assumed that the distribution function does not depend on frequency, or equivalently that the energy and momentum are related to each other so that the energy is not an independent variable. Let us now choose the simplest spectral function possible, namely the one corresponding to the mean-field approximation 4.72. In the present case

$$A(\mathbf{p}, \omega; \mathbf{R}, T) = \delta(\hbar\omega - \epsilon_p), \qquad (4.152)$$

with

$$\epsilon_p = \frac{p^2}{2m} + H'(\mathbf{R}, T). \qquad (4.153)$$

If we replace ansatz 4.151, with the spectral function 4.152, into Eq. 4.150 we obtain

$$A(\mathbf{p}, \omega; \mathbf{R}, T) \left(\frac{\partial}{\partial T} + \frac{\mathbf{p}}{m} \cdot \nabla_\mathbf{r} - \nabla_\mathbf{R} H'(\mathbf{R}, T) \cdot \nabla_\mathbf{p} \right) f(\mathbf{p}, \mathbf{R}, T)$$
$$= -i \frac{A(\mathbf{p}, \omega; \mathbf{R}, T)}{\hbar} \left[f(\mathbf{p}, \mathbf{R}, T) \Sigma^>(\mathbf{p}, \omega; \mathbf{R}, T) \right.$$
$$\left. + (1 - f(\mathbf{p}, \mathbf{R}, T)) \Sigma^<(\mathbf{p}, \omega; \mathbf{R}, T) \right]. \qquad (4.154)$$

If we now integrate in frequency and use the properties of the δ-function, we get

$$\left(\frac{\partial}{\partial T} + \frac{\mathbf{p}}{m} \cdot \nabla_\mathbf{r} - \nabla_\mathbf{R} H'(\mathbf{R}, T) \cdot \nabla_\mathbf{p} \right) f(\mathbf{p}, \mathbf{R}, T)$$
$$= -\frac{i}{\hbar} \left[f(\mathbf{p}, \mathbf{R}, T) \Sigma^>(\mathbf{p}, \omega = \epsilon_p/\hbar; \mathbf{R}, T) \right.$$
$$\left. + (1 - f(\mathbf{p}, \mathbf{R}, T)) \Sigma^<(\mathbf{p}, \omega = \epsilon_p/\hbar; \mathbf{R}, T) \right]$$
$$\equiv \mathbf{I}[f]. \qquad (4.155)$$

This is precisely the Boltzmann equation 2.130 with H' replacing V_{ext}, and the collision integral given by

$$\mathbf{I}[f] = -\frac{i}{\hbar} \left[f(\mathbf{p}, \mathbf{R}, T) \Sigma^>(\mathbf{p}, \omega = \epsilon_p/\hbar; \mathbf{R}, T) \right.$$
$$\left. + (1 - f(\mathbf{p}, \mathbf{R}, T)) \Sigma^<(\mathbf{p}, \omega = \epsilon_p/\hbar; \mathbf{R}, T) \right]. \qquad (4.156)$$

Note that at equilibrium $\hat{\Sigma}^>$ and $\hat{\Sigma}^<$ are given by Eqs. 4.89, with $f = f^{eq}$ the global equilibrium Fermi–Dirac distribution 4.23. If we replace these expressions in 4.156, with $f = f^{eq}$, we immediately see that the collision integral is identically zero, as one would expect in the equilibrium case.

As we have explicitly demonstrated for non-interacting electrons (see Eq. 4.122), $\Sigma^</\hbar$ is proportional to the rate for an electron in the central region to tunnel (scatter) elastically into the left and right electrodes.

Similarly, in the interacting case, by looking at its role in the collision integral 4.156, we can interpret $\Sigma^<(\mathbf{p},\omega = \epsilon_p/\hbar;\mathbf{R},T)/\hbar$ as the collision rate for a particle to scatter *into* the state with momentum $\mathbf{p}$ and energy ϵ_p at the phase-space point $\mathbf{R}$, T, provided that state is unoccupied. Since that state is unoccupied according to the factor $[1-f(\mathbf{p},\mathbf{R},T)]$, the product $[1-f(\mathbf{p},\mathbf{R},T)]\Sigma^<(\mathbf{p},\omega=\epsilon_p;\mathbf{R},T)/\hbar$ counts how many particles can actually scatter into the above state at $\mathbf{R}$, T.

On the other hand, $\Sigma^>(\mathbf{p},\omega = \epsilon_p/\hbar;\mathbf{R},T)/\hbar$ is the collision rate for a particle in a state with momentum $\mathbf{p}$ and energy ϵ_p to scatter *out* of that state at position $\mathbf{R}$ and at a time T. This rate, multiplied by the distribution function $f(\mathbf{p},\mathbf{R},T)$, counts how many particles scatter out of the above state at a point $\mathbf{R}$, T in phase space.

The right-hand side of Eq. 4.156 is then the *net* rate of change of particles with momentum $\mathbf{p}$ and energy ϵ_p, at position $\mathbf{R}$ and time T. This is precisely the definition of collision integral of the semiclassical Boltzmann equation (see discussion after Eq. 2.129).

Finally, if we consider small departures from equilibrium so that we can approximate $\hat{\Sigma}^>$ and $\hat{\Sigma}^<$ with their equilibrium value, with $f = f^{eq}$ in Eqs. 4.89, but retain the non-equilibrium distribution function f in 4.156 we get

$$\mathbf{I}[f] \simeq \frac{2\mathrm{Im}\{\Sigma^+(\epsilon_p)\}}{\hbar}[f(1-f^{eq}) - f^{eq}(1-f)]$$
$$= -\frac{[f(\mathbf{p},\mathbf{R},T)) - f^{eq}(\mathbf{p},\mathbf{R}))]}{\tau_p}, \qquad (4.157)$$

where in the last step I have defined the scattering lifetime (see, e.g., Eq. 4.69)

$$\tau_p = -\hbar/2\mathrm{Im}\{\Sigma^+(\epsilon_p)\}. \qquad (4.158)$$

Equation 4.157 has the same form as the relaxation-time approximation 2.134 we have introduced in Sec. 2.7, where here, as in the semiclassical case, the lifetime can be due to both elastic and inelastic effects among particles.

If one is interested in the current, from the solution of Eq. 4.155 one obtains the non-equilibrium function f from which the current can be determined as in Eq. 2.137, which completes our search for the quantum analogue of the Boltzmann equation.

Summary and open questions

In this chapter I have introduced the non-equilibrium Green's function formalism. This approach pertains to the study of the dynamics of closed quantum systems subject to a time-dependent deterministic perturbation, and for which many-body perturbation theory applies. All physical properties of interest are then calculated from appropriate single-particle Green's functions.

The application of the formalism to steady-state transport requires similar assumptions and approximations to those used in the scattering formulation of Chapter 3. Under these assumptions one can obtain a closed set of equations of motion for the non-equilibrium Green's functions from which the current can be determined. All interactions are included in a self-energy function, which, in principle, contains all possible scattering events that a single particle experiences in the presence of all other particles. However, it is no simple task to compute the current in realistic systems when interactions beyond mean field are included, since the self-energy, and hence the Green's functions in the presence of electron flow, are generally unknown.

Needless to say, if the system consists of non-interacting particles (or particles interacting at a mean-field level) we recover the current 3.229 we have derived in Chapter 3 using single-particle non-interacting scattering theory.

As an example of application of the NEGF I have discussed the phenomenon of Coulomb blockade, which is a *non-equilibrium many-body* effect whereby the addition of an extra electron to a central region weakly coupled to two electrodes may be energetically unfavorable despite the presence of a bias. We have used a simple "impurity" Hamiltonian to study this effect, and I am not aware of a full time-dependent first-principles calculation of this phenomenon in nanoscale systems. Such a study is desirable since it would clarify the microscopic dynamics, and self-consistent charge and current distributions during the onset of single-electron tunneling, and would shed new light both on the assumptions of the "orthodox" theory of Coulomb blockade, and on the limits of applicability of the NEGF approach to transport phenomena in nanoscale systems.

In the same vein, the effect of interactions on the formation and microscopic properties of steady states is still an open problem, and is not at all addressed within the NEGF. Its study would help us better understand the basic assumptions of the static formulation of transport.

Finally, I have derived the quantum analogue of the semiclassical Boltzmann equation, which reveals the link between self-energies and collision integral.

Exercises

4.1 Derive Eq. 4.22.

4.2 Use the definitions of retarded and advanced Green's functions 4.40 and 4.41 to prove that the spectral function 4.52 is positive definite. Show that Eq. 4.56 is satisfied. (*Hint:* you will need to use the commutation relation A.21 of field operators.)

4.3 **Fluctuation-dissipation theorem.** Write the lesser and greater Green's functions 4.44 and 4.45 in terms of the *exact* eigenstates of the Hamiltonian $\hat{H}_0$. Prove Eq. 4.59 which is at the basis of the fluctuation-dissipation relation 4.59.

4.4 **Equation of motion for $\hat{G}^<$.** Starting from the matrix notation for Green's functions and self-energies as in Eqs. 4.75 and 4.84, and the equations of motion for the contour-ordered Green's function **G**, Eqs. 4.82 and 4.83, derive Eqs. 4.91 and 4.92.

4.5 Prove Eq. 4.112.

4.6 Derive Eq. 4.124.

5
Noise

So far I have discussed the most obvious quantity in a transport problem: the average current. In reality, as I have anticipated in Sec. 1.2, electrical current continually fluctuates in time so that it carries *noise*.

There are several sources of noise in a conductor that can be classified as *external* and *internal*. Two of the most common sources of external noise are known as $1/f$ and *telegraph noise*. The first takes its name from the fact that it shows a spectrum at small frequencies f, which scales as $1/f$. No general theory exists on this noise, and it is believed to be of extrinsic origin, for instance due to distributions of defects in the conductor. The interested reader should consult the review by Dutta and Horn (1981) where $1/f$ noise is discussed at length.

On the other hand, telegraph noise is due to rapid fluctuations between states of the conductor that are very close in energy, but that carry different currents. This noise is also believed to be of extrinsic nature and has a typical Lorentzian power spectrum.[1]

In addition to the above fluctuations, one has to remember that our nanoscale system (sample), whose noise properties we want to determine, is connected to a real external circuit with its own external resistance, call it R_{ext} (Fig. 5.1). Therefore, due to the presence of this extra resistance, even if the voltage source is noiseless (call the corresponding potential drop V_{source}), the voltage drop across the sample (call it V_{sample}) is different from V_{source} and would manifest fluctuations. This is easy to see using the following simple arguments.

If I is the average current flowing in the system at steady state, the above two voltages are related via (assuming no quantum-mechanical interference occurs between the external resistance and the resistance of the nanoscale

---

[1] It could be due, e.g., to fluctuations of atomic configurations induced by thermal effects.

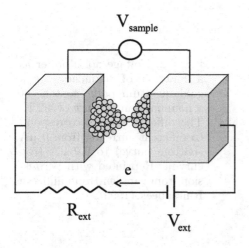

Fig. 5.1. Due to the presence of the resistance of the external circuit, the voltage at the sample V_{sample} is different than the external applied voltage, causing fluctuations in V_{sample}.

system)

$$V_{\text{source}} = R_{\text{ext}} I + V_{\text{sample}}. \tag{5.1}$$

Therefore, even if we could control the source voltage with arbitrary precision (i.e., its fluctuations are zero), from Eq. 5.1 we see that if the current fluctuates in time as $\Delta I(t)$, the voltage across the sample fluctuates as $\Delta V_{\text{sample}}(t) = -R_{\text{ext}} \Delta I(t)$, which allows us to relate current fluctuations to *voltage fluctuations* at the sample.

In this chapter I will not discuss these types of noise. Instead, I will focus only on noise that cannot be eliminated even in clean, "impurity-free" samples, embedded in a circuit with zero external resistance. These are fluctuations of the current due to both the discreteness of charge present even at zero temperature (*shot* or *partition noise*), and random fluctuations of the electron distribution at finite temperatures and close to equilibrium (*thermal* or *Johnson–Nyquist noise*). I will show that the latter is related to the conductance of the system via the fluctuation-dissipation theorem (see Sec. 2.3.3), while the first provides more information on the electron dynamics than the conductance itself. Both have a power spectrum that is independent of frequency (white noise).

Figure 5.2 shows a typical noise power measurement for a gold quantum point contact (like the one represented in Fig. 5.1). The curves correspond to voltage fluctuations, as a function of frequency, with increasing current, from 0 µA (bottom curve) to 0.9 µA (top curve) (van den Brom and van Ruitenbeek, 1999). Even at zero average current, noise is observed and is precisely due to thermal fluctuations (the nominal background temperature of the system is 4.2 K). This noise is independent of frequency up to a

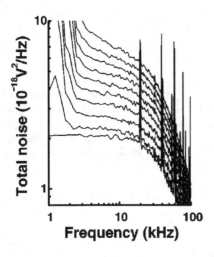

Fig. 5.2. Voltage noise power as a function of frequency for a gold quantum point contact at a nominal temperature of 4.2 K. The different curves correspond to different currents, from 0 µA (bottom curve) to 0.9 µA (top curve). Reprinted with permission from van den Brom and van Ruitenbeek (1999).

cut-off frequency imposed by the capacitance of the external circuit (about 50 kHz in this particular experiment). The spikes in the spectrum are attributed to electromagnetic interference with the outside environment. By increasing the current, one observes the $1/f$ behavior at small frequencies, which saturates fast to some value that contains both shot and thermal noise contributions.

In the following, I will limit our discussion to noise in the coherent transport regime within a mean-field approximation. A discussion of noise in systems where electron interactions are important (such as in the Coulomb blockade regime of Sec. 4.5) can be found in the review by Blanter and Büttiker (2000). A brief discussion of the effect of electron-phonon scattering on noise properties of nanoscale systems in the quasi-ballistic regime is given in Appendix H (see also Chen and Di Ventra, 2005).

In addition, in Chapter 8 I will show that, under certain conditions, the electron liquid may undergo a transition from laminar to turbulent flow when crossing a nanojunction. If this occurs, turbulent current noise will also be present, and will correlate with the other sources of internal noise. The analytical description of turbulence is a notoriously difficult problem (see, e.g., Landau and Lifshitz, 1959b), so that I do not attempt to describe this type of noise here, even if it is of internal origin.

Finally, I mention that a comprehensive experimental study of all these sources of noise is still lacking for nanoscale systems. This study is desirable since the noise characteristics of a conductor provide much more information on electron flow than the average current, and for nanoscale systems it may even be used to determine their detailed microscopic configuration, since internal noise (shot noise in particular) is much more sensitive to atomic details than the average current (Chen and Di Ventra, 2003; Lagerqvist et al., 2004; Yao et al., 2006).

5.1 The moments of the current

Since the current fluctuates in time, and the knowledge of this observable is only statistical, it is what we call a *stochastic process*. Its value $I(t)$ at any given instant of time t is a *realization* of the process (or *sample function*) and the stochastic process is the *ensemble* of these realizations.

From the theory of classical stochastic processes we then know that we can obtain information on the *statistical distribution* of this process (namely, how its properties are distributed in time) by calculating all its associated *moments* (see, e.g., van Kampen, 1992). In quantum mechanics we can do the same.

Assume the state of a many-body system is described by a statistical operator $\hat{\rho}(t)$. The *first moment* of the current is simply its average 1.86

$$e\langle \hat{I} \rangle_t = e\text{Tr}\{\hat{\rho}(t)\hat{I}\}. \tag{5.2}$$

The *second moment* is the current *autocorrelation function*

$$S(t,t') = \frac{e^2}{2}\text{Tr}\left\{\hat{\rho}(t=0)[\hat{I}(t)\hat{I}(t') + \hat{I}(t')\hat{I}(t)]\right\}, \tag{5.3}$$

which we have defined in Sec. 2.3.3 when discussing the fluctuation-dissipation theorem, where the current operator is in the Heisenberg picture.[2]

At an ideal steady (stationary) state, the average current 5.2 does not depend on time, $\langle \hat{I} \rangle_t = \langle \hat{I} \rangle$, and it is convenient to calculate the second moment with respect to the average current by defining the operator $\Delta \hat{I}(t) \equiv \hat{I}(t) - \langle \hat{I} \rangle$. In this case, the autocorrelation function can depend only on time differences and is

$$S(t - t') = \frac{e^2}{2}\text{Tr}\left\{\hat{\rho}(t=0)[\Delta\hat{I}(t)\Delta\hat{I}(t') + \Delta\hat{I}(t')\Delta\hat{I}(t)]\right\}$$
$$\equiv \frac{e^2}{2}\langle \Delta\hat{I}(t)\Delta\hat{I}(t') + \Delta\hat{I}(t')\Delta\hat{I}(t)\rangle. \tag{5.4}$$

This will be our starting point to determine shot and thermal noise.

---

[2] Recall that the symmetrization of this quantity (which is not necessary in classical stochastic processes) is because operators in quantum mechanics do not necessarily commute at different times. Also, in certain literature the term autocorrelation function is often used only for Eq. 5.4, and Eq. 5.3 is simply called second moment. I do not make this distinction here, and will use these terms interchangeably.

In general, the *m-th moment* can be similarly defined as

$$S(t_1, t_2, \ldots, t_m) = \frac{e^m}{m!} \text{Tr} \left\{ \hat{\rho}(t=0) \, \text{P} \left[\prod_j \hat{I}(t_j) \right] \right\}, \tag{5.5}$$

where P indicates all possible permutations of the argument.

In an ideal stationary state, the above moments do not change by a time translation, namely

$$S(t_1 + \tau, t_2 + \tau, \ldots, t_m + \tau) = S(t_1, t_2, \ldots, t_m), \quad \text{ideal steady state,} \tag{5.6}$$

with τ an arbitrary interval of time.

These higher moments give rise to what is now called *counting statistics* (Levitov and Lesovik, 1993). I will briefly discuss the information we can extract from these higher moments in Sec. 5.3.

Autocorrelation time and stationary processes

Since a steady state is just a theoretical idealization, one is naturally led to ask if there is a condition that allows us to *approximate* a given stochastic process as stationary. This condition is related to the existence of an *autocorrelation time*.

This is defined as the finite time, call it τ_c, after which the autocorrelation of the stochastic process is negligibly small. This is because a process that does *not* have a finite autocorrelation time will always "remember" its history and its initial conditions, and thus cannot be approximated as an ideal stationary process.

In the present case, if

$$S(t, t') = \frac{e^2}{2} \langle \hat{I}(t)\hat{I}(t') + \hat{I}(t')\hat{I}(t) \rangle - e^2 \langle \hat{I}(t) \rangle \langle \hat{I}(t') \rangle \tag{5.7}$$

is the autocorrelation of the current, we can approximate the process as a stationary process if

$$|S(t, t')| \simeq 0, \quad \text{for } |t - t'| > \tau_c, \quad \text{approximate steady state,} \tag{5.8}$$

or, if we take $|t - t'|$ to infinity

$$\lim_{|t-t'| \to +\infty} \frac{\langle \hat{I}(t)\hat{I}(t') + \hat{I}(t')\hat{I}(t) \rangle}{2} = \langle \hat{I}(t) \rangle \langle \hat{I}(t') \rangle. \tag{5.9}$$

By comparing this condition with Eq. 2.76, we see that an ideal steady state is based on the assumption of ergodicity (Sec. 2.3.3.1): it is equivalent to assuming that the system's representative point in phase space explores, during time evolution, *any* neighborhood of any point of the relevant phase space. This is again an idealization, and it is probably not true for many physical systems considered experimentally. Nonetheless, we will make this approximation in this chapter; namely, we will assume that the stochastic current has a finite autocorrelation time so that the stationary condition 5.8 holds.

Approximations 1 to 5 of the Landauer approach

Here I also remark that, as for the average current, the above moments cannot be calculated exactly for a general many-body system, and we need approximations to describe the system dynamics.

From now on I will make the *Approximations* 1 − 5 of the Landauer approach (Sec. 3.1); namely, I will work with an ideal steady state for the average current of a closed but infinite system, and electrons experience at most mean-field interactions even in the nanostructure. I want to determine the internal noise properties of this simplified problem.

5.2 Shot noise

Shot noise is the term employed to define electrical current fluctuations at steady state and zero temperature due to the discreteness of charge.[3] Its discovery in vacuum tubes dates back to experiments by Schottky who observed that after eliminating all other (extrinsic) sources of noise there would still be a contribution due to the random motion of the charge caused by thermal effects of the conducting part of the circuit (thermal noise), and a contribution due to the irregular arrival of (independent) electrons at the anode across the vacuum region that separates the anode from the cathode of the tube (Schottky, 1918).

Since this latter type of noise is due to the discreteness of charge it is not just a prerogative of electrons, but it is observed, for instance, in quantum optics due to the corpuscular nature of light, or in any classical problem in which one measures the fluctuations of a given steady particle flux about its average (e.g., raindrops on a given surface).

The quantum theory of shot noise can be developed similarly for fermions and bosons. In this chapter we just focus on the first class of particles. A brief review of shot noise for bosons is presented in Blanter and Büttiker (2000).

---

[3] In literature this is sometimes called *partition noise*. This nomenclature comes, as we will see in a moment, from the random "partition" of the electron flow due to, e.g., a potential energy barrier. I will use these two terms interchangeably to represent the same type of noise.

5.2.1 The classical (Poisson) limit

Before describing the quantum-mechanical theory of shot noise, let us begin by considering the case – also analyzed by Schottky – in which electrons are emitted from a source (e.g., a cathode) at random intervals. The electrons are also considered *independent* (no Coulomb interaction between them) and these emission events *uncorrelated* from each other. We can then approximate the time-dependent current that crosses an arbitrary surface perpendicular to the electron flow as a succession of charge spikes, which can be represented as delta functions centered at different random times

$$I(t) = \sum_i e\,\delta(t - t_i). \qquad (5.10)$$

If T_p is an arbitrary *finite* interval of time, on average, at steady state, N electrons cross a given surface in the interval T_p, i.e., the electron flow has average current $\bar{I}_{T_P} = eN/T_p$ over that interval. By taking the limit of $T_p \to \infty$, N also goes to infinity, and we write

$$\langle I \rangle = \lim_{T_p \to \infty} \frac{1}{T_p} \int_{-T_p/2}^{T_p/2} dt\, I(t) = \lim_{T_p \to \infty} \bar{I}_{T_P}, \qquad (5.11)$$

where, once again, I have made use of the ergodic hypothesis (Sec. 2.3.3.1) to transform the time average on the right-hand side into an ensemble average over replicas of the system on the left-hand side of Eq. 5.11, *irrespective* of the initial conditions.

With these definitions and assumptions we can now calculate the current fluctuations about the average current. We can proceed in two ways: (i) we can formally compute the noise using the current autocorrelation function 5.3, or (ii) we realize that the ensemble of current events follows a Poisson distribution and evaluate from that distribution the fluctuations of the number of carriers that cross a given surface in time. I will employ the first method since it is also the starting point for the quantum theory of noise we will develop later, and leave the second approach as an exercise for the reader (Exercise 5.1).

It is convenient to calculate the fluctuations about the average by defining $\Delta I(t) = I(t) - \langle I \rangle$, and writing the current autocorrelation function as

$$S(t') = \lim_{T_p \to \infty} \frac{1}{T_p} \int_{-T_p/2}^{T_p/2} \Delta I(t)\, \Delta I(t+t')\, dt = \langle \Delta I(t)\, \Delta I(t+t') \rangle, \qquad (5.12)$$

where the dependence on a single time comes from the fact that we are considering a steady state.

5.2 Shot noise

If we replace 5.10 into 5.12 we get

$$S(t') = \lim_{T_p \to \infty} \frac{e^2}{T_p} \sum_i \sum_{i'} \int_{-T_p/2}^{T_p/2} \delta(t - t_i)\delta(t - t_{i'} + t')dt - \langle I \rangle^2$$

$$= \lim_{T_p \to \infty} \frac{e^2}{T_p} \sum_i \sum_{i'} \delta(t_i - t_{i'} + t') - \langle I \rangle^2, \quad (5.13)$$

where in the last expression I have used the properties of the δ-function. Now, there is a theorem of statistical mechanics (the *Wiener–Khinchin theorem*, see, e.g., van Kampen, 1992) that relates the *noise spectrum* $S(\omega)$ – or *spectral density of fluctuations* – of a stationary process to the autocorrelation function via[4]

$$S(\omega) = 2 \int_{-\infty}^{+\infty} dt' \, e^{i\omega t'} S(t'). \quad (5.14)$$

Note that the validity of this theorem requires the existence of a finite autocorrelation time, which I have implicitly assumed by treating the transport problem as stationary.

Before replacing 5.10 into 5.14 we recognize that the δ-function terms with $t_i \neq t_{i'}$ do not contribute on average to $S(t')$. We are then left with N δ-functions for $t_i = t_{i'}$. The noise power spectrum is thus

$$S(\omega) = 2 \lim_{T_p \to \infty} \frac{e^2}{T_p} \sum_{i=1}^{N} \int_{-\infty}^{+\infty} dt' \, e^{i\omega t'} \delta(t') - 2\langle I \rangle^2 \int_{-\infty}^{+\infty} dt' \, e^{i\omega t'}$$

$$= 2e \lim_{T_p \to \infty} \frac{eN}{T_p} - 2\langle I \rangle^2 \delta(\omega) = 2e\langle I \rangle - 2\langle I \rangle^2 \delta(\omega), \quad (5.15)$$

where I have used the Fourier transform of the δ-function, $\delta(\omega) = \int dt \, e^{i\omega t}$. In the limit of zero frequency we then obtain a constant value of noise proportional to the average current[5]

$$S_P \equiv \lim_{\omega \to 0} S(\omega) = 2e\langle I \rangle. \quad (5.16)$$

I will refer to this classical result as the *Poisson limit* of shot noise.

It is called poissonian because it can be derived by assuming that the number of charges that are transmitted across a given surface of the conductor follows a Poisson distribution (Exercise 5.1). This is also known as a *counting process* with *independent increments* (see, e.g., Ross, 1996), in the

---
[4] The factor of 2 in 5.14 is due to the symmetry $S(-\omega) = S(\omega)$.
[5] Recall that $\lim_{\omega \to 0} \delta(\omega) = 0$. This limit must be interpreted as follows: consider the δ-function in any of its representations. For clarity, use the representation in Eq. 3.126. Take the limit of that representation for $\epsilon \to 0$, with frequency $\omega > \epsilon$. Then take the limit of $\omega \to 0$.

sense that it "counts" the number of events that occur in a given interval of time, and these events are statistically *independent* from those occurring in a different and disjoint interval of time.

5.2.2 Quantum theory of shot noise

We are now ready to derive the shot noise expression in the quantum case. Let us begin by considering an idealized case of a single electron in a single one-dimensional channel. The electron scatters on a potential energy barrier. It thus has a probability T to be transmitted through the barrier and a probability $R = 1 - T$ to be reflected.

From Eq. 3.56 we know that, in an interval of energy dE, the current (per spin) that is transmitted across the barrier is $dI = \frac{e}{2\pi\hbar} f(E)T(E)dE$, i.e., it is proportional to the occupation $n = fT$ of the current-carrying states. In the zero-frequency limit the current fluctuations are thus related to fluctuations in the occupation number: $S(0) \propto \langle \Delta n \Delta n \rangle$.

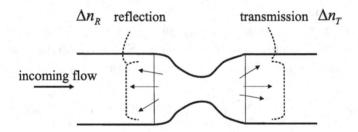

Fig. 5.3. Shot noise is due to the *statistical correlation* between transmitted and reflected states caused by their fluctuating occupation numbers, Δn_T and Δn_R, respectively.

A simple way to calculate these fluctuations is as follows (see also Blanter and Büttiker, 2000). Assume that the average occupation number for the incoming flux is $\langle n_i \rangle = f$. It is then $\langle n_T \rangle = fT$ and $\langle n_R \rangle = fR$ for the transmitted and reflected states, respectively. The fluctuations in the transmitted and reflected states can be easily calculated by recognizing that in a given experimental measurement of the current the electron is either transmitted or reflected so that the ensemble average $\langle n_T n_R \rangle = 0$. In addition, we are considering the statistical properties of an ideal Fermi gas, so that the second moment of occupation fluctuations at equilibrium is equal to its average: $\langle n^2 \rangle = \langle n \rangle$. By defining $\Delta n_T = n_T - \langle n_T \rangle$ and $\Delta n_R = n_R - \langle n_R \rangle$ we therefore find

$$\langle (\Delta n_T)^2 \rangle = \langle n_T \rangle - \langle n_T \rangle^2 = fT(1 - fT), \qquad (5.17)$$

$$\langle(\Delta n_R)^2\rangle = \langle n_R\rangle - \langle n_R\rangle^2 = fR(1-fR), \qquad (5.18)$$

$$\langle(\Delta n_T)(\Delta n_R)\rangle = -\langle n_T\rangle\langle n_R\rangle = -f^2 TR. \qquad (5.19)$$

If we assume the incident state is populated with probability 1 (zero temperature, $f=1$), then the above expressions reduce to

$$\langle(\Delta n_T)^2\rangle = \langle(\Delta n_R)^2\rangle = -\langle(\Delta n_T)(\Delta n_R)\rangle = TR = T(1-T). \qquad (5.20)$$

Recalling that shot noise is $S(0) \propto \langle \Delta n \Delta n \rangle$, we see from Eq. 5.20 that this is due to the statistical correlation between transmitted and reflected states caused by their respective occupation number fluctuations (Fig. 5.3). In addition, unlike the current, which can be expressed in terms of transmission probabilities only (see Eq. 3.56), shot noise depends on *both* the transmission and reflection probabilities.

From Eq. 5.20 the term *partition noise* should now be clear: it refers to the separation of the incoming electron flow into two flows: transmitted and reflected. It is also clear that this noise is maximum when $T = 1/2$, and there is no shot noise if $T = 0$ (complete reflection) or $T = 1$ (complete transmission). The latter result can be interpreted by thinking that a "uniform" flow of electrons (or of any other particle for that matter), which is embodied in the condition $T = 1$, does not produce any second moment for the current.

The single-channel case

Let us now derive the above results more rigorously using the scattering theory of Chapter 3. For simplicity, we again limit ourselves to the case of a single channel i (in both left and right leads) in a two-probe geometry (see, e.g., Fig. 3.4). The general scattering wave-function, $\Psi(\mathbf{r}, E)$, is given by Eq. 3.200, where only one channel is considered. We will generalize later to the multi-channel case.

I proceed by using the second-quantization formalism (Appendix A), which allows us to handle the ensemble averages appearing in the moments of the current in a much easier way. Instead of the wave-function $\Psi(\mathbf{r}, E)$ we then have the field operator (see Eqs. A.19 and A.20)

$$\psi(\mathbf{r}, t) = \psi^L(\mathbf{r}, t) + \psi^R(\mathbf{r}, t), \qquad (5.21)$$

where the field operators $\psi^L(\mathbf{r}, t)$ and $\psi^R(\mathbf{r}, t)$ are (see Chen and Di Ventra, 2003; Lagerqvist *et al.*, 2004)

$$\psi^{L(R)} = \int dE\, D(E)\, \Psi^+_{i(\pm k_i)}(\mathbf{r})\, \hat{a}_E^{L(R)}\, e^{-iEt/\hbar} \equiv \int dE\, \Psi_E^{L(R)}(\mathbf{r})\, \hat{a}_E^{L(R)}\, e^{-iEt/\hbar}, \qquad (5.22)$$

with $D(E)$ the channel density of states, Eq. 3.48.[6] In Eq. 5.22 the wavefunctions $\Psi^+_{i(\pm k_i)}(\mathbf{r})$ correspond to processes with an incident wave from the L electrode $(+|k_i|)$ and those with a wave incident from the R electrode $(-|k_i|)$, as in Sec. 3.5.1. We have also defined the annihilation operators $\hat{a}_E^{L(R)}$ associated with these processes.[7] Since we are dealing with fermions and assume the left-moving and right-moving electrons uncorrelated, these operators satisfy the anti-commutation relations (from Eq. A.17)

$$\{\hat{a}_E^i, \hat{a}_{E'}^{j\dagger}\} = \delta_{ij}\delta(E - E'), \qquad i,j = R, L. \tag{5.23}$$

We have also assumed that electrons injected from the left and from the right of the sample have local equilibrium distributions 3.14 and 3.15, respectively. From Eq. 4.24, the occupation of the single-particle states is thus

$$\langle \hat{a}_E^{i\dagger} \hat{a}_{E'}^j \rangle = \delta_{ij}\delta(E - E') f_E^i, \qquad i,j = R, L, \tag{5.24}$$

where, for brevity, I have used the symbol $f_E^i = f_i(E)$ for the Fermi–Dirac distribution.

Let us then define the following current operator

$$\hat{I}(x,t) = \frac{\hbar}{2mi} \int d\mathbf{r}_\perp \left[\psi^\dagger(\mathbf{r},t) \left(\frac{\partial}{\partial x}\psi(\mathbf{r},t)\right) - \left(\frac{\partial}{\partial x}\psi^\dagger(\mathbf{r},t)\right)\psi(\mathbf{r},t) \right], \tag{5.25}$$

where $d\mathbf{r}_\perp = dy\,dz$ is a two-dimensional coordinate element in the transverse direction.

The average current

Before calculating the noise let us show that the average current calculated with the above field operators is indeed the same as we have derived in Chapter 3 (Eq. 3.56). If we replace Eq. 5.21 into 5.25 we get

$$\hat{I}(x,t) = \frac{\hbar}{2mi} \int dE \int dE' \int d\mathbf{r}_\perp e^{i(E-E')t/\hbar} \left[(\hat{a}_E^L)^\dagger \hat{a}_{E'}^L \tilde{I}_{E,E'}^{LL} \right.$$
$$\left. + (\hat{a}_E^R)^\dagger \hat{a}_{E'}^R \tilde{I}_{E,E'}^{RR} + (\hat{a}_E^L)^\dagger \hat{a}_{E'}^R \tilde{I}_{E,E'}^{LR} + (\hat{a}_E^R)^\dagger \hat{a}_{E'}^L \tilde{I}_{E,E'}^{RL} \right], \tag{5.26}$$

---

[6] Note also that the field operator 5.21 is quite general, i.e., the derivation that follows can be done by assuming this operator written in terms of any type of single-particle states, not necessarily scattering states (Chen and Di Ventra, 2003).

[7] Alternatively, we could have defined operators associated with the incoming, $\tilde{a}_{L,R}$, and outgoing, $\tilde{b}_{L,R}$, *amplitudes* of the scattering processes (see Fig. 3.14). The derivation of shot noise would follow similarly (see, e.g., Blanter and Büttiker, 2000).

where we have defined the functions[8]

$$\tilde{I}^{ij}_{E,E'} \equiv (\Psi^i_E)^* \frac{\partial \Psi^j_{E'}}{\partial x} - \frac{\partial (\Psi^i_E)^*}{\partial x} \Psi^j_{E'}, \qquad i,j = R, L. \qquad (5.27)$$

Let us now calculate the ensemble average of the current operator using the relations 5.24

$$\langle I \rangle = e \langle \hat{I}(x) \rangle = \frac{e\hbar}{2mi} \int dE \int d\mathbf{r}_\perp \left[f^L_E \tilde{I}^{LL}_{E,E} + f^R_E \tilde{I}^{RR}_{E,E} \right]$$

$$= \frac{e\hbar}{2mi} \int dE \int d\mathbf{r}_\perp \left[f^L_E - f^R_E \right] \tilde{I}^{LL}_{E,E}, \qquad (5.28)$$

where in the last identity I have used the property $\tilde{I}^{LL}_{E,E} = -\tilde{I}^{RR}_{E,E}$, a direct consequence of flux conservation (see discussion in Sec. 3.3.3).

First, we note that this expression does not depend on time. Second, by comparing it with Eq. 3.37, we see that it does not depend on the position x either. We can thus evaluate it at any point in space. Let us for instance calculate it at $x \to +\infty$ for leads that are described by translationally invariant Hamiltonians.

In this case we know from Sec. 3.3.2 that the asymptotic behavior of the wave-functions $\Psi^+_{i(\pm k_i)}(\mathbf{r})$ is given by Eqs. 3.197 and 3.198.[9] Substituting this limit into 5.28, and carrying out the spatial integral we finally get

$$\langle I \rangle = \frac{e}{2\pi\hbar} \int dE \left[f^L_E - f^R_E \right] T(E), \qquad (5.29)$$

which is precisely the current (per spin), Eq. 3.56, for the one-channel case.

The second moment of the current

We can now turn to the derivation of shot noise. From the general definition 5.3 of second moment of the current, and the noise spectrum 5.14 we write[10]

$$S(\omega; x_1, x_2) = e^2 \int_{-\infty}^{+\infty} dt \, e^{i\omega t} \langle \Delta \hat{I}(x_1, t) \Delta \hat{I}(x_2, 0) + \Delta \hat{I}(x_2, 0) \Delta \hat{I}(x_1, t) \rangle, \qquad (5.30)$$

with $\Delta \hat{I}(x, t) = \hat{I}(x, t) - \langle \hat{I} \rangle$.

We can now proceed as we did for the calculation of the current, i.e., we replace Eqs. 5.26 and 5.28 into Eq. 5.30 and perform the averages using

---

[8] Note that for $E = E'$, the functions $\tilde{I}^{ij}_{E,E}$ contain only one density of states $D(E)$.
[9] Clearly, the summations in Eqs. 3.197 and 3.198 do not appear in the one-channel case.
[10] Again, since we are considering an ideal steady state, the second moment of the current can depend on time differences only.

the properties 5.24 of the creation and annihilation operators. In this case, however, the calculation is lengthier and cumbersome as one comes across terms of the form $\langle a_1^\dagger a_2 a_3^\dagger a_4 \rangle$, where a_i's are operators corresponding to different energies and satisfying different statistics, i.e., left and right Fermi distributions. The evaluation of these averages is simplified by the use of Wick's theorem (see, e.g., Giuliani and Vignale, 2005), which allows us to write $(i,j,k,l = R, L)$

$$\left\langle a_{E_1}^{i\dagger} a_{E_2}^{j} a_{E_3}^{k\dagger} a_{E_4}^{l} \right\rangle = \left\langle a_{E_1}^{i\dagger} a_{E_2}^{j} \right\rangle \left\langle a_{E_3}^{k\dagger} a_{E_4}^{l} \right\rangle$$
$$+ \delta_{il}\delta_{jk}\delta(E_1 - E_4)\delta(E_2 - E_3) f_{E_1}^{i} \left[1 - f_{E_2}^{j}\right]. \quad (5.31)$$

I give here just the final result of this long calculation (Blanter and Büttiker, 2000; Lagerqvist et al., 2004)

$$S(\omega; x_1, x_2) = -\frac{e^2 \hbar^3}{2m^2} \int dE \int d\mathbf{r}_{1\perp} \int d\mathbf{r}_{2\perp} \sum_{i,j=L,R} f_{E+\hbar\omega}^{i}(1 - f_E^{j})$$
$$\times \left[\tilde{I}_{E,E+\hbar\omega}^{ij}(\mathbf{r}_1) \times \tilde{I}_{E+\hbar\omega,E}^{ji}(\mathbf{r}_2) \right]. \quad (5.32)$$

We are interested in the zero-temperature and zero-frequency limits of Eq. 5.32. If the electrochemical potential on the left μ_L is higher than the electrochemical potential on the right μ_R (with $\mu_L - \mu_R = eV$), then the shot noise is[11]

$$S(0; x_1, x_2) = -\frac{e^2 \hbar^3}{2m^2} \int_{\mu_L}^{\mu_R} dE \int d\mathbf{r}_{1\perp} \int d\mathbf{r}_{2\perp} \left[\tilde{I}_{E,E}^{LR}(\mathbf{r}_1) \times \tilde{I}_{E,E}^{RL}(\mathbf{r}_2) \right]. \quad (5.33)$$

I note that if one knows the single-particle states self-consistently anywhere in space (using, e.g., ground-state DFT in combination with the Lippmann–Schwinger equation, as discussed in Chapter 3), then one can employ Eq. 5.33 directly to estimate shot noise without the need to extract transmission probabilities (Chen and Di Ventra, 2003).

It is, however, common to express shot noise in terms of scattering amplitudes and this is what I set out to do in the following.

Expression 5.33 can be evaluated at any pair of points in space (giving rise to different *cross-correlations*). Let us then evaluate $S(0; x_1, x_2)$ for, say, $x_1 \to \infty$ and $x_2 \to \infty$ and call this quantity S_{RR}. Substituting the asymptotic behavior of the wave-functions $\Psi_{i(\pm k_i)}^{+}(\mathbf{r})$ (Eqs. 3.197 and 3.198)

---

[11] In this case $f_E^i(1 - f_E^j) = 0$, except for $i = L$, $j = R$ and $\mu_R < E < \mu_L$.

into Eq. 5.32 and carrying out the integrals we get (Exercise 5.2)

$$S_{RR} = \frac{e^2}{\pi\hbar}\int_{\mu_R}^{\mu_L} dE\, T(E)\, R(E) = \frac{e^2}{\pi\hbar}\int_{\mu_R}^{\mu_L} dE\, T(E)\,[1 - T(E)]. \quad (5.34)$$

We readily see that we have recovered the form of shot noise in terms of transmission and reflection probabilities we have derived with simple arguments in 5.20. We can also perform the same calculation for $x_1 \to -\infty$ and $x_2 \to -\infty$ and call this noise S_{LL}, and for $x_1 \to -\infty$ and $x_2 \to \infty$ (or equivalently $x_1 \to \infty$ and $x_2 \to -\infty$) and call the result S_{LR} (S_{RL}). We then find the relations

$$S_{RR} = S_{LL} = -S_{LR} = -S_{RL}, \quad (5.35)$$

as in Eq. 5.20.[12]

The Poisson limit

Let us now discuss the limits under which the shot noise 5.34 reduces to the Poisson limit 5.16. If $T(E) \ll 1$, from Eq. 5.34 we get

$$S_{RR} \approx \frac{e^2}{\pi\hbar}\int_{\mu_L}^{\mu_R} dE\, T(E). \quad (5.36)$$

If we also assume the transmission coefficient is independent of energy ($T(E) \equiv T$) we can take it out of the energy integral so that[13]

$$S_{RR} \approx \frac{e^2}{\pi\hbar}T\int_{\mu_L}^{\mu_R} dE = \frac{e^3 V}{\pi\hbar}T, \quad (5.37)$$

where in the last equation I have made use of the fact that $\mu_L - \mu_R = eV$. Within the *same* approximations, the average current 5.29 is

$$\langle I \rangle \approx \frac{e^2 V}{2\pi\hbar}T, \quad (5.38)$$

so that Eq. 5.37 can be written as

$$S_{RR} \approx 2e\frac{e^2 V}{2\pi\hbar}T = 2e\langle I\rangle = S_P, \quad (5.39)$$

which is the classical value 5.16 of independent electrons.

---

[12] The condition $\sum_i S_{ij} = 0$, with i,j either L or R, is a direct consequence of current conservation (Exercise 5.2).

[13] We may make this approximation when the bias is small and away from resonances. In this case, the transmission coefficient is evaluated at the Fermi level (see discussion in Sec. 3.3.3).

The multi-channel case

The derivation of shot noise we have done so far can be easily extended to the case in which there are N_c^L independent incoming channels i and N_c^R independent outgoing channels f (still in a two-probe geometry).[14] I give here the final result for S_{RR}, since the other cross-correlations S_{ij} follow the relations 5.35. By summing up all contributions from independent channels one finds

$$S_{RR} = \frac{e^2}{\pi\hbar} \sum_{ijkl} \int_{\mu_L}^{\mu_R} dE\, r^*_{kf} r_{ki} \tau^*_{li} \tau_{lf} = \frac{e^2}{\pi\hbar} \int_{\mu_L}^{\mu_R} dE\, \text{Tr}\{r^\dagger r \tau^\dagger \tau\}$$

$$= \frac{e^2}{\pi\hbar} \sum_n \int_{\mu_L}^{\mu_R} dE\, T_n(E)[1 - T_n(E)], \quad (5.40)$$

where the matrices r and τ are the block matrices defined in Eq. 3.208. The last equality in 5.40 is for the particular choice of eigenchannel basis in which the matrix $\tau^\dagger \tau$ is diagonal with eigenvalues T_n ($1 \leq n \leq N_c^L$), and the matrix $r^\dagger r$ is diagonal with eigenvalues $1 - T_n$ (see Eq. 3.211).

As noted before, shot noise cannot be written in terms of transmission probabilities only. In addition, from Eq. 5.40, we see that it is proportional to the sum of terms of the type $r^*_{kf} r_{ki} \tau^*_{li} \tau_{lf}$, with i, f, l, k indexes describing different channels. If $i \neq f$, these products are not real-valued so that they depend on the relative *phase* of the transmission and reflection amplitudes, i.e., they describe *interference* among carriers in different channels. This is not the case for the current (see Eq. 5.42 below).

If the coefficients $T_n(E)$ are again assumed independent of energy ($T_n(E) \equiv T_n$) Eq. 5.40 becomes

$$S_{RR} \approx \frac{e^3 V}{\pi\hbar} \sum_n T_n(1 - T_n), \quad (5.41)$$

that generalizes the result 5.37.

---

[14] Note that the assumption of the existence of independent channels (*Approximation 5* of the Landauer approach) is even more important for the noise than for the current, owing to the fact that shot noise is related to the current autocorrelation function: any channel mixing may result in a completely different noise form. For instance, if we allow for mixing of the channels we would have noise terms of the type (assuming $N_c^L = N_c^R$)

$$S = \frac{e^2}{\pi\hbar} \int_{\mu_R}^{\mu_L} dE\, \text{Tr}\{r^\dagger \tau\} \text{Tr}\{\tau^\dagger r\},$$

which would be zero whenever expression 5.40 is zero, but would show different noise structure in all other cases. In addition, this mixing, being of quantum-mechanical origin, would not lead to the Poisson limit 5.16 even at small transmissions.

5.2 Shot noise

Similarly, I have shown (Sec. 3.6) that in the eigenchannel basis the average current is

$$\langle I \rangle = \frac{e}{2\pi\hbar} \sum_n \int dE \left[f_E^L - f_E^R \right] T_n(E) = \frac{e}{2\pi\hbar} \int dE \left[f_E^L - f_E^R \right] \text{Tr}\{\tau\tau^\dagger\}, \tag{5.42}$$

which in the limit of zero temperature, and assuming the coefficients $T_n(E)$ do not depend on energy, reduces to

$$\langle I \rangle \approx \frac{e^2 V}{2\pi\hbar} \sum_n T_n. \tag{5.43}$$

The multi-channel expression for the Poisson noise then reads

$$S_P \approx 2e\langle I \rangle = \frac{e^3 V}{\pi\hbar} \sum_n T_n. \tag{5.44}$$

It is customary to define a dimensionless quantity, called the *Fano factor*, that quantifies the departure of shot noise from its Poisson value S_P. The Fano factor is defined as

$$F = \frac{S_{RR}}{S_P}, \tag{5.45}$$

which, inserting Eq. 5.40 and Eq. 5.42, becomes

$$F = \frac{S_{RR}}{S_P} = \frac{\int_{\mu_R}^{\mu_L} dE \, \text{Tr}\{r^\dagger r \tau^\dagger \tau\}}{\int_{\mu_R}^{\mu_L} dE \, \text{Tr}\{\tau\tau^\dagger\}}. \tag{5.46}$$

For energy-independent transmission coefficients (and in the eigenchannel basis) Eq. 5.46 reduces to

$$F \approx \frac{\sum_n T_n[1 - T_n]}{\sum_n T_n}. \tag{5.47}$$

Since $0 \leq T_n \leq 1$, the Fano factor F is always smaller than unity, so that from the above we would conclude that shot noise is always suppressed with respect to its Poisson value (the noise is called *sub-poissonian*). However, this result is not generally true. Let us indeed recall that we have assumed all along non-interacting particles. For interacting particles one can indeed obtain *super-poissonian* values of noise, i.e., $F > 1$ (see, e.g., Iannaccone et al., 1998).

Finally, I just mention that, under the same approximations and assumptions we have employed above, the generalization of Eq. 5.40 to the case in which multiple terminals are present (Sec. 3.9) follows similarly. I leave this case as an exercise for the reader (Exercise 5.3).

5.3 Counting statistics

We can now perform the same calculation for higher moments of the current, Eq. 5.5. This is quite complicated and I will not report it here. Instead, I will proceed in a somewhat simpler way, which shows the underlying concept (see, e.g., Lee et al., 1995).

Assume we know the probability density $P_m(t)$ that m particles[15] cross a given surface of the sample in a given interval of time t much larger than the characteristic time $e/\langle I \rangle$ for electrons to pass through that same surface.[16]

Given the above probability distribution, we can construct its *characteristic function*

$$G(k) = \sum_{m=0}^{N} e^{imk} P_m(t), \qquad (5.48)$$

which is the Fourier transform of $P_m(t)$ in the range of values of the distribution. In the above, N is the average number of "attempts" to cross the surface in the interval t. The function $G(k)$ is also known as the *moment generating function*, because its series expansion provides *all* moments of the distribution

$$G(k) = \sum_{0}^{\infty} \frac{(ik)^m}{m!} \frac{S_m}{e^m}, \qquad (5.49)$$

with S_m the m-th moment 5.5, and $S_0 = 1$.

The logarithm of the characteristic function generates the *cumulants* of the distribution[17]

$$\ln G(k) = \sum_{1}^{\infty} \frac{(ik)^m}{m!} \langle\langle n_m(t) \rangle\rangle, \qquad (5.50)$$

---

[15] I am implicitly assuming that this number is discrete so that I am employing a semiclassical description of electron motion.

[16] The distribution $P_m(t)$ is usually assumed to be of the *Bernoulli* type (also known as the *binomial distribution*). For a single channel with transmission probability T this is

$$P_m(t) = \binom{N}{m} T^m (1-T)^{N-m} = \frac{N!}{m!(N-m)!} T^m (1-T)^{N-m}$$

with $m \leq N$ (both integers). The physical reason for this choice is that it represents the distribution of finding m non-interacting particles out of N contained in two separate but communicating volumes. When one of the two volumes is set to go to infinity (and hence it represents a reservoir), while the particle density in the other stays constant, this distribution approaches the Poisson distribution.

[17] Here I am considering the correlations defined by subtracting the averages as in Eq. 5.4.

where the m-th cumulant is

$$\langle\langle n_m\rangle\rangle = \frac{1}{e^m}\int_0^t dt_1\ldots dt_m \langle\langle I(t_1)\ldots I(t_m)\rangle\rangle. \qquad (5.51)$$

For an ideal steady state, and under the same assumptions we have used to derive shot noise, the first cumulant is proportional to the average current $\langle\langle n(t)\rangle\rangle = \langle I\rangle t/e$, while the second, to the shot noise, $\langle\langle n_2(t)\rangle\rangle = S_{RR}t/2e^2$.

By knowing *all* cumulants of the distribution, one can evaluate the characteristic function $G(k)$, from which we can back-Fourier transform Eq. 5.48 to obtain the distribution function $P_m(t)$. For instance, a gaussian distribution has zero cumulants for $m > 2$, while a poissonian distribution has all cumulants equal to its average value. Therefore, if we could measure cumulants we would know the full (counting) statistics of our transport problem.

The actual measurement of these cumulants is, however, not an easy task. The reason is because the n-th cumulant depends on sums and differences of functions of all the $(n-1)$-th moments (see, e.g., van Kampen, 1992). Since measurements are done on time scales much longer than the characteristic time $e/\langle I\rangle$, the *central limit theorem* makes the lower-order moments the dominant contribution to the cumulants, so that the latter are difficult to distinguish beyond the second order.[18]

Recently, the third cumulant has been measured in the case of electronic transport across a tunnel junction, showing that in this case the distribution is likely to be poissonian (Bomze et al., 2005).

5.4 Thermal noise

As anticipated, thermal noise (also called *Johnson–Nyquist noise*) is due to thermal fluctuations of the electrons in a conductor close to equilibrium. From the analysis that led us to Eqs. 5.17, 5.18 and 5.19 we see that it originates from the fluctuations in the occupation number Δn_i of the incident state.

We can derive the expression for thermal noise using a similar approach to that we have used to derive shot noise. We first calculate the frequency dependence of the noise, and then set the frequency to zero *after* setting the equilibrium condition, i.e., by imposing the same distribution functions in the left and right electrode (at finite temperature).

Our starting point is therefore Eq. 5.32 and we evaluate it for say $x_1 \to$

---

[18] The central limit theorem states that if $y_1, y_2, \ldots, y_N$ is a set of *independent* stochastic variables whose probability distribution has *finite* variance, the scaled sum $(y_1 + y_2 + \ldots + y_N)/\sqrt{N}$ has a gaussian distribution for $N \to \infty$. The gaussian distribution has zero cumulants beyond the second.

∞ and $x_2 \to \infty$. Call this quantity $S_{RR}(\omega)$. For the one-channel (and two-terminal) case, and assuming that the transmission amplitudes do not depend on energy, we then find (Exercise 5.4)[19]

$$S_{RR}(\omega) = \frac{e^2}{2\pi\hbar}\left\{T(1-T)\int dE\left[f_{LR}(E,\omega) + f_{RL}(E,\omega)\right]\right.$$
$$\left. + T^2\int dE[f_{LL}(E,\omega) + f_{RR}(E,\omega)]\right\}, \qquad (5.52)$$

where

$$f_{ij}(E,\omega) = f_E^i[1 - f_{E+\hbar\omega}^j] + f_{E+\hbar\omega}^j[1 - f_E^i], \qquad (5.53)$$

for i,j either L or R. In the multi-channel case (and in the eigenchannel basis) Eq. 5.52 generalizes to

$$S_{RR}(\omega) = \frac{e^2}{2\pi\hbar}\left\{\sum_n T_n(1-T_n)\int dE\left[f_{LR}(E,\omega) + f_{RL}(E,\omega)\right]\right.$$
$$\left. + \sum_n T_n^2\int dE[f_{LL}(E,\omega) + f_{RR}(E,\omega)]\right\}. \qquad (5.54)$$

Since the transmission probabilities are outside the energy integrals, the energy integration can be performed exactly to give (Exercise 5.5)

$$S_{RR}(\omega) = \frac{e^2}{2\pi\hbar}\left\{\sum_n T_n(1-T_n)\left[(\hbar\omega - eV)\coth\left(\frac{\hbar\omega - eV}{2k_B\theta}\right)\right.\right.$$
$$\left.+ (\hbar\omega + eV)\coth\left(\frac{\hbar\omega + eV}{2k_B\theta}\right)\right]$$
$$\left.+ 2\hbar\omega\coth\left(\frac{\hbar\omega}{2k_B\theta}\right)\sum_n T_n^2\right\}. \qquad (5.55)$$

This equation contains *both* contributions to noise: thermal and partition. For instance, if we set the frequency to zero and *then* the temperature to zero in Eq. 5.54, we recover the shot noise formula Eq. 5.41 (Exercise 5.6). If instead we assume the system to be at equilibrium, $f_L = f_R$ ($V = 0$), we get

$$S_{RR}(\omega) = \frac{e^2\omega}{\pi}\coth\left(\frac{\hbar\omega}{2k_B\theta}\right)\sum_n T_n, \qquad (5.56)$$

---

[19] Since we are now working with a finite frequency, we are also implicitly assuming that this frequency, temperature and bias are all smaller than a characteristic (internal) frequency of the system in such a way that the transmission amplitudes are frequency-independent and thus the *particle* current is conserved. If this were not the case, then only the *total* current would be conserved, i.e., the particle current plus the displacement current. In this case, the condition $\sum_i S_{ij}(\omega) = 0$, with i,j either L or R, is not satisfied (see Footnote [12]).

which in the limit of zero frequency is[20]

$$S_{RR}(0) = \frac{4e^2 k_B \theta}{h} \sum_n T_n = 4k_B \theta \frac{\langle I \rangle}{V} = 4k_B \theta G, \qquad (5.57)$$

where in the last equation I have used the definition of two-probe linear-response conductance (per spin), $G = \langle I \rangle / V$, and used the result 5.43.

Equation 5.57 is the classical result for thermal noise. Similar to the Poisson limit of shot noise, we could have derived it with classical arguments (Exercise 5.7), and it simply states that the fluctuations in the occupation number at finite temperature and zero bias are accompanied by dissipative effects, proportional to the conductance of the sample. This is one form of the more general *fluctuation-dissipation theorem* for systems close to equilibrium I have introduced in Sec. 2.3.3 in the context of the Kubo formalism. We also see that thermal noise does not provide more information on the carrier dynamics than the conductance itself.

Summary and open questions

I have discussed here two types of intrinsic noise: shot and thermal. I have derived, within a single-particle scattering picture, closed forms for both. I have also discussed, within a simplified semiclassical picture, the type of information one could extract from a measurement of the higher moments of the current (counting statistics).

It is worth pointing out that noise properties of nanoscale systems have received much less attention than the average current. This is somewhat unfortunate since noise may provide more information on the microscopic properties of nanoscale structures than the average current itself.

In addition, effects of many-body interactions (beyond mean field) on noise properties of nanoscale conductors are still poorly understood. Along these lines, a comprehensive study of the effect of electron-ion interactions on noise would also be desirable.

Exercises

5.1 **The Poisson limit of shot noise.** Consider electrons crossing a given surface at a rate $n(t) = I(t)/e$, where $I(t)$ is the current. Define the number of electrons that pass in an interval of time T_p as

$$N = \int_0^{T_p} dt\, n(t), \qquad (E5.1)$$

---

[20] Recall that $\lim_{x \to 0} \coth(x) \approx 1/x + O(x)$.

with ensemble average $\langle N \rangle$ (see Eq. 5.11). Define the variations $\Delta N = N - \langle N \rangle$ with respect to the average. Assume that N is a Poisson process, so that its moments are all equal to its average. Using the Wiener–Khinchin theorem 5.14, show that the zero-frequency limit of the noise is $S(0) = 2e\langle I \rangle$.

5.2 Use Eqs. 3.197 and 3.198 to derive Eq. 5.34, and prove the conditions 5.35, which are a direct consequence of flux conservation.

5.3 **Multi-terminal noise.** Consider the current operator 5.25 at a terminal $\hat{I}_\alpha(t)$ with average steady-state current $\langle \hat{I}_\alpha \rangle$, so that $\Delta \hat{I}(t) = \hat{I}_\alpha(t) - \langle \hat{I}_\alpha \rangle$. Define the correlation function between two terminals α and β as (from Eq. 5.4)

$$S_{\alpha\beta}(t - t') \equiv \frac{e^2}{2} \langle \Delta \hat{I}_\alpha(t) \Delta \hat{I}_\beta(t') + \Delta \hat{I}_\alpha(t') \Delta \hat{I}_\beta(t) \rangle. \qquad (E5.2)$$

Within the single-particle scattering picture, generalize the result 5.40 to multiple terminals by calculating the zero-frequency and zero-temperature limit of Eq. E5.2. Show that the correlations are negative for $\alpha \neq \beta$, and positive otherwise.

5.4 Starting from the noise power, Eq. 5.32, derive the relation 5.52 assuming that the transmission amplitudes do not depend on energy.

5.5 Prove that Eq. 5.55 obtains from Eq. 5.54.

5.6 **Order of limits.** Start from the power spectrum, Eq. 5.54, and show that if we take the limit of the frequency to zero *before* the zero-temperature limit we recover the shot noise formula 5.41. Take instead the equilibrium limit first (zero bias) in Eq. 5.54, and *then* the zero-frequency limit. Show that this order of limits leads to Eq. 5.57.

5.7 **Thermal noise from a Langevin approach.** Consider a general RC-circuit with resistance R and capacitance C. The charge Q on the capacitor obeys a *stochastic differential equation*

$$\frac{dQ}{dt} = -\frac{Q}{RC} + L(t) \qquad (E5.3)$$

where $L(t)$ is a *stochastic process* with zero average and δ-autocorrelation:

$$\langle L(t) \rangle = 0, \qquad (E5.4)$$

$$\langle L(t) L(t') \rangle = \Gamma \delta(t - t'), \qquad (E5.5)$$

where $\langle \ldots \rangle$ indicates classical canonical average, and Γ is some positive constant. Equation E5.3 is a type of *Langevin equation*.

Using equilibrium statistical mechanics, determine the constant Γ and show that the current autocorrelation function is

$$\langle \Delta I(t)\Delta I(t')\rangle = \frac{2k_B\theta}{R}\delta(t-t'), \qquad (E5.6)$$

from which the thermal noise result 5.57 follows.

Show also that the *voltage* autocorrelation function is

$$\langle \Delta V(t)\Delta V(t')\rangle = 2k_B\theta R\,\delta(t-t'). \qquad (E5.7)$$

6
Electron-ion interaction

Electrical current is affected by the interaction between electrons and ions. Due to this interaction electrons may undergo *inelastic* transitions between states of different energy, even if the electrons themselves are considered non-interacting with each other. These transitions appear as discontinuities (steps) in the current (conductance) at biases corresponding to the phonon spectrum of the structure. In reality, the phonon spectrum is *renormalized* by both the electron-phonon interaction at equilibrium, and by the current itself. The latter fact makes the concept of phonons under current flow fundamentally less obvious. I will discuss this point in Sec. 6.5.

An example of inelastic features in nanoscale systems is illustrated in Fig. 6.1 where the conductance of a gold point contact is measured as a function of bias. The conductance shows a step in the range between 10 and 20 meV corresponding to the energy of the vibrational modes of the whole system – gold point contact plus electrodes – that couple more effectively with electrons.

Via the same inelastic mechanism, electrons can exchange energy with the ions and thus heat up the nanostructure while they propagate across it. As we will see later, this phenomenon, called *local ionic heating*, may have dramatic effects on the stability of nanostructures.

Finally, in a current-carrying system ions may be displaced by local current-induced rearrangements of the electronic distribution – the local resistivity dipoles I discussed in Sec. 3.2 – without the intervention of inelastic processes. The forces responsible for such displacements are known as *current-induced forces*. Despite many studies, past and present, these forces challenge our understanding of non-equilibrium phenomena, starting from their basic definition for a current-carrying system to their, yet unsolved, conservative character.

In this chapter I will discuss all these effects. I will present simple argu-

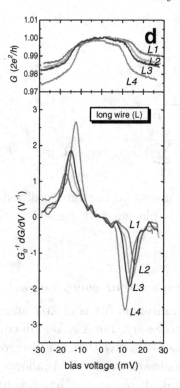

Fig. 6.1. Experimental differential conductance of a gold point contact of different nominal lengths. The numerical derivative of the differential conductance as a function of bias is also shown. Reprinted with permission from Agraït et al. (2002).

ments on the bias dependence of these phenomena, as well as their quantum description within a set of approximations that render their solution tractable.

6.1 The many-body electron-ion Hamiltonian

Since now the ions enter the transport problem explicitly, we need to find the most appropriate Hamiltonian to describe the combined dynamics of electrons and ions *in the presence of current*. Here, I will just focus on the steady-state transport problem and will briefly mention dynamical effects in the "Summary and open questions" section of this chapter.

Let us start from the general many-body Hamiltonian[1]

$$\hat{\mathbf{H}} = \hat{\mathbf{H}}_{el} + \hat{\mathbf{H}}_{ion} + \hat{\mathbf{H}}_{el-ion}, \qquad (6.1)$$

---

[1] As in Chapter 4, operators in bold face indicate many-body quantities. We also neglect relativistic effects.

where

$$\hat{H}_{el} = -\sum_i \frac{\hbar^2}{2m} \nabla_i^2 + \frac{1}{2}\sum_{i\neq j} \frac{e^2}{|\mathbf{r}_i - \mathbf{r}_j|},$$

$$\hat{H}_{ion} = -\sum_i \frac{\hbar^2}{2M_i} \nabla_{\mathbf{R}_i}^2 + \frac{1}{2}\sum_{i\neq j} \frac{Z_i Z_j e^2}{|\mathbf{R}_i - \mathbf{R}_j|},$$

$$\hat{H}_{el-ion} = -\sum_i \sum_j \frac{Z_j e^2}{|\mathbf{r}_i - \mathbf{R}_j|} \equiv \sum_i \sum_j V_{el-ion}(\mathbf{r}_i - \mathbf{R}_j). \quad (6.2)$$

In Eq. 6.2 $Z_i|e|$, M_i and $\mathbf{R}_i$ are the charge, mass and coordinate of ion i, respectively. The symbol $\nabla_{\mathbf{R}_i}^2$ indicates the Laplacian with respect to the ionic coordinates.

6.1.1 The adiabatic approximation for a current-carrying system

Direct diagonalization of the many-body Hamiltonian 6.1 is clearly an impossible task, so further approximations are necessary. For a system in equilibrium, we know that the ions, being much more massive than the electrons ($m/M_i \sim 10^{-4}$), move slower than the electronic subsystem. This allows us to conjecture that the electrons, for each atomic displacement, have enough time to "readjust" themselves before the atoms move again. We can therefore solve the electronic part of the Hamiltonian at fixed atomic positions, the latter treated as parameters.

This is called the *adiabatic* or *Born–Oppenheimer approximation* – see, e.g., Bransden and Joachain (1983), for a complete derivation. In this approximation the ionic Hamiltonian $\hat{H}_{ion}$ can be written as an effective Hamiltonian where the ion-ion interaction potential, at given ionic positions, is "mediated" by the electronic configuration corresponding to those ionic positions. Quite generally, this new effective Hamiltonian can be written as

$$\hat{H}_{ion}^{eff} = -\sum_i \frac{\hbar^2}{2m} \nabla_{\mathbf{R}_i}^2 + \sum_{ij} V_{ion}(\mathbf{R}_i - \mathbf{R}_j). \quad (6.3)$$

The energy manifold that is spanned by the nuclear motion – the energy manifold defined by the potential $\sum_{ij} V_{ion}(\mathbf{R}_i - \mathbf{R}_j)$ – is called the *Born-Oppenheimer surface*.

The question now is whether this approximation is valid for a system out of equilibrium, and in particular if an electronic current is present. The first comment I make is that out of equilibrium – as well as in equilibrium – this approximation may neglect correlated effects of coupled electron-ion dynamics when the ion wave-functions cannot be simply approximated as δ-functions centered at the classical ion positions. I will neglect these effects and adopt the adiabatic approximation, which in the case of current-carrying systems simply states that the electrons can quickly readjust their *non-equilibrium* charge distribution in response to the ionic motion. In other words I define

Adiabatic approximation for a current-carrying system: *The electronic subsystem is assumed to remain infinitesimally close to an instantaneous current-carrying steady state, corresponding to the instantaneous set of ionic positions.*

We therefore assume that for fixed ionic positions, $\mathbf{R}_i^0$, we can solve the electronic part of the Hamiltonian 6.1.

Single-particle approximation

In practice, I will assume in the following discussions that using, for instance, the approximations of the Landauer approach (Sec. 3.1), we have been able to determine the single-particle scattering states associated with an effective single-particle Hamiltonian that replaces Hamiltonian $\hat{H}_{el} + \hat{H}_{el-ion}$ at fixed ionic positions. These are the states we have determined in Chapter 3, where the effective single-particle Hamiltonian could be, e.g., the Hartree one, Eq. 3.10, or the ground-state DFT Hamiltonian 3.12.

In the context of a one-electron theory, it is also customary to define the electron-ion interaction in terms of an effective screened potential $V^{ps}(\mathbf{r}; \mathbf{R}_i)$ so that the one-particle electron-ion Hamiltonian can be written as

$$\hat{H}_{el-ion} = \sum_i V^{ps}(\mathbf{r}; \mathbf{R}_i). \qquad (6.4)$$

In practical calculations, $V^{ps}(\mathbf{r}; \mathbf{R}_i)$ could be for instance a *pseudopotential*, an effective potential where the energetically deep electrons of the ions (*core* electrons) are considered "frozen", and only *valence* electrons participate directly in the bonding (see, for instance, Harrison, 1989).[2]

This is a reasonable assumption also in the transport problem, since the

---

[2] Note that pseudopotentials are not necessarily local in the electronic coordinates. Here, we assume them local to simplify the notation. Also, while I am not aware of an "all-electron" calculation of inelastic scattering in nanostructures, (i.e., one in which also core electrons are explicitly taken into account), the reader should keep in mind that such a calculation is in principle possible following similar considerations as the ones we make in this section.

core electrons are highly localized and tightly bound to the nucleus. In addition, their energies often fall well below the bottom of the conduction bands of both electrodes, and definitely well below the electrochemical potentials even at large voltages (Di Ventra and Lang, 2002).

The core electrons plus the positive bare-ion charge thus form an effective ion of charge Z_i^*. In the following, I will implicitly adopt this pseudopotential picture, but the discussions that follow do not depend on the chosen form of such potential.

6.1.2 The phonon subsystem

Let us now determine the effect of ionic displacements on the Hamiltonians 6.2. Let us start with $\hat{H}_{ion}$.

As we will see in detail in Sec. 6.5, at finite current the ions may be displaced by local rearrangements of the electronic distribution without the intervention of inelastic processes. These current-induced forces, however, are proportional to the current – at least in the linear response, see Eq. 6.105 – and for relatively small biases we can therefore assume their effect on the ionic displacement to be negligible.[3]

Let us then apply the following (small) displacement to the ions

$$\mathbf{R_i} = \mathbf{R}_i^0 + \mathbf{Q}_i \tag{6.5}$$

and expand $V_{ion}(\mathbf{R}_i - \mathbf{R}_j)$ in Taylor series about the equilibrium positions to second order. The first-order term in the expansion is zero because we assume the ions at equilibrium. Higher-order terms give rise to phonon-phonon scattering processes. I will consider these later when dealing with heat dissipation (see Sec. 6.3). We thus find

$$V_{ion}(\mathbf{R}_i - \mathbf{R}_j) \approx V_{ion}(\mathbf{R}_i^0 - \mathbf{R}_j^0) + \frac{1}{2}\sum_{\mu,\nu}(\mathbf{Q}_i - \mathbf{Q}_j)_\mu (\mathbf{Q}_i - \mathbf{Q}_j)_\nu F_{i\mu;j\nu}, \tag{6.6}$$

where $\mu, \nu = x, y, z$, are the coordinates of each ionic position, and

$$F_{i\mu;j\nu} = \left.\frac{\partial^2 V_{ion}(\mathbf{R}_i - \mathbf{R}_j)}{\partial R_{i\mu}\, \partial R_{j\nu}}\right|_{\mathbf{R}_i = \mathbf{R}_i^0; \mathbf{R}_j = \mathbf{R}_j^0} \tag{6.7}$$

are known as the elements of the *dynamical response matrix*, and we have defined $R_{i\mu}$, the μ-th component of the vector $\mathbf{R}_i$. Since the term $V_{ion}(\mathbf{R}_i^0 - \mathbf{R}_j^0)$ only redefines the zero of energy, we assume it zero in Eq. 6.6.

---

[3] For metallic nanojunctions, such as gold point contacts, at biases of tens of millivolts these forces are of the order of $10^{-3} - 10^{-2}$ eV Å^{-1}, and therefore negligible (see also Sec. 6.5).

6.1 The many-body electron-ion Hamiltonian

Let us now introduce "normal" coordinates, $\{q_{i\mu}\}$, for the N ions such that the μ-th component ($\mu = x, y, z = 1, 2, 3$) of $\mathbf{Q}_i$ is

$$(\mathbf{Q}_i)_\mu = \sum_{j=1}^{N} \sum_{\nu=1}^{3} A_{i\mu,j\nu} q_{j\nu}. \tag{6.8}$$

The transformation matrix, $\mathbf{A} = \{A_{i\mu,j\nu}\}$, which contains information on the geometry of the system, satisfies the "orthonormality" relation[4]

$$\sum_{i,\mu} M_i A_{i\mu,j\nu} A_{i\mu,j'\nu'} = \delta_{j\nu,j'\nu'}. \tag{6.9}$$

In the following it will be convenient to use a single index j to label the index pair $i\mu$. Using relation 6.9 and the definition 6.8 it is then easy to prove (Exercise 6.1) that the Hamiltonian $\hat{\mathbf{H}}_{ion}^{eff}$ can be written as

$$\hat{\mathbf{H}}_{ion}^{eff} = \frac{1}{2} \sum_i \dot{q}_i^2 + \frac{1}{2} \sum_i \omega_i^2 q_i^2, \tag{6.10}$$

where ω_i is the frequency of the *normal mode i*, and the summation runs over all possible normal modes of the system.[5] We have also used the symbol $\dot{q}_i \equiv dq_i/dt$ to indicate the momentum conjugated with the normal coordinate q_i. If we now use the canonical transformation

$$\begin{cases} \tilde{Q}_i = \sqrt{\dfrac{\omega_i}{2\hbar}}\, q_i \\ \tilde{P}_i = \sqrt{\dfrac{1}{2\hbar\omega_i}}\, \dot{q}_i, \end{cases} \tag{6.11}$$

with $\left[\tilde{Q}_i, \tilde{P}_j\right] = i\hbar \delta_{ij}$, we can define the operators

$$\begin{cases} \hat{b}_i = \tilde{Q}_i + i\tilde{P}_i \\ \hat{b}_i^\dagger = \tilde{Q}_i - i\tilde{P}_i, \end{cases} \tag{6.12}$$

which satisfy boson commutation relations $\left[\hat{b}_j, \hat{b}_{j'}^\dagger\right] = \delta_{j,j'}$ and $[\hat{b}_j, \hat{b}_{j'}] =$

---

[4] With this definition the ion mass appears implicitly in the transformation matrix $A_{i\mu,j\nu}$. One could alternatively introduce it explicitly in the electron-phonon coupling constant.

[5] For N ions there are $3N$ normal modes so that, if we consider all the atoms of the junction plus the electrodes, N is essentially infinite. However, for our scope only the modes associated with the nanojunction – which may include quite a few atoms of the bulk electrodes – couple most strongly with the electrons, and therefore need to be considered explicitly. The modes of the bulk electrodes form a continuum and, therefore, all information is included in their density of states.

$[\hat{b}_j^\dagger, \hat{b}_{j'}^\dagger] = 0$ (Appendix A). With these operators Eq. 6.10 can be written as

$$\hat{\mathbf{H}}_{ion}^{eff} \equiv \hat{\mathbf{H}}_{ph} = \sum_j \left(\hat{b}_j^\dagger \hat{b}_j + \frac{1}{2} \right) \hbar\omega_j. \tag{6.13}$$

With the above approximations, we have thus written the ionic Hamiltonian as the one corresponding to a set of independent harmonic oscillators.

Let us now call $|N_j\rangle$ the orthonormal eigenvectors corresponding to the modes with energy $\hbar\omega_j$.[6] For later use, we also mention that the operators 6.12 when acting on the phonon wave-functions satisfy (see Appendix A)

$$\begin{cases} \hat{b}_i|N_j\rangle = \sqrt{N_j}\,|N_j - 1\rangle \\ \hat{b}_i^\dagger|N_j\rangle = \sqrt{N_j + 1}\,|N_j + 1\rangle \\ \hat{b}_i|0_j\rangle = 0, \end{cases} \tag{6.14}$$

where we have defined with $|0_j\rangle$ the state with zero phonons in the corresponding mode.[7]

In the absence of electron-phonon or phonon-phonon interactions, the average number of phonons in a given mode at a given temperature θ_{ph} is determined by the Bose–Einstein statistics 1.81, i.e.,

$$N_j = \langle \hat{b}_j^\dagger \hat{b}_j \rangle = \frac{1}{e^{\hbar\omega_j/k_B \theta_{ph}} - 1}, \tag{6.15}$$

where the symbol $\langle \cdots \rangle$ indicates average over the canonical equilibrium statistical operator $\hat{\rho}_{ph} = e^{-\beta\hat{\mathbf{H}}_{ph}}/\text{Tr}\{e^{-\beta\hat{\mathbf{H}}_{ph}}\}$ of independent phonons (see also Sec. 1.4.6).

In the presence of current and interactions among phonons, the above statistical operator is not necessarily the equilibrium one, but some arbitrary $\hat{\rho}_{ph}(t)$. For an arbitrary operator $\hat{A}$ acting on the phonon subsystem, the average of this operator on the phonon degrees of freedom is

$$\langle \hat{A} \rangle_t = \text{Tr}_{ph}\{\hat{\rho}_{ph}(t)\hat{A}\}, \tag{6.16}$$

where Tr_{ph} means the trace is done in the space of the phonons.

In what follows, any average involving the phonon degrees of freedom will mean an average of the type 6.16. However, I will assume that the *form* of the phonon distribution is the equilibrium 6.15 with at most an effective local temperature.

I also emphasize here that the temperature θ_{ph} appearing above is the

---

[6] This is a simplified notation to indicate the many-body state, in the Fock space of phonons that has N_j phonons in the mode with energy $\hbar\omega_j$ (see Appendix A).
[7] The *vacuum state* is defined as the many-body state with zero phonons in *all* modes.

temperature of the *ions*, i.e., the temperature associated with the ionic kinetic energy. As we will discuss in Sec. 6.3 this temperature can be quite different, locally in the junction, from the temperature θ of the electrons which appears in the Fermi–Dirac distribution (Eq. 1.80).

The propagator of non-interacting phonons

For later use, let us also write the "free" phonon propagator: the Green's function associated with the ionic vibrations, possibly screened by electrons, but not interacting with other ionic vibrations. We will need it when we treat the electron-phonon interaction within the NEGF approach (Sec. 6.2.2).

We start from the general definition of time-ordered Green's functions, Eq. 4.26. However, instead of replacing the field operators with the phonon operator b_j we choose the combination $\psi_j = \hat{b}_j^\dagger + \hat{b}_j$. This is a convenient choice so that the resultant Green's function describes propagation of phonons both forward and backward in time (Mattuck, 1976).

Let us then define the time-ordered single-phonon Green's function (which at equilibrium depends on the time difference only) for each mode j

$$D_j^0(t-t') = -\frac{i}{\hbar}\langle T[\psi_j(t)\psi_j^\dagger(t')]\rangle, \qquad (6.17)$$

where $\psi_j(t)$ is the operator $\psi_j = \hat{b}_j^\dagger + \hat{b}_j$ in the Heisenberg picture with Hamiltonian 6.13 (see Eq. 4.25), T is the time-ordering operator for bosons (see Footnote [8] in Chapter 4), and the average is done according to the equilibrium statistical operator $\hat{\rho}_{ph}$.

By carrying out the calculations explicitly and Fourier transforming $D_j^0(t-t')$ to the frequency domain, we finally obtain the "free" phonon propagator (see, e.g., Mahan, 1990)

$$D_j^0(\hbar\omega) = \frac{1}{\hbar\omega - \hbar\omega_j + i\delta} - \frac{1}{\hbar\omega + \hbar\omega_j - i\delta} = \frac{2\hbar\omega_j}{(\hbar\omega)^2 - (\hbar\omega_j)^2 + 2i\hbar\omega_j\delta}, \qquad (6.18)$$

where δ is an infinitesimal positive number.

We will also need the lesser phonon Green's function $D_j^<$, which according to Eq. 4.44 is defined as

$$D_j^<(t;t') = \frac{i}{\hbar}\langle \psi_j^\dagger(t)\psi_j(t')\rangle. \qquad (6.19)$$

At equilibrium it assumes the form (Mahan, 1990)

$$D_j^<(t-t') = -\frac{i}{\hbar}\left[(N_j+1)e^{-i\omega_j(t-t')} + N_j e^{i\omega_j(t-t')}\right]. \qquad (6.20)$$

This expression can be easily Fourier transformed to give

$$D_j^<(\hbar\omega) = -2\pi i\left[(N_j + 1)\,\delta(\hbar\omega + \hbar\omega_j) + N_j\delta(\hbar\omega - \hbar\omega_j)\right]. \quad (6.21)$$

Interacting phonons

Similar to the electron case, if one considers phonon interactions, and these processes can be described within many-body perturbation theory, the phonon propagator satisfies the Dyson equation 4.33, as well as the Keldysh equation 4.95. In equilibrium, if we call $\Pi_j(\omega)$ the irreducible self-energy associated with these interactions, the Dyson equation for phonons reads (from 4.35)

$$D_j(\hbar\omega) = \frac{D_j^0(\hbar\omega)}{1 - D_j^0(\hbar\omega)\Pi_j(\hbar\omega)} = \frac{2\hbar\omega_j}{(\hbar\omega)^2 - (\hbar\omega_j)^2 + 2i\hbar\omega_j\delta - 2\hbar\omega_j\Pi_j(\omega)}. \quad (6.22)$$

The self-energy $\Pi_j(\hbar\omega)$ is often called the *polarization* contribution to the phonon propagator as it describes distortion (polarization) of the lattice around each phonon vibration ("dressing" of the phonon).

I finally point out that the phonon frequencies ω_j should, in principle, be calculated in the presence of current. In practice, this is never done as it would require a self-consistent calculation of the dynamical response matrix 6.7 evaluated not at the ground-state density but at the non-equilibrium one. As we will see in Sec. 6.5 this is not simply a numerical problem but a fundamental one as the definition of phonons in the presence of current is not at all obvious.

6.1.3 Electron-phonon coupling in the presence of current

We are now left to determine the Hamiltonian 6.4 that describes the electron-phonon interaction. In addition, we want to do this when the electronic system carries a current. As in the case of shot noise it is more convenient to work using second quantization. The reason is that it is easier to keep track of both electron and phonon statistics when we perform the averages over both the electronic and phonon states. This is very important because when an electron is scattered by a phonon it can transition into a state only if the latter is unoccupied. The Pauli exclusion principle therefore needs to be taken into account explicitly.

In what follows I will assume that the averaging over electronic and

6.1 The many-body electron-ion Hamiltonian

phonon degrees of freedom can be done independently for each subsystem: the electron and phonon "reservoirs" do not interact with each other and they are at their own respective local temperature.

As we did for $\hat{\mathbf{H}}_{ion}$, we first expand Eq. 6.4 about the equilibrium positions $\mathbf{R}_i^0$

$$\hat{\mathbf{H}}_{el-ion} \approx \sum_i \left[V^{ps}(\mathbf{r}; \mathbf{R}_i^0) + \mathbf{Q}_i \cdot \nabla_{\mathbf{R}_i} V^{ps}(\mathbf{r}; \mathbf{R}_i)|_{\mathbf{R}_i = \mathbf{R}_i^0} \right], \qquad (6.23)$$

and keep only first-order terms, since the higher-order ones are assumed to be much smaller (see also discussion below).

The first term can be absorbed into the electronic Hamiltonian $\hat{\mathbf{H}}_{el}$, so we are interested in the second one, which we relabel

$$\hat{\mathbf{H}}_{e-ph} = \sum_i \mathbf{Q}_i \cdot \nabla_{\mathbf{R}_i} V^{ps}(\mathbf{r}; \mathbf{R}_i)|_{\mathbf{R}_i = \mathbf{R}_i^0}. \qquad (6.24)$$

Using the convention we have adopted of replacing a single index j to label the index pair $i\mu$, Eq. 6.8 can be written as

$$\mathbf{Q}_i = \sum_{j=1}^{3N} A_{i,j} q_j = \sum_{j=1}^{3N} A_{i,j} \sqrt{\frac{\hbar}{2\omega_j}} (\hat{b}_j + \hat{b}_j^\dagger), \qquad (6.25)$$

where in the last equality we have made use of the operators 6.12, and the notation $A_{i,j}$ is an abbreviation for $A_{i\mu,j\nu}$.

Replacing 6.25 in 6.24 we then get

$$\hat{\mathbf{H}}_{e-ph} = \sum_{i=1}^{3N} \sum_{j=1}^{3N} A_{i,j} \sqrt{\frac{\hbar}{2\omega_j}} (\hat{b}_j + \hat{b}_j^\dagger) \frac{\partial V^{ps}(\mathbf{r}; \mathbf{R}_i)}{\partial R_i} \bigg|_{\mathbf{R}_i = \mathbf{R}_i^0}. \qquad (6.26)$$

This expression is in second-quantized form in the phonon degrees of freedom but not in the electronic ones.

The electrons carry current between left and right electrodes, the latter ones being kept at a given bias. As in Eq. 5.21 we then define the field operator[8]

$$\psi(\mathbf{r}) = \psi^L(\mathbf{r}) + \psi^R(\mathbf{r}) = \sum_E \Psi_E^L(\mathbf{r}) \hat{a}_E^L + \sum_E \Psi_E^R(\mathbf{r}) \hat{a}_E^R, \qquad (6.27)$$

with the wave-functions $\Psi_E^{L,R}(\mathbf{r}) = \Psi_{i(\pm k_i)}^+(\mathbf{r})$ corresponding to processes with an incident wave from the L electrode $(+|k_i|)$, and the ones with a wave incident from the R electrode $(-|k_i|)$, as calculated in Sec. 3.5.1. The operators $\hat{a}_E^{L(R)}$ satisfy the relations 5.23.

---

[8] The only difference with Eq. 5.21 is that we do not consider any time dependence here. We also use a discrete notation for the energy. We will take the continuum limit later.

As explained in Appendix A, given the above field operators, the Hamiltonian $\hat{H}_{e-ph}$ – which in this form is a one-body operator in the electronic coordinates – can be written in second-quantization form as

$$\hat{H}_{e-ph} = \int d\mathbf{r}\, \psi^\dagger(\mathbf{r}) \left[\sum_{i=1}^{3N}\sum_{j=1}^{3N} A_{i,j} \sqrt{\frac{\hbar}{2\omega_j}} (\hat{b}_j + \hat{b}_j^\dagger) \frac{\partial V^{ps}(\mathbf{r};\mathbf{R}_i)}{\partial R_i} \bigg|_{\mathbf{R}_i = \mathbf{R}_i^0} \right] \psi(\mathbf{r}). \tag{6.28}$$

By introducing 6.27 into 6.28 we finally get

$$\hat{H}_{e-ph} = \sum_{\alpha,\beta=L,R} \sum_{E_1,E_2} \sum_{i,j} \sqrt{\frac{\hbar}{2\omega_j}} A_{i,j}\, J^{i,\alpha\beta}_{E_1,E_2}\, a^{\alpha\dagger}_{E_1} a^{\beta}_{E_2} \left(\hat{b}_j + \hat{b}_j^\dagger\right), \tag{6.29}$$

where I have defined the coupling constant

$$J^{i,\alpha\beta}_{E_1,E_2} = \int d\mathbf{r}\, \Psi^{\alpha*}_{E_1}(\mathbf{r}) \frac{\partial V^{ps}(\mathbf{r};\mathbf{R}_i)}{\partial R_i}\bigg|_{\mathbf{R}_i=\mathbf{R}_i^0} \Psi^{\beta}_{E_2}(\mathbf{r}). \tag{6.30}$$

Recalling that the field operator in 6.27 describes both carriers moving from left to right, $\Psi^L_E(\mathbf{r})$, and carriers moving from right to left, $\Psi^R_E(\mathbf{r})$, we see from Eq. 6.30 that the electron-phonon interaction couples these two types of scattering states. In addition, the coupling constant 6.30 is a non-trivial function of the scattering states, and therefore there is no reason to expect it to be, for a given ion, constant in energy or, for that matter, any simple function of the energy and/or bias.

6.2 Inelastic current

We now have all the tools to calculate the inelastic (electron-phonon) contribution to the current. I will first derive it using the scattering approach we have developed in Chapter 3. I will then discuss a way to calculate it using the Keldysh formalism developed in Chapter 4. In both cases we will need approximations to derive a closed form for the inelastic current.

Since typical dimensions of a nanoscale junction are much smaller than the inelastic electron-phonon mean-free path λ_{ph}[9] – the average distance an electron travels between two successive scattering events with phonons – we can use perturbation theory in the electron-phonon coupling to determine the dominant contributions to inelastic scattering (Montgomery et al., 2003; Chen et al., 2003). The use of perturbation theory seems also to be justified by the fact, known from experiments, that the observed inelastic features in nanoscale structures are very small, in the sense that the extra current

---

[9] This is typically of the order of thousands of Å (see Sec. 6.2.1).

due to inelastic effects is much smaller than the current due to elastic effects (see, e.g., Agraït et al., 2003).

However, the most important approximation we will have to make in order to have a closed form for the current is that inelastic scattering occurs at energies away from electronic resonances and/or for phonon modes that are not localized in the junction: we do not consider modes that may have very weak coupling with the phonons of the adjacent bulk electrodes (see also Sec. 6.3). The reason for these assumptions is that, in both instances, the electron-phonon scattering rates vary in a non-trivial way due to an increased (electronic and/or vibrational) local density of states at the relevant energies for scattering. As I will show later, this makes the calculation of the current quite difficult – if not impossible – to carry out.

6.2.1 Inelastic current from standard perturbation theory

Equation 6.29 is the desired expression to apply standard perturbation theory in the electron-phonon coupling. We thus seek corrections to the unperturbed wave-functions due to this coupling.[10] We use as unperturbed states of the full system – electron plus phonons – the states $|\Psi_E^{L(R)}; N_j\rangle \equiv |\Psi_E^{L(R)}(\mathbf{r})\rangle |N_j\rangle$.

Let us then expand the wave-functions to second order in the electron-phonon coupling 6.30

$$\left|\Phi_E^{L(R)}; N_j\right\rangle = \left|\Psi_E^{L(R)}; N_j\right\rangle + \left|\delta\Psi_E^{L(R)}; N_j\right\rangle_1 + \left|\delta\Psi_E^{L(R)}; N_j\right\rangle_2, \quad (6.31)$$

where $|\delta\Psi_E^{L(R)}; N_j\rangle_1$ and $|\delta\Psi_E^{L(R)}; N_j\rangle_2$ are the first-order and second-order corrections, respectively, to the wave-function. Once these corrections are calculated one then introduces 6.31 in the current operator

$$\hat{I}(x) = \frac{e\hbar}{2mi} \int d\mathbf{r}_\perp \left[\psi^\dagger(\mathbf{r}) \left(\frac{\partial}{\partial x}\psi(\mathbf{r})\right) - \left(\frac{\partial}{\partial x}\psi^\dagger(\mathbf{r})\right)\psi(\mathbf{r})\right], \quad (6.32)$$

with the field operator given by Eq. 6.27 and the single-particle electronic states as modified in 6.31.

If we only retain terms up to second order in the electron-phonon coupling we recover the unperturbed current 5.28 plus two types of inelastic

---

[10] Note that this approach neglects any possible *dynamical* self-consistent modification of the electronic Hamiltonian due to the inelastic electron-phonon interaction.

contributions. The first one contains terms proportional to

$$\int d\mathbf{r}_\perp \langle \delta\Psi_E^{L(R)}; N_j|\mathbf{r}\rangle_1 \left(\frac{\partial}{\partial x}\langle \mathbf{r}|\delta\Psi_E^{L(R)}; N_j\rangle_1\right) - \left(\frac{\partial}{\partial x}\langle \delta\Psi_E^{L(R)}; N_j|\mathbf{r}\rangle_1\right)\langle \mathbf{r}|\delta\Psi_E^{L(R)}; N_j\rangle_1. \quad (6.33)$$

These are explicitly written in Appendix H in the case in which there is weak elastic backscattering.

The second type of inelastic corrections to the current contains terms proportional to

$$\int d\mathbf{r}_\perp \langle \Psi_E^{L(R)}; N_j|\mathbf{r}\rangle \left(\frac{\partial}{\partial x}\langle \mathbf{r}|\delta\Psi_E^{L(R)}; N_j\rangle_2\right) - \left(\frac{\partial}{\partial x}\langle \Psi_E^{L(R)}; N_j|\mathbf{r}\rangle\right)\langle \mathbf{r}|\delta\Psi_E^{L(R)}; N_j\rangle_2. \quad (6.34)$$

These terms describe the *interference* between the zero-order wave-function $\left|\Psi_E^{L(R)}; N_j\right\rangle$ and the second-order correction $\left|\delta\Psi_E^{L(R)}; N_j\right\rangle_2$. They involve the emission and absorption of a "virtual" phonon by an electron with initial and final state at the same energy, i.e., they describe the "elastic" contribution to the inelastic current. I will write this contribution explicitly in Sec. 6.2.2 when I discuss the calculation of the inelastic current using the NEGF approach (see Eq. 6.71).

Inelastic transition probabilities

Here I want to give a qualitative account of the inelastic features we expect due to electron-phonon scattering. In particular, I want to estimate the inelastic mean-free path λ_{ph}, and determine which modes contribute the most to electron-phonon scattering.

This information is included in the terms 6.33 and 6.34. We can however extract it much more easily as follows. For a general perturbation $\hat{\mathbf{H}}_p$, the transition probability per unit time between an initial state $|i\rangle$ and a final state $|f\rangle$ is given by the following perturbation expression[11]

$$P_{i\to f} = \frac{2\pi}{\hbar}|\langle f|\hat{\mathbf{H}}_p|i\rangle|^2 \delta(E_f - E_i). \quad (6.35)$$

In the present case, the initial and final states are of the type $\left|\Psi_E^{L(R)}; N_j\right\rangle \equiv \left|\Psi_E^{L(R)}(\mathbf{r})\right\rangle |N_j\rangle$, and the perturbation is Hamiltonian 6.29. We consider two

---

[11] This form is also known as the *Fermi golden rule* (see, e.g., Sakurai, 1967).

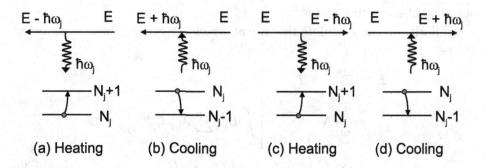

Fig. 6.2. Schematic of the first-order electron-phonon scattering processes contributing to the correction of the wave-function in the quasi-ballistic regime. Diagram (a) corresponds to a process of an electron propagating from right to left and emitting a phonon of energy $\hbar\omega_j$. The corresponding phonon subsystem makes a transition from state $|N_j\rangle$ to state $|N_j+1\rangle$. Similarly for the other diagrams.

types of processes due to electron-phonon scattering: (i) electrons (moving from left to right or from right to left) can absorb phonons, (ii) electrons can emit phonons (see schematic in Fig. 6.2). During these processes electrons have to satisfy the Pauli exclusion principle.

Let us consider for simplicity a single mode with energy $\hbar\omega_j$. If we calculate the matrix elements appearing in 6.35 explicitly (Exercise 6.2), the total rate at which electrons *emit* such a phonon is

$$\mathcal{I}_j^+ = \mathcal{I}_j^{R,+} + \mathcal{I}_j^{L,+}$$

$$= \frac{2\pi}{\omega_j}(1+N_j)\sum_{\alpha,\beta=L,R}\left[\int dE \left|\sum_i A_{i,j} J_{E-\hbar\omega_j,E}^{i,\alpha\beta}\right|^2\right.$$

$$\left. \times f_E^\alpha\left(1-f_{E-\hbar\omega_j}^\beta\right) D_{E-\hbar\omega_j}^\alpha D_E^\beta\right]. \qquad (6.36)$$

This corresponds to the scattering diagrams (a) and (c) in Fig. 6.2. A factor of 2 is included in Eq. 6.36 due to spin degeneracy, and we have explicitly introduced the density of states of the two types of current-carrying states, $D_E^{R(L)}$:

$$D_E^\alpha = \int d\mathbf{r}\, D_\alpha(\mathbf{r}, E) \qquad (6.37)$$

where $D_\alpha(\mathbf{r}, E)$ is the local density of states defined in Eq. 3.127 for only one type of scattering states.[12]

Similarly, we can calculate the rate at which the same phonon is *absorbed*. We get

$$\mathcal{I}_j^- = \mathcal{I}_j^{R,-} + \mathcal{I}_j^{L,-}$$

$$= \frac{2\pi}{\omega_j} N_j \sum_{\alpha,\beta=L,R} \left[\int dE \left| \sum_i A_{i,j} J_{E+\hbar\omega_j,E}^{i,\alpha\beta} \right|^2 \right.$$

$$\left. \times f_E^\alpha \left(1 - f_{E+\hbar\omega_j}^\beta\right) D_{E+\hbar\omega_j}^\alpha D_E^\beta \right]. \qquad (6.38)$$

In the presence of many independent phonon modes the total rate is simply the sum

$$\mathcal{I} = \sum_j (\mathcal{I}_j^- + \mathcal{I}_j^+). \qquad (6.39)$$

We see that the coefficients entering in Eqs. 6.36 and 6.38 contain terms of the type $f_E^\alpha \left(1 - f_{E\pm\hbar\omega_j}^\beta\right)$ which appear precisely because of the Pauli exclusion principle. Note also that in 6.36 there are terms proportional to $1 + N_j$. This means that even if all modes have zero average occupation, i.e., $N_j = 0$, $\forall j$ (this corresponds to the ionic temperature $\theta_{ph} = 0$ in Eq. 6.15), there is still a contribution to the electron-phonon scattering.

This contribution, known as *spontaneous phonon emission*, is due to the energy $\frac{1}{2} \sum_j \hbar\omega_j$ appearing in Hamiltonian 6.13.[13] This term, which is called the *zero-point* energy of the phonons, is therefore not zero even at zero ionic temperature. The scattering processes at $\theta_{ph} = 0$ can be thus interpreted as due to the zero-point energy (vacuum) fluctuations of the ions.

Inelastic mean-free path

Let us discuss the case in which $\theta_e = 0$ and $\theta_{ph} = 0$, namely both the electronic and the ionic temperatures are zero. Therefore, only processes involving spontaneous emission contribute. By setting these temperatures

---

[12] Explicitly,

$$D_{L,R}(\mathbf{r}, E) = \sum_{i=1}^{N_c^{L,R}} |\Psi_{ik_i}^+(\mathbf{r})|^2 D_i(E),$$

where $D_i(E)$ is the density of states per spin per direction of motion (Eq. 3.48). The sum $D_E^L + D_E^R = D(E)$ is the total density of states.

[13] This energy contains, in principle, an infinite sum of terms.

in Eq. 6.39 to zero we get the total rate

$$\mathcal{I} = \sum_j \frac{2\pi}{\omega_j} \int_{\mu_R+\hbar\omega_j}^{\mu_L} dE \left| \sum_i A_{i,j} J^{i,LR}_{E-\hbar\omega_j,E} \right|^2 D^L_{E-\hbar\omega_j} D^R_E, \qquad (6.40)$$

which is non-zero for biases $V = (\mu_L - \mu_R)/e > \min\{\hbar\omega_j\}/e$.

In addition, let us assume for simplicity that the bias is small but still $eV \gg \hbar\omega_j$, namely the bias is much larger than the energy per unit charge of the phonon modes that contribute to the rate $\mathcal{I}$. Let us also assume the density of states, $D^{L(R)}_E$, and coupling constant $J^{i,\alpha\beta}_{E_1,E_2}$ are slowly varying functions of energy – call these values, evaluated at the equilibrium Fermi level, $D^{L(R)}$ for the density and $J^{i,\alpha\beta}$ for the coupling.

We can then approximate the rate 6.40 as

$$\mathcal{I} \simeq eV \sum_j \frac{2\pi}{\omega_j} \left| \sum_i A_{i,j} J^{i,LR} \right|^2 D^L D^R, \qquad (6.41)$$

which states that, at low bias and under the above conditions, the electron-phonon scattering rate varies linearly with bias.

We can employ Eq. 6.41 to estimate the inelastic scattering rate $1/\tau_{ph}$ and consequent mean-free path λ_{ph} in nanojunctions. Consider just one mode of energy $\hbar\omega$ and a bias V some constant times $\hbar\omega/e$, with this constant of order one.

For typical metallic quantum point contacts we can take $J^{i,LR} \sim 0.1$ eV/Å, $D^L \simeq D^R \simeq 1$ eV^{-1}. By taking the ion mass $M \simeq 10^4\, m_e$ (which enters via the matrix $A_{i,j}$, see Eq. 6.9), the scattering rate $\mathcal{I} \equiv 1/\tau_{ph}$ calculated from Eq. 6.41 is about 10^{13} s^{-1}.

If we assume the single electrons travel at a typical velocity $v_F \sim 10^8$ cm/s, we get

$$\lambda_{ph} = v_F\, \tau_{ph} \simeq 1000\,\text{Å}. \qquad (6.42)$$

This distance is much longer than the typical length of a nanojunction.

Therefore, this result confirms our initial hypothesis that electrons remain quasi-ballistic (in the sense discussed in Sec. 3.7.4) even in the presence of inelastic effects and justifies *a posteriori* the use of perturbation theory for electron-phonon scattering.[14]

---

[14] If we assume a finite ionic temperature θ_{ph} (but still zero electronic temperature) Eq. 6.41 needs to be multiplied by a factor $\sim (1+2\bar{N}_j)$ (from Eq. 6.39). At the local ionic temperatures one expects in these systems (see Sec. 6.3) this factor is of order one, so that the estimate for the inelastic mean-free path holds.

"Selection rules" for inelastic scattering

We can now answer the question of which modes contribute the most to electron-phonon scattering. The structural information on these modes is contained in the transformation matrix defined in 6.8. From Eq. 6.29 we then see that it is the product $A_{i,j} J^{i,\alpha\beta}_{E_1,E_2}$ of this matrix with the coupling constant 6.30 that determines the magnitude of the inelastic scattering.

In a perfectly one-dimensional system we know, from simple considerations on conservation of momentum, that only longitudinal modes can couple to the current – these are modes with an oscillation pattern in the direction of current flow; transverse modes would have an oscillation pattern perpendicular to current flow.[15]

In a general three-dimensional structure like a nanojunction there is no such thing as a purely longitudinal or a purely transverse mode. However, we expect that if a mode has large "longitudinal" character, namely it has an oscillation pattern that is mostly in the direction of current flow, then this mode couples easily with electrons.[16] Conversely, if the mode has large "transverse" character then its coupling with the current is small. These are the "selection rules" that determine the strength of the electron-phonon scattering in nanostructures.

6.2.2 Inelastic current from the NEGF

Let us now calculate the inelastic contribution to the current using the Keldysh formalism I discussed in Chapter 4. Here too we will need a series of approximations to derive a closed set of equations for the single-particle Green's functions. Some of these approximations are similar to the ones we have used in perturbation theory, others are specific to the present approach.

Our starting point is again the static Hamiltonian 6.1. I then assume that we have made the adiabatic approximation and thus separated the problem into an electronic one described by a Hamiltonian $\hat{H}_{el}$ for fixed ionic positions, a phonon Hamiltonian $\hat{H}_{ph}$ (Eq. 6.13) and the Hamiltonian 6.29 describing the electron-phonon interaction.

I also make the same physical assumptions we have discussed before, namely I assume that the electronic problem, in the absence of phonons, can be solved in an effective *single-particle approximation*.

---

[15] In one dimension, if the electron has wave-vector k and the longitudinal phonon mode has wave-vector q, then momentum conservation requires $q = 2k$.

[16] In molecular structures making contact to bulk electrodes, due to the presence of the contacts, there is a large number of "longitudinal" modes, otherwise absent in the isolated nanostructure, at very low energies - typically energies of a few meV (see, e.g., Chen et al., 2003; Agraït et al., 2003).

6.2 Inelastic current

To make things even more specific let us assume that, as done in Sec. 3.5.1, we can divide the system into a central region representing the nanostructure – and possibly some atoms of the electrodes – which is coupled via spatially finite potentials to two infinite electrodes decoupled from each other (see Fig. 3.11), each at equilibrium with its own local distribution.

Referring to Fig. 3.11 the single-electron Hamiltonian can be again written as in Eq. 3.137

$$\hat{H} = \hat{H}_L + \hat{H}_R + \hat{H}_C + \hat{V}_{LC} + \hat{V}_{LC}^\dagger + \hat{V}_{CR} + \hat{V}_{CR}^\dagger. \quad (6.43)$$

In the absence of electron-phonon interaction the retarded Green's function associated with Hamiltonian 6.43 is 3.156, which we rewrite here for convenience

$$\hat{G}^+(E) = \frac{\hat{1}}{E - \hat{H}_C - \hat{\Sigma}_L^+(E) - \hat{\Sigma}_R^+(E)}. \quad (6.44)$$

The current, in the absence of phonons, is again given by Eq. 3.229. Up to this point there is *no need* of the NEGF.

Due to the confined geometry, we expect the effect of electron-phonon scattering to be stronger in the junction than in the bulk. We can then assume that the electron-phonon interaction is confined to the central region only. This interaction takes the form of the Hamiltonian 6.29 if we know the scattering wave-functions everywhere in space.

Here, instead, we want to describe the interaction between the "isolated" states of the central region and the phonons in it. Later on, we will combine this interaction with the "interaction" with the electrodes. The interaction Hamiltonian – to be added to 6.43 – thus has a general second-quantized form (cf. Eq. 6.30)

$$\hat{H}_{e-ph} = \sum_{n,n'} \sum_j \Lambda_{n,n'}^j \hat{a}_n^\dagger \hat{a}_{n'} \left(\hat{b}_j + \hat{b}_j^\dagger \right), \quad (6.45)$$

where $\Lambda_{n,n'}^j$ is some effective electron-phonon coupling constant – for each mode j – to be determined. To simplify the notation I will also assume that this coupling is energy independent, $\Lambda_{n,n'}^j \to \Lambda^j$. The results can be easily generalized to an energy-dependent coupling constant by simply summing over the states in the central region.

Note that the operators a_n in Eq. 6.45 act on the states with energies ϵ_n of the central region – Hamiltonian H_C – not of the full Hamiltonian 6.43 (see also discussion in Sec. 4.4). The phonon Hamiltonian is instead the same as before, Eq. 6.13.

Electron-phonon vs. elastic scattering in the leads

Now, following the Keldysh approach I have described in Sec. 4.4, I cannot simply apply the electron-phonon perturbation to the single-particle states of the Hamiltonian 6.43. The reason is that the electron-phonon scattering may *correlate* with the electron scattering into the leads, namely the processes of elastic scattering into the leads may be a complicated function of the electron-phonon scattering, and vice versa, the latter may be strongly influenced by the elastic scattering.

So, the correct way to apply many-body perturbation theory is the following: at a given initial time, we switch on adiabatically *both* the interaction between the electrons and phonons, *and* the elastic scattering of the electrons in the central region into the electrodes, the latter containing non-interacting electrons – otherwise no closed equation for the current can be obtained (see discussion in Sec. 4.4).

In addition, if we want to describe a static transport problem, we need to assume that *both* the electron and phonon subsystems have reached, in the long-time limit, an ideal steady state independent of the initial correlations, and therefore solve for the lesser Green's function of the electronic subsystem *with* the electron-phonon interaction included. We would like to solve this problem to *all orders* in the electron-phonon perturbation. In practice, this cannot be done exactly and we therefore need further approximations, some of which may indeed neglect important effects in nanostructures. Let me expand on this.

First of all, we need to decide whether we solve first for the electrode coupling with no phonons involved, and then include the phonons, or the reverse: calculate first the electron-phonon self-energy with no electrode coupling, and then introduce the electrode coupling. The two procedures *do not necessarily lead to the same physical state*, and when the two coupling effects are comparable, *correlated* electron-phonon states originate (see discussion below) and one, in principle, has to rely on the full many-body perturbation calculation, which is in general intractable.

We therefore first make the approximation that the electron-phonon coupling does not affect the coupling potentials $\hat{V}_{LC}$ and $\hat{V}_{CR}$, and corresponding self-energies $\hat{\Sigma}_L(E)$ and $\hat{\Sigma}_R(E)$, Eqs. 3.149 and 3.150, respectively.

This approximation is not valid if the coupling potentials describe the connection of the nanostructure with the immediately adjacent atoms of the electrodes. This is because, as already mentioned, for nanoscale systems phonon modes localized at the interface between the nanostructure and the electrodes may be present.

I thus assume that the central region contains enough atoms of the bulk electrodes so that the potentials $\hat{V}_{LC}$ and $\hat{V}_{CR}$ are localized to a spatial region where the electron-phonon coupling is negligible compared to the larger coupling in the nanostructure.

We then calculate the contribution of the electron self-energy due to the electron-phonon interaction, call this $\hat{\Sigma}^+_{el-ph}(E)$, *in the presence* of the electrodes, namely in the presence of tunneling. The self-energy $\hat{\Sigma}^+_{el-ph}(E)$ will then be constructed from electron Green's functions *in the absence* of phonons.[17] The above procedure is easy to justify if $|\hat{\Sigma}^+_L(E)|, |\hat{\Sigma}^+_R(E)| \gg |\hat{\Sigma}^+_{el-ph}(E)|$, at all electronic energies.

Whenever $|\hat{\Sigma}^+_L(E)|, |\hat{\Sigma}^+_R(E)| \simeq |\hat{\Sigma}^+_{el-ph}(E)|$, then one cannot in principle apply perturbation theory separately on the electron-phonon or electrode coupling. In this case, the "renormalization" of the electron energy and its "lifetime" due to electron-phonon scattering are comparable to the equivalent effect due to elastic scattering into the bulk electrodes. In other words, *coherent* effects between electrons and phonons need to be taken into account and the global current-carrying states may have a non-trivial phonon contribution – even though we have assumed the phonon coupling occurs in the central region only. One would then need to resort to the full many-body perturbation theory with the interaction with the phonons *and* electrodes treated on the same footing.

The opposite regime, $|\hat{\Sigma}^+_L(E)|, |\hat{\Sigma}^+_R(E)| \ll |\hat{\Sigma}^+_{el-ph}(E)|$, represents an electron strongly coupled to the phonons. The latter would then "slow down" the electron motion (by increasing its effective mass) and, if $E_F \simeq |\hat{\Sigma}^+_{el-ph}(E)|$, with E_F the ground state Fermi energy, the electron will move carrying a *polarization field* with it.

This new *coherent* electron-phonon excitation is called a *polaron* and it may be more stable than a free electron state. I stress however that strong electron-phonon coupling is not enough for a polaron to form: the resultant electron-phonon self-energy has to be comparable to the Fermi energy of the system, or the relevant energy for transport.

In general, the absolute value of electron-phonon self-energies is of the order of the phonon energies, i.e., tens of meV. For a nanostructure connected to bulk metallic leads $E_F \sim 4-5$ eV, so that we do not expect this excitation to be present for most of these systems, and we will therefore not discuss this case any further.[18]

---

[17] I will outline a self-consistent procedure that allows some degree of correction to the electron Green's functions due to phonons, without however changing the *type* of self-energy, namely the type of scattering processes.
[18] In heavily doped semiconductors, or in polymers, the energy scale of interest is determined

For quasi-ballistic systems considered here we expect

$$|\hat{\Sigma}_L^+(E)|, |\hat{\Sigma}_R^+(E)| \gg |\hat{\Sigma}_{el-ph}^+(E)| \; \forall E, \quad \text{quasi-ballistic systems}, \quad (6.46)$$

so that, for this case, we can safely start by assuming the electron states, *with* the elastic scattering into the leads included, are unperturbed by the electron-phonon interaction. (I will discuss later some important exceptions to this condition.)

All the above analysis allows us to say that, if we can treat the electrode coupling *independently* of the electron-phonon one, the contribution of all these interactions to the total self-energy is simply additive

$$\hat{\boldsymbol{\Sigma}}^+(E) \simeq \hat{\Sigma}_L^+(E) + \hat{\Sigma}_R^+(E) + \hat{\Sigma}_{el-ph}^+(E), \quad \text{non-crossing approximation}, \quad (6.47)$$

which is called the *non-crossing approximation* precisely because one neglects any possible scattering event that contains simultaneously elastic scattering in the leads and electron-phonon scattering.

Let us then assume we are in a regime in which the approximation 6.47 holds. Using the Dyson equation 4.33, with the "unperturbed" (or "free") Green's function $\hat{G}_C^+$ associated with the central region Hamiltonian $\hat{H}_C$, the single-electron retarded Green's function in the presence of electron-phonon coupling is

$$\begin{aligned}\hat{\mathbf{G}}^+(E) &= \hat{G}_C^+(E) + \hat{G}_C^+(E)\, \hat{\boldsymbol{\Sigma}}^+(E)\, \hat{\mathbf{G}}^+(E) \\ &= \hat{G}^+(E) + \hat{G}^+(E)\, \hat{\Sigma}_{el-ph}^+(E)\, \hat{\mathbf{G}}^+(E) \\ &= \frac{\hat{1}}{E - \hat{H}_C - \hat{\Sigma}_L^+(E) - \hat{\Sigma}_R^+(E) - \hat{\Sigma}_{el-ph}^+(E)}. \end{aligned} \quad (6.48)$$

The corresponding advanced Green's function is simply $\hat{\mathbf{G}}^-(E) = [\hat{\mathbf{G}}^+(E)]^\dagger$ (Eq. 3.84).

If the couplings between the central region and the two electrodes are proportional (see Sec. 4.4) this is enough to calculate the current: we can use Eq. 4.114 where the spectral function is given by Eq. 4.54. If not, we need to employ the full expression Eq. 4.113, and thus determine the lesser Green's function $\hat{\mathbf{G}}^<(E)$.

Since our unperturbed Hamiltonian describes non-interacting electrons, from the Keldysh equation 4.95 the lesser Green's function is (see also Eq. 4.121)

$$\mathbf{G}^<(E) = \mathbf{G}^+(E)\, \boldsymbol{\Sigma}^<(E)\, \mathbf{G}^-(E), \quad (6.49)$$

by the number of thermally activated carriers, and is thus of the order of $k_B \theta$. This thermal energy is comparable to the magnitude of the electron-phonon self-energy, and thus polarons can easily form in these systems.

6.2 Inelastic current

where, within the same approximation 6.47,

$$\Sigma^<(E) \simeq \Sigma_L^<(E) + \Sigma_R^<(E) + \Sigma_{el-ph}^<(E). \tag{6.50}$$

The lesser self-energy $\Sigma_L^<(E) + \Sigma_R^<(E)$ is given in Eq. 4.122, while $\Sigma_{el-ph}^<(E)$ has yet to be determined. We are then just left to determine the self-energies $\hat{\Sigma}_{el-ph}^+(E)$ and $\hat{\Sigma}_{el-ph}^<(E)$. An exact general solution is not available – even at equilibrium, let alone out of equilibrium – and we need further approximations. Let us first start from the electron system at equilibrium coupled with the phonons (themselves at equilibrium). We will generalize later to the non-equilibrium case.

6.2.2.1 Electron-phonon self-energies at equilibrium

Let us take a step back, and consider the system at equilibrium, $f_L = f_R \equiv f$. Let us then follow a single electron in time and see which scattering mechanisms due to electron-phonon coupling we need to consider, when interactions with other electrons are absent – or included only at the mean-field level.

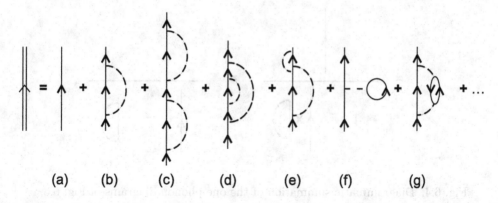

Fig. 6.3. Many-body perturbation expansion of the full Green's function of a single electron interacting with phonons but *not* with other electrons. Diagrams (f) and (g) involve the interaction of phonons with the electron gas.

As in Sec. 3.4.1, we stipulate the convention that a solid line represents the "free" – with no interactions – electron Green's function 6.44, a dashed line the phonon Green's function 6.18. The sum of all these scattering mechanisms is what determines the full electron Green's function, the one determined by the Dyson equation 4.33. We represent this full Green's function as a double vertical arrow (see Fig. 6.3). This is an example of the many-body perturbation expansion I have invoked in Chapter 4 but never discussed explicitly.

The electron can first evolve in time without scattering, Fig. 6.3(a). Or, at a given time, it can emit a phonon of frequency ω. Both the phonon and the electron propagate, and after some time the electron absorbs the phonon back, Fig. 6.3(b). This is called a *one-phonon diagram*.

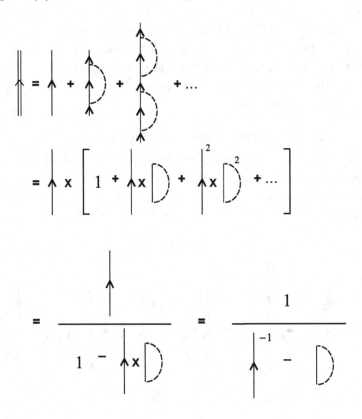

Fig. 6.4. Diagrammatic summation of the one-phonon diagrams of Fig. 6.3.

Another possible mechanism is the one in which the electron scatters with one phonon and then scatters again with another phonon in a succession of two one-phonon diagrams, Fig. 6.3(c).

Other diagrams may correspond to the emission and absorption of several phonons by the same electron, for instance, diagram (d) in Fig. 6.3, where the electron scatters a phonon, then scatters another phonon – before absorbing the first one – absorbs the latter one and later on re-absorbs the first phonon. Diagram (e) is even more complicated and is called a *vertex correction* since it *renormalizes* the vertex between the electron line and the phonon line. Other diagrams can be as complicated as one can imagine!

The summation of all these diagrams cannot be done exactly. However, the summation of diagrams (b), (c) and all (infinite number) other diagrams that are simple successions of one-phonon diagrams can be summed up exactly. This is easy to see in diagrammatic form in Fig. 6.4. The resulting self-energy is then represented by the left diagram in Fig. 6.5. The corresponding approximation which neglects all other scattering mechanisms except this one, is called *Born approximation* (BA).

A further step is to replace (*renormalize*) the free electron line with the one corresponding to the (unknown) full propagator, as if all possible scattering processes occur in between the emission and absorption of a phonon. The solution to this problem can only be obtained self-consistently: one starts with the free propagator, obtains the perturbed Green's function, and replaces this propagator back into the self-energy, and repeats the process till self-consistency, thus the name *self-consistent Born approximation* (SCBA), right diagram of Fig. 6.5.

This procedure takes care of diagrams of the type (d) in Fig. 6.3 but not of diagrams of type (f) or (g) in the same figure. Diagram (g), in particular, takes into account the fact that the emitted phonon, before being re-absorbed, can generate an *electron-hole pair* in the electron gas.

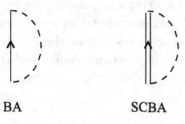

Fig. 6.5. Left: one-phonon contribution to the electron-phonon self-energy (Born approximation). Right: the electron free propagator is replaced by the full propagator (self-consistent Born approximation).

The Migdal theorem

At equilibrium and for small electron-phonon coupling, one can argue that all other higher-order processes involving many phonons (for each electron line) can be neglected. In bulk metals, one can also prove a theorem due to Migdal (Migdal, 1958) which states that vertex corrections (diagram (e)) are proportional to $\sqrt{m/M}$, with m the electron mass and M the ion mass, and are therefore small.

There are however cases, specific to transport in nanostructures, where these approximations may fail. We will discuss these cases later. For the moment we will assume these approximations can be done, and work with either the BA or the SCBA.

If one considers explicitly electron-electron interactions, to the diagrams in Fig. 6.3 one has to add similar diagrams due to the Coulomb interaction, making the problem much more complicated. In Sec. 8.6 I will show that electron-electron interactions beyond mean field *are* important in nanoscale systems, and contribute non-trivially to the heating of a nanojunction.

Here we will keep on working with single electrons in a mean-field approximation, namely with the electron Green's function 6.44.

"Bubble" diagrams in nanostructures

Unlike the case of translationally invariant systems, in the present case there are certain diagrams involving the electron gas that are of the same order – in the electron-phonon coupling – as the one-phonon diagrams. These are represented in Fig. 6.3(f). These diagrams – also called *Hartree diagrams* – correspond to an electron that scatters with a phonon of zero wave-vector and zero frequency, creating a *bubble* in the electron gas: another electron is scattered but instantaneously returns to its initial state.

In a translationally invariant crystal this corresponds to a rigid movement of the *whole* crystal. The contribution of this process is thus zero on the basis that an external force keeps the whole crystal in place.

In our problem, instead, diagram (f) would correspond to a rigid motion of all the atoms *inside* the central region with respect to the ions of the nearby electrodes (kept fixed). Its effect on the electron-phonon self-energy is thus not zero (Galperin *et al.*, 2004).

From diagrams to equations

We can now transform from the diagram language to equations. For the diagrams in Fig. 6.5, this is done by replacing the electron line with the "free" electron Green's function 6.44 (or the full Green's function in the SCBA), the phonon line with the "free" phonon propagator 6.17, and each vertex with the interaction strength Λ^j. The result is multiplied by a factor i and integrated over the frequency of the *virtual* phonons.[19]

The self-energy function on the Keldysh contour (see Sec. 4.3) thus has the form

$$\Sigma_{el-ph}(t-t') = i\hbar \sum_j \Lambda^j \, G(t-t') D_j^0(t-t') \Lambda^j. \qquad (6.51)$$

One then replaces the contour variable with the real time one (*analytic*

---

[19] The scattering conserves frequency, $E/\hbar$, and momentum. However, the particle energy is not conserved, hence the name *virtual processes* described by the diagrams in Fig. 6.3.

6.2 Inelastic current

continuation) and finds for the Fourier transforms of the self-energies (see, e.g., Haug and Jauho, 1996)

$$\hat{\Sigma}^+_{el-ph}(E) = i\hbar \sum_j \int \frac{d\omega}{2\pi} \left[\Lambda^j\, G^<(E-\hbar\omega) D^0_j(E)\, \Lambda^j \right.$$
$$+ \Lambda^j\, G^+(E-\hbar\omega) D^<_j(E)\, \Lambda^j$$
$$\left. + \Lambda^j\, G^+(E-\hbar\omega) D^0_j(E)\, \Lambda^j \right], \quad \text{Born approximation,} \quad (6.52)$$

and

$$\hat{\Sigma}^<_{el-ph}(E) = i\hbar \sum_j \int \frac{d\omega}{2\pi} \Lambda^j\, G^<(E-\hbar\omega) D^<_j(E)\, \Lambda^j, \quad \text{Born approximation.} \quad (6.53)$$

In these equations, $G^+(E)$ is defined in Eq. 6.44, and $G^<(E)$ has been defined in Eq. 4.123 for the non-interacting system with two different Fermi distributions. Here we consider equilibrium conditions, $f_L = f_R \equiv f$.

As discussed above, in addition to the self-energy of Fig. 6.5 one needs to add the self-energy corresponding to the Hartree diagram (f) in Fig. 6.3. This is done by replacing the "bubble" with the canonical average

$$\langle \hat{a}^\dagger_n \hat{a}_n \rangle = -i \int \frac{dE}{2\pi} \{G^<(E)\}_{nn}, \quad (6.54)$$

where the equality comes from the Fourier transform of definition 4.101.

As before, the phonon line is replaced by the "free" phonon propagator D^0_j, but this time only its value at zero frequency is necessary. Each vertex is replaced by the interaction strength Λ^j. The result is multiplied by a factor -1 and summed over the states of the central region. The contribution of the Hartree term to the retarded electron-phonon self-energy is thus

$$i \int \frac{dE}{2\pi} \sum_j \sum_n \Lambda^j \{G^<(E)\}_{nn} D^0_j(\omega=0) \Lambda^j. \quad (6.55)$$

The total retarded self-energy in the Born approximation is therefore

$$\hat{\Sigma}^+_{el-ph}(E) = i\hbar \sum_j \int \frac{d\omega}{2\pi} \left[\Lambda^j\, G^<(E-\hbar\omega) D^0_j(E)\, \Lambda^j \right.$$
$$+ \Lambda^j\, G^+(E-\hbar\omega) D^<_j(E)\, \Lambda^j$$
$$\left. + \Lambda^j\, G^+(E-\hbar\omega) D^0_j(E)\, \Lambda^j \right]$$
$$+ i \int \frac{dE}{2\pi} \sum_j \sum_n \Lambda^j \{G^<(E)\}_{nn} D^0_j(\omega=0) \Lambda^j. \quad (6.56)$$

Together with the self-energy 6.53 it allows the calculation of the electron-phonon self-energies at equilibrium.

If the SCBA is used, instead of $G^+(E)$ and $G^<(E)$ as defined in Eqs. 6.44 and 4.123, one would replace in Eqs. 6.53 and 6.56 the full Green's functions 6.48 and 6.49, and solve the whole set of equations self-consistently.

An even more refined calculation would include the *polarization* contribution to the phonon propagator we have introduced in Eq. 6.22.[20] In this case one makes the substitution $D_j^0 \to D_j$. This problem also needs to be solved self-consistently, since the polarization contribution to the phonon propagator is not generally known analytically.

The Hartree term in 6.52 is easy to understand, the others are less clear. A simplification to the problem, which helps us understand these terms, occurs when one assumes $\hat{\Sigma}_L^+ = \hat{\Sigma}_R^+ \equiv \hat{\Sigma}^+$ and $\Gamma_L(E) = \Gamma_R(E) \equiv \Gamma(E) = -2\mathrm{Im}\{\Sigma^+(E)\}$ so that $G^<(E)$ is (from Eq. 4.59)

$$G^<(E) = 2\pi i\, f(E)\, A(E),\qquad(6.57)$$

with $A(E)$ the spectral function. If we approximate $A(E)$ with a δ-function, $A(E) = \delta(E - \epsilon_n)$, which is appropriate for a free particle (Eq. 4.61), we can carry out the integral in Eq. 6.52 exactly and obtain (Exercise 6.3)

$$\hat{\Sigma}_{el-ph}^+(E) = \sum_j \Lambda^j \left[\frac{N_j + 1 - f(\epsilon_n)}{E - \hbar\omega_j - \epsilon_n + i\eta} + \frac{N_j + f(\epsilon_n)}{E + \hbar\omega_j - \epsilon_n + i\eta} \right] \Lambda^j,\qquad(6.58)$$

with η an infinitesimal positive number.[21]

Expression 6.58 as it stands does not seem very transparent, in particular with respect to the statistics of electrons and phonons. However, one can prove that it is similar to the energy correction to the single-particle state of energy ϵ_n one would obtain from second-order perturbation theory with terms proportional to[22]

$$\frac{\Lambda^j (1 + N_j)\, f(\epsilon_{n'})[1 - f(\epsilon_n)]\, \Lambda^j}{\epsilon_{n'} - \epsilon_n - \hbar\omega_j}, \quad \text{phonon emission,}\qquad(6.59)$$

---

[20] More generally, the phonons in the central region acquire a finite lifetime due to the (elastic or inelastic) interaction with the bulk phonons of the electrodes (see discussion in Sec. 6.3.1), or due to electron-hole pair production (diagram (g) in Fig. 6.3).
[21] In this example the eigenvalues ϵ_n are in reality $\epsilon_n + \mathrm{Re}\{\Sigma^+(E)\}$ (see Exercise 6.3).
[22] We urge the interested reader to read Mahan (1990), p. 176-178, for such a proof. One can also show (Mahan, 1990, p. 171-173) that for a given frequency ω_j, $\mathrm{Im}\Sigma_{el-ph}^+(E) = 0$ in the energy range $[E_F - \hbar\omega_j, E_F + \hbar\omega_j]$. This means that the electron (or hole for energies below the Fermi level) does not acquire an electron-phonon lifetime unless its energy is large enough to scatter a phonon, namely large enough so that the Pauli exclusion principle is not violated.

and
$$\frac{\Lambda^j N_j \, f(\epsilon_{n'})[1 - f(\epsilon_n)] \, \Lambda^j}{\epsilon_{n'} - \epsilon_n - \hbar\omega_j}, \quad \text{phonon absorption,} \qquad (6.60)$$

where the terms $f(\epsilon_{n'})[1-f(\epsilon_n)]$ are necessary to satisfy the Pauli exclusion principle.

This shows that the electron-phonon self-energy is a correction to the energy of the free particle, proportional to the occupation of the phonon modes and the strength of the coupling, whenever the scattering process is allowed (initial state occupied, final state empty), similar to what we have found for the inelastic transition probabilities (cf. Eqs. 6.36 and 6.38).

State lifetime due to electron-phonon scattering

From all this we can then define the "broadening" of the electron states due to the electron-phonon interaction as (see Eq. 4.67)

$$\hat{\Gamma}_{el-ph}(E) = -2\mathrm{Im}\{\hat{\Sigma}^+_{el-ph}(E)\}, \qquad (6.61)$$

and corresponding lifetime (Eq. 4.69)

$$\tau_{el-ph}(E) = -\frac{\hbar}{\mathrm{Im}\{\hat{\Sigma}^+_{el-ph}(E)\}}. \qquad (6.62)$$

6.2.2.2 Electron-phonon self-energies out of equilibrium

We are now ready to move on to the non-equilibrium case. Under the same assumption 6.46 of quasi-ballistic electrons, and within the Born approximation (or SCBA) we can generalize the above results to the case in which $f_L \neq f_R$ by simply replacing the equilibrium $G^<(E)$ with the non-equilibrium one defined in Eq. 4.123. If we employ, for instance, the Born approximation, we can then use this Green's function in Eq. 6.56 and carry out the calculation of the self-energy as in the equilibrium case.

In order to take into account the non-equilibrium nature of the phonons one can replace in Eq. 6.56 the "free" phonon propagator D^0_j with a modified one D_j, which includes an effective phonon lifetime due to the scattering of phonons into the bulk electrodes (see also Sec. 6.3.1), much like the propagator D_j one would calculate due to polarization effects (see Eq. 6.22). The problem could then be solved self-consistently with respect to the phonons as well.

Electronic resonances and localized phonons

However, in the non-equilibrium case we need to be more careful about the approximations we employ to calculate $\hat{\Sigma}^+_{el-ph}(E)$. For instance, in the presence of electronic resonances, multiple electron reflections inside the central region can reduce the values of $\hat{\Sigma}^+_L(E)$ or $\hat{\Sigma}^+_R(E)$ around the resonance energies considerably (see discussion in Sec. 3.7.1.1).

At these energies, the magnitude of $\hat{\Sigma}^+_L(E) + \hat{\Sigma}^+_R(E)$ may be thus comparable to the magnitude of $\hat{\Sigma}^+_{el-ph}(E)$, rendering the BA (and SCBA) discussed above less justified. Physically, this is because an electron may emit or absorb several phonons in one process before "escaping" from the central region, and therefore one would need to sum diagrams that are not included in the BA (or SCBA).

Another case that needs particular care is when there are phonons that are weakly coupled to the phonons of the electrodes – *localized phonons*. In this case, replacing the "free" phonon propagator D^0_j with a modified phonon one D_j, which includes an effective phonon lifetime does not make much sense.

The issue is related to the fact that these localized phonons can be in a highly non-equilibrium configuration due to the continuous interaction with non-equilibrium electrons: these modes are continually "pumped" by the incoming electrons and may actually lead to a structural instability of the nanojunction and, eventually, to breaking of atomic bonds (see also discussion in Sec. 6.3). This again cannot be captured by the BA or even the SCBA.

Total current of independent electrons interacting with phonons

We therefore exclude these cases and just focus on quasi-ballistic systems for which the electron self-energy can be calculated as in Eq. 6.56. We have considered all along independent electrons (interacting with each other at most at a mean-field level) but interacting with phonons in the central region. The current for this case is again given by 4.113 which I re-write here for convenience

$$I = 2\frac{ie}{2\hbar} \int \frac{dE}{2\pi} \text{Tr} \left\{ [\mathbf{\Gamma}_L(E) - \mathbf{\Gamma}_R(E)] \mathbf{G}^<(E) \right. \\ \left. + [f_L(E)\mathbf{\Gamma}_L(E) - f_R(E)\mathbf{\Gamma}_R(E)](\mathbf{G}^+(E) - \mathbf{G}^-(E)) \right\}. \tag{6.63}$$

The Green's functions are now propagators with both elastic scattering into the electrodes *and* inelastic scattering due to phonons. All quantities depend on the external bias and, in general, need to be determined self-consistently at any given bias.

We can gain more insight into the physics of the inelastic current by doing the following algebraic manipulations. From Eq. 6.48 we get

$$[\hat{\mathbf{G}}^+]^{-1} - [\hat{\mathbf{G}}^-]^{-1} = \left(\hat{\mathbf{\Sigma}}_L^- - \hat{\mathbf{\Sigma}}_L^+\right) + \left(\hat{\mathbf{\Sigma}}_R^- - \hat{\mathbf{\Sigma}}_R^+\right) + \left(\hat{\mathbf{\Sigma}}_{el-ph}^- - \hat{\mathbf{\Sigma}}_{el-ph}^+\right)$$
$$= i\left[\hat{\mathbf{\Gamma}}_L + \hat{\mathbf{\Gamma}}_R + \hat{\mathbf{\Gamma}}_{el-ph}\right], \qquad (6.64)$$

that can be equivalently written as

$$i\left[\hat{\mathbf{G}}^+ - \hat{\mathbf{G}}^-\right] = \hat{\mathbf{G}}^+\left[\hat{\mathbf{\Gamma}}_L + \hat{\mathbf{\Gamma}}_R + \hat{\mathbf{\Gamma}}_{el-ph}\right]\hat{\mathbf{G}}^-. \qquad (6.65)$$

By replacing this expression (and using 6.50) in Eq. 6.63, the latter can be separated in two parts (Exercise 6.4)

$$I = I_{el} + I_{in}, \qquad (6.66)$$

where

$$I_{el} = \frac{e}{\pi\hbar}\int_{-\infty}^{+\infty} dE\,[f_L(E) - f_R(E)]\,\text{Tr}\{\hat{\mathbf{\Gamma}}_R(E)\,\hat{\mathbf{G}}^+(E)\,\hat{\mathbf{\Gamma}}_L(E)\,\hat{\mathbf{G}}^-(E)\}, \qquad (6.67)$$

and

$$I_{in} = \frac{ie}{2\hbar}\int_{-\infty}^{+\infty}\frac{dE}{2\pi}\text{Tr}\left\{[\Gamma_L(E) - \Gamma_R(E)]\hat{\mathbf{G}}^+(E)\,\hat{\mathbf{\Sigma}}_{el-ph}^<(E)\,\hat{\mathbf{G}}^-(E) \right.$$
$$\left. - i[f_L(E)\Gamma_L(E) - f_R(E)\Gamma_R(E)]\hat{\mathbf{G}}^+(E)\,\hat{\mathbf{\Gamma}}_{el-ph}(E)\,\hat{\mathbf{G}}^-(E)\right\}. \qquad (6.68)$$

We call the first term *elastic* because it has the same form as the current 3.229 for non-interacting electrons without phonon scattering. However, this terminology is deceiving because the Green's functions appearing in Eq. 6.67 *contain* contributions from the phonons.

We also note that if the coupling of the central region with the left electrode or with the right electrode is zero (either $\mathbf{\Gamma}_L(E) = 0$ or $\mathbf{\Gamma}_R(E) = 0$), then *both* I_{el} and I_{in} are zero. This is easy to see for I_{el}, less so for I_{in}. We leave this demonstration as an exercise (Exercise 6.4).

We can further manipulate the expression of I_{el} using the Dyson equation 6.48. By introducing 6.48 into 6.67 we get

$$I_{el} = I_{el}^0 + \delta I_{el}, \qquad (6.69)$$

where

$$I_{el}^0 = \frac{e}{\pi\hbar} \int_{-\infty}^{+\infty} dE \, [f_L(E) - f_R(E)] \, \text{Tr}\{\hat{\Gamma}_R(E) \, \hat{G}^+(E) \, \hat{\Gamma}_L(E) \, \hat{G}^-(E)\}, \quad (6.70)$$

and

$$\begin{aligned}
\delta I_{el} = \frac{e}{\pi\hbar} \int_{-\infty}^{+\infty} & dE \, [f_L(E) - f_R(E)] \\
\times \Big[& \text{Tr}\{\hat{\Gamma}_R \, \hat{G}^+ \, \hat{\Sigma}_{el-ph}^+ \, \hat{G}^+ \, \hat{\Gamma}_L \, \hat{G}^- \\
& + \hat{\Gamma}_R \, \hat{G}^+ \, \hat{\Gamma}_L \, \hat{G}^- \, \hat{\Sigma}_{el-ph}^- \, \hat{G}^- \\
& + \hat{\Gamma}_R \, \hat{G}^+ \, \hat{\Sigma}_{el-ph}^+ \, \hat{G}^+ \, \hat{\Gamma}_L \, \hat{G}^- \, \hat{\Sigma}_{el-ph}^- \, \hat{G}^-\} \Big] .
\end{aligned} \quad (6.71)$$

The term I_{el}^0 is precisely the current *in the absence* of electron-phonon interaction (see Eq. 3.229). The total inelastic contribution to the current is therefore $\delta I_{el} + I_{in}$. As anticipated in Sec. 6.2.1 the term δI_{el} is precisely the one that originates from the *interference* between the zero-order wavefunction and its second-order correction, and corresponds to the "elastic" emission and absorption of a virtual phonon.

Conductance increase or decrease due to phonon scattering

In Appendix H I show that for weak elastic backscattering the current is reduced by inelastic effects. An experimental example of this case is shown in Fig. 6.1 where a drop in conductance is observed at specific biases. Here I show that the sum $\delta I_{el} + I_{in}$ may also be positive: the conductance may increase due to inelastic scattering. The reason for this can be understood quite easily despite the complicated structure of both δI_{el} and I_{in}.

In Sec. 3.4.2.1 we saw that the retarded and advanced Green's functions – in a positional basis – of a free particle in one dimension are oscillating functions of position (Eqs. 3.131 and 3.134), while for bound states they are the same real and exponentially decaying functions (Eq. 3.136).[23] In addition, one can show that at zero temperature the imaginary part of the retarded self-energy is an even function of the energy (the zero being the equilibrium Fermi energy), while the real part is an odd function (see Mahan, 1990, p. 596).

For simplicity let us consider the proportional coupling case, $\Gamma_L(E) = \Gamma_R(E) \equiv \Gamma(E), \forall E$, and work at a low bias, but large enough that electron-phonon scattering is allowed. Let us then assume that the central region has

---

[23] Similar considerations apply in 3D.

very few electronic states that are weakly coupled to the states of the electrodes. The electron Green's functions, in a positional basis, are therefore exponentially small in the central region at those energies.

To first order in the Dyson equation 6.48 we can then write

$$\hat{\mathbf{G}}^+(E) \simeq \hat{G}^+(E) + \hat{G}^+(E)\,\hat{\mathbf{\Sigma}}^+_{el-ph}(E)\,\hat{G}^+(E), \quad \text{weak coupling}, \quad (6.72)$$

and similarly for $\hat{\mathbf{G}}^-(E)$. By replacing these into Eqs. 6.68 and 6.71 and neglecting cubic (and higher) terms in the Green's functions, we see that δI_{el} is zero and I_{in} is expression 6.68 with $\hat{\mathbf{G}}^+(E)$ and $\hat{\mathbf{G}}^-(E)$ replaced by $\hat{G}^+(E)$ and $\hat{G}^-(E)$, respectively:

$$I_{in} \simeq \frac{e}{2\hbar} \int_{-\infty}^{+\infty} \frac{dE}{2\pi} [f_L(E) - f_R(E)] \text{Tr}\left\{\Gamma(E)\hat{G}^+(E)\,\hat{\mathbf{\Gamma}}_{el-ph}(E)\,\hat{G}^-(E)\right\}. \quad (6.73)$$

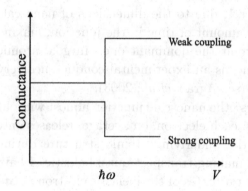

Fig. 6.6. Schematic of the conductance as a function of bias in the presence of electron-phonon scattering. If the electronic states of the central region are weakly coupled to the electronic states of the electrodes, an increase of conductance corresponding to the phonon energy may occur. If these states are strongly coupled to the states of the electrodes, a decrease of conductance occurs.

The term in the trace contains the square of real functions times $\hat{\mathbf{\Gamma}}_{el-ph}(E) = -2\text{Im}\{\hat{\mathbf{\Sigma}}^+_{el-ph}(E)\}$, which is a positive quantity (see discussion in Sec. 4.2.4). Therefore, under these conditions, the contribution I_{in} is positive: the current *increases* due to electron-phonon scattering, as if extra "channels" are introduced by the interaction with phonons. One would then observe a step-like increase in conductance corresponding to the energy of those modes that couple strongly with electrons (Fig. 6.6).

If, on the other hand, we consider many electron states in the central region, and these states are strongly coupled to the states of the electrodes, then the electron Green's functions are oscillating functions of position with coefficients given by a mixture of $\text{Re}\{\hat{\mathbf{\Sigma}}^+_{el-ph}(E)\}$ and $\text{Im}\{\hat{\mathbf{\Sigma}}^+_{el-ph}(E)\}$. Since these functions are odd and even in energy, respectively, the sums in

Eqs. 6.68 and 6.71 can be viewed as combinations of *constructive* and *destructive interference* patterns of scattering amplitudes with the cumulative effect being a small contribution to the elastic current. When the destructive interference dominates over the constructive one, the current is reduced by electron-phonon scattering (Fig. 6.6).[24]

The above analysis is clearly over-simplified, and in a real nanojunction one may have *both* types of couplings at different energies. The increase or decrease of the current due to electron-phonon coupling is thus a function of the microscopic details of both the self-consistent electronic distribution under current flow and the atomic configuration.

6.3 Local ionic heating

Let us now discuss another important effect that occurs due to electron-phonon scattering: *current-induced local ionic heating* of nanoscale systems (Todorov, 1998; Chen et al., 2003; Segal and Nitzan, 2002).

As I have anticipated in Chapter 1, due to the dimensions of nanoscale structures, electrons spend a small amount of time in the junction, making the elastic contribution to the current the dominant one. In gold atomic point contacts, for instance, this means an experimental conductance very close to the quantized one $(2e^2/h)$ (see Agraït et al., 2003).

Every electron, therefore, "crosses" the nanoscale junction almost without inelastic scattering. In other words, each electron, on average, releases only a small fraction of its energy to the underlying atomic structure during the time it spends in the junction, making transport quasi-ballistic. I have corroborated this point with simple estimates of the inelastic electron mean-free path in Sec. 6.2.1.

However, due to the much smaller number of atoms in a nanoscale system compared to the bulk of the electrodes, the current *density* is much larger in the junction than in the bulk. As a consequence, the number of scattering events per unit volume and unit time, or equivalently the *power per atom*, is much larger in the junction compared to the bulk. This power must then be dissipated into the bulk electrodes via some efficient mechanism of lattice heat conduction (see discussion later in this section). At steady state we thus expect that the ions in the junction are at a different local temperature than those in the bulk. This is illustrated schematically in Fig. 6.7.

---

[24] The fact that this contribution is negative in the weak backscattering case is not so easy to determine and requires explicit calculations of δI_{el} and I_{in}. Physically one can explain this fact by noting that an additional electron momentum change occurs due to electron-phonon scattering, thus increasing the resistance.

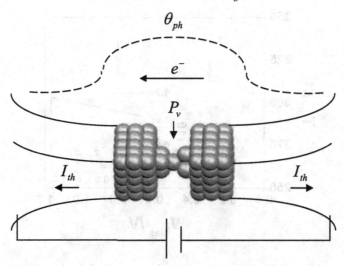

Fig. 6.7. Schematic of an atomic junction with current flowing across it. The electrons transfer some power P_v to the ions. This power has to be dissipated away to the bulk electrodes via lattice heat conduction I_{th}. At steady state the ions have thus an effective temperature locally in the junction higher than in the bulk.

As in Sec. 2.4, here by "local temperature" I mean the one *ideally measured* by a *local probe* at equilibrium with the system at the given point in space.

This is clearly an idealization and one needs different tools to extract the local temperature experimentally. This has been done in atomic and molecular junctions demonstrating the phenomenon of local ionic heating (Huang *et al.*, 2006; Tsutsui *et al.*, 2007; Wang *et al.*, 2007). In Huang *et al.* for instance, the force necessary to break the junction has been measured as a function of bias. From this force one can extract an effective local temperature. This experimental temperature as a function of bias is shown in Fig. 6.8 for an octanedithiol molecule in between two gold electrodes. From the figure it is clear that an effective temperature larger than the nominal background (room) temperature is found. I anticipate here that this temperature also contains an important contribution from electron-electron interactions (see Sec. 8.6).

In this section I want to determine this temperature and its bias dependence from a microscopic theory.

Local ionic heating: simple considerations

In a nanojunction there is a discrete set of vibrational modes – with some degree of coupling to the electrode bulk modes – so that local heating of the

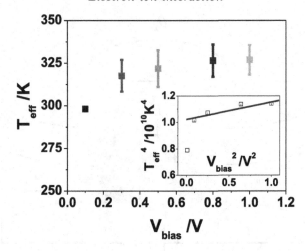

Fig. 6.8. Experimental local temperature of an octanedithiol molecule in between two bulk electrodes as a function of the bias. The inset shows its fourth-power dependence on the square of the bias. Reprinted with permission from Huang *et al.* (2006).

ions of the junction occurs when the bias is large enough to excite the mode with lower energy.

Let us call the energy of this mode $\hbar\omega_c$ and the corresponding "critical" bias V_c. At steady state, the energy exchanged between electrons and ions, and the energy dissipated from the junction into the bulk make the net power exchanged between electrons and ions zero, so that the ions of the junction experience an effective local temperature θ_{eff} higher than the one associated with the ions in the bulk (see Fig. 6.7).

Here, the phonons may have a local distribution different in the junction compared to the bulk. At steady state, however, as an approximation we can take the phonon distribution to have the same *form* as the equilibrium one 6.15 but with a spatially dependent temperature, $\theta_{eff}(\mathbf{r})$. While this assumption may not be exact, it is a good starting point to determine the effective local temperature in the junction.[25]

Before discussing the quantum theory of local ionic heating let us make simple considerations on the bias dependence of this effect. Let us begin by assuming that the electrons are at zero temperature and the ions deep into the bulk electrodes also have zero background temperature, $\theta_0 = 0$.

The power generated in the nanostructure due to the electron flow, has

---

[25] Determining the actual *form* of the non-equilibrium distribution would require, for each nanostructure, a self-consistent theory of coupled phonon *and* electron flow. To the best of my knowledge such a theory has not been developed yet.

6.3 Local ionic heating

to be a small fraction of the total power of the circuit V^2/R (V is the bias, R is the resistance). Let us define this fraction $P_v = \Theta(V - V_c)\gamma V^2/R$, with γ a positive constant to be determined from a microscopic theory, and I have explicitly introduced the step function Θ so that for $V < V_c$ there is no transfer of energy between electrons and ions of the junction.

As discussed above, at steady state this power has to balance the thermal current I_{th} (heat per unit time) carried away into the bulk electrodes. For a nanostructure, lattice heat conduction may be a complicated function of the coupling between the modes of the junction and the modes of the bulk electrodes. For instance, I have anticipated that a nanojunction may support local phonon modes that are weakly coupled to the bulk modes of the electrodes. The heat associated with these modes would thus be difficult to dissipate and the structure may become unstable. I will come back later to this point.

For the moment let us suppose that the lattice heat conduction follows a bulk law. I will show later that this law can be recovered under conditions of *elastic* phonon scattering, i.e., as for electrons, scattering processes that allow phonons to scatter from the nanojunction out to the bulk region without changing their energy.

In the bulk, we know that lattice energy is carried away at low temperatures at a rate $I_{th} = \alpha\,\theta_{eff}^4$ with α the lattice heat conductance (see, e.g., Ashcroft and Mermin, 1975).

By equating the power in and the power out the junction – steady-state condition – we then get[26]

$$P_v = I_{th} \Longrightarrow \theta_{eff} = \Theta(V - V_c)\left(\frac{\gamma}{\alpha R}\right)^{1/4}\sqrt{V}, \quad \text{zero background } \theta. \quad (6.74)$$

We therefore see from this simple argument that, under the assumption of bulk lattice heat conductance, the bias dependence of the temperature of the junction follows a square-root law (Todorov, 1998; Chen et al., 2003), if the resistance, as well as the other parameters in 6.74, does not vary appreciably with bias – reasonable in linear response.

Let us now take the background temperature far away from the junction to be some some finite temperature θ_0. As before, let us also consider the bulk form for lattice heat conductance. In addition to the energy generated in the junction, which is carried away to the bulk at an *out*-rate $I_{th}^{out} = \alpha\,\theta_{eff}^4$,

---

[26] The estimate I make of the effective temperature of the junction corresponds to an "average" temperature associated with the atoms of the nanostructure and possibly a few atoms of the nearby electrodes. As discussed in Footnote [25], the calculation of the actual profile $\theta_{ph}(\mathbf{r})$ would also require a self-consistent calculation of the non-equilibrium phonon distribution in the presence of current.

we also have the energy that is carried *from* the bulk to the junction at an *in*-rate $I_{th}^{in} = \alpha \theta_0^4$.[27]

The steady-state condition now reads

$$P_v = I_{th}^{in} - I_{th}^{out} \implies \theta_{eff} = (\theta_0^4 + \theta_v^4)^{1/4}, \quad \text{finite background } \theta, \quad (6.75)$$

where θ_v is now the contribution to the local temperature due to the bias only, namely $\theta_v = \Theta(V - V_c) \left(\frac{\gamma}{\alpha R}\right)^{1/4} \sqrt{V}$.

Heating from localized modes

These are simple considerations and are based on the assumption that the heat conduction mechanism is the same as in the bulk. As mentioned above this may not be the case when there are modes that have weak coupling with the modes of the nearby bulk electrodes. In this case, the rate at which electrons excite these modes may be considerably faster than the rate at which their associated energy can be dissipated in the bulk. This *electron-pumping* effect can thus give rise to extremely high temperatures. We can estimate these temperatures with the following argument.

In a current-carrying system the electrons are not in their ground state. Therefore, even if their background temperature is zero, some of them are above the ground-state Fermi level simply due to the presence of the bias V. This provides an effective energy window for inelastic electron-phonon scattering of width of order eV. The localized vibrational modes therefore "see" this excited electronic system as if the latter has an effective finite temperature of the order of eV/k_B. The effective ionic temperature is then simply

$$\theta_{eff} \propto eV/k_B, \quad \text{localized phonons}. \quad (6.76)$$

Note that this argument is *independent* of the strength of the electron-phonon coupling: the effective temperature of a localized phonon mode is related to the "departure" of the electron distribution from equilibrium, not to the energy transferred from the electrons to the ions. We therefore expect the proportionality constant to be of order one.

If this is the case, then a bias of 0.1 V would give rise to a temperature of about 1000 K! At these temperatures the nanostructure is likely to undergo structural instabilities and fracture of bonds. If these instabilities occur, they are likely to *nucleate* at the atomic bonds with reduced coordination,

---

[27] This reasoning is similar to one I have used to determine the total electrical current, which can be written as the difference of the current from right to left and the current from left to right, Sec. 3.3.3.

for instance the atoms at the contact region between the nanostructure and the bulk electrodes.

Quantum theory of local ionic heating

We can now move on to the calculation of the local heating effect using the microscopic theory of electron-phonon coupling we have developed in the previous sections. Our starting point is the coupling Hamiltonian 6.29. We want to determine the energy per unit time transferred among electrons and ions coupled via this Hamiltonian. This power can be calculated using time-dependent perturbation theory as follows.

If we refer again to Eq. 6.35 with $|i\rangle$ and $|f\rangle$ states corresponding to any of the processes represented in Fig. 6.2, where a phonon of energy $\hbar\omega_j$ is exchanged, then the energy per unit time transferred between electrons and ions for that particular process is simply

$$W_j = \hbar\omega_j P_{i\to f} = \frac{2\pi}{\hbar}\hbar\omega_j |\langle f|\hat{\mathbf{H}}_\mathbf{p}|i\rangle|^2 \delta(E_f - E_i). \quad (6.77)$$

The matrix elements appearing in Eq. 6.77 are the ones we have computed for Eqs. 6.36 and 6.38. We therefore get for each mode with energy $\hbar\omega_j$

$$W_j^\mp = W_j^{R,\mp} + W_j^{L,\mp}$$

$$= 2\pi\hbar(\delta + N_j) \sum_{\alpha,\beta=L,R} \left[\int dE \left| \sum_i A_{i,j} J_{E\pm\hbar\omega_j,E}^{i,\alpha\beta} \right|^2 \right.$$

$$\left. \times f_E^\alpha \left(1 - f_{E\pm\hbar\omega_j}^\beta\right) D_{E\pm\hbar\omega_j}^\alpha D_E^\beta \right]. \quad (6.78)$$

In the last equality, $\delta = 1$ and "$-$" sign are for the scattering diagrams (a) and (c) in Fig. 6.2, and $\delta = 0$ and "$+$" sign for diagrams (b) and (d). A factor of 2 due to spin degeneracy also appears in Eq. 6.78.

The total thermal power generated in the junction is therefore the sum over all vibrational modes for the four processes of Fig. 6.2

$$P = \sum_j \left(W_j^{R,+} + W_j^{L,+} - W_j^{R,-} - W_j^{L,-} \right), \quad (6.79)$$

with processes (a) and (c) contributing to *heating* of the structure, and processes (b) and (d) contributing to *cooling*.

Due to the spontaneous emission terms in Eq. 6.79, by increasing the ionic temperature θ_{ph}, heating is at first the dominant term (see Fig. 6.9). Cooling processes have zero contribution at $\theta_{ph} = 0$, but slowly pick up

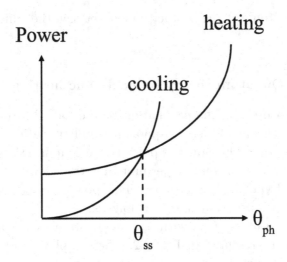

Fig. 6.9. Power as a function of temperature for heating and cooling processes. When these two balance each other, a steady-state temperature θ_{ss} is achieved.

with temperature, until at a given (steady-state) temperature θ_{ss} they exactly compensate the heating ones and make the power P in 6.79 zero (see Fig. 6.9).

In reality, from the power in Eq. 6.79 we need to subtract the energy per unit time that is transferred from the junction to the bulk electrodes. This is an extra cooling process which lowers the temperature even further: $\theta_{ss} \to \theta_{eff}$.

Before discussing this point, let us note that if we set $\theta_e = 0$ and $\theta_{ph} = 0$ – both the electronic and the ionic temperatures equal to zero – then only processes of type (c) involving spontaneous emission contribute. Setting these temperatures in Eq. 6.78 to zero we get the total power

$$P = 2\pi\hbar \sum_j \int_{\mu_R+\hbar\omega_j}^{\mu_L} dE \left| \sum_i A_{i,j} J_{E-\hbar\omega_j,E}^{i,LR} \right|^2 D_{E-\hbar\omega_j}^L D_E^R. \qquad (6.80)$$

As anticipated with the simple model 6.74, we thus see that at zero background temperature – for both electrons and ions – a bias larger than $V_c = \min\{\hbar\omega_j\}/e$ is necessary to generate heat locally in the junction.

As we have done for the scattering rate calculation in Sec. 6.2.1, let us also assume $eV \gg \hbar\omega_j$ and the density of states, $D_E^{L(R)}$, and coupling constant, $J_{E_1,E_2}^{i,\alpha\beta}$, slowly varying with energy (with $D^{L(R)}$ and $J^{i,\alpha\beta}$ their values, respectively, evaluated at the equilibrium Fermi level). The power 6.80 can then be approximated as

$$P \simeq 2\pi\hbar eV \sum_j \left| \sum_i A_{i,j} J^{i,LR} \right|^2 D^L D^R. \qquad (6.81)$$

The quantity $2\pi\hbar \sum_j \left| \sum_i A_{i,j} J^{i,LR} \right|^2 D^L D^R$ "counts" the number of scattering events per unit time that occur locally in the volume of the junction. Since in linear response the current $I \propto V$, we therefore see from Eq. 6.81 that the power P is proportional to the current density in the junction, as we have anticipated at the beginning of this section.

I finally note that for a current of about 10 µA at a bias of 0.1 V, the power of the total circuit is about 1 µW. For systems that carry a similar current at that bias we can take again $J^{i,LR} \simeq 0.1$ eV/Å, $D^L \sim D^R \simeq 1$ eV^{-1}. With $M \simeq 10^4\, m_e$, the power for a single mode calculated from Eq. 6.81 is about 10^{-2} µW at 0.1 V.

As anticipated, the power dissipated to a single mode in the junction is a small fraction of the total power of the circuit, even though the power *per atom* is larger in the junction.

6.3.1 Lattice heat conduction

We are now left to consider the mechanisms that carry the power generated in the junction away into the bulk. I stress here that a complete theory of heat transport in nanostructures is not yet available. However, we can argue about a plausible form of heat conduction along the following lines.

At the length scales we consider here and for quasi-ballistic transport, the heat needs to be dissipated very fast for the nanostructure to remain stable. For instance, for a typical current of 1µA, electrons "cross" the nanojunction at a rate of about 10^{13} s^{-1}, and in the process they excite modes of the junction. Therefore, this frequency sets the minimum rate at which heat needs to be dissipated in the bulk to avoid structural instabilities.

In bulk systems, we know that if we consider ionic temperatures, θ_{ph}, smaller than the Debye temperature, θ_D, of the bulk (typically of the order of few hundred K), then the rate at which inelastic phonon processes – transitions between different modes of different energy – occur is of the order of $e^{-\theta_D/\theta_{ph}}$, i.e., it is exponentially suppressed (see, e.g., Ashcroft and Mermin, 1975). This is due to the fact that in a perfect crystal there is a small probability of *Umklapp processes*, those that destroy crystal momentum and thus contribute to a finite heat conductivity.

In the absence of such processes, the total crystal momentum would be conserved even if *anharmonic effects* – originating from higher-order terms in the expansion 6.6 – are present[28] and the thermal conductivity would be infinite: the system cannot reach either local or global thermodynamic equilibrium.

A nanostructure breaks the translational invariance of the bulk electrodes, and therefore we expect these processes to be of limited use in proximity to the junction (provided they can be defined at all in the absence of translation invariance).

We therefore expect that *elastic scattering* of the phonons of the junction into the bulk (continuum) phonon modes of the electrodes to be the most efficient way to dissipate heat and restore local thermal equilibrium.[29]

Similar to what we have done in the case of electron transport within the Landauer approach of Chapter 3, this elastic phonon scattering can be thought of as originating from an "imbalance" of phonon distributions – or, equivalently, a thermal gradient – between two *reservoirs* of phonons (see Fig. 6.10). Here I describe the following simple model that captures the main physics of this problem.

Let us consider two reservoirs of phonons connected via a weak mechanical link of length L – our junction, which we will later assume to be a spring with a given stiffness. A phonon of energy $E = \hbar\omega$ thus has a probability T_{ph} to carry its energy across the mechanical link from one reservoir to the other. Let us call N_R the local phonon distribution 6.15 of the right reservoir – with effective temperature θ_R higher than the left one – and N_L the distribution of the left one.

The net thermal current carried by this phonon in an interval of energy dE must thus have the form (see discussion in Sec. 3.3.3 and Rego and Kirczenow, 1998; Patton and Geller, 2001)[30]

$$dI_{\hbar\omega} \propto dE\, E\, T_{ph}(E)\left[N_R(E) - N_L(E)\right], \qquad (6.82)$$

which is similar to the Landauer equation for electrons (Eq. 3.56).[31]

In order to take into account all possible energies of independent phonon

---

[28] Processes that conserve total crystal momentum are called *normal*. The terminology of normal and Umklapp processes applies to electrons as well. As we will discuss in Chapter 7, the first cannot contribute to inelastic relaxation effects, and thus to the resistance.

[29] Inelastic phonon processes would again be of importance away from the junction where Umklapp processes are also expected to contribute.

[30] I am here implicitly assuming that all the energy is carried away by the phonons. Also, that the only contribution to heat dissipation is from *longitudinal* phonons, so that, as in the case of electron transport, the 1D density of states cancels exactly the velocity of the phonon (see discussion in Sec. 3.3.3).

[31] Following a similar argument one can also show that in ballistic (with respect to phonons) quasi-1D wires supporting one phonon channel, when heat is carried only by phonons (*dielectric media*), the thermal conductance is quantized with quantum $\pi k_B^2 \theta / 6\hbar$ (Rego and Kirczenow, 1998; Schwab *et al.*, 2001).

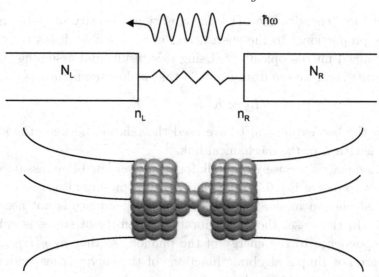

Fig. 6.10. Schematic of two reservoirs of phonons with different distributions (N_L and N_R) connected by a weak mechanical link, represented by a spring of given stiffness. The energy per unit time carried by a phonon of energy $\hbar\omega$ depends on the local vibrational density of states at the contact (n_L and n_R) between the link and the left and right reservoirs.

modes then one simply integrates in energy Eq. 6.82 from zero – the lowest possible mode energy – to infinity:[32]

$$I_{th} \propto \int_0^\infty dE\, E\, T_{ph}(E)\, [N_R(E) - N_L(E)]. \tag{6.83}$$

We just need to determine the proportionality constant and the transmission probability. A microscopic model of the link between the nanostructure and the bulk electrodes is thus necessary. If we describe the junction as a weak mechanical link of a given stiffness K[33] and denote with $n_R(E)$ and $n_L(E)$ the corresponding local vibrational density of states of phonon modes at the *interface* between the right reservoir and the mechanical link, and the latter and the left reservoir (see Fig. 6.10), then one can show along the lines of the transfer Hamiltonian (Exercise 3.21) that at temperatures $\theta_{ph} \ll \theta_D$ (Patton and Geller, 2001)

$$I_{th} = \frac{4\pi K^2}{\hbar} \int_0^\infty dE\, E\, n_R(E)\, n_L(E)\, [N_R(E) - N_L(E)]. \tag{6.84}$$

---

[32] A similar argument allows us to derive an equivalent expression for electron heat conduction (see Sec. 8.6.1).

[33] The stiffness $K = \pi d^2 Y/4L$, where Y is the Young's modulus of the junction, and d (L) is its diameter (length) (Patton and Geller, 2001).

At very low energies the 2D local vibrational density of states $n(E)$ is typically proportional to the energy, i.e., $n(E) = bE$ with b a constant to be determined microscopically.[34] Using this result, and assuming the bulk temperature zero, we see from Eq. 6.84 that at low temperatures

$$I_{th} \propto K^2 \theta^4 \propto L^{-2} \theta^4, \qquad (6.85)$$

where in the last expression I have used the relation between the stiffness and the length L of the mechanical link.[33]

Equation 6.85 is precisely the bulk fourth-power law I have assumed from the outset to derive Eq. 6.74. This microscopic picture also shows why, for a localized phonon mode, this fourth-power dependence is not necessarily satisfied. In this case, the local vibrational density of states is enhanced in correspondence to the energy of the phonon, so that $n_{L,R}(E)$ and also $T_{ph}(E)$ are not simple algebraic functions of the energy, thus leading to a different temperature dependence of the thermal current.

Neglecting the above case, and considering now the length dependence of the thermal current I have just derived, Eq. 6.85, I can redo the simple estimates of the local temperature, and find its dependence on the length of the junction.

If the power generated locally in the nanostructure is length independent[35]

$$P_v = I_{th} \implies \theta_{eff} \propto \Theta(V - V_c) \sqrt{L} \sqrt{V}, \quad \text{zero background } \theta, \qquad (6.86)$$

which shows a weak dependence on length.

For molecular structures the current decays exponentially with length, and the resistance increases accordingly. If $R = R_0 e^{\beta L}$ is the change of the resistance with length from a fundamental value R_0 (β is a characteristic inverse decay length) the power generated locally in the structure is $P_v = \Theta(V - V_c)\gamma V^2/R = \Theta(V - V_c)\gamma e^{-\beta L} V^2/R_0$. Equating this power to Eq. 6.85 we find

$$P_v = I_{th} \to \theta_{eff} \propto \Theta(V - V_c) \sqrt{L}\, e^{-\beta L/4} \sqrt{V}, \quad \text{zero background } \theta, \qquad (6.87)$$

so that the effective temperature shows a stronger dependence on length (Chen et al., 2005).[36]

---

[34] This dependence is due to the presence of *acoustic* phonon branches. Quite generally, the relation is of the type $n(E) = bE^\alpha$, with α some positive number. For instance, $n(E)$ scales as E^2 in 3D. $\alpha = 1$ at the planar surface of a semi-infinite isotropic elastic continuum, which is what I am assuming here. A different power law gives rise to a different heat dissipation behavior.

[35] This is, for instance, fairly well satisfied by metallic quantum point contacts.

[36] This analysis is clearly a simplification of the real physical situation of heat generation and transport. For instance, for molecular structures, the phonon wave-function "mismatch" between the vibrational modes inside the structure and the bulk ones may favor non-trivial elastic scattering of the former, possibly leading to a different local temperature dependence on length.

The fact that local heating decreases with the length of the junction in molecular structures has been recently verified experimentally on alkanethiols of different lengths (Huang *et al.*, 2007) (see Sec. 8.6).

Referring again to Fig. 6.10, if we consider the average effective temperature in the junction, θ_{eff}, to first approximation the latter can be assumed to be the average between the left and right phonon reservoir temperatures.[37] In reality, the nanojunction has a temperature θ_{eff} higher than the temperature of both left and right electrodes, θ_0. We can thus modify the above arguments by realizing that the thermal current from the junction with temperature θ_{eff} dissipated to the left electrode with temperature θ_0 is equivalent to the thermal current of a weak mechanical link between heat "reservoirs" with temperatures $2\theta_{eff} - \theta_0$ and θ_0 (Chen *et al.*, 2003). Analogously for the thermal current into the right electrode. The total thermal energy carried away from the junction due to elastic phonon scattering would be the sum of these two contributions.

6.4 Thermopower

Let us conclude the discussion of heat transport with a brief introduction to the *thermoelectric power* or simply *thermopower*. This effect is of great technological importance – for an extensive account see, e.g., the book by Pollock (1985). It has received much less attention at the nanoscale (Sivan and Imry, 1986; Bogachek *et al.*, 1996; Ludolph and van Ruitenbeek, 1999; Paulsson and Datta, 2003; Reddy *et al.*, 2007) but it may yield a lot of information on energy transport mechanisms and their relation to charge transport.

In Sec. 2.4 I have shown that electrons respond to both an electric field **E** and to a chemical potential gradient as if they are driven by a "field" $\mathcal{E} = \mathbf{E} - \nabla\bar{\mu}/e$ (Eq. 2.92).

The thermopower (or *Seebeck coefficient*) is defined as the proportionality constant Q between this "field" and the temperature gradient

$$\mathcal{E} = Q\nabla\theta. \tag{6.88}$$

The above relation shows that if there is a temperature gradient applied at the two ends of an open circuit a voltage difference has to develop between

---

[37] This argument neglects any spatial variation of the local temperature, and it assumes that the main temperature "drop" occurs at the contacts between the nanostructure and the electrodes. This may not always be the case.

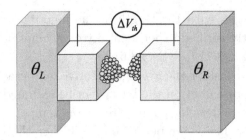

Fig. 6.11. If two sides of a nano-junction initially in global equilibrium are heated up at different temperatures, θ_L and θ_R, a current flows. If the circuit is open, in the long-time limit the current goes to zero, and a thermoelectric voltage ΔV_{th} develops.

the high- and low-temperature regions of the sample (Fig. 6.11).[38] The charge carriers *diffuse* from the hot region to the cold region, similar to the behavior of a classical gas (see also discussion in Sec. 2.5). This creates a separation of charges and, therefore, a *thermoelectric voltage* difference ΔV_{th}. This voltage difference can be evaluated by integrating Eq. 6.88 – at zero current – between two points $\mathbf{r}_L$ and $\mathbf{r}_R$ whose temperature difference is $\Delta \theta = \theta_L - \theta_R$[39]

$$\Delta V_{th} = -\int_{\mathbf{r}_R}^{\mathbf{r}_L} \mathcal{E} \cdot d\mathbf{r} = Q \int_{\mathbf{r}_R}^{\mathbf{r}_L} (-\nabla \theta) \cdot d\mathbf{r} = Q(\theta_R - \theta_L) = -Q \Delta \theta. \quad (6.89)$$

In bulk metals, if $Q < 0$ then the current is carried by electrons, if $Q > 0$ it is carried by holes. This is easy to see.

Consider the system at initial zero bias, and global chemical potential $\bar{\mu}$. Now let $\theta_L = 0$ on the left boundary of the metal and θ_R some finite temperature on the right boundary. In the right region where we have a finite temperature, the Fermi distribution broadens (see, e.g., Fig. 1.7). Electrons in the metal with energy in the neighborhood of the chemical potential have thus access to states on the left boundary, while electrons on the far left have a lower probability to move to the right (they have lower average kinetic energy than those on the right boundary). Electrons will then *diffuse* to the left, and if the circuit is open, they accumulate on the left boundary of the metal leaving a depletion of electrons on the right boundary. The accumulated electrons on the left have thus a higher potential energy compared to those on the right boundary, and hence a lower electrostatic potential. The difference $\Delta V_{th} = V_L - V_R < 0$, and thus $Q < 0$. The above argument is clearly reversed if we consider holes as charge carriers, and hence $Q > 0$ for positive carriers.

---

[38] If the circuit is closed via some resistance, a voltage difference is produced at the two ends of this resistance, and current flows in the circuit.

[39] In an actual experiment, the voltage probe that is connected to the system in order to measure the thermopower is necessarily invasive, since the applied thermal gradient would induce, locally at the voltage probe contact, an extra voltage difference. This extra effect needs to be subtracted to get the actual thermopower of the nanojunction.

6.4 Thermopower

As I will discuss below, the above sign assignment is not so straightforward, and does not have the same meaning in nanoscale systems. Note also that thermopower is a non-equilibrium phenomenon: after the transient dynamics that leads to the charge imbalance between the two opposite ends of the conductor, there is still a *continuous flow of energy* between the right and left boundaries. In the stationary state the thermopower may not change in time (in quantum mechanics we say that the macro-state of the system is time-independent), but energy continually flows in from the high-temperature contact out into the low-temperature contact. If we removed this external temperature gradient, the system would cancel *internal* temperature differences, so that a *global* equilibrium state can be established in the long-time limit.

Relation between thermopower and scattering probabilities

As we did for the electrical current within the single-particle Landauer approach, we can determine an approximate form of the thermoelectric voltage within the same approximations as follows (Butcher, 1990). Consider the system of Fig. 6.11 in *global* equilibrium with temperature θ_0 and chemical potential $\bar{\mu}$. Let us then raise slightly the local temperature *and* electrochemical potential in the left electrode

$$\theta_L = \theta_0 + \Delta\theta/2, \tag{6.90}$$

$$\mu_L = \bar{\mu} + |e|V/2. \tag{6.91}$$

At the same time, we also lower the local temperature and electrochemical potential in the right electrode

$$\theta_R = \theta_0 - \Delta\theta/2, \tag{6.92}$$

$$\mu_R = \bar{\mu} - |e|V/2. \tag{6.93}$$

Let us now Taylor-expand the local Fermi distributions as we have done in Sec. 3.3.3 (Eqs. 3.57 and 3.60). This yields

$$f_L(E) = f(E) - \left.\frac{\partial f(E)}{\partial E}\right|_{\bar{\mu}} \left[\frac{eV}{2} - (E - \bar{\mu})\frac{\Delta\theta}{2\theta_0}\right] \tag{6.94}$$

where $f(E)$ is the global equilibrium Fermi distribution 1.80 with temperature θ_0 and chemical potential E_F.

Similarly, we have

$$f_R(E) = f(E) + \left.\frac{\partial f(E)}{\partial E}\right|_{\bar{\mu}} \left[\frac{eV}{2} - (E - \bar{\mu})\frac{\Delta\theta}{2\theta_0}\right]. \tag{6.95}$$

If we replace the above expansions into the Landauer formula 3.56 we get

$$\begin{aligned}I &\simeq \frac{e^2 V}{\pi\hbar}\int_{-\infty}^{+\infty} dE \left(-\left.\frac{\partial f(E)}{\partial E}\right|_{\bar{\mu}}\right)\mathcal{T}(E) \\ &\quad -\frac{e}{\pi\hbar}\frac{\Delta\theta}{\theta_0}\int_{-\infty}^{+\infty} dE \left(-\left.\frac{\partial f(E)}{\partial E}\right|_{\bar{\mu}}\right)(E-\bar{\mu})\mathcal{T}(E).\end{aligned} \tag{6.96}$$

If the equilibrium temperature θ_0 is very low, we can approximate $\bar{\mu} \simeq E_F$ and the first term gives rise to the current 3.67, with the transmission coefficient evaluated at the equilibrium Fermi level.

The second integral can be calculated by expanding the transmission coefficient to first order as in Eq. 3.65, provided we are away from electronic resonances so that its variation at the equilibrium Fermi level is not sharp. We can then use the well-known *Sommerfeld expansion* to evaluate it (see, e.g., Appendix C of Ashcroft and Mermin, 1975).[40] Since the thermal energies we consider here are much smaller than the equilibrium Fermi energy, we can stop at the first order of that expansion so that the current 6.96 is

$$\begin{aligned}I &\simeq \frac{2e^2}{h}\mathcal{T}(E_F)V - \frac{e}{\pi\hbar}\frac{\pi^2 k_B^2}{3}\theta_0 \left.\frac{\partial \mathcal{T}(E)}{\partial E}\right|_{E_F}\Delta\theta \\ &= G(E_F)V + e\, L_{eE}\Delta\theta.\end{aligned} \tag{6.97}$$

The *response coefficient* L_{eE} is related to the fact that a temperature gradient can induce both a particle flow *and* energy flow (see Footnote [43] in Chapter 2). If we finally want to determine the thermoelectric voltage we simply set the current equal to zero in the above expression to get

$$\Delta V_{th} = -\frac{\pi^2 k_B^2}{3e}\theta_0 \frac{1}{\mathcal{T}(E)}\left.\frac{\partial \mathcal{T}(E)}{\partial E}\right|_{E_F}\Delta\theta = -\frac{\pi^2 k_B^2}{3e}\theta_0 \left.\frac{\partial \ln \mathcal{T}(E)}{\partial E}\right|_{E_F}\Delta\theta, \tag{6.98}$$

---

[40] Given the Fermi distribution $f(E)$ and some function $L(E)$ which vanishes at $E \to -\infty$, and is bound at $E \to +\infty$ or, if not, it diverges no more rapidly than a power of E, the Sommerfeld expansion relates to the following integral

$$\int_{-\infty}^{\infty} dE\, L(E) f(E) = \int_{-\infty}^{\bar{\mu}} dE\, L(E) + \frac{\pi^2}{6}(k_B\theta_0)^2 \left.\frac{\partial L(E)}{\partial E}\right|_{\bar{\mu}} + O\left(\frac{k_B\theta_0}{\bar{\mu}}\right)^4.$$

so that the thermopower is

$$Q = -\frac{\pi^2 k_B^2}{3|e|}\theta_0 \left.\frac{\partial \ln \mathcal{T}(E)}{\partial E}\right|_{E_F}. \tag{6.99}$$

Notice that if the transmission coefficient is a slowly varying function of the energy around the Fermi level, then from Eq. 6.99 we find that the thermopower is close to zero. Also, anything else being equal, a larger transmission coefficient implies lower thermopower. This is consistent with experimental results on metallic quantum point contacts where it has been found that the thermopower is suppressed for contacts with larger conductance (Ludolph and van Ruitenbeek, 1999).

Transient response and dynamical effects in thermopower

We have thus related the thermopower to single-particle properties, such as the transmission coefficient, in the linear – with respect to both thermal gradient and bias – regime. However, the above result is clearly an idealization and lacks important physics at the nanoscale.

First of all, we have derived it by assuming that a voltage source is *already* present together with a temperature gradient. This is contrary to the experimental situation in which a thermal gradient is applied and, *as a consequence*, a voltage difference is created.

To be more specific, if a thermal gradient is applied to the junction, the transient dynamics is fundamental in establishing this voltage difference since the *dynamical* formation of local resistivity dipoles creates strong local fields at the junction. These fields influence the electron motion in a self-consistent way, and thus influence the long-time behavior of the carrier dynamics. This is particularly important *away* from linear response.

It is the self-consistent formation of these fields that makes the thermopower very sensitive to atomic details, and thus to the contact geometry between the nanostructure and bulk electrodes; a fact that has also been demonstrated experimentally (Ludolph and van Ruitenbeek, 1999). The sign of the thermopower is therefore very sensitive to these microscopic details, precluding an easy interpretation in terms of "electron" or "hole" excitations as in bulk metals.

Finally, interactions among electrons – and electrons and ions – may also play an important role during the formation of the thermoelectric voltage. For instance, during transient dynamics part of the electron energy is transferred to both ions and electrons, thus creating a transient local heating of the junction. The ionic temperature is a "loss" of energy for electrons while the effective local electron temperature (see Sec. 8.6) reduces the external thermal gradient locally in the junction, creating an effective local energy barrier for diffusion. These effects may thus have a large bearing in the thermopower generation in nanoscale systems, and have not been explored at all in the literature. To this aim, since the system is in *dynamical* interaction with two different baths, one needs to go beyond the Landauer approximations that have led us to Eq. 6.99, and consider an open quantum system approach like the one described in Sec. 1.4.5 (see also Appendices C and G).

6.5 Current-induced forces

In addition to the local ioninc heating effect discussed in the previous sections, there is yet another phenomenon associated with the interaction between electrons and ions in the presence of current: *current-induced forces*. These forces are due to the momentum that electrons flowing in a conductor transfer to the ions. Or alternatively, from a static point of view, the steady-state electron density in a current-carrying conductor is microscopically different from that in the absence of current. In either picture, current flow generates additional forces on ions. This may result in atomistic rearrangements and diffusion of atoms (Di Ventra *et al.*, 2004b).

In microelectronics this phenomenon is known as *electro-migration*, and it is the major failure mechanism in metallic interconnects used in solid-state circuits.[41] Once more, since nanoscale conductors may carry current densities much larger than their macroscopic counterparts, electro-migration may also be a key issue in their stability.

6.5.1 Elastic vs. inelastic contribution to electro-migration

Before moving on to the (tricky) definition of forces in the presence of current, let us say a few words on what we mean by "momentum transfer". Electro-migration may occur due to two processes: (i) electrons transfer only momentum to the ions and no energy is exchanged (elastic processes); (ii) electrons transfer momentum *and* energy to the ions (inelastic processes). I give here a simple argument to show that inelastic processes contribute negligibly to current-induced forces, at least in quasi-ballistic conditions (Yang and Di Ventra, 2003).

---

[41] In microelectronics, electro-migration generally refers to the current-induced motion of vacancies or other defects. I will instead use this term to indicate current-induced forces on *any* type of atoms.

6.5 Current-induced forces

Let us apply a bias V between left and right electrodes. Due to inelastic processes an electron moving from left to right with energy equal to the left electrochemical potential loses, at most, all the energy eV due to an inelastic collision with a given ion in the junction – and emerges in an unoccupied state. In the process, the electron transfers a momentum Δp to the ion of the order of $\sqrt{2meV}$.

Now, if the electron inelastic mean-free path due to electron-phonon collisions is λ_{ph}, and the electron moves at a Fermi velocity v_F, the time between collisions is of the order of $\tau_{ph} = \lambda_{ph}/v_F$ (from Eq. 6.42). If, say, the average current in the system is I, we can estimate the average number N_{in} of electrons that experience collisions with that atom in the interval of time τ_{ph}, by multiplying the latter by the current I, namely $N_{in} = I\tau_{ph}/e$.

The total force exerted on the ion due to these inelastic collisions is therefore $F_{in} = N_{in}\Delta p/\tau_{ph}$.

In reality, for quasi-ballistic systems an electron does not lose all its energy via inelastic effects, only a fraction γ. In Sec. 6.3, we have estimated the fraction of power P exchanged between electrons and phonons in the junction. In the interval of time τ_{ph} the electron thus loses on average an energy $E_{in} \simeq P\tau_{ph}$. The transfer of momentum to the ion Δp is therefore of the order of $\sqrt{2mE_{in}}$. As previously discussed, in typical quasi-ballistic systems we expect $\lambda_{ph} \sim 1000$ Å, and $v_F \sim 10^8$ cm/s. In Sec. 6.3 we have seen that for a typical current of 10 µA, at a bias $V = 0.1$ V, $P \simeq 0.01$ µW, so that $E_{in} \simeq 0.5 \times 10^{-2}$ eV, and $N_{in} \sim 6$. We then get $F_{in} = N_{in}\Delta p/\tau_{ph} = N_{in}\sqrt{2mE_{in}}/\tau_{ph} \sim 0.001$ eV/Å ~ 0.002 nN (1 eV/Å $\simeq$ 1.602 nN).

From the above simple arguments we can also estimate the bias dependence of this force – at least in the linear-response regime. Let us then consider a linear relation between current and bias, and recall that for slowly varying density of states and electron-phonon coupling constant, $\tau_{ph} \propto V^{-1}$ (see Eq. 6.41). Within the same approximations the power fraction $P \propto V$ (Eq. 6.81) and N_{in} does not vary with bias. We therefore find the bias – and current – dependence

$$|F_{in}| \propto |V| \propto |I|, \quad \text{inelastic force, linear response,} \qquad (6.100)$$

where I have introduced the absolute value, because the sign of the force cannot be easily determined from the above arguments.

If there are N atoms in the nanostructure, then the average force per atom would be F_{in}/N. As we will see later, this force is at least an order of magnitude smaller than the corresponding elastic contribution. Also, a typical bond strength is of the order of 1 nN or more, so that the inelastic contribution to electro-migration does not affect bond breaking substantially, and can therefore be neglected even at large biases. This is what I will do in the following and focus only on the elastic part of current-induced forces.

6.5.2 One force, different definitions

We now need to define forces on ions in the presence of current. There are several ways to do this. I will first discuss the approaches used in the past. These approaches will clarify the physical origin of these forces. I will then derive a general form for forces out of equilibrium in the adiabatic approximation. This force has a simple form within the single-particle scattering picture of conduction.

Direct and wind forces

Historically, current-induced forces have been understood via the separation of the total force on an ion into two contributions, one known as the *direct force*, and the other as the *wind force* (see, e.g., Sorbello, 1998). The direct force is often defined as the direct electrostatic force – due to the macroscopic electrostatic field **E** that drives the electron current – on the bare positive charge of the ion. The wind force is defined as the net momentum transfer from the electron current due to the scattering of electrons on the ion.

The separation into these two terms is clearly arbitrary, and in literature one can find the definition of direct force as the one acting not on the bare charge of the ion, but on its charge partially screened by some neutralizing electron charge – typically, the core electron charge if a pseudopotential picture is used, see Sec. 6.2. If $Z_i|e|$ is the bare charge of the ion, then the direct force is simply

$$\mathbf{F}_d = -Z_i\, e\, \mathbf{E}, \qquad \text{direct force}, \qquad (6.101)$$

or

$$\mathbf{F}_d = -Z_i^*\, e\, \mathbf{E}, \qquad \text{direct force}, \qquad (6.102)$$

with $Z_i^*|e|$ an effective screened charge of the ion.

Since the wind force is proportional to the rate of momentum transfer from electrons to the ions, one can estimate it in the following way within a single-particle scattering picture.

Consider, as usual, a current present due to a bias $V = (\mu_L - \mu_R)/e$. An electron from the left electrode can scatter from a state with momentum **k** to a state in the right electrode with momentum **k'** due to an elastic collision with the ion. The distribution of incoming electrons on the left is $f_L(E(\mathbf{k}))$ and the distribution of available states on the right, $1 - f_R(E(\mathbf{k}'))$, where I have chosen an arbitrary dependence of the energy on momentum.[42] If we consider very small temperatures $k_B\theta \ll eV$, we can neglect the small contribution from the opposite processes corresponding to electrons moving from the right electrode with occupation $f_R(E(\mathbf{k}))$, scattering into the left electrode in unoccupied states with weight $1 - f_L(E(\mathbf{k}'))$.

The only quantity that is left to determine is the number of possible transitions per unit time for an electron to elastically scatter from a state with momentum $\mathbf{k}$ to a state with momentum $\mathbf{k}'$. Similar to the local heating problem (Sec. 6.3), we determine this rate using time-dependent perturbation theory (Eq. 6.35). If we employ pseudopotentials to represent electron-ion interactions we get

$$P_{\mathbf{k}\to\mathbf{k}'} = \frac{2\pi}{\hbar}|\langle\mathbf{k}'|V^{ps}(\mathbf{r};\mathbf{R}_i)|\mathbf{k}\rangle|^2\delta(E(\mathbf{k}') - E(\mathbf{k})), \qquad (6.103)$$

where I have indicated with $|\mathbf{k}\rangle$ the single-particle state with momentum $\mathbf{k}$. Putting all together, the wind force can be approximated as

$$\mathbf{F}_w \simeq 2\pi \int dE \sum_{\mathbf{k}}\sum_{\mathbf{k}'} (\mathbf{k} - \mathbf{k}')|\langle\mathbf{k}'|V^{ps}(\mathbf{r};\mathbf{R}_i)|\mathbf{k}\rangle|^2$$
$$\times f_L(E(\mathbf{k}))[1 - f_R(E(\mathbf{k}'))]D_E^L D_E^R. \qquad (6.104)$$

The above expression for the force is clearly valid whenever perturbation theory is applicable. This may not be the case, for instance, if electronic resonances are present.

At very low biases, we can assume a slowly varying density of states for both left and right electrodes, and constant – in energy – overall momentum change as obtained from the sum $\sum_{\mathbf{k}}\sum_{\mathbf{k}'}(\mathbf{k}-\mathbf{k}')|\langle\mathbf{k}'|V^{ps}(\mathbf{r};\mathbf{R}_i)|\mathbf{k}\rangle|^2$. In the limit of zero temperature, the term $f_L(E(\mathbf{k}))[1 - f_R(E(\mathbf{k}'))]$ can then be integrated in energy to give the energy window eV. In linear response this is proportional to the current I.

We thus see from Eq. 6.104 that, within the above conditions, the absolute value of the wind force is proportional to the absolute value of the current

$$|\mathbf{F}_w| \propto |I| \propto |V|, \quad \text{wind force, linear response.} \qquad (6.105)$$

As in the case of the inelastic contribution to the force, I have emphasized the absolute values because contrary to naive expectations the wind force does *not* always drive the atom in the direction of the electron particle current (opposite to the standard sign convention for the current), and thus

---

[42] For instance, in the case of translationally invariant electrodes this relation is given by Eq. 3.23.

opposite to the applied electric field. I will show in a moment that the sign of this force depends on the details of the scattering between electrons and ion, the chemical nature of the latter, and its environment. All this requires a self-consistent calculation of the electron distribution under current flow. Out of linear response we therefore expect strong deviations from the simple linear relation 6.105 (Yang and Di Ventra, 2003; Yang et al. 2005) – see, e.g., Fig. 6.12.

Forces from Kubo linear-response theory

An alternative way of calculating these forces is via the expectation value of the electron-ion potential variation induced by an external (oscillating) electric field acting on the many-body Hamiltonian 6.1. This can be done in linear response by using the Kubo formalism I have introduced in Sec. 2.3 (see, e.g., Sorbello, 1998). A simplification occurs if we make the adiabatic approximation as discussed in Sec. 6.2, and use pseudopotentials to represent the electron-ion interaction. In this case, the observable $\hat{A}$ whose expectation value we are interested in is $\nabla_{\mathbf{R}_i} V^{ps}(\mathbf{r}; \mathbf{R}_i)$, with $\mathbf{R}_i$ the position of ion i (Kumar and Sorbello, 1975).

If the oscillating field has the form $\mathbf{E}e^{i\omega t}$, and couples to the electronic positions as $\mathbf{r} \cdot \mathbf{E}e^{i\omega t}$ (Eq. 2.69), the force on an ion at position $\mathbf{R}_i$ is from Eq. 2.46 (including the direct force)

$$\mathbf{F}_i = -Z_i^* e\mathbf{E} + \lim_{\omega \to 0} \frac{ie}{\hbar\omega} \int_0^\infty dt\, \mathrm{Tr}\left\{\hat{\rho}_{eq}^C \left[\nabla_{\mathbf{R}_i} V^{ps}(\mathbf{r}; \mathbf{R}_i, t), \hat{\mathbf{j}}(0)\right]\right\} \cdot \mathbf{E}\, e^{i(\omega+i\epsilon)t}, \tag{6.106}$$

where $\nabla_{\mathbf{R}_i} V^{ps}(\mathbf{r}; \mathbf{R}_i, t)$ is the Heisenberg representation of the operator $\nabla_{\mathbf{R}_i} V^{ps}(\mathbf{r}; \mathbf{R}_i)$, and $\hat{\mathbf{j}}(t)$ is defined in Eq. 2.71. In Eq. 6.106, the parameter ϵ is a positive infinitesimal quantity that arises from the adiabatic switch-on of the external field and guarantees the convergence of the integral (see Sec. 2.3), and the trace is again over the equilibrium canonical statistical operator of the electronic degrees of freedom.

The first term on the right-hand side of Eq. 6.106 is the direct force (Eq. 6.102), and the second term can be interpreted as the wind force.

However, the Kubo approach has two practical disadvantages; namely, it is limited to the linear-response regime and, most importantly, is computationally very demanding as it requires the self-consistent calculation of response functions. In addition, the above Kubo formula has been obtained using a uniform applied field. In reality, the field at the atom contains a self-consistent contribution due to the electron scattering, and this needs to be taken into account when calculating the force from Eq. 6.106. This contribution can be understood as follows.

6.5.3 Local resistivity dipoles and the force sign

If we go back to the single-particle scattering approach of Chapter 3, we can think of a single atomic defect in the presence of current as an obstacle for the electrons. Much like a nanojunction (see Fig. 3.5), this obstacle creates a local charge excess on one side of the atom and a charge deficit on the other side: a local resistivity dipole (Landauer, 1957; Landauer and Woo, 1974). This dipole gives rise to an electric field. This extra field, which is absent at equilibrium, is then responsible for a force on the ion.

The atom we consider is however in a current-carrying wire with other atoms or defects around. The resistivity dipole at that atom will be thus influenced by the other resistivity dipoles that may form around the other atoms or defects. The global effect of these resistivity dipoles is to change the macroscopic field via a "polarization field" that can be thought of as arising from the polarization charge at the surfaces of the sample (Landauer and Woo, 1974).

In addition, due to the positive charge of the ion, extra charge accumulates around the atom to partially neutralize it. This charge accumulation effectively reduces the local field at the ion self-consistently. The total force on the given atom is then a sum of all the above effects. Its sign is thus not so obvious to predict *a priori*.

The above analysis is even more complicated in the case of nanoscale structures. For these systems the self-consistent local charge inhomogeneities at the junction can be quite substantial due to partial electronic screening. We therefore need a definition of forces that contains all these effects and is computationally tractable. This is what I will derive in the following sections.

6.5.4 Forces at equilibrium

Separating the force into a direct- and a wind-force contribution is clearly a matter of choice. The only important quantity is the total net force on the ion. We therefore seek here a general quantum-mechanical definition of forces. To simplify the discussion let us work in the adiabatic approximation.[43]

---

[43] The concept of forces without the adiabatic approximation is more complicated since one has to deal with ion wave-functions explicitly, i.e., treat the ionic charge as a *distribution* not as

Before proceeding with the definition of forces in the presence of current let us recall their definition for a closed (not interacting with an environment), and finite (with a finite number of electrons and ions) system *at equilibrium*. In this case, forces are obtained from the *Hellmann–Feynman theorem* (Hellmann, 1937; Feynmann, 1939).

This theorem states that if the electronic system is described by the many-body Hamiltonian $\hat{H}$ which depends parametrically on the ion coordinates $\mathbf{R}_i$, and is in an *exact* eigenstate $|\Psi\rangle$ of the Hamiltonian ($\hat{H}|\Psi\rangle = E|\Psi\rangle$), the force on an ion at position $\mathbf{R}_i$ is defined as (Exercise 6.5)[44]

$$\mathbf{F}_i = -\frac{\langle \Psi | \nabla_{\mathbf{R}_i} \hat{H} | \Psi \rangle}{\langle \Psi | \Psi \rangle}$$

$$= -\nabla_{\mathbf{R}_i} \frac{\langle \Psi | \hat{H} | \Psi \rangle}{\langle \Psi | \Psi \rangle}$$

$$\equiv -\nabla_{\mathbf{R}_i} U(\mathbf{R}_i), \quad \text{closed and finite system at equilibrium,} \quad (6.107)$$

where I have explicitly introduced the normalization constant $\langle \Psi | \Psi \rangle$ for the state $|\Psi\rangle$, and defined the average energy of the system $U(\mathbf{R}_i)$, which depends on the ionic coordinates.[45] The equivalence between the first and the second equality in Eq. 6.107 is precisely the Hellmann–Feynman theorem.

In a single-particle picture, by introducing the single-particle number density $n(\mathbf{r})$, and assuming that the only explicit dependence on $\mathbf{R}_i$ is contained in the single-particle electron-ion potential $V^{ps}(\mathbf{r};\mathbf{R}_i)$ – namely, the wavefunctions do not depend explicitly on ionic positions – Eq. 6.107 becomes

$$\mathbf{F}_i = -\int d\mathbf{r}\, \nabla_{\mathbf{R}_i} V^{ps}(\mathbf{r};\mathbf{R}_i) n(\mathbf{r}), \quad \text{closed and finite system at equilibrium.} \quad (6.108)$$

Since there are generally other ions in the system, we then need to add to 6.108 the repulsive force due to the other ions (with effective charge $Z_j^*|e|$)

$$\mathbf{F}_i^{ion} = \sum_{j \neq i} \frac{Z_j^* Z_i^* e^2}{|\mathbf{R}_j - \mathbf{R}_i|^3}(\mathbf{R}_j - \mathbf{R}_i), \quad \text{ion-ion force.} \quad (6.109)$$

Within a single-particle picture, the total force on the ion in a closed and

---

a δ-function centered at the classical ion position. In addition, correlated motion of electrons and ions needs to be taken into account.

[44] If the states are not exact eigenstates of the Hamiltonian, as in numerical calculations, or they depend explicitly on the atomic positions (e.g., a localized basis set), extra terms – known as *Pulay forces* (Pulay, 1977) – appear in Eq. 6.107, and one has to calculate forces with some care (see Exercise 6.5).

[45] Note that square-integrable wave-functions are a necessary condition for the demonstration of the theorem (see Exercise 6.5).

finite system at equilibrium is therefore

$$\mathbf{F}_i = -\int d\mathbf{r}\, \nabla_{\mathbf{R}_i} V^{ps}(\mathbf{r}; \mathbf{R}_i) n(\mathbf{r}) + \sum_{j \neq i} \frac{Z_j^* Z_i^* e^2}{|\mathbf{R}_j - \mathbf{R}_i|^3}(\mathbf{R}_j - \mathbf{R}_i). \quad (6.110)$$

6.5.5 Forces out of equilibrium

Let us now go back to the transport – or, generally, a non-equilibrium – problem. Here, the definition of forces via the Hellmann–Feynman theorem does not hold (Di Ventra and Pantelides, 2000). This is because the equivalence between the first and the second equality in Eq. 6.107 cannot be generally proved for a system out of equilibrium.

General definition

In order to proceed, let us work, as in the equilibrium case, within the adiabatic approximation and consider the electronic Hamiltonian $\hat{H}$ (not necessarily time-independent) which depends parametrically on the ionic coordinates.

Consider the system in a general (normalizable) state $|\Psi(t)\rangle$. The force on an ion i due to the electron subsystem can be determined via the most general definition of forces in quantum mechanics, namely as the total time derivative of the average over this state of the ion momentum operator $\hat{\mathbf{P}}_i = -i\hbar \nabla_{\mathbf{R}_i}$ (Kumar and Sorbello, 1975; Di Ventra and Pantelides, 2000)

$$\boxed{\mathbf{F}_i = \frac{d\langle \Psi(t)|\hat{\mathbf{P}}_i|\Psi(t)\rangle}{dt} = -\frac{i}{\hbar}\left\langle \Psi(t)\left|\left[\hat{\mathbf{P}}_i, \hat{H}\right]\right|\Psi(t)\right\rangle = -\langle \Psi(t)|\nabla_{\mathbf{R}_i}\hat{H}|\Psi(t)\rangle} \quad (6.111)$$

where the second equality comes from the general equation of motion of the expectation value of any operator that does not depend explicitly on time, and the last from the value of the commutator between the momentum and the Hamiltonian (see, e.g., Messiah, 1961).[46]

From the above we immediately see the difference with the Hellmann-Feynman theorem 6.107: the quantity $\langle \Psi(t)|\nabla_{\mathbf{R}_i}\hat{H}|\Psi(t)\rangle$ is generally *different* from $\nabla_{\mathbf{R}_i}\langle \Psi(t)|\hat{H}|\Psi(t)\rangle$, to which it reduces only if the state $|\Psi(t)\rangle$ is square-integrable, stationary and corresponds to a closed finite system (Di Ventra and Pantelides, 2000).

The definition 6.111 need not to be limited to pure states, but can be extended to mixed states defined by some statistical operator $\hat{\rho}(t)$. It can

---

[46] In the presence of an external vector potential $\mathbf{A}(\mathbf{r}, t)$, the ion momentum needs to be changed into $\hat{\mathbf{P}}_i \to \hat{\mathbf{P}}_i + Z_i^* e\, \mathbf{A}(\hat{\mathbf{R}}_i, t)/c$.

also be extended to systems that are open to an environment, and thus their statistical operator is the solution of some non-unitary equation of motion, e.g., the Lindblad equation 1.73.

If $\hat{\rho}(t)$ is the statistical operator describing our system (whether closed or open), the force on an ion – still in the adiabatic approximation – is then

$$\mathbf{F}_i = \frac{d}{dt}\text{Tr}\{\hat{\rho}(t)\,\hat{\mathbf{P}}_i\} = -\frac{i}{\hbar}\text{Tr}\left\{\hat{\rho}(t)\left[\hat{\mathbf{P}}_i,\hat{H}\right]\right\} = -\text{Tr}\left\{\hat{\rho}(t)\,\nabla_{\mathbf{R}_i}\hat{H}\right\}, \tag{6.112}$$

which, again, cannot be derived from the Hellmann–Feynman theorem.

The single-particle case

If $\hat{H}$ is an effective single-particle Hamiltonian[47] and $V^{ps}(\mathbf{r};\mathbf{R}_i)$ represents the single-particle electron-ion interaction, we have the relation

$$\left[\hat{\mathbf{P}}_i,\hat{H}\right] = -i\hbar\nabla_{\mathbf{R}_i}\,V^{ps}(\mathbf{r};\mathbf{R}_i). \tag{6.113}$$

If we also work with pure states, by replacing this relation into 6.111 we get

$$\mathbf{F}_i = -\frac{\langle\Psi(t)|\nabla_{\mathbf{R}_i}\,V^{ps}(\mathbf{r};\mathbf{R}_i)|\Psi(t)\rangle}{\langle\Psi|\Psi\rangle}. \tag{6.114}$$

In the above equation I have emphasized the normalization of the wave-function in anticipation of its application to the single-particle scattering problem.

As discussed in Sec. 3.3.2 the scattering wave-functions do not belong to the Hilbert space of the system since they are not square-integrable. As we did for the continuum states of the current operator (Eq. 1.32) we can solve this issue by constructing eigendifferentials (Messiah, 1961). In the present case, this is done as follows.

If $|\Psi_E\rangle$ is a general continuum eigensolution of the Hamiltonian $\hat{H}$ with energy E, the eigendifferential $|\Psi_{\delta E}\rangle$ associated with it is

$$|\Psi_{\delta E}\rangle = \frac{1}{\sqrt{\delta E}}\int_E^{E+\delta E}dE\,|\Psi_E\rangle, \tag{6.115}$$

with δE a small interval of energy in the continuum.

Then one can prove (Messiah, 1961) (see also Eq. 1.33)

$$\lim_{\delta E\to 0}\langle\Psi_{\delta E}|\Psi_{\delta E}\rangle = 1, \tag{6.116}$$

---

[47] In the single-particle scattering problem this Hamiltonian describes an infinite system (see Sec. 3.1).

6.5 Current-induced forces

i.e., eigendifferentials are square-integrable, and thus belong to the Hilbert space.

With the normalization of wave-functions in the continuum solved, we can then define the force on an ion in the presence of an ideal steady-state current within the single-particle scattering theory of conduction (Chapter 3). In this case, the force is simply the sum of all single-electron states, both discrete (call them $|\Psi_k\rangle$) and continuum global scattering states (call these $|\Psi_E\rangle$, given in Eq. 3.200) (Di Ventra and Pantelides, 2000)

$$\mathbf{F}_i = -2\sum_k \langle \Psi_k | \nabla_{\mathbf{R}_i} V^{ps}(\mathbf{r};\mathbf{R}_i) | \Psi_k \rangle$$

$$-\lim_{\delta E \to 0} 2 \int_\sigma dE D(E) \langle \Psi_{\delta E} | \nabla_{\mathbf{R}_i} V^{ps}(\mathbf{r};\mathbf{R}_i) | \Psi_{\delta E} \rangle, \quad (6.117)$$

where I have used the normalization condition $\langle \Psi_k | \Psi_k \rangle = 1$ for the discrete states, and Eq. 6.116 for the continuum states. I have also explicitly included the density of states $D(E)$. The integral in the continuum is over the range σ of occupied states. At zero temperature and for $\mu_L > \mu_R$ this corresponds to the energy window between μ_L and the conduction band bottom of the right electrode (see, e.g., Fig. 3.12). The factor of 2 accounts for the spin degree of freedom.

At finite temperature, if we call $n_{neq}(\mathbf{r})$ the total non-equilibrium number density (which may include fully occupied bound states) and using the local density of states of left-moving $D_R(E,\mathbf{r})$ and right-moving $D_L(E,\mathbf{r})$ states (Eq. 3.127), we have the relation (see also Eq. 3.16)

$$n_{neq}(\mathbf{r}) = 2\sum_k |\langle \mathbf{r}|\Psi_k\rangle|^2 + \lim_{\delta E \to 0} 2 \int_\sigma dE D(E) |\langle \mathbf{r}|\Psi_{\delta E}\rangle|^2$$

$$= 2\sum_k |\langle \mathbf{r}|\Psi_k\rangle|^2 + 2 \int_{-\infty}^{+\infty} dE\, f_L(E)\, D_L(E,\mathbf{r})$$

$$+ 2 \int_{-\infty}^{+\infty} dE\, f_R(E)\, D_R(E,\mathbf{r}). \quad (6.118)$$

We can then rewrite Eq. 6.117 as

$$\mathbf{F}_i = -\int d\mathbf{r}\, \nabla_{\mathbf{R}_i} V^{ps}(\mathbf{r};\mathbf{R}_i)\, n_{neq}(\mathbf{r}). \quad (6.119)$$

Due to the presence of the other ions, to 6.119 we need to add the force 6.109 – which, being purely classical, is the same whether the system is at equilibrium or not – so that the total force on an ion in a current-carrying system at steady state is (within the single-particle picture)

$$\mathbf{F}_i = -\int d\mathbf{r} \nabla_{\mathbf{R}_i} V^{ps}(\mathbf{r};\mathbf{R}_i) n_{neq}(\mathbf{r}) + \sum_{j\neq i} \frac{Z_j^* Z_i^* e^2}{|\mathbf{R}_j - \mathbf{R}_i|^3}(\mathbf{R}_j - \mathbf{R}_i). \quad (6.120)$$

Note the equivalent *form* of 6.120 with the corresponding force at equilibrium 6.110. However, I stress again that despite this formal analogy the force 6.120 *cannot be derived from the Hellmann–Feynman theorem*.

We can now calculate the *extra* force due to the current by first calculating the force for the system at equilibrium (zero bias, $V = 0$). From Eq. 6.120

$$\mathbf{F}_i(V=0) = -\int d\mathbf{r} \nabla_{\mathbf{R}_i} V^{ps}(\mathbf{r};\mathbf{R}_i) n_{eq}(\mathbf{r}) + \sum_{j\neq i} \frac{Z_j^* Z_i^* e^2}{|\mathbf{R}_j - \mathbf{R}_i|^3}(\mathbf{R}_j - \mathbf{R}_i),$$
$$(6.121)$$

where I have indicated with $n_{eq}(\mathbf{r})$ the electron number density at equilibrium.

By taking the difference between Eq. 6.120 and Eq. 6.121 we finally get the extra contribution due to the non-equilibrium charge distribution

$$\delta\mathbf{F}_i(V) = -\int d\mathbf{r} \nabla_{\mathbf{R}_i} V^{ps}(\mathbf{r};\mathbf{R}_i)[n_{neq}(\mathbf{r}) - n_{eq}(\mathbf{r})]$$
$$\equiv -\int d\mathbf{r} \nabla_{\mathbf{R}_i} V^{ps}(\mathbf{r};\mathbf{R}_i)\delta n(\mathbf{r}), \quad (6.122)$$

where the density has to be evaluated self-consistently.

An example

In Fig. 6.12 I show an example of current-induced forces calculated using Eq. 6.120 for an aluminum chain of four atoms between aluminum electrodes represented by an electron gas compensated by a uniform positive background (also known as *jellium model*, in this example with $r_s \simeq 2a_0$, with a_0 the Bohr radius). The different quantities appearing in 6.120 have been calculated using static density-functional theory (Appendix D). This calculation thus has the same limitations as those I have discussed in Sec. 3.11, namely it misses dynamical many-body effects of the true electron dynamics. How important these effects are on current-induced forces has never been studied.

The electron-ion interaction is represented using pseudopotentials. From Fig. 6.12 it is clear that at any given bias forces on single atoms are not necessarily oriented in the same direction. Some are *against* the current

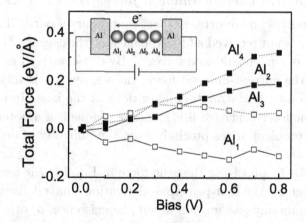

Fig. 6.12. Current-induced forces as a function of bias in an atomic wire containing four aluminum atoms between aluminum electrodes represented by an electron gas compensated by a uniform positive background ($r_s \simeq 2a_0$). The inset shows a schematic of the wire. The left electrode is positively biased so that electrons flow from right to left. Positive force pushes the atom against electron flow. Adapted with permission from Yang et al. (2005).

flow, some *along* the current flow due to the detailed self-consistent charge distribution around the ions.

In addition, the largest force is on average on the edge contact atoms, suggesting that if electro-migration occurs it would nucleate from the bonds of these atoms. As I have discussed in Sec. 6.5.2, we also see that forces on ions are a non-linear function of the bias, and thus of the current.

The magnitude of the average force per atom is about 0.01 eV/Å at 0.1 V (for this system and at this bias the current is about 10 μA). For comparison, the inelastic contribution to the force we have estimated in Sec. 6.5.1 for this bias and comparable current is about 0.001 eV/Å, which has to be distributed over at least four atoms. On average, the inelastic contribution to the force per atom is ~ 0.00025 eV/Å; more than an order of magnitude smaller than the elastic one, confirming that the former can reasonably be neglected.

6.5.6 Are current-induced forces conservative?

Let us finally address an important aspect that deserves particular attention, and that so far has not received a full explanation: the conservative character – or lack of it – of current-induced forces.[48] By conservative I mean that the work done by the current-induced forces on a given ion is independent of the path between two arbitrary configurations of the ions. In other words, can we write the force 6.117 (or 6.111) as the gradient of a potential energy function? The problem arises precisely from the difficulty of defining forces out of equilibrium.

Let us start by emphasizing that the force in Eq. 6.117 is *not necessarily conservative* because the first-principles derivation I have followed to derive it does not contain any assumption about the existence, or otherwise, of an energy functional, of which this force might be the derivative. In a closed and finite system at equilibrium the existence of such a functional is provided by the Hellmann–Feynman theorem – the quantity $U(\mathbf{R}_i)$ in Eq. 6.107.[49]

In the scattering problem, the density $n_{neq}(\mathbf{r})$ depends parametrically, via self-consistency, on the ionic positions, so that in general

$$\int d\mathbf{r}\, \nabla_{\mathbf{R}_i} V^{ps}(\mathbf{r}; \mathbf{R}_i) n_{neq}(\mathbf{r}) \neq \nabla_{\mathbf{R}_i} \int d\mathbf{r}\, V^{ps}(\mathbf{r}; \mathbf{R}_i) n_{neq}(\mathbf{r}). \qquad (6.123)$$

It is thus not at all clear if an energy functional exists in this case.

The reader may ask at this point: Why bother with this issue? I just mention two reasons, one fundamental and one practical.

Can phonons be defined in the presence of current?

If current-induced forces are not conservative, their cross-derivatives with respect to ionic positions are not necessarily equal and hence the dynamical response matrix 6.7 for a current-carrying structure is not necessarily symmetric

$$\left.\frac{\partial^2 V_{ion}(\mathbf{R}_i - \mathbf{R}_j)}{\partial R_{i\mu}\, \partial R_{j\nu}}\right|_{\mathbf{R}_i=\mathbf{R}_i^0;\mathbf{R}_j=\mathbf{R}_j^0} \neq \left.\frac{\partial^2 V_{ion}(\mathbf{R}_i - \mathbf{R}_j)}{\partial R_{j\mu}\, \partial R_{i\nu}}\right|_{\mathbf{R}_i=\mathbf{R}_i^0;\mathbf{R}_j=\mathbf{R}_j^0}. \qquad (6.124)$$

Since this symmetry is fundamental to uniquely define phonons (see, e.g., Ashcroft and Mermin, 1975), it implies that the conventional notion of

---

[48] Clearly, I refer here to the case in which the inelastic part of the force can be completely neglected, otherwise the force is by definition non-conservative.
[49] Forces on ions need not be conservative even for a closed, finite system at equilibrium along closed trajectories that span a point of intersection of two distinct Born–Oppenheimer surfaces. In the present discussion, I will assume that no such pathologies are present.

phonons would be lost in the presence of current. In other words, we would not be able to talk about phonons in a current-carrying system at all!

Current-induced forces and topology

A practical reason – with impact in nanoscale electronics – of the importance of the conservative character of current-induced forces is the following. Consider a perfect, defect-free quasi-2D wire and identify two points in it, along the direction of current flow, in such a way that a straight line connecting them separates the wire into two equivalent regions. If current-induced forces are non-conservative, then an atom can move from one point to the other along different paths in the two equivalent regions such that the work done by the current-induced force is different for the different paths. As a consequence, it is in principle possible to break *topologically equivalent* regions of a wire differently! While this effect may not be important in macroscopic wires, it could be relevant in nanoscale conductors made of a relatively small number of atoms.

A *gedanken* experiment

Let us now turn to the question of whether or not these forces are conservative. There exist specific arguments that suggest that current-induced forces cannot be conservative. These arguments consist in *gedanken* experiments, aimed at showing that the work done by current-induced forces along specific closed paths is not zero and that, therefore, these forces are not conservative. Below I refer to one notable such construction (Sorbello, 1998), but other similar arguments can be formulated. I choose this one because it is very simple and is particularly convincing. The argument goes like this.

Consider a single defect atom in a current-carrying wire between bulk electrodes under an applied bias V (Fig. 6.13).[50] First, let the atom move, quasi-statically, from a point A to a point B inside the wire under the influence of an external force. The current-induced force on the defect in the wire does some non-zero work on the way. Now imagine taking the atom from A to B along a path that goes *outside* the wire (see Fig. 6.13). Since outside the wire there apparently is no current, no current-induced force should be present. Hence, the work done by current-induced forces on the atom along the path from A to B inside the wire cannot be the same

---

[50] For simplicity, we may here imagine a wire made of electrons compensated by a uniform positive background charge (jellium model).

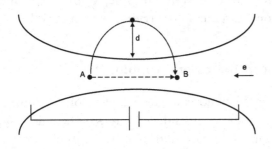

Fig. 6.13. Schematic of a current-carrying wire between two bulk electrodes. An atom is moved from a point A to a point B along two different paths, one inside, one outside the wire. d is the largest distance of the atom from the wire along the "vacuum" path.

as along the path outside the wire. This would be sufficient to prove that current-induced forces are not conservative.

However, something is missing in this argument: the quantum-mechanical nature of the current *outside* the wire (Di Ventra *et al.*, 2004). Consider again the atom at a point in space outside the wire. Let us also work within a single-particle picture of conduction.

Let $|\Psi\rangle$ be the global current-carrying electron states for the system. If there always is *some* (however small) quantum-mechanical coupling between the atom and the wire, there will always be some non-zero tunneling probability for electrons to get from the wire onto the atom, and back. The atom then acts as a *resonant center* and there will always exist an energy window where the states $|\Psi\rangle$ contain *resonant paths* that emerge from the wire, pass through the atom, and go back into the wire. Due to the scattering of electrons at the atom, these resonant paths may be expected to create a charge dipole, centered at the atom, much like the local resistivity dipole I have discussed in Sec. 6.5.3.

Increasing the distance d between the atom and the wire will make the resonant energy window narrower, while the resonance will remain pinned against the energy window eV, due to self-consistency. Now, the total current carried by the resonance is given by the energy integral, over the energy window eV, of a transmission function $T(E)$ (Eq. 3.56). As the resonance gets narrower in energy, so does $T(E)$.

The function $T(E)$ is always bounded and hence, as d increases, the total current carried by the resonance will decrease, since the integral window eV remains constant. On the other hand, the charge dipole across the resonant center originates from the density of states $D(E)$ due to the populated current-carrying electron states in the energy window eV.

However, unlike $T(E)$, the density of states $D(E)$ is *unbounded*: if d increases, the resonant peak in $D(E)$ gets narrower but taller in such a way that its energy integral remains constant (no charge is lost). Since current-induced forces are related to the integral of the density of states (see, e.g., Eq. 6.117), we expect the force on the atom due to the resonant dipole to be constant with d. In other words, there is *always* a current-induced force on the atom, even outside the wire, and this force does not vanish with increasing d.[51]

The existence of such a resonant dipole has been numerically shown – within a static DFT calculation – using the definition of forces Eq. 6.117 in the simplest possible resonant configuration: a single atom in the vacuum between two bulk electrodes (Di Ventra *et al.*, 2004). In that case, it was shown that this resonant dipole gives rise to a finite current-induced force that does not vanish with increasing electrode separation, even when the current dies out.

Clearly, these results do not prove that current-induced forces are conservative but they give extra support in favor of such a conclusion.

6.6 Local ionic heating vs. current-induced forces

I conclude by mentioning that structural instabilities and breaking of bonds in nanostructures are due to a combined effect of *both* local ionic heating *and* current-induced forces. I have derived expressions for each of these quantities assuming the other is negligible. This is clearly an idealization, since at any given bias they coexist and influence each other. The relation between these two phenomena – or which one dominates for a given system and bias – is still an open subject.

Using the approaches discussed in this chapter, theoretical work on metallic quantum point contacts suggests that local ionic heating promotes structural instabilities at very low biases if heat dissipation into the bulk electrodes is not efficient, while current-induced forces are mainly responsible for the junction breakup at large biases (Yang *et al.*, 2005).

Instead, in systems with large resistance – e.g., molecular junctions – similar calculations show that current-induced forces are negligible even at very large biases (Di Ventra *et al.*, 2002), and thus the instability of these large-resistance systems is mainly due to local ionic heating.

These results, however, neglect the mutual effect between ionic heating and forces. In particular, the intermediate regime in which these two phe-

---

[51] It is clear that this "vacuum resonance" will become more and more sensitive to external perturbations as the distance d increases and the resonance gets narrower and narrower. This simply implies that, in practice, this resonance may be destroyed easily by external perturbations and be difficult to detect in experiments.

nomena are of comparable importance has not been studied. More work to understand this issue is highly desirable.

Summary and open questions

In this chapter I have discussed many effects related to the interaction of electrons and ions, such as inelastic current, local ionic heating, and forces.

In the case of the inelastic current I have described how to calculate it using standard perturbation theory. I have also given an explicit example of many-body perturbation theory in the electron-phonon coupling, and have shown how to determine the electron-phonon self-energy within the Born approximation.

I have also underlined the difficulty in defining forces out of equilibrium. Such difficulty, and the related open question of whether current-induced forces are conservative or not, has both fundamental and practical consequences which have yet to be explored fully.

I finally note that I have worked out all the calculations in this chapter by considering independent electrons interacting at most at the mean-field level. The effect of interactions beyond mean field on all the phenomena described here is still poorly understood.

In particular, I have left aside the issue of correlated electron-ion dynamics, which may be of importance when electronic resonances and/or localized modes are present. Due to the large current densities at nanojunctions, and consequent increase of electron-electron and electron-phonon scattering rates, this issue cannot be addressed within a non-interacting electron picture, and would thus require a totally different approach than the ones presented in this chapter. This approach needs to take into account the dynamical properties of the electron liquid in interaction with quantum-mechanical ions. Also this line of research has not been fully explored yet.

Exercises

6.1 Use Eqs. 6.9 and 6.8 to prove that the effective ion Hamiltonian can be written as 6.10.

6.2 Start from the Fermi golden rule 6.35, and use the definitions of creation and annihilation operators for both electrons and phonons (Eqs. A.8, A.9, A.29, and A.30) to derive Eq. 6.36.

6.3 Calculate the self-energy 6.58 explicitly.

6.4 **Inelastic current from the NEGF.** Prove that the total current 6.63 of independent electrons interacting with phonons can be separated into an "elastic" component 6.67 and an "inelastic" component 6.68. Show that if the coupling of the central region with the left electrode or with the right electrode is zero (either $\Gamma_L(E) = 0$ or $\Gamma_R(E) = 0$), both the elastic and the inelastic components are zero.

6.5 **The Hellmann–Feynman theorem.** (i) Prove first the Hellmann-Feynman relation 6.107 in the case in which the wave-functions do not depend parametrically on the ion coordinates. (ii) Show that extra terms, known as *Pulay forces*, appear in the definition of forces, if the wave-functions depend explicitly on ionic positions. (iii) Show that a necessary condition for the theorem to hold is that the wave-functions be square-integrable.

7
The micro-canonical picture of transport

So far, we have dealt mostly with the electron system *already* at an ideal steady state. We never questioned whether this state *exists* at all, and if so, how a many-body system does actually *reach* that steady state, whether the steady state we *impose* via single-particle scattering boundary conditions is actually what the electrons want to realize when they flow across a nanojunction, or even if it is *unique*.

In addition, we have mostly worked with electrons interacting at a mean-field level (see discussion in Sec. 4.2.4). We discussed in Chapter 4 how, in principle, one can introduce electron-electron interactions beyond mean field using the non-equilibrium Green's function formalism. However, except for simple model systems, it is computationally demanding – and most of the time, outright impossible – to use the interacting version of the NEGF practically in transport calculations (and not only in transport).

A simpler and more efficient way to treat electron-electron interactions in a transport problem would be thus desirable.

In this chapter I will introduce an alternative picture of transport – I will name it *micro-canonical* (Di Ventra and Todorov, 2004) – that does not rely on the approximations of the Landauer approach. In addition, this formulation does not require partitioning the system into leads and a central region, or assuming the leads contain non-interacting electrons in order to have a closed form for the current.

In fact, within this picture I will prove several theorems of dynamical density-functional theory (Appendices E, F and G) that, in principle, allow us to calculate the *exact* current – namely with *all* many-body effects included – within an *effective single-particle picture*.

I will then discuss what fundamental processes allow a nanoscale system to reach a steady state, analyze the properties of the latter, formulate a principle on the formation of steady states, and make a connection with the static Landauer approach.

7.1 Formulation of the problem

As I have discussed in Sec. 1.2, we know that when a nanojunction between two macroscopic electrodes is connected to an electron source, electrical current flows across it. The source provides and, most importantly, *maintains* the charge imbalance between the electrode surfaces needed to *sustain* steady-state conduction in the junction.

I have also discussed several times in this book – see for instance Sec. 3.2 – that a net electron current from one electrode into another across a nanojunction is accompanied by an excess of electrons in one electrode, and a deficit of electrons in the other.

These charges take the form of *surface charge densities*, present within a screening length of the electrode surface. Thus, electron transport goes hand in hand with surface charges on each side of the junction: a negative sheet of charge on one side and a positive sheet on the other, as if two capacitor plates were present. The transport problem in the presence of electron sources can then be viewed as a *continuous* attempt by the electrons to *passivate* the surface charges on each side of the junction. Similarly, the ideal steady-state transport problem described by the Landauer approach of Chapter 3 can be interpreted as the continuous discharge of an *infinite* capacitor.

This reasoning is exactly in reverse order compared to the one I have made in Sec. 1.2, where I have argued that the discharge of a *finite* capacitor generates a current, which in the limit of *infinite* electrodes may realize the steady-state condition 1.13.

7.1.1 Transport from a finite-system point of view

Therefore, there seems to be no conceptual difference between the continuous flow of charges across a nanojunction as maintained by an electron source – such as a battery – or the discharge of an infinite capacitor.

In real life there is no such thing as an infinite capacitor, and indeed the term "infinite" really means with characteristic time longer than some time scale dictated by the nature of the problem.

Quasi-steady states of open systems

If, as in Sec. 1.2, we consider the discharge of a finite, though possibly large,

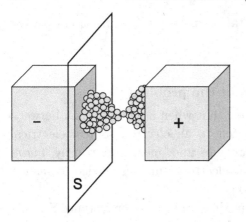

Fig. 7.1. Schematic of the system considered in the micro-canonical picture of transport: the electron flow is studied as the discharge of two large but finite charged electrodes across a nanoscale junction.

capacitor (Fig. 7.1), which is not, however, closed – namely its degrees of freedom are coupled with an environment – then the above time scale is the following. Call C the capacitance and R the resistance across which the capacitor discharges. According to classical circuit theory the discharge will then take place over a characteristic time RC. The larger C, for a given R, the more "stationary" the properties of the system will appear, in a temporally local sense, at any one stage of the discharge (see Fig. 1.4 and related discussion).

This ultimately transient, but very slowly varying, conducting state, which I call a *quasi-steady state*, indeed encapsulates the intuitive picture of the d.c. steady state I have discussed in many parts of this book. As discussed in Sec. 1.2, we finally expect this system to reach a truly time-independent steady state with zero current in the course of its evolution.[1]

Here, I will focus mainly on d.c. (direct current) bias conditions. The micro-canonical picture, however, applies also to the a.c. (alternating current) case, that is, periodic time-varying disturbances of the charge. It can also be extended to the multi-probe conductance measurement case as discussed in Sec. 3.9 via the addition of extra electrodes connected somewhere along the main structure (see Fig. 3.20).

Quasi-steady states of closed systems

If the capacitor plus the nanojunction across which it discharges form a truly closed and finite system, then we expect the above quasi-steady state – in the sense discussed above – to persist until the time when multiple electron reflections off the far boundaries of the system begin to develop.

---

[1] If the system is infinite, then this could be a current-carrying steady state.

Within a single-particle picture, we can obtain an order-of-magnitude estimate of the electrode size necessary to observe the quasi-steady state as follows. Let us call t_d the time it roughly takes for an electron to travel at the Fermi velocity v_F along the characteristic length L of one electrode and back, i.e., $t_d \approx 2L/v_F$. Therefore, one should observe a quasi-steady state if the time necessary for this to form during the transient dynamics, τ_c, is less than t_d, namely $L > \tau_c v_F/2$. Since, as I will discuss later, in nanoscale systems τ_c is of the order of femtoseconds (see Sec. 7.6), L is of the order of nanometers or less.

From then on, the electronic system may return to the neighborhood of its initial conditions within the Poincaré recurrence time – see discussion in Sec. 1.2.

As shown in Eq. 1.15, for an interacting many-body system, this time increases exponentially with the number of particles, so that even a relatively small system has a macroscopically large recurrence time (see estimate in Footnote [16] of Chapter 1).

For a finite and closed system, however, we still need to answer the question of whether a quasi-steady state actually forms at all. I will show in Sec. 7.6 that in nanoscale systems a quasi-steady state may indeed develop in the *absence* of inelastic effects and in extremely short time scales. This is, however, not true for all initial conditions.

7.1.2 Initial conditions and dynamics

Let us focus on a given finite system (electrodes plus nanojunction) with a finite number of electrons and ions whose many-body Hamiltonian is $\hat{H}$ (Fig. 7.1). This Hamiltonian could be for instance the one in Eq. 6.1. It could also contain an external deterministic field, such as an electromagnetic field.

Let us prepare the system in an initial state $|\Psi_0\rangle$. From a physical point of view it is a state that represents a charge imbalance between the two electrodes that sandwich our nanojunction, an imbalance that can be generated in many ways. The most important point is that $|\Psi_0\rangle$ is *not an eigenstate* of $\hat{H}$, and thus is not a stationary state. This many-body state may describe the initial electronic configuration at a given static ionic configuration, or it may describe the many-body state of electrons *and* ions, the latter also treated as quantum-mechanical entities. For later use, it is convenient to think of this initial state as the *ground state* of some Hamiltonian $\hat{H}_{\text{initial}} \neq \hat{H}$.

The time evolution of $|\Psi_0\rangle$ is described by a many-body state vector

$|\Psi(t)\rangle$, governed by the time-dependent Schrödinger equation 1.16 which I re-write here for convenience

$$i\hbar \frac{d|\Psi(t)\rangle}{dt} = \hat{H} |\Psi(t)\rangle, \quad |\Psi(t_0)\rangle = |\Psi_0\rangle. \tag{7.1}$$

The above dynamics pertains to a system prepared in a pure state. If the system is in a macro-state, with initial statistical operator $\hat{\rho}(t_0)$, we solve the Liouville–von Neumann equation 1.60.

With the solution of either the Schrödinger equation 7.1 or the Liouville–von Neumann equation 1.60 the current flowing in the system across a given surface S is given by the average of the current operator over the state of the system (Eq. 1.86)

$$I_S(t) = e\langle \hat{I} \rangle_t = e\mathrm{Tr}\{\hat{\rho}(t)\hat{I}_S\}. \tag{7.2}$$

It is now clear what is the conceptual difference with the time-dependent approaches I have discussed previously, namely the Kubo (Sec. 2.3) and NEGF (Chapter 4) approaches. In these, the system is prepared in a *global equilibrium state*. At a given time a *perturbation* is switched on adiabatically and the expectation value of a given observable – e.g., the current – is calculated, whether in linear response (Kubo) or beyond linear response (NEGF). This expectation value determines the *deviations* from global equilibrium.

In the micro-canonical picture, the system is prepared in some arbitrary state which is *not* the ground state of the Hamiltonian, and then let to evolve in time to find its own dynamics – and possible quasi-steady state – under the many-body Schrödinger equation 7.1 (or the Liouville–von Neumann equation 1.60). In this respect, no perturbation to some Hamiltonian is applied – even if the Hamiltonian $\hat{H}$ may contain an external field – and thus no adiabatic switch-on is necessary as in the above two approaches.

The above picture is also close to a real experimental realization in the presence of an electron source, such as a battery. This performs just that: it prepares the quantum system in a given initial state which corresponds to the charging of the electrodes that sandwich the nanojunction. However, it does so *continually*, and very likely with *different* initial conditions at different times. I will come back to this point later.

The approach I have described here does not fall in either of the two viewpoints I have discussed in Sec. 1.6. In the present case, the current discharge is driven by the electric field that is *dynamically* built by the two electrodes, while the magnitude of the current flux – and hence the resistance – is determined by the self-consistent charge pileup, and consequent local field, at the nanoscale junction.

7.2 Electrical current theorems within dynamical DFTs

All of the above analysis regarding the micro-canonical picture of transport would be worthless if in order to do time-dependent transport calculations one had to solve the many-body time-dependent Schrödinger equation 7.1.

Here I formulate a series of theorems on the total current which provide solid ground to perform electrical current calculations within an *effective single-particle picture*. These theorems are based on the dynamical formulations of density-functional theory (DFT).

At this point, if the reader is not familiar with dynamical density-functional theories I suggest a quick stop to take a look at the Appendices E, F and G, where time-dependent DFT (TDDFT; the main variable is the *density*), time-dependent current DFT (TD-CDFT; the main variable is the *current density*), and stochastic time-dependent current DFT (stochastic TD-CDFT; the main variable is the *ensemble-averaged current density*) are briefly introduced.

I will discuss both the case in which the system is closed, and the case in which it is in dynamical interaction with an environment – open system.[2]

7.2.1 *Closed and finite quantum systems in a pure state*

Within TDDFT the dynamics of the *many-body* system can be mapped into the dynamics of effective single particles – known as the Kohn–Sham (KS) system. In particular, the density $n(\mathbf{r},t)$ extracted from the *exact* many-body wave-function (see Sec. 2.8.3) is the same as the one obtained from the KS equations E.4. If we know the *exact* dynamical functional (see Eq. E.5),

$$n^{KS}(\mathbf{r},t) = n(\mathbf{r},t). \qquad (7.3)$$

I show here that under the same conditions of validity of TDDFT we can prove the following (Di Ventra and Todorov, 2004)

Theorem T1: *Given an initial condition, the many-body total current of a finite and closed non-equilibrium system is, at any given time, given exactly by the one-electron total current, obtained from time-dependent density-functional theory.*

The rigorous connection provided by the above theorem T1 is independent

---

[2] For this case, the term "micro-canonical" does not seem very appropriate, but for lack of a better word, I will use the same name for this case as well.

of whether the system has reached a quasi-steady state or not. Let us then prove it.

Proof. Let us suppose that we have solved the time-dependent Kohn–Sham equations of TDDFT (Eqs. E.4) and obtained a set of KS time-dependent one-electron orbitals $\phi_k^{KS}(\mathbf{r},t)$ (here k includes also spin degrees of freedom). The KS current $\mathbf{j}^{KS}(\mathbf{r},t)$ is the sum of expectation values of the current-density operator 1.22 in the populated (say N) KS orbitals (see Eq. 2.61)

$$\mathbf{j}^{KS}(\mathbf{r},t) = \frac{\hbar}{m} \sum_{k=1}^{N} \operatorname{Im}\left\{[\phi_k^{KS}(\mathbf{r},t)]^* \nabla \phi_k^{KS}(\mathbf{r},t)\right\}. \qquad (7.4)$$

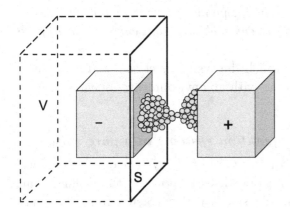

Fig. 7.2. The volume V is bounded by the surface S across which the current is calculated, and which completely envelops one of the electrodes.

Referring to Fig. 7.2, we define the KS total current I_S^{KS} across a surface S as

$$I_S^{KS}(t) = \int_S d\mathbf{S} \cdot \mathbf{j}^{KS}(\mathbf{r},t) = \int_V d\mathbf{r}\, \nabla \cdot \mathbf{j}^{KS}(\mathbf{r},t). \qquad (7.5)$$

In the above equation we close S in the vacuum, as indicated by the dashed box in Fig 7.2, sufficiently far from the boundaries of the system to enable us to ignore any contributions to the surface integral over the surface of the dashed box.[3] In Eq. 7.5 I have also invoked Gauss' theorem with V the volume bounded by S, which I take to completely envelop one of the electrodes, as shown in Fig. 7.2.

From Eq. 7.3, we know that the KS charge density is equal to the exact density. Furthermore, $n^{KS}(\mathbf{r},t)$ and $\mathbf{j}^{KS}(\mathbf{r},t)$ satisfy the continuity equation 1.11, just like the density $n(\mathbf{r},t)$ and current density $\mathbf{j}(\mathbf{r},t)$ of the true

---

[3] Here I assume that all electrons are bound within the electrode-junction-electrode finite system, so that the charge and current densities are exponentially small into the vacuum, at all times.

many-body system. From 7.3, we then have

$$\nabla \cdot \mathbf{j}^{KS}(\mathbf{r},t) = \nabla \cdot \mathbf{j}(\mathbf{r},t), \tag{7.6}$$

even though $\mathbf{j}^{KS}(\mathbf{r},t)$ and $\mathbf{j}(\mathbf{r},t)$ need not be equal.[4]
Hence,

$$\boxed{I_S^{KS}(t) = \int_V d\mathbf{r}\, \nabla \cdot \mathbf{j}^{KS}(\mathbf{r},t) = \int_V d\mathbf{r}\, \nabla \cdot \mathbf{j}(\mathbf{r},t) = I_S(t),} \tag{7.7}$$

where $I_S(t)$ is the *true* total many-body electron current across the surface S.

The above proof is valid only for a *finite* system because for an infinite system we could not have made the transition between surface and volume integrals above.

We can bypass this problem by working with time-dependent current DFT (TD-CDFT).

7.2.2 Closed quantum systems in a pure state with arbitrary boundary conditions

In TD-CDFT, our basic variable is the current density. Solving the TD-CDFT KS equations F.1 with the *exact* exchange-correlation *vector* potential would give the exact current density $\mathbf{j}(\mathbf{r},t)$. In this case, however, the KS current density contains also an effective diamagnetic contribution (see Eq. 2.61)

$$\mathbf{j}^{KS}(\mathbf{r},t) = \sum_{k=1}^{N} \left[\frac{\hbar}{m} \mathrm{Im}\left\{ [\phi_k^{KS}(\mathbf{r},t)]^* \nabla \phi_k^{KS}(\mathbf{r},t) \right\} \right.$$
$$\left. - \frac{e}{mc} |\phi_k^{KS}(\mathbf{r},t)|^2\, \mathbf{A}_{eff}(\mathbf{r},t) \right]$$
$$= \mathbf{j}(\mathbf{r},t), \tag{7.8}$$

where the effective vector potential is (considering also an arbitrary external vector potential $\mathbf{A}(\mathbf{r},t)$ acting on the system)

$$\mathbf{A}_{eff}(\mathbf{r},t) = \mathbf{A}_{ext}(\mathbf{r},t) + \mathbf{A}_{xc}(\mathbf{r},t), \tag{7.9}$$

with $\mathbf{A}_{xc}(\mathbf{r},t)$ an exchange-correlation *vector* potential that is a *functional* of the initial condition $|\Psi_0\rangle$ *and* the current density $\mathbf{j}(\mathbf{r},t')$ at all *previous* times

---

[4] In general, the current density is not V-representable, i.e., the mapping between current density and external scalar potential is not invertible (D'Agosta and Vignale, 2005). Therefore, the KS current density in TDDFT is not guaranteed to be exact; only its divergence (Eq. 7.6).

$t' \leq t$, namely it is dependent on the *history* of the system. It describes both the correlations due to the Pauli exclusion principle, and all other quantum correlation effects among electrons.

The theorem on the total current is then

Theorem T2: *Given an initial condition, the many-body total current of a closed non-equilibrium system is, at any given time, given exactly by the one-electron total current, obtained from time-dependent current density-functional theory.*

Proof. Let us refer again to Fig. 7.1. The KS total current I_S^{KS} across an arbitrary surface S is simply

$$\boxed{I_S^{KS}(t) = \int_S d\mathbf{S} \cdot \mathbf{j}^{KS}(\mathbf{r},t) = \int_S d\mathbf{S} \cdot \mathbf{j}(\mathbf{r},t) = I_S(t),} \qquad (7.10)$$

where I have used 7.8 and $I_S(t)$ is again the exact total many-body current across the surface S.

Note that the above demonstration holds for an arbitrary surface and irrespective of the boundary conditions on our system. It thus holds for truly infinite systems as well.

7.2.3 Current in open quantum systems

Finally, let us consider a system open to an environment, and assume that its dynamics can be described by a quantum master equation such as the equation 1.73 with arbitrary Hamiltonian and bath operators. I work here with the *stochastic* Schrödinger equation (see Appendix G)

$$\partial_t |\Psi(t)\rangle = -\frac{i}{\hbar} \hat{H}(t) |\Psi(t)\rangle - \frac{1}{2} \hat{V}^\dagger \hat{V} |\Psi(t)\rangle + \ell(t) \hat{V} |\Psi(t)\rangle, \qquad (7.11)$$

where $\hat{V}$ is an operator representing the interaction of our system with a bath, and $\ell(t)$ is a stochastic process chosen such that it has both zero ensemble average and δ-autocorrelation

$$\overline{\ell(t)} = 0; \qquad \overline{\ell(t)\ell(t')} = \delta(t - t'), \qquad (7.12)$$

where the symbol $\overline{\cdots}$ indicates the average over a statistical ensemble of identical systems all prepared in the same initial quantum state $|\Psi_0\rangle$. The

7.2 Electrical current theorems within dynamical DFTs

results can be easily generalized to a mixed initial quantum state, and to many statistical operators $\hat{V}_\alpha$ as discussed in Appendix G.

The reason for working with a stochastic equation and not with an equation of motion for the statistical operator (e.g., Eq. 1.72) is that we are dealing with KS Hamiltonians, which depend on the density and/or current density and are thus different for each element of the ensemble. In general, this prevents us from writing a closed equation of motion for the statistical operator (see discussion in Sec. 1.4.3). An additional complication arises when using a quantum master equation of the type 1.72 with time-dependent Hamiltonians and/or bath operators: the statistical operator may lose positivity (Sec. 1.4) during time evolution, thus leading to un-physical solutions (see Appendix C). These issues do not appear in a formulation in terms of the stochastic Schrödinger equation 7.11.

From the solutions of Eq. 7.11, we define a many-body statistical operator

$$\hat{\rho}(t) = \overline{|\Psi(t)\rangle\langle\Psi(t)|} \equiv \sum_i p_i(t) |\Psi_i(t)\rangle\langle\Psi_i(t)|, \qquad (7.13)$$

with $p_i(t)$ the probability that a state $|\Psi_i(t)\rangle$ occurs in the ensemble with $\sum_i p_i(t) = 1$ (see also discussion in Sec. 1.4).

From the theorem of stochastic TD-CDFT (Appendix G), we can map the many-body problem into an effective single-particle problem where we solve a set of KS equations G.20 with an effective vector potential $\mathbf{A}_{eff}(\mathbf{r},t)$, defined as in Eq. 7.9, sum of the external vector potential and the exchange-correlation vector potential $\mathbf{A}_{xc}(\mathbf{r},t)$. The latter is a functional of the operator $\hat{V}$ (or operators $\hat{V}_\alpha$) in addition to being a functional of the initial conditions, and the *ensemble-averaged* current density $\overline{\mathbf{j}(\mathbf{r},t')}$, for $t' \leq t$. The latter is simply the ensemble average of the current density 7.8.

The theorem of stochastic TD-CDFT guarantees that

$$\overline{\mathbf{j}^{KS}(\mathbf{r},t)} = \overline{\mathbf{j}(\mathbf{r},t)} = \mathrm{Tr}\{\hat{\rho}(t)\hat{\mathbf{j}}(\mathbf{r})\}, \qquad (7.14)$$

namely, the ensemble-averaged current density operator in the populated KS states is equal to the corresponding quantity of the many-body system, with the many-body statistical operator given in Eq. 7.13.

We then also have (Di Ventra and D'Agosta, 2007)

Theorem T3: *Given an initial condition, and a set of external baths, the ensemble-averaged many-body total current of a non-equilibrium system is, at any given time, given exactly by the ensemble-averaged one-electron total current, obtained from stochastic time-dependent current density-functional theory.*

Proof. If we refer once more to Fig. 7.1, the *ensemble-averaged* KS total current $\overline{I_S^{KS}(t)}$ across an arbitrary surface S is simply

$$\overline{I_S^{KS}(t)} = \int_S d\mathbf{S} \cdot \overline{\mathbf{j}^{KS}(\mathbf{r},t)} = \int_S d\mathbf{S} \cdot \overline{\mathbf{j}(\mathbf{r},t)} = \overline{I_S(t)}, \qquad (7.15)$$

where $\overline{I_S(t)} = \text{Tr}\{\hat{\rho}(t)\hat{I}\}$ is the *exact* ensemble-averaged total many-body current across the surface S.[5]

7.2.4 Closure of the BBGKY hierarchy

The above theorems are very important formal results, because, unlike the static transport problem discussed in Chapter 3, they guarantee that, if we knew the *exact* functionals of dynamical density-functional theories, we would obtain the *exact* current of the many-body system with *all* electron-electron interactions included, whether the system is closed or open.

This is because the above functionals are universal functionals of the density, current density, or ensemble-averaged current density. Therefore, we are guaranteed that the BBGKY hierarchy of equations of motion one generates, e.g., in the NEGF (see discussion in Sec. 4.2.1.1 and Appendix B) can be formally *closed* to *all orders* in the electron-electron interactions!

These interactions pertain to *all* contributions from free and bound charges and localized currents (see Sec. 1.2), so that polarization and magnetization currents are described at the same level as particle currents.

Therefore, these theorems put on a rigorous footing the calculation of *dynamical* transport properties – the static transport problem being just a

---

[5] Note that a particular case of theorem T3 is when the operator $\hat{V} = \hat{1}$. In this case, Eq. 7.11 provides the same physics as the Liouville–von Neumann equation 1.60. So the theorem on the current holds also for a closed system in a mixed state. In addition, I am here assuming that the bath operators are local in space (see Sec. G.2 in Appendix G). Otherwise, to prove the theorem, one would need to determine the density, and from this quantity calculate the total current as the rate of change of the total charge in one of the electrodes (see Footnote [22] in Chapter 1).

subclass of these in the limit of zero frequency – in nanoscale systems by the use of effective one-electron time-dependent Schrödinger equations, without the need to impose the physical approximations of the Landauer approach in order to obtain the current.

Many-body effects (e.g, the Coulomb blockade discussed in Sec. 4.5) which are generally described with model Hamiltonians could be thus studied *dynamically*, with effective single-particle equations.[6]

7.2.5 Functional approximations and loss of information

Unfortunately, we do not know the exact functionals, and in actual calculations approximations must be used, some of which are described in the Appendices. However – and this is the most important point – unlike the static transport problem where ground-state DFT is employed, by improving the *dynamical* functionals one is *guaranteed* by the above theorems that the approximate value of the KS current one obtains from these improved approximate functionals must "converge" to the exact many-body value. By "improved" I mean functionals that contain more and more physical properties.[7] This is a fundamental guiding principle similar to the one that guides the calculations of the ground-state total energy using ground-state DFT.

In the case of mixed states, if approximate Kohn–Sham Hamiltonians are used in the calculations of the current, these Hamiltonians are different for every element of the statistical ensemble, namely they become *stochastic Hamiltonians* (Sec. 2.8.6). As I have discussed in Sec. 2.8.6, the loss of information associated with these Hamiltonians necessarily increases with time due to the concavity property of the entropy (Eq. 2.155). This means that the very fact that we make approximations to the exact functional renders the problem intrinsically irreversible, even though the underlying equations of motion are time-reversal invariant.

---

[6] Note that the generalization of the above theorems to spin-dependent Hamiltonians may open the possibility to describe other many-body problems, e.g., the dynamical formation of the Kondo effect, using effective single-particle equations.

[7] Note that, contrary to naive expectations, even in the simplest adiabatic local-density approximation (ALDA) to the exchange-correlation functional (Appendix E) one is *not* guaranteed to obtain the same current as from the ground-state DFT, within the LDA, combined with the Landauer approach (Sec. 3.11). The reason is that in the dynamical calculation one imposes initial conditions on the state of the system, which may not – in the thermodynamic limit of electrodes size and long-time limit (in this order) – be represented by scattering boundary conditions.

7.3 Transient dynamics

An interesting application of the the micro-canonical picture and the above theorems is the study of the transient dynamics of nanoscale systems. This dynamics is still poorly understood.

Here I show an illustrative example (Sai et al., 2007b) using TDDFT within the most common approximation for the exchange-correlation scalar potential: the adiabatic local-density approximation (Appendix E).

Let us consider a narrow constriction that separates two large but finite electrodes (Fig. 7.3). Both the electrodes and the constriction are quasi-2D structures and contain electrons compensated by a uniform background charge (jellium model) with a density close to that of bulk gold ($r_s \simeq 3a_0$).

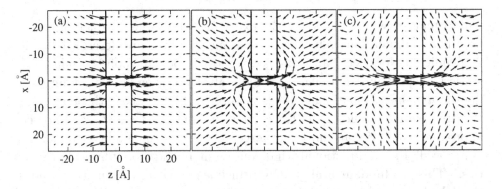

Fig. 7.3. Current density for a series of times in a nanoscale constriction. Electrons flow from right to left, and the current density vector is oriented along the usual convention of the current (i.e., opposite to electron flow). At $t < 0$ the system is prepared with an applied bias $\Delta V = 0.2$ V. The field lines in each panel depict the direction and amplitude of the current density vectors. (a) $t = 0.4$ fs ; (b) $t = 0.8$ fs ; (c) $t = 1.6$ fs. Reprinted with permission from Sai et al. (2007).

The simulations are initiated by preparing the system with a charge imbalance. This is done by applying a step-function-like electric bias across the junction such that the two electrodes bear equal and opposite potential offsets relative to the potential at the center of the junction. The ground state of this system is then calculated, which provides the initial condition $|\Psi_0\rangle$. Equations E.4 of TDDFT are then solved with this initial condition for the same electron system in the absence of the above bias.

Due to the initial bias offset there is the dynamical formation of resistivity dipolar layers at the electrode surfaces. This is shown in Fig 7.4 where a time series of the $x-y$ plane-averaged excess charge density along the z axis is shown.[8]

---
[8] For the definition of planar average, see Sec. 2.4.

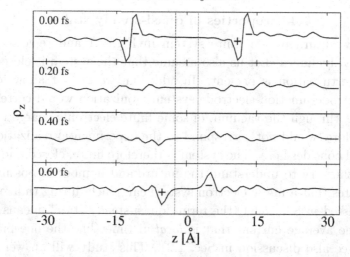

Fig. 7.4. Planar average of the charge density at different times for the nanojunction of Fig. 7.3. Thin dashed lines mark the edges of the positive uniform background of the electrodes. The sign of the surface charges indicates that electron charges accumulate on the right and deplete on the left of the junction. Reprinted with permission from Sai *et al.* (2007).

As a result, the initial current flow is uniform on both sides as shown in Fig. 7.3(a). After this, there is a period of adjustment during which the dominant flow is in the lateral direction, i.e., *parallel* to the facing surfaces of the electrodes: the electrons try to *passivate* the extra dipolar charges (Fig. 7.3(b)). This passivation is continuous, but after the initial transient the net current density converges, for this particular initial condition, toward the center of the nanojunction (Fig. 7.3(c)). This adjustment occurs on very fast time scales related to the fast momentum relaxation in nanojunctions (see discussion in Sec. 2.2.3 and in Sec. 7.6).

The physical situation described above confirms the discussion I have made in the preceding section. Electrons moving across a nanostructure form local resistivity dipoles at the junction, or in its interior. The formation of these dipoles is dynamical, and indeed the discharge of a finite capacitor can be viewed as a continuous attempt of the electrons to passivate the positive charges of these dipoles.

The above example also shows that the electron flow is very similar to a *classical liquid flow* across a constriction: the electrons move like a liquid in going from a "bucket" – the negatively charged electrode – to another "bucket" – the positively charged electrode – across a narrow aperture. I will formalize this notion in the next chapter within a hydrodynamic theory of conduction.

7.4 Properties of quasi-steady states

Let us now return to our finite system in Fig. 7.1 and refer to the true many-body dynamics. Let us also assume that the system is closed and no external perturbation is present. In addition, we do not allow ionization of the electrode-junction-electrode system. Ionization would correspond to the escape, through the vacuum, of some finite electronic charge to infinity and, therefore, would not correspond to the experimental realization of d.c. transport I consider here. The system is therefore finite, closed and isolated.

Let us then try to understand the nature and number of possible quasi-steady states that a many-electron system can reach during time evolution. As I discussed in Sec. 7.1.1, the term "quasi-steady state" means a steady state of the average current that is *local* in time (like the one pictured in Fig. 7.6; see also discussion in Sec. 1.2). This study will answer a fundamental question:

How many states realize a given average total current?

Above, by "average" I mean in a temporal sense, as will become clear in a moment. Also, here "states" may be either pure or mixed, and may differ in terms of the microscopic density and/or current density.

For simplicity, I will work with pure states but the following discussions can be generalized to mixed states as well, by replacing $|\Psi\rangle \to \hat{\rho}$ everywhere. Let us then release the electrons from an arbitrary but definite initial state $|\Psi_0\rangle$, characterized by a charge imbalance between the electrodes. We thenceforth allow the electrons to propagate dynamically with the time-dependent Schrödinger equation 7.1.

As clearly shown in the example of Fig. 7.3, after an initial transient time – related to the initial state of the electrons and to the electron-electron relaxation time inside the capacitor plates – during which electrons first start to traverse the nanojunction, a quasi-steady state may be established. If it is established, we expect it to persist until multiple electron reflections off the far boundaries of the system develop. We are interested in the intermediate quasi-steady state before the occurrence of these reflections.

7.4.1 Variational definition of quasi-steady states

Let us first show that when the system has reached a quasi-steady state this property can be expressed in variational terms.[9] The advantage of a variational procedure is that it will allow us to discuss the *robustness* of

---

[9] This derivation follows closely the one in Di Ventra and Todorov (2004).

7.4 Properties of quasi-steady states

steady states with respect to variations, which will guide us in formulating a principle regarding their existence and formation.

Our electrons are described by a many-body state vector $|\Psi(t)\rangle$, governed by the time-dependent Schrödinger equation 7.1.

For the moment, I regard the initial condition $|\Psi(t=0)\rangle = |\Psi_0\rangle$ as a parameter. As usual, the electron number density and current density are given by

$$n(\mathbf{r},t) = \langle \Psi(t)|\hat{n}(\mathbf{r})|\Psi(t)\rangle,$$
$$\mathbf{j}(\mathbf{r},t) = \langle \Psi(t)|\hat{\mathbf{j}}(\mathbf{r})|\Psi(t)\rangle, \quad (7.16)$$

where $\hat{n}(\mathbf{r})$ and $\hat{\mathbf{j}}(\mathbf{r})$ are the number density (Eq. 1.24) and current density (Eq. 1.25) operators, respectively.

Referring to Fig. 7.2 the current $I_S(t)$, through a surface S across the electrode-junction-electrode system, is once again given by

$$I_S(t) = \int_S \mathbf{j}(\mathbf{r},t) \cdot d\mathbf{S} = \int_V d\mathbf{r}\, \nabla \cdot \mathbf{j}(\mathbf{r},t). \quad (7.17)$$

The time variation of the charge density is related to the current density via the continuity equation

$$\nabla \cdot \mathbf{j}(\mathbf{r},t) + e \frac{\partial n(\mathbf{r},t)}{\partial t} = 0. \quad (7.18)$$

By differentiating once again Eq. 7.18 in time, and replacing the result into Eq. 7.17 we find

$$\frac{dI_S(t)}{dt} = -e \int_V d\mathbf{r}\, \frac{\partial^2 n(\mathbf{r},t)}{\partial t^2}. \quad (7.19)$$

It is then evident that when

$$\frac{\partial^2 n(\mathbf{r},t)}{\partial t^2} = 0, \quad \forall \mathbf{r}, \quad \forall t, \quad (7.20)$$

an *exact* ideal steady state is enforced, namely

$$\frac{dI_S(t)}{dt} = 0, \quad \forall S, \quad \forall t, \quad \text{exact steady state.} \quad (7.21)$$

However, in our closed and finite system an exact steady state can never be realized and a *local* definition – both in time and space – of steady states needs to be introduced, and, as before, we call these *quasi-steady states*.

By appealing to the intuitive concept of a steady state in the case of a macroscopic classical capacitor, considered earlier, we therefore adopt the view that a quasi-steady state is one in which the temporal variation of *local properties* is minimal. Guided by the condition 7.20 we see that a measure of temporal variations is provided by the functional of the density

$$A[n] = \int_{t_1}^{t_2} dt \int_{\text{all space}} d\mathbf{r} \left(\frac{\partial n(\mathbf{r}, t)}{\partial t} \right)^2, \tag{7.22}$$

where (t_1, t_2) is some time interval. In this interval one thus has the average current (cf. Eq. 1.12)

$$\langle I_S \rangle = \frac{1}{t_2 - t_1} \int_{t_1}^{t_2} dt \, I_S(t). \tag{7.23}$$

Let us first perform an unconstrained variational minimization of A with respect to n, for a given $n(\mathbf{r}, t_1)$ and $n(\mathbf{r}, t_2)$. The result is

$$\frac{\partial^2 n(\mathbf{r}, t)}{\partial t^2} = 0, \quad \forall \mathbf{r}, \quad \forall t \in (t_1, t_2). \tag{7.24}$$

This result is indeed the condition 7.20 for an exact steady state with the exception that it is now limited to an interval of time (t_1, t_2). In fact, from Eq. 7.19 we see that, under the conditions expressed by Eq. 7.24,

$$\frac{dI_S(t)}{dt} = 0, \quad \forall S, \quad \forall t \in (t_1, t_2). \tag{7.25}$$

It therefore seems natural to stop here and use the functional 7.22 to define quasi-steady states in our closed and finite system. In reality, Eq. 7.25 by itself says nothing about the actual value of $I_S(t)$ or about the dependence of $I_S(t)$ on S. Equation 7.25 simply describes a *generic* type of conducting state in which matter (in this case electrons) is being transferred from region to region *at a steady rate*. Such a state is a true *global* steady state.

We have seen that such states are minima of the quantity A in Eq. 7.22. If we write $t_2 - t_1 = \delta t > 0$ and use the above procedure for smaller and smaller δt, then in the limit $\delta t \to 0$ we obtain a weaker version of Eq. 7.24, with reference to a single instant in time. This instantaneous form of Eq. 7.24 defines a type of conducting state that I call a true *instantaneous* global steady state.

However, the minimization procedure applied to A above treats $n(\mathbf{r}, t)$ as a freely adjustable function of time, without regard for the actual physical dynamical laws governing our system. In fact, the time-dependent Schrödinger Eq. 7.1 may not permit the system to attain a true instantaneous global steady state, as defined above.

My next task, therefore, is to seek the dynamical state *closest* to a true instantaneous global steady state, permitted by the laws of motion that govern our system. I do this by *constraining* the minimization as follows.

7.4 Properties of quasi-steady states

Let us first replace $(\partial n(\mathbf{r},t)/\partial t)^2$ in Eq. 7.22 with $(\nabla \cdot \mathbf{j}(\mathbf{r},t))^2$ (from Eq. 7.18).

I now define the instantaneous dynamical state closest to a true steady state by the following variational procedure. Consider all possible evolutionary paths allowed by the dynamics of the system, in a small time interval (t_1, t_2). Of these, consider those that have a given total energy $E = \langle \Psi(t)|\hat{H}|\Psi(t)\rangle$ and a given total number of electrons N. In our microcanonical picture, if the Hamiltonian does not depend on time, E is a *constant of the motion* (N is a constant of the motion even if $\hat{H}$ is time dependent).

Of those paths, consider the evolutionary paths that produce a given average current $\langle I_S \rangle$, during the interval (t_1, t_2), across a given open surface S (Eq. 7.23). Of those, we seek that evolutionary path that minimizes A in Eq. 7.22.

To do this, let us define B as $A = B\,\delta t$, $\delta t = t_2 - t_1$. Then, in the limit of vanishing δt, the above search is equivalent to minimizing (with the same constraints) the *instantaneous* functional, at time $t = t_1 = t_2$,

$$B = B[|\Psi(t)\rangle] = \int_{\text{all space}} d\mathbf{r}\, (\nabla \cdot \mathbf{j}(\mathbf{r},t))^2 \qquad (7.26)$$

with $\mathbf{j}(\mathbf{r},t)$ given in Eq. 7.16.

Let us now develop the state of the system as

$$|\Psi(t)\rangle = \sum_i c_i |\Psi_i\rangle, \qquad (7.27)$$

where $\{|\Psi_i\rangle\}$ are all N-electron *bound* states of the system in Fig. 7.2, with eigenenergies $\{E_i\}$, and $\{c_i\}$ some complex coefficients. The reason for restricting the expansion to the bound part of the spectrum of $\hat{H}$ is again that we do not allow for ionization of the electrode-junction-electrode system.

Let us substitute the expansion Eq. 7.27 into Eq. 7.16, and then into Eq. 7.26. Our minimization procedure then corresponds to seeking minima, with respect to $\{c_i\}$, of the functional

$$B[\{c_i\}] = \sum_{i,i',i'',i'''} c_i^* c_{i'} c_{i''}^* c_{i'''}$$

$$\times \int_{\text{all space}} d\mathbf{r}\, (\nabla \cdot \mathbf{j}_{ii'}(\mathbf{r}))(\nabla \cdot \mathbf{j}_{i''i'''}(\mathbf{r})) \qquad (7.28)$$

where $\mathbf{j}_{ii'}(\mathbf{r}) = \langle \Psi_i | \hat{\mathbf{j}}(\mathbf{r}) | \Psi_{i'} \rangle$, subject to the constraints

$$\langle I_S \rangle = \lim_{t_2 \to t_1} \frac{1}{t_2 - t_1} \int_{t_1}^{t_2} dt \, I_S(t) = \sum_{i,i'} c_i^* c_{i'} \int_S d\mathbf{S} \cdot \mathbf{j}_{ii'}(\mathbf{r}), \qquad (7.29)$$

$$E = \sum_i c_i^* c_i E_i, \qquad (7.30)$$

$$\sum_i c_i^* c_i = 1. \qquad (7.31)$$

Each solution for $|\Psi\rangle = \sum_i c_i |\Psi_i\rangle$ is an *instantaneous* many-body state, which we label $|\Psi(E, I_S, t)\rangle$, containing only electrons bound within the electrode-junction-electrode system in Fig. 7.2, that produces a given average current $\langle I_S \rangle$ and a given total energy E, while globally minimizing the divergence of the current density.

We call the solution $|\Psi(E, I_S, t)\rangle$ an *instantaneous quasi-steady state*: the best our closed finite system can do to "mimic" a true instantaneous global steady state (i.e., the one for which condition 7.25 holds). Once these solutions are found, we may then let the electrodes size go to infinity. We expect that in this "thermodynamic limit" the instantaneous quasi-steady states become true global steady states.[10]

7.4.2 Dependence of quasi-steady states on initial conditions

We may now back-track in time and look for the initial condition required to obtain a given instantaneous quasi-steady state. From Eq. 1.17, when the Hamiltonian does not depend explicitly on time, this initial condition is

$$|\Psi_0(E, I_S)\rangle = e^{i\hat{H}t/\hbar} |\Psi(E, I_S, t)\rangle, \qquad (7.32)$$

where I have chosen $t = 0$ as our arbitrary initial time.

We can now make several important points regarding the above minimization procedure.

(i) For a given total energy E and average current $\langle I_S \rangle$ we may obtain *more than one* quasi-steady state solution $|\Psi(E, I_S, t)\rangle$. In other words, there may be different – in terms of charge and current densities – microscopic realizations of the *same* steady-state current.[11]

---

[10] Note that this may not be necessarily true for all the instantaneous quasi-steady states. The reason is that a local (in time) definition of such states may still not allow for a global evolutionary steady-state path (see discussion in Sec. 7.5). If one has only *local stability*, compared to *global stability*, of quasi-steady states we expect that perturbations/fluctuations in the system drive it away from a global steady state.

[11] If we work with mixed states and not pure states, the above is equivalent to stating that there

(ii) There may be combinations of E and $\langle I_S \rangle$ for which *no solution* for $|\Psi(E, I_S, t)\rangle$ exists: the system may *never* reach a quasi-steady state of specified current.

(iii) There may be initial conditions $|\Psi_0\rangle$ that *do not ever lead* to a quasi-steady state, i.e., that cannot be reached by back propagation from any $|\Psi(E, I_S, t)\rangle$.[12]

(iv) Finally, there may exist quasi-steady state solutions that have the same E but different I_S. If these solutions describe quasi-steady states of different currents at the same time t corresponding to the *same* voltage drop, they correspond to chaotic transport and may occur in systems with intrinsic non-linear dynamics.[13] In Sec. 8.5 I will discuss such an instance when the system dynamics is turbulent.[14]

7.5 A non-equilibrium entropy principle

In addition to the above points we can also draw the following conclusion which is a direct consequence of the variational nature of quasi-steady states.

At a quasi-steady state $|\Psi(E, I_S, t)\rangle$, the functional $B[|\Psi\rangle]$ is, by construction, *stationary* against variations of $|\Psi\rangle$ about $|\Psi(E, I_S, t)\rangle$, compatible with the constraints. But B is a measure of the magnitude of the divergence of the current density, at least in a macroscopically averaged sense.

Thus, we may expect the quasi-steady state flow pattern itself to be relatively *robust* against variations about $|\Psi(E, I_S, t)\rangle$, at least on a "coarse-grained" scale. However, after back-propagation to our initial time $t = 0$, the

---

are two or more statistical operators $\hat{\rho}$ that give the same total current at time t, but whose trace over the density and current-density operators is different. This is reminiscent of the case encountered in the ground state of a many-body system, when the latter is degenerate thus giving rise to different microscopic densities.

[12] Note that it was argued in Stefanucci and Almbladh (2004), that *any* initial condition would lead to a steady state in an infinite system. In reality, in that publication the authors consider a system that in the long-time limit allows for a *stationary* solution, i.e., they *assume* from the outset that the system will indeed reach a steady state. We ask the opposite question: if one starts with an arbitrary initial condition does the system actually reach an instantaneous steady state that in the thermodynamic limit may develop into a true global steady state?

[13] A further consequence of all the above discussion is the following. If we consider the quasi-steady state solutions in an ensemble $\{\langle I_S \rangle\}$ of average currents and form a linear combination of many-body wave-functions out of their respective sets of initial conditions, the system with this new initial condition could possibly evolve in time into the quasi-steady state of yet *another* average current $\langle I_S \rangle$, that does *not* belong to the original ensemble $\{I_S\}$. If this happens, then the system can fluctuate *coherently* between microscopic quasi-steady states with *different* currents and steady-state noise is produced (Di Ventra and Todorov, 2004). This additional noise has nothing to do with the ordinary shot noise we have discussed in Chapter 5; instead, it could be due to possible realizations of a steady state as a linear combination of microscopic states corresponding to different currents. This noise would be sensitive to decoherence effects and is thus likely to be reduced if the latter are strong.

[14] Initial conditions that do not lead to a quasi-steady state may still do so if other inelastic effects are included, such as electron-phonon interactions.

corresponding spread of initial conditions, about $|\Psi_0(E, I_S)\rangle$, may contain *large* variations in microscopic quantities such as the charge density.

In other words, there may be a large set of initial conditions – in the Hilbert space of the system – which I denote symbolically by $P_0(|\Psi(E, I_S, t)\rangle)$, that differ in their microscopic properties but that produce the same, or nearly the same, quasi-steady state flow pattern at some later time t. This is illustrated schematically in Fig. 7.5.

This conclusion supports the intuitive notion that a steady state, if formed, should be relatively insensitive to the microscopic details in the initial conditions.

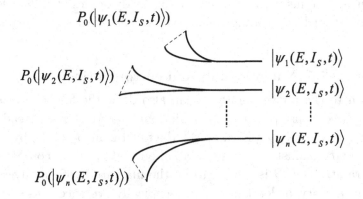

Fig. 7.5. A quasi-steady state $|\Psi_i(E, I_S, t)\rangle$ at time t is robust against local variations but may originate from a large set $P_0(|\Psi_i(E, I_S, t)\rangle)$ of different initial conditions in the Hilbert space of the system.

This conclusion also suggests a link with the notion of entropy I have discussed in Sec. 2.8. In fact, the "likelihood" of a system with a given total energy E attaining a steady state, $|\Psi(E, I_S, t)\rangle$, with a given total current $\langle I_S \rangle$ at time t, and the stability of this steady state against small perturbations, is measured by the relative weight (in the Hilbert space) of the set $P_0(|\Psi(E, I_S, t)\rangle)$, among all initial conditions that lead to the current $\langle I_S \rangle$ at time t.

Call $p_i(|\Psi(E, I_S, t = 0)\rangle)$ the probability that a *given* initial condition occurs in the set $P_0(|\Psi(E, I_S, t)\rangle)$. From Eq. 2.142, we can then define the following measure of disorder for that particular set:

$$S[P_0(|\Psi(E, I_S, t)\rangle)] = -k_B \sum_{i=1}^{M} p_i(|\Psi(E, I_S, t = 0)\rangle) \ln[p_i(|\Psi(E, I_S, t = 0)\rangle)], \quad (7.33)$$

where M is the number of accessible states in the set $P_0(|\Psi(E, I_S, t)\rangle)$.[15]

Now, suppose that for a given total energy E and average current $\langle I_S \rangle$ there are several *distinct* quasi-steady states $|\Psi_i(E, I_S, t)\rangle$ (Fig. 7.5). For *each* of them we can define an entropy as in Eq. 7.33. The question is: which state would be realized, or preferentially realized, in a macroscopic experiment on the system at time t?

In equilibrium statistical mechanics we are guided by the *maximum statistical entropy principle* which states that *among all possible statistical operators $\hat{\rho}$ compatible with the constraints on the system, the latter is represented by the macro-state that has the largest statistical entropy* 2.142 (see, e.g., Balian, 1991). From this principle we are able to derive all laws of thermodynamics.

In the present non-equilibrium problem we cannot rely on this principle. However, from the above discussion it seems natural to formulate the following (Di Ventra and Todorov, 2004)

Initial-state maximum entropy principle: *A dynamical system reaches the steady state whose ensemble of initial conditions has maximum disorder.*

In other words, the system is driven towards a specific steady state at time t – if the dynamics supports steady states – by a "maximum-entropy principle" where the entropy here does not measure the disorder of the state at time t, but rather the number of different initial conditions that realize the given steady state. The more disordered (larger) that set of initial conditions is, the more likely the steady state is to be realized.[16]

The above principle also seems to encompass the situation in which the time variation of the entropy (*entropy production*) is *minimal* at the quasi-steady state that develops from the set of initial conditions with largest entropy. This is reminiscent of *Prigogine's principle of minimal entropy production* at a steady state close to equilibrium, by which a steady state is characterized by minimal entropy variation (Prigogine, 1967). However, unlike the latter principle which has a bearing only for states close to equilibrium and has no relation to the actual dynamics that *leads* to a steady state, the one formulated above has applicability even far from equilibrium, and contains a statistical criterion for a given steady state to actually form.

The last point is particularly important because dynamical systems go

---

[15] If we choose to work within the micro-canonical ensemble, the probabilities $p_i(|\Psi(E, I_S, t = 0)\rangle)$ are all equal to $1/M$ (see Sec. 2.8.2).
[16] In this respect, this steady-state may be called an *attractor* of the dynamical system, and the set of initial conditions that lead to it a *basin of attraction*.

through several evolutionary paths, some of which may contain large "barriers" that may force the system in some arbitrary state, not necessarily a steady state. Therefore, by focusing only on the final result – the steady state – without regard to the *dynamics* of the system, we cannot answer the question of how the latter overcomes these evolutionary "barriers".

It is only when all evolutionary paths are taken into account with all possible initial conditions that the system may overcome those "barriers" and reach a steady state. It thus seems natural that the more "attempts" – initial conditions – that allow the system to reach a given steady state, the more likely the latter is to be realized.

Finally, the principle I have formulated above is not necessarily limited to steady states in transport, but may have a bearing in the formation of dynamical steady states of any physical system – close to or far from equilibrium – with these dynamical steady states belonging to some other observable of the system. An example of this is the approach to equilibrium of ultra-cold gases in optical lattices (Kinoshita et al., 2006). It was shown experimentally that, according to the initial conditions, these systems may reach steady states that are not necessarily canonical (namely, they are not equilibrium states), and that are strongly dependent on the initial state of the system. In other words, the final state has *memory* of the initial state it originates from.

Electron sources revisited

At this point, let us refer to the case in which the system is connected to an electron source, such as a battery. At any given time, the latter prepares the many-body system in some state which then evolves in time. The only constraint on this state is that it represents a fixed electrostatic potential difference V. This constraint, however, is a "mean-field property" of the many-body system (Eq. 2.94) that can be realized by many states: several many-body wave-functions can be constructed that produce different single-particle densities but same voltage difference. According to the above analysis, some of these states may lead to steady states, some to no steady state at all.

Those states that *do* lead to a steady state may be grouped into different final steady states. The above principle suggests that the one that is likely to be realized is the one that has the largest number of initial states associated with it: the one with maximum initial-state disorder.

7.6 Approach to steady state in nanoscale systems

I have discussed the nature and number of quasi-steady states, and their dependence on initial conditions, but I have yet to discuss the microscopic physical processes that help the system establish a steady state.

Electron-phonon interaction

If the many-body electron system is open to a phonon bath with a dense spectrum,[17] the dynamics of the system may be described by an equation of motion of the Lindblad type (Eq. 1.73). As discussed in the example of Sec. 2.2.1, it thus seems natural that electron-phonon interactions help the system evolve in one of the (possibly many) steady states compatible with the equation of motion 1.73. This happens within relaxation times in the range of $10^{-14} - 10^{-13}$ s with the smaller times corresponding to the system at room temperature or higher (see Secs. 2.2.2 and 6.2.1). This is, however, not the only process that helps establish a steady state.

Electron-electron interaction

In fact, let us consider our electron system closed and finite. Electron-electron interactions are responsible for screening, which keeps the electron density macroscopically constant, a condition we call *charge neutrality*. This creates an effective "incompressibility" of the electron liquid in the electrodes, which makes current flow somewhat similar to a classical liquid flow: local disturbances in the density are not tolerated and "heal" fast (I will expand further on this hydrodynamic connection in Chapter 8).

However, another important effect of electron-electron interactions, in particular *Umklapp processes* – processes that destroy crystal momentum (Ashcroft and Mermin, 1975) – is to produce relaxation of the *total electron momentum* in the electrodes, which is not a constant of the motion in our finite system. Even though these processes are well defined in strictly periodic systems – and thus not in our electrode-nanojunction-electrode system – deep into the electrodes, far away from the boundaries, we can argue (as we did in Sec. 6.3.1 for the heat conduction) that such processes exist, and can be described similarly to the case of translationally invariant systems.[18]

In bulk materials, relaxation times for electron-electron interactions are typically larger than electron-phonon scattering times for a wide range of

---

[17] And very short correlation times compared to the electron dynamics.
[18] In the same region of the electrodes, there would be *normal* electronic processes which conserve total electron momentum and therefore cannot contribute to the resistance.

temperatures (Ashcroft and Mermin, 1975). In nanoscale systems the rate of electron-electron interactions actually increases locally at the nanojunction with respect to the bulk, thus leading to the electron heating phenomenon I will discuss in Sec. 8.6. Electron-electron interactions may thus help the formation of steady states.

An elastic relaxation mechanism

There is, however, yet another intrinsic mechanism that facilitates momentum relaxation in the crucial region of the junction and which is particularly effective for systems of nanoscale dimensions. I have anticipated this mechanism in Sec. 2.2.3 where I discussed the role of momentum relaxation in driving the system to local equilibrium.

This mechanism – which changes electron momentum but not energy – is provided simply by the geometrical constriction experienced by electron wave-packets as they approach the nanojunction (Di Ventra and Todorov, 2004; Bushong et al., 2005). It is thus due to the wave properties of the electron wave-functions and the resultant uncertainty principle, and has nothing to do with electron-electron interactions.

It is indeed this relaxation mechanism that generates most of the (elastic) resistance for electrons scattering at impurities. However, unlike an impurity in a bulk material that can be avoided by electrons moving around it, a nanojunction cannot be avoided by the carriers. In addition, in nanojunctions this relaxation mechanism turns out to be generally much faster than the other inelastic processes described above.

Since momentum relaxation drives the system to local equilibrium (Sec. 2.2.3), the above mechanism, combined with the effective incompressibility of the electron liquid, may also drive the system to a quasi-steady state. In the thermodynamic limit of large electrodes this quasi-steady state may then turn into a true global steady state, but the infinite number of degrees of freedom provided by the electrodes is *not necessary* for the quasi-steady state to develop, only to *maintain* the developed quasi-steady state indefinitely in time.[19] Let us expand on this point.

I follow a similar argument as in Sec. 2.2.3 but I now take advantage of the single-particle scattering theory I developed in Chapter 3. Let us assume that the nanojunction has cross section w and a single electron wave-packet moves into it. The wave-packet has to adjust to the motion appropriate to

---

[19] This is clear if one considers turbulent or chaotic transport. In this case, a true global steady state does not develop, even if the system is infinite.

7.6 Approach to steady state in nanoscale systems

the given junction geometry in a time

$$\tau_c \sim \frac{\hbar}{\Delta E}, \qquad (7.34)$$

where ΔE is the typical energy spacing of lateral modes in the constriction. If we assume a rectangular lateral confining potential, the energy spacing of lateral modes is of the order of (from Eq. 3.25)[20]

$$\Delta E \sim \frac{\pi^2 \hbar^2}{m_e w}. \qquad (7.35)$$

We therefore find

$$\tau_c = \frac{1}{\nu_c} \sim \frac{m_e w}{\pi^2 \hbar}. \qquad (7.36)$$

For later use I have also defined a frequency ν_c associated with this process.

If $w = 1$ nm^2, τ_c is of the order of 1 fs, i.e., orders of magnitude smaller than typical electron-electron, or even electron-phonon scattering times for a wide range of temperatures. As anticipated, this shows that the elastic collisions at the nanostructure contribute largely to momentum relaxation.

If we call τ_{in} the scattering time due to inelastic effects, and assume these processes independent of the above elastic scattering with the nanojunction, then the total scattering time τ_{tot} is

$$\frac{1}{\tau_{tot}} \approx \frac{1}{\tau_c} + \frac{1}{\tau_{in}} \approx \nu_c, \qquad (7.37)$$

where the last approximation comes precisely from the assumption that the elastic scattering time is much faster than the other relaxation times.

The relaxation time τ_{tot} is the one that would enter, for instance, in the collision integral 4.157 of the quantum kinetic equations 4.155. Thus, this effect would seem to suggest that, even *without* electron-phonon or electron-electron inelastic scattering, a steady state could be reached in a nanoconstriction, with at most mean-field interactions.

Clearly, the result in Eq. 7.37 is just a crude estimate in the spirit of the relaxation-time approximation I have discussed in Sec. 4.6. In reality, the scattering rate due to the presence of the nanojunction must be dependent on the many-body configuration of all electrons. Also, in truly mesoscopic/macroscopic systems we expect the above elastic relaxation mechanism to be of less importance in establishing a quasi-steady state: inelastic effects are likely to dominate the approach to steady state.

---

[20] The electron mass entering Eq. 7.35 could be an *effective mass* m_e^* if one uses an effective Schrödinger equation to describe transport (see Chapter 8).

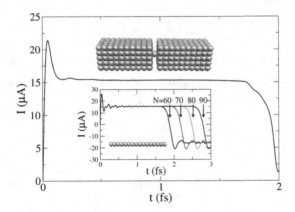

Fig. 7.6. Current as a function of time through a model 3D finite nanojunction (schematic is shown) calculated with a single-particle tight-binding model. The inset shows the corresponding current for a finite linear chain of N = 60, 70, 80 and 90 atoms. Reprinted with permission from Bushong et al. (2005).

A numerical example

An example that the above relaxation mechanism may indeed lead to a quasi-steady state is shown in Fig. 7.6, where electronic transport is studied within the micro-canonical picture for a finite three-dimensional model of a nanojunction, and a finite quasi-one-dimensional atomic wire (Bushong et al., 2005).[21]

The dynamics of the system is described by the N-site, nearest-neighbor, tight-binding Hamiltonian

$$\hat{H} = \sum_{i=1}^{N} \epsilon_i |r_i\rangle \langle r_i| + t \sum_{i=1}^{N} \left(|r_i\rangle \langle r_{i+1}| + \text{H.c.} \right), \qquad (7.38)$$

where there is one orbital state $|r_i\rangle$ per atomic site with energy ϵ_n and hopping matrix element t connecting nearest-neighbor sites (the symbol "H.c." means Hermitian conjugate).

We then prepare the system such that one half has a different local configuration than the other, i.e., the initial state is not the ground state of the above Hamiltonian. For this non-interacting case, this can be done by increasing the on-site energies ϵ_i on one side of the system by an energy

---

[21] For another example of the use of the micro-canonical picture of transport in molecular structures see, e.g., Cheng et al. (2006).

"barrier" E_B.[22] This seems a natural choice as it "mimics" the role of the electron source. However, as I have discussed above it is just one of the many (essentially infinite) initial conditions one can choose to initiate current flow; some of these initial conditions may not lead to a quasi-steady state.

Taking this state as the initial state of the system, we then remove E_B, and let the electrons propagate according to the time-dependent Schrödinger equation 7.1 with the time-independent Hamiltonian 7.38.

Due to the closed and finite nature of the system, the total current can be calculated by differentiating in time the charge accumulated on one side of the system, i.e. (the factor of 2 is for spin)

$$I(t) = -2e\frac{d}{dt}\sum_{n=1}^{N/2}\sum_{i=1}^{N_L}\langle\psi^n(t)|r_i\rangle\langle r_i|\psi^n(t)\rangle. \qquad (7.39)$$

Here $|\psi^n(t)\rangle$ are the occupied single-electron states that are solutions of the time-dependent Schrödinger equation, and N_L is the number of sites on the left of the junction interface.

The onset of a quasi-steady state is clear in Fig. 7.6, where the current Eq. 7.39 as a function of time is plotted for $E_B = 0.2$ eV. In the inset, it is shown that a similar quasi-steady-state current develops in 1D wires of different lengths, where the initial-time energy barrier forces electrons to change the spread of electron momentum, and hence plays a role similar to that of the geometric constriction in the 3D case.

In all cases, the current initially rises rapidly, but quickly settles to a quasi-constant value $I_{\rm ss}$.[23] The quasi-steady state lasts for a time t_d during which the electron waves propagate to the ends of the structure and back.

As is clear from the inset of Fig. 7.6, the duration of a quasi-steady-state – when developed – can be made as long as one pleases by simply increasing the size of the electrodes. This again confirms our previous discussion, namely that a quasi-steady state current with a finite lifetime can develop even in the *absence* of dissipative effects. In addition, it shows that the *formation* of quasi-steady states is related to properties of the nanojunction, while its *lifetime* is related to the size of the electrodes.[24] In the thermodynamic limit of electrode size, this lifetime can be made infinitely long.

---

[22] For the 1D wire (see inset of Fig. 7.6), the interface between the two regions separated by the barrier defines the nanoscale "junction".

[23] In the 1D structures, small oscillations (related to the finite size of the system) are observed that decay in time.

[24] As discussed in Sec. 7.1, $t_d \approx 2L/v_F$, with L the characteristic length of one electrode. Taking into account Eq. 7.36, in order to observe a quasi-steady state one needs a characteristic length for the electrodes $L > v_F m_e w/2\pi^2\hbar$.

7.7 Definition of conductance in the micro-canonical picture

I finish this chapter with considerations on how to define a conductance in this finite-system approach. I recall that the conductance is a derived quantity, and, therefore, it does not have a unique quantitative definition (Sec. 3.3.4).

Let us consider the case in which a quasi-steady state has formed from some initial state. Let $I_S(t)$ be the total current associated with this quasi-steady state, across a chosen surface S. Now that we have dispensed with the infinite reservoirs of the Landauer approach, we may no longer appeal to the definition of conductance we have described in Chapter 3, with respect to bulk electrochemical potentials. We may, however, fall back to its *non-invasive* definition, with respect to the *electrostatic potential drop* in the system, similar in spirit to the definition we have employed in Sec. 3.8.

The electrostatic potential $\phi(\mathbf{r}, t)$ (Eq. 2.94), subject to the boundary condition $\phi(\mathbf{r}, t) \to 0$ as $r \to \infty$ appropriate to our isolated finite system, is a functional of the electron density $n(\mathbf{r}, t)$ and is thus unambiguously known in the quasi-steady state, or in any other state for that matter.

It is physically plausible – and numerical calculations confirm this assumption (Bushong *et al.*, 2005) – that, for large enough electrodes, in a quasi-steady state the potential ϕ tends – possibly within microscopic Friedel-like oscillations – to well-defined values $\phi_L(t)$ and $\phi_R(t)$ in the interior of the left and right electrodes, respectively, at least for times much smaller than the time for the electrons to traverse the electrodes and come back.

This enables us to define a potential difference $V(t) = \phi_L(t) - \phi_R(t)$, with respect to which we may then define a conductance as

$$\sigma(t) = \frac{dI_S(t)}{dV(t)}. \tag{7.40}$$

It has been shown numerically (Bushong *et al.*, 2005) that for metallic quantum point contacts the conductance evaluated according to Eq. 7.40 shows the quantization properties I have discussed in Sec. 3.3.4 in the context of the Landauer approach.[25]

We could, however, measure the above electrostatic potentials anywhere along the electrodes (see Fig. 3.19). With an appropriate macroscopic average, we could define another set $\bar{\phi}_L(t)$ and $\bar{\phi}_R(t)$ of electrostatic potentials

---

[25] In a single-particle picture one can also project the occupation of the dynamical states $|\psi(t)\rangle$ onto the eigenstates of the Hamiltonian $\hat{H}$, within the left and right regions of the electrodes (Bushong *et al.*, 2005). This gives rise to two local electron distributions. It was shown numerically that these distributions approach two local Fermi-like distributions centered at different local energies, which can be interpreted as local electrochemical potentials (Bushong *et al.*, 2005). In this case, the conductance can be defined in terms of the difference between these two local electrochemical potentials.

7.7 Definition of conductance in the micro-canonical picture

locally in those regions of the electrodes where a *non-invasive* measurement is ideally performed, from which a new electrostatic potential difference $\bar{V}(t) = \bar{\phi}_L(t) - \bar{\phi}_R(t)$ could be defined. The corresponding conductance

$$\bar{\sigma}(t) = \frac{dI_S(t)}{d\bar{V}(t)}, \tag{7.41}$$

would thus have, at a quasi-steady state, all the characteristics of the four-probe conductance 3.293.

Summary and open questions

In this chapter I have introduced an alternative picture of transport I named micro-canonical. In this picture electrical current is studied via the discharge of a large but finite capacitor across a nanojunction. This system may be either closed or open.

This picture allows us to prove theorems on the total current within dynamical density-functional theories. These theorems guarantee that, if we know the exact dynamical functional, *all* many-body dynamical effects can be evaluated using effective single-particle equations. Such theorems cannot be proved for the static formulations of transport I have described in previous chapters.

The micro-canonical picture also allows for a variational definition of quasi-steady states, and suggests a principle regarding their formation. This principle does not refer to the entropy – or information – content of the steady state, as traditional principles for systems close to equilibrium do. Rather, it gives a prescription on the likelihood that a steady state be formed based on the disorder of the *initial states* that lead to such a steady state.

This principle is not necessarily limited to the transport problem I have discussed here, but may have applications to the more general class of non-equilibrium steady states.

Interesting effects that have not been studied yet within this approach relate to electron-ion interactions and noise properties of the current. Studies in this direction may help to better understand the approach to steady state, and memory effects on the electron dynamics.

8
Hydrodynamics of the electron liquid

In the preceding Chapter I have discussed an alternative picture of transport, which I called micro-canonical. From this picture we have learned that electrons flow from one electrode to the other across a nanojunction like a liquid would do in confined geometries. This was illustrated in Fig. 7.3.

The junction is an *unavoidable* obstacle for the electrons, which need to change their momenta while crossing it. This forces the system to reach local equilibrium fast, greatly helping the effect of other inelastic processes whose role is also to force the system towards local equilibrium.

To the above properties we need to add the known fact that, due to Coulomb interactions, the electron liquid is also *viscous*, namely electron-electron interactions create an internal *friction* for an electron to propagate through the electron liquid, much like the friction experienced by an object moving across a viscous liquid or gas.

To be more specific, the shear viscosity of the electron liquid at the density of a typical metal, such as gold, is about 10^{-7} Pa s (from Eq. F.8 with $r_s = 3a_0$). For comparison the viscosity of water at room temperature is about 10^{-3} Pa s.

The viscosity of the electron liquid is thus very small but not zero. In fact, the smaller the viscosity, the less stable the flow with respect to perturbations, especially those provided by obstacles or constrictions along the path.

All of the above properties suggest an intriguing analogy of the electron flow in a nanojunction with the flow of a classical liquid in restricted geometries. In this chapter I show that this notion is not far-fetched, and it is based on the fact that the time-dependent many-body Schrödinger equation can be cast *exactly* in a "hydrodynamic" form in terms of the single-particle density and the *velocity field*, i.e., the ratio between the current density and density.

The equivalence between the Schrödinger equation and hydrodynamics goes back to the early studies by Madelung and Bloch (Madelung, 1926; Bloch, 1933) at the beginning of quantum mechanics; later on put on a formal basis by Martin and Schwinger for a general many-body system (Martin and Schwinger, 1959).[1] However, as I will show here, the complete equivalence between these two formulations rests on the theorems of dynamical density-functional theories.

The above also shows that the dynamics of a quantum system can be described *exactly* using "collective" variables – precisely the density and velocity field – much like one does for classical liquids. Indeed, due to the continuous nature of the wave-functions, a hydrodynamic description is much more "natural" in quantum mechanics than in classical physics. In the latter case particles are objects of finite size, and the density and velocity field are associated with volumes of the liquid small compared to the other relevant length scales (such as variations of any external potential) but large to contain "enough" particles so that a continuum mechanics can be developed. In quantum mechanics the above coarse-grained requirement is not necessary.

However, despite its great physical appeal, the hydrodynamic formulation of the Schrödinger equation – in its exact form – has the same range of applicability as the latter: for a true many-body problem it is very limited. This is due to the difficulty in identifying the analytical structure of the quantity that describes electron-electron interactions (in hydrodynamics this quantity is the *stress tensor*).

I will show that, due to the fast relaxation mechanism I have discussed in the previous chapter (Sec. 7.6), the electron flow at nanostructures can be described by the quantum analogue of the Navier–Stokes equations (D'Agosta and Di Ventra, 2006). This opens up the possibility to observe *turbulence* in nanoscale systems as in in the classical case. This phenomenon is purely dynamical, and thus cannot be captured by static formulations of transport.

Finally, due to the viscous nature of the electron liquid, and the large current density locally in the nanojunction, electrons "heat up" in the junction region (D'Agosta *et al.*, 2006), similar to the local heating of the ions due to electron-phonon interactions I have discussed in Sec. 6.3. The hydrodynamic formulation comes in handy for this problem, since it allows us to estimate the magnitude of this phenomenon without the explicit calculation of matrix elements between non-equilibrium many-body states of different configurations.

---

[1] A similar derivation has been repeated in Fröhlich (1967), where the Navier–Stokes equations have also been obtained.

8.1 The Madelung equations for a single particle

Let us start by showing that the time-dependent Schrödinger equation can be written in a "hydrodynamic" form in the simplest case possible: the dynamics of a single particle under the action of a potential $V(\mathbf{r}, t)$. This equivalence was formulated by Madelung in 1926 (Madelung, 1926), immediately after Schrödinger published his famous equation.

Let us consider the single-particle Hamiltonian

$$H = -\frac{\hbar^2}{2m}\nabla^2 + V(\mathbf{r}, t), \tag{8.1}$$

which, assigned an initial condition, gives rise to the dynamics

$$i\hbar \frac{\partial \Psi(\mathbf{r}, t)}{\partial t} = \left(-\frac{\hbar^2}{2m}\nabla^2 + V(\mathbf{r}, t)\right)\Psi(\mathbf{r}, t). \tag{8.2}$$

Let us now write the wave-function as[2]

$$\Psi(\mathbf{r}, t) = \sqrt{n(\mathbf{r}, t)}\, e^{iS(\mathbf{r}, t)/\hbar}, \tag{8.3}$$

in terms of the probability density $n(\mathbf{r}, t)$ and the phase $S(\mathbf{r}, t)$.

Let us insert this form into Eq. 8.2 and divide by $\Psi(\mathbf{r}, t)$ (assuming it is not zero). This generates two coupled equations of motion: the real and imaginary parts of a complex equation. These are (Exercise 8.1)

$$\frac{\partial n}{\partial t} = -\nabla \cdot (n\mathbf{v}), \tag{8.4}$$

$$\frac{\partial S}{\partial t} = -V(\mathbf{r}, t) - U(\mathbf{r}, t) - \frac{1}{2m}(\nabla S)^2, \tag{8.5}$$

where I have defined the *velocity field*

$$\mathbf{v}(\mathbf{r}, t) = \frac{\nabla S(\mathbf{r}, t)}{m}, \tag{8.6}$$

and the "internal potential"

$$U(\mathbf{r}, t) = -\frac{\hbar^2}{2m}\frac{\nabla^2(\sqrt{n})}{\sqrt{n}}. \tag{8.7}$$

Equation 8.4 is the continuity equation 1.11. The second, Eq. 8.5, can be written in a more transparent way by differentiating it once more, and by defining the *stress tensor* ($i, j = 1, 2, 3 = x, y, z$)

$$P_{ij}(\mathbf{r}, t) = -\left(\frac{\hbar^2}{4m}\right) n(\mathbf{r}, t) \frac{\partial^2 (\ln n(\mathbf{r}, t))}{\partial x_i \partial x_j}, \tag{8.8}$$

---
[2] This is sometimes referred to as the *de Broglie ansatz*.

to get (Exercise 8.1)

$$m\,n(\mathbf{r},t)\frac{\partial v_j(\mathbf{r},t)}{\partial t} + m\,n(\mathbf{r},t)\,(\mathbf{v}\cdot\nabla)v_j = -n(\mathbf{r},t)\,\nabla_j V(\mathbf{r},t) - \nabla_i P_{ij}(\mathbf{r},t), \tag{8.9}$$

where the symbol $\nabla_i P_{ij}$ is short for the repeated-index summation $\sum_i \nabla_i P_{ij}$. By defining the *convective derivative*

$$D_t \equiv \frac{\partial}{\partial t} + \mathbf{v}\cdot\nabla, \tag{8.10}$$

we finally get

$$\boxed{m\,n(\mathbf{r},t)\,D_t \mathbf{v}_j(\mathbf{r},t) = -n(\mathbf{r},t)\,\nabla_j V(\mathbf{r},t) - \nabla_i P_{ij}(\mathbf{r},t)\,, \quad j = x,y,z\,.} \tag{8.11}$$

Equations 8.11 are the Newton equations of motion of a fluid of density n and velocity $\mathbf{v}$ (Landau and Lifshitz, 1959b). They simply state that all forces external to the fluid, $-\nabla V$, and "internal", $-\nabla_i P_{ij}$, change its velocity. In the present context, they are often referred to as the *Madelung equations*.

I have thus shown that the time-dependent Schrödinger equation 8.2 for a single particle in an external potential can be written *exactly* in terms of the density and the velocity field.

What about the reverse? Namely, by starting from the continuity equation 8.4 and the Madelung equations 8.11, do we recover full information on the wave-function whose dynamics is given by the time-dependent Schrödinger equation 8.2?

It has been argued (see, e.g., Wallstrom, 1994) that, since the phase S in the wave-function 8.3 is generally multi-valued, this step requires an *ad hoc* quantization condition on the velocity field 8.6. However, this is not necessary if we recall, from the theorem of time-dependent density-functional theory (Appendix E), that given an initial condition one can determine uniquely – apart from a trivial time-dependent constant – the external scalar potential that generates the density,[3] and hence the associated wave-function (see also discussion in Sec. 2.8.3). It is from the given initial conditions and this correspondence that quantization must arise. The above thus guarantees the exact equivalence between the two formulations. It is then a practical difficulty, not a conceptual one, to back-track from the solution of the Madelung equations 8.11 to the solution of the Schrödinger equation 8.2.

---

[3] If the external potential cannot be written as a scalar one only, namely one has a vector potential, then the theorem of time-dependent *current* density-functional theory (Appendix F) states that given an initial condition one can determine the external vector potential that generates the velocity field.

Finally, I mention that in this single-particle case, since the velocity field can be written – in the absence of a vector potential – as the gradient of a function (Eq. 8.6), the associated fluid flow is *irrotational*, namely the curl of the velocity field is zero everywhere in space and at any given time (Ghosh and Deb, 1982). This is not necessarily true for an interacting many-body system, where electron-electron interactions produce *dissipative* effects in the system dynamics. This is the topic of the next sections.

8.2 Hydrodynamic form of the Schrödinger equation

For a many-body interacting system the above equivalence is a bit more difficult to derive. Nonetheless, it can be done exactly (Martin and Schwinger, 1959). Here I simply state and discuss the results, and refer the reader to Appendix I for some points of the derivation.

More specifically, let us consider a closed many-body electron system whose Hamiltonian has the general form

$$\hat{\mathbf{H}}(t) = \hat{\mathbf{T}} + \hat{\mathbf{W}} + \hat{\mathbf{V}}_{ext}(t) \qquad (8.12)$$

describing electrons interacting via the Coulomb potential (described by the operator $\hat{\mathbf{W}}$) subject to a time-dependent external potential represented by the operator $\hat{\mathbf{V}}_{ext}(t)$, and $\hat{\mathbf{T}}$ is the sum of all kinetic energies of the electrons (see, e.g., Eq. 6.1). The external potential may represent, for instance, the electron-ion interaction potential 6.4, and any other external potential.

With this Hamiltonian we solve either the time-dependent Schrödinger equation 1.16 or the Liouville–von Neumann equation 1.60 if the system is in a mixed state.

I then define the *velocity field*[4]

$$\mathbf{v}(\mathbf{r},t) = \frac{\mathbf{j}(\mathbf{r},t)}{n(\mathbf{r},t)} = \frac{\langle \hat{\mathbf{j}}(\mathbf{r},t) \rangle}{\langle \hat{n}(\mathbf{r},t) \rangle}, \qquad (8.13)$$

the ratio between the average of the current density operator 1.25 and the number density operator 1.24. The average is done over the many-body state of the system, whether the latter is in a pure or a mixed state.[5] The above definition holds whenever the average density is not zero, and reduces to Eq. 8.6 for a single particle if we make the de Broglie ansatz 8.3 (Exercise 8.3). Note also that, unlike the classical case, since the wave-functions

---

[4] In this chapter the current density is defined as the expectation value of the current density operator 1.25. It thus does not contain a charge dimension.

[5] Note that the velocity field 8.13 is *not* the expectation value of some velocity operator over the state of the system. It is thus not an observable in the proper sense. Therefore, the properties associated with this field must be interpreted as those derived from the density and current density, which are observables.

8.2 Hydrodynamic form of the Schrödinger equation

decay exponentially in the vacuum outside the sample region, both the density and the current density decay exponentially away from the sample. However, their ratio 8.13 may remain constant, thus leading to a finite velocity field even outside the sample; a fact which may be used to interpret the Aharonov–Bohm effect (a purely quantum-mechanical effect) in a hydrodynamic picture (see, e.g., Casati and Guarnieri, 1979, for a different derivation of this effect from quantum hydrodynamics).

Similar to the single-particle case of the preceding section, all "internal interactions" are included in a *stress tensor* of the type

$$\mathbf{P}_{ij}(\mathbf{r},t) = \mathbf{W}_{ij}(\mathbf{r},t) + \mathbf{T}_{ij}(\mathbf{r},t), \qquad (8.14)$$

the sum of a Coulomb interaction part $\mathbf{W}_{ij}(\mathbf{r},t)$ and a kinetic contribution $\mathbf{T}_{ij}(\mathbf{r},t)$ (see Appendix I for the explicit definition of these tensors).

In the present many-body case, however, the word "interactions" really means all interactions among particles, and therefore the above stress tensor cannot be easily written in terms of the single-particle density as in Eq. 8.8.

Finally, the equations of motion for the density and velocity field that, if solved with given initial conditions, are equivalent to solving the many-body time-dependent Schrödinger equation are (see Appendix I)

$$i\hbar \frac{d|\Psi(t)\rangle}{dt} = \hat{\mathbf{H}}(t)|\Psi(t)\rangle$$

$$\updownarrow$$

$$\begin{cases} D_t n(\mathbf{r},t) + n(\mathbf{r},t)\nabla \cdot \mathbf{v}(\mathbf{r},t) = 0 \\ mn(\mathbf{r},t) D_t v_j(\mathbf{r},t) + \nabla_i \mathbf{P}_{ij}(\mathbf{r},t) + n(\mathbf{r},t)\nabla_j V_{ext}(\mathbf{r},t) = 0, \quad j = x, y, z. \end{cases}$$

$$(8.15)$$

Equations 8.15 have the same form as the Madelung equations 8.11 with the notable exception that the stress tensor does not have the same analytical structure as in the single-particle case. These equations are also *exact*, namely no approximation has been made to derive them.[6]

Once again, they are similar to the equations of motion of a classical liquid of density $n(\mathbf{r},t)$ and velocity $\mathbf{v}(\mathbf{r},t)$ subject to an external field, where all interactions among particles of the liquid are described by a stress tensor $\mathbf{P}_{ij}$ (Landau and Lifshitz, 1959b).

As in the single-particle case, given appropriate boundary conditions, if we know the density and velocity of the system, then according to the theorems of dynamical density-functional theories, we can determine the external

---

[6] Note also that while we always discuss about electrons, this exact hydrodynamic equivalence is also valid for bosons (Fröhlich, 1967).

(scalar or vector) potential that generates these quantities, and hence calculate the state of the system via the Schrödinger equation. This guarantees that solving 8.15 is equivalent to solving the many-body time-dependent Schrödinger equation.

From a practical point of view, however, in order to solve 8.15 one needs to know the equation of motion for the two-particle stress tensor $\mathbf{P}_{ij}$ which turns out to be dependent on the three-particle stress tensor, i.e., the one describing interactions among three particles (see Appendix I). The equation of motion of the latter depends on the four-particle stress tensor, and so forth, thus generating a BBGKY hierarchy of infinite equations (Appendix B).

It is true that, thanks to the theorems of dynamical density-functional theories, the stress tensor $\mathbf{P}_{ij}$ can be written, in principle, as a functional of the density or velocity field (Appendices E and F). In practice, we do not know such a functional.

It is then a question of what approximations we need to make to solve the hydrodynamic equations for the single-particle variables $n(\mathbf{r},t)$ and $\mathbf{v}(\mathbf{r},t)$. I will show in the next section that transport in nanoscale systems allows us to approximate these equations to the equivalent Navier−Stokes equations of hydrodynamics (D'Agosta and Di Ventra, 2006).

In addition, the form of the obtained stress tensor describing interactions among particles is similar to the one derived from linear-response theory of the electron liquid with an effective viscosity (Vignale et al., 1997; Ullrich and Vignale, 2002). This allows us to use, as a first approximation, the calculated viscosity of the electron liquid in the estimates of many phenomena I will describe in this chapter.

8.2.1 Quantum Navier−Stokes equations

By using what we have learned from the micro-canonical picture of transport of the previous chapter, I now show that in nanoscale systems the set of equations 8.15 can be closed, so that we can derive equations similar to the Navier−Stokes ones that describe the hydrodynamics of classical liquids.[7]

In order to do this we need to know the dependence of the stress tensor $\mathbf{P}_{i,j}$ on collisions among particles. An approximate form of this dependence can be derived from the quantum Boltzmann equation 4.155 I derived in Sec. 4.6. From the definition of non-equilibrium distribution function we

---

[7] An alternative derivation based on the assumption of local equilibrium is reported in Fröhlich (1967).

8.2 Hydrodynamic form of the Schrödinger equation

know that (see Eqs 2.136 and 2.137)[8]

$$n(\mathbf{r}, t) = \sum_{\mathbf{p}} f(\mathbf{r}, \mathbf{p}, t) \qquad (8.16)$$

and

$$\mathbf{v}(\mathbf{r}, t) = \frac{1}{m\,n(\mathbf{r}, t)} \sum_{\mathbf{p}} \mathbf{p} f(\mathbf{r}, \mathbf{p}, t). \qquad (8.17)$$

In the language of stochastic processes I have discussed in Sec. 5.1 we thus see that the density and velocity field are the first two *moments* of the distribution – the density is the zero moment, the velocity is the first. In general, the N-th moment of the distribution function is defined as

$$\mathbf{P}^{(N)}_{i_1...i_N}(\mathbf{r}, t) = \frac{1}{m^{N-1}} \sum_{\mathbf{p}} p_{i_1} \cdots p_{i_N} f(\mathbf{r}, \mathbf{p}, t), \qquad (8.18)$$

and provides information on all possible scattering processes among particles of the fluid. Our goal is to find an equation of motion for the second moment $\mathbf{P}^{(2)}_{i_1 i_2} \equiv \mathbf{P}_{ij}$, the stress tensor entering Eq. 8.15.

Let us then start from Eq. 4.155 which in this case reads

$$\frac{\partial f(\mathbf{r}, \mathbf{p}, t)}{\partial t} + \frac{\mathbf{p}}{m} \nabla_\mathbf{r} f(\mathbf{r}, \mathbf{p}, t) - \nabla_\mathbf{r} V_{ext}(\mathbf{r}, t) \cdot \nabla_\mathbf{p} f(\mathbf{r}, \mathbf{p}, t) = \mathbf{I}[f], \qquad (8.19)$$

where the collision integral

$$\mathbf{I}[f] = \mathbf{I}_{el}[f] + \mathbf{I}_{in}[f] \qquad (8.20)$$

can be separated into an elastic contribution due to the scattering of the electrons with the nanojunction, and any other inelastic contribution.

In Sec. 7.6 I have shown that, in true nanoscale systems, the elastic contribution dominates the collision processes at a total frequency approximately given by ν_c, Eq. 7.37. This implies that the system reaches a local equilibrium fast due to the electron "squeezing" in the nanojunction. Since in Eq. 8.20 the non-equilibrium distribution is the same for both the elastic and inelastic processes, if the first ones force the system into a local equilibrium, the second type of processes will follow with the same distribution.

In Appendix J I guide the reader through the derivation of the equation of motion for $\mathbf{P}_{ij}$. All calculations are easier to do in a reference frame *co-moving* with the liquid, one in which the current density is zero. Clearly, all results are independent of the chosen reference frame.

---

[8] Strictly speaking, $\mathbf{r}$ and t are the Wigner coordinates 4.140.

Here I just give the final result (summation over repeated indexes is implied)

$$D_t \mathbf{P}_{ij} + \mathbf{P}_{ij} \nabla \cdot \mathbf{v} + \mathbf{P}_{ik} \nabla_k v_j + \mathbf{P}_{kj} \nabla_k v_i + \nabla_k \mathbf{P}^{(3)}_{ijk} = \frac{1}{m} \sum_{\mathbf{p}} p_i p_j \mathbf{I}[f]. \quad (8.21)$$

As anticipated, we see from Eq. 8.21 that the equation of motion for $\mathbf{P}_{ij}$ involves the third moment $P^{(3)}_{ijk}$. Using a similar approach, we could derive the equation of motion for $P^{(3)}_{ijk}$ that would include an equation of motion for the fourth moment $P^{(4)}_{ijkl}$, and so forth, so that we again obtain an infinite hierarchy of equations.

However, we note that $P^{(3)}_{ijk}$ enters in Eq. 8.21 only through its spatial derivative. If the latter is small compared to the other terms, then the hierarchy can be truncated.[9] The moment $P^{(3)}_{ijk}$ varies mostly within an effective length L where the inhomogeneities of the liquid involving interactions among three particles occur. The derivative of $P^{(3)}_{ijk}$ is thus proportional to $1/L$. The length L may depend on the current density and, in general, is much larger than the atomic length of the nanojunction itself.[10]

The velocity field may depend on the frequency ω of the external field (in the d.c. limit $\omega \to 0$) or on the frequency ν_c at which the system reaches equilibrium (Eq. 7.36), so that from the continuity equation in Eq. 8.15 we see that the spatial derivative of the velocity field $\mathbf{v}$ is of the order of ω or ν_c, whichever is larger. The first four terms in Eq. 8.21 are therefore of the order of the larger between ω and ν_c.

Finally, the collision integral is proportional to the frequency ν_c. This is easy to see if we choose the relaxation-time approximation, Eq. 4.157, for the collision integral as discussed in Sec. 4.6 (with $1/\tau_{tot} \approx \nu_c$)[11]

$$\mathbf{I}[f] = -\nu_c \left[f(\mathbf{r}, \mathbf{p}, t) - f^{eq}(\mathbf{r}, \mathbf{p}, t) \right], \quad (8.22)$$

where $f^{eq}(\mathbf{r}, \mathbf{p}, t)$ is the local equilibrium Fermi distribution function 2.135.

Putting all this together we see that the spatial derivative of $P^{(3)}_{ijk}$ can be neglected with respect to the other terms in Eq. 8.21 whenever

$$g = \frac{\bar{v}}{L \max\{\omega, \nu_c\}} \ll 1, \quad (8.23)$$

---

[9] Clearly, if $P^{(3)}_{ijk}$ and all higher-order moments are small, the equations of motion can also be closed. However, this is a much more restrictive condition than the smallness of the spatial derivative of $P^{(3)}_{ijk}$.

[10] For quasi-ballistic structures, this length can be identified as the average length of electron-electron collisions. In typical systems it is of the order of 1000 Å or larger.

[11] The conclusions, however, are independent of the chosen form of collision integral (see Footnote [13] in this chapter).

8.2 Hydrodynamic form of the Schrödinger equation

with $\bar{v}$ the average electron velocity, and the symbol $\max\{\omega, \nu_c\}$ indicates the maximum of the two quantities. In what follows we will be interested only in the d.c. limit, $\omega \to 0$.

The parameter g is indeed small for transport in nanostructures. In Sec. 7.6 we have seen that for typical nanoscale systems $\nu_c \sim 10^{15}$ Hz. Taking $\bar{v} \sim 10^8$ cm/s (typical Fermi velocity), then condition 8.23 is satisfied whenever $L \gg 1$ nm. This is true for many nanostructures.

To first approximation we can thus neglect the term $\nabla_k \mathbf{P}^{(3)}_{ijk}$ in Eq. 8.21, which allows us to derive the form of the stress tensor $\mathbf{P}_{ij}$. In order to do this let us write the stress tensor – as we do in hydrodynamics, (see Landau and Lifshitz, 1959b) – quite generally as

$$\mathbf{P}_{ij}(\mathbf{r}, t) = P(\mathbf{r}, t)\delta_{ij} - \pi_{ij}(\mathbf{r}, t), \qquad (8.24)$$

where the diagonal part gives the *pressure* of the liquid, and π_{ij} is a traceless tensor ($\text{Tr}\{\pi\} = 0$) that describes the *shear* effect on the liquid, i.e., the internal *friction* experienced by electrons when moving one against the other at different velocities.

In d dimensions ($d > 1$), the pressure is related to the local equilibrium Fermi distribution as

$$P(\mathbf{r}, t) = \frac{1}{dm} \sum_{\mathbf{p}} |\mathbf{p}|^2 f^{eq}(\mathbf{r}, \mathbf{p}, t), \qquad (8.25)$$

which can be alternatively used as the definition of local equilibrium.

Replacing the definition 8.24 into Eq. 8.21 and neglecting the term $\nabla_k \mathbf{P}^{(3)}_{ijk}$, we then find that Eq. 8.21 is equivalent to the coupled equations (see Appendix J)

$$D_t P + \left(1 + \frac{2}{d}\right) P \nabla_k v_k - \frac{2}{d} \pi_{kj} \nabla_k v_j = 0, \qquad (8.26)$$

$$- D_t \pi_{ij} - \left[\nabla_k v_i \pi_{jk} + \nabla_k v_j \pi_{ik} + \nabla_k v_k \pi_{ij} - \frac{2}{d} \pi_{kl} \nabla_k v_l \delta_{ij}\right]$$

$$+ P\left[\nabla_k v_i \delta_{kj} + \nabla_k v_i \delta_{ik} - \frac{2}{d} \delta_{ij} \nabla_k v_k\right] = \frac{1}{m} \sum_{\mathbf{p}} p_i p_j \mathbf{I}[f]. \qquad (8.27)$$

From Eqs. 8.22, 8.24 and 8.25 we have

$$\frac{1}{m}\sum_{\mathbf{p}} p_i p_j \mathbf{I}[f] = -\frac{\nu_c}{m}\sum_{\mathbf{p}} p_i p_j \left[f(\mathbf{r},\mathbf{p},t) - f^{eq}(\mathbf{r},\mathbf{p},t)\right]$$

$$= -\nu_c \left\{\sum_{\mathbf{p}} p_i p_j f(\mathbf{r},\mathbf{p},t) - \frac{1}{m}\sum_{\mathbf{p}} p_i p_j f^{eq}(\mathbf{r},\mathbf{p},t)\right\}$$

$$= -\nu_c \left\{\mathbf{P}_{ij}(\mathbf{r},t) - \frac{1}{dm}\delta_{ij}\sum_{\mathbf{p}} \mathbf{p}^2 f^{eq}(\mathbf{r},\mathbf{p},t)\right\}$$

$$= -\nu_c \left\{\mathbf{P}_{ij}(\mathbf{r},t) - P(\mathbf{r},t)\delta_{ij}\right\} = \nu_c \pi_{ij}(\mathbf{r},t). \tag{8.28}$$

Following the same arguments that led us to neglect the contribution of the third moment $P^{(3)}_{ijk}$, in the zero-frequency limit – or generally for $\omega/\nu_c \ll 1$ – we can neglect all terms on the left-hand side of Eq. 8.27[12] except the one containing the pressure.

Using Eq. 8.28, the first-order solution of Eq. 8.27 is thus

$$\pi_{ij}(\mathbf{r},t) = \eta(\mathbf{r},t)\left(\nabla_j v_i(\mathbf{r},t) + \nabla_i v_j(\mathbf{r},t) - \frac{2}{d}\delta_{i,j}\nabla_k v_k(\mathbf{r},t)\right), \tag{8.29}$$

where $\eta(\mathbf{r},t) = P(\mathbf{r},t)/\nu_c$ is a real, positive coefficient which can be identified with the *shear viscosity* of the liquid. It is associated with the internal resistance to flow experienced by the liquid under *shear* stress.[13] From a microscopic point of view it is related to the *transfer of momentum* from a layer of the liquid with a given velocity to an adjacent layer with a different velocity.

Without proof, I just say that if we included higher-order terms in Eq. 8.27 one would then get corrections to Eq. 8.29 related to the compression of the liquid beyond the simple pressure term. The new stress tensor – which is no longer traceless – would then have the form

$$\sigma_{ij}(\mathbf{r},t) = \eta(\mathbf{r},t)\left(\nabla_j v_i(\mathbf{r},t) + \nabla_i v_j(\mathbf{r},t) - \frac{2}{d}\delta_{i,j}\nabla_k v_k(\mathbf{r},t)\right)$$
$$+ \zeta(\mathbf{r},t)\nabla \cdot v(\mathbf{r},t)\delta_{ij}, \tag{8.30}$$

with the positive constant ζ called the *bulk viscosity* of the liquid (see also Appendix K). In liquids (and this is also true for the electron liquid, see next section) the bulk viscosity is generally much smaller than the shear viscosity, so that we can safely neglect this term in the following discussions.

---

[12] As well as the term $(2/d)\pi_{kj}\nabla_k v_j$ in Eq. 8.26.
[13] The use of the relaxation-time approximation 8.22 to derive Eq. 8.29 is not necessary since in collisional dynamics the second moment $\frac{1}{m}\sum_{\mathbf{p}} p_i p_j \mathbf{I}[f]$ is always a linear function of the stress tensor π_{ij}.

8.2 Hydrodynamic form of the Schrödinger equation

Finally, using the form 8.29 of the traceless part of the stress tensor in Eq. 8.15 we obtain the generalized Navier–Stokes equations for the electron liquid in nanoscale systems $(j = x, y, z)$[14]

$$D_t n(\mathbf{r}, t) + n(\mathbf{r}, t) \nabla \cdot \mathbf{v}(\mathbf{r}, t) = 0, \qquad (8.31)$$

$$mn(\mathbf{r}, t) D_t v_j(\mathbf{r}, t) + \nabla_j P(\mathbf{r}, t) - \nabla_i \pi_{ij}(\mathbf{r}, t)$$
$$+ n(\mathbf{r}, t) \nabla_j V_{ext}(\mathbf{r}, t) = 0, \qquad (8.32)$$

which need to be solved together with the continuity equation 8.31. Equations 8.32 are formally equivalent to the classical Navier–Stokes equations (Landau and Lifshitz, 1959b) with quantum-mechanical parameters entering via the shear viscosity of the electron liquid, its pressure, velocity and density.

Important. It goes without saying that there may be nanoscale systems in which condition 8.23 is not satisfied, e.g., those in which inelastic relaxation mechanisms compete in importance with elastic effects. In addition, in many mesoscopic structures g is not necessarily a small parameter. In those instances, one cannot neglect the third moment of the distribution and the equations of motion for the density and velocity field may not have the simple form 8.32. Nonetheless, one could still derive those equations for high enough frequencies and/or low enough densities.

In the atomic systems considered here it is convenient to include in $V_{ext}(\mathbf{r}, t)$ the electron-ion potential energy Eq. 6.4 (which is time-independent) plus the self-consistent time-dependent Hartree potential 3.11[15] so that

$$V_{ext}(\mathbf{r}, t) = \sum_i V^{ps}(\mathbf{r}; \mathbf{R_i}) + V_H(\mathbf{r}, t). \qquad (8.33)$$

This is what we will consider in the following. This way, we just need to know the self-consistent *form* of $V_{ext}(\mathbf{r}, t)$ which enters as a "geometric" confinement for the liquid in the above Navier–Stokes equations. If we know this form we can then use those equations to predict the electron flow *with* electron-electron interactions included both in a pressure term and in a shear viscosity. In particular, as in the classical case, the above equations predict that the electron liquid at nanojunctions may undergo a transition from *laminar* to *turbulent* flow.

---

[14] The Navier–Stokes equations would be the same if one used the stress tensor σ_{ij}, Eq. 8.30. It is enough to replace π_{ij} with σ_{ij} in Eq. 8.32. Also, in the absence of viscous terms Eqs. 8.32 are called *Euler equations*.

[15] If the Hartree term is added into $V_{ext}(\mathbf{r}, t)$ then it must be subtracted from the stress tensor $\mathbf{W}_{ij}(\mathbf{r}, t)$ (Eq. I.13).

Before discussing this phenomenon, however, let us derive some known results from the above equations and estimate the contribution of the viscosity to the resistance of a nanojunction.

8.3 Conductance quantization from hydrodynamics

Let us consider Eqs. 8.32 and assume the liquid is incompressible, namely the density and viscosity are constant everywhere in space, even in the presence of an external time-dependent field: $n(\mathbf{r},t) = n$, $\eta = \eta(n)$.[16]

For an incompressible fluid then the continuity and Navier–Stokes equations 8.32 reduce to

$$D_t n = 0, \quad \nabla \cdot v(\mathbf{r},t) = 0, \quad (8.34)$$

$$m n D_t v_i(\mathbf{r},t) + \nabla_i P(\mathbf{r},t) - \eta \nabla^2 v_i(\mathbf{r},t)$$
$$+ n \nabla_i V_{ext}(\mathbf{r},t) = 0. \quad (8.35)$$

Given the external potential V_{ext}, one then needs to solve these equations to find the density and velocity profiles. This generally requires a numerical procedure, but it can be done analytically in the following case.[17]

The single-component liquid

Let us consider an *ideal* ($\eta = 0$) infinite 1D liquid along the x axis that we may conceptually assume adiabatically connected to two infinite electrodes, with $\mu_{L(R)}$ the electrochemical potential deep into the left (right) electrode, and $\mu_L - \mu_R = eV$, with V the bias, which we take to be very small so that we can work in the linear approximation. The electrodes here play the role of the reservoirs of the Landauer approach (Chapter 3).

From Eq. 8.35 we can derive Bernoulli's equation, which simply states the conservation of energy (see Appendix L for its derivation)

$$\frac{v^2}{2} + h(n) + \frac{V_{ext}}{m} = \text{constant}, \quad (8.36)$$

where v is the fluid velocity and $h(n) = P/n = E_F/m$, the ratio between the pressure and the density, or equivalently, the Fermi energy and the mass, is the *enthalpy*.

From a hydrodynamic point of view the fluid can be seen as composed of two types of carriers: those moving from left to right at velocity v_L,

---

[16] Incompressibility is a reasonable approximation for metallic nanojunctions at low bias. However, it neglects the formation of local resistivity dipoles at the junction.

[17] This discussion follows closely the one in D'Agosta and Di Ventra (2006).

and those from right to left with velocity v_R (as measured by a laboratory frame, and I choose the positive axis for velocities pointing from left to right). These two velocities are the *local Fermi velocities* on the left and right electrodes where the fluid "originates" and gets "absorbed". If we take the external bias small enough that the velocities of the particles in the fluid are distributed uniformly between the above two extreme values, v_L and v_R, the fluid velocity is nothing other than

$$v = \frac{1}{2}(|v_L| - |v_R|), \tag{8.37}$$

which is the velocity of a reference frame co-moving with the fluid.

Within the same approximation of low bias, I can define a global Fermi velocity (which is an "incoherent" component of the fluid velocity) as the average between the left and right local Fermi velocities

$$v_F = \frac{1}{2}(|v_L| + |v_R|). \tag{8.38}$$

This is the velocity with respect to which the density of the fluid is defined. In this 1D, stationary case $V_{ext} = 0$, since the liquid is confined to move in just one direction with no external "barrier", so that Bernoulli's equation 8.36 is

$$\frac{v^2}{2} + \frac{v_F^2}{2} = \frac{1}{2}\left(\frac{v_L^2}{2} + \frac{v_R^2}{2}\right) = \frac{1}{2}\left(\frac{\mu_L}{m} + \frac{\mu_R}{m}\right). \tag{8.39}$$

Deep into the left electrode the liquid velocity merges with the velocity of carriers on the left so that from the above equation

$$\frac{v_L^2}{2} + \frac{v_F^2}{2} = \frac{\mu_L}{m}, \qquad x \to -\infty. \tag{8.40}$$

Similarly for the liquid on the infinite far right

$$\frac{v_R^2}{2} + \frac{v_F^2}{2} = \frac{\mu_R}{m}, \qquad x \to +\infty. \tag{8.41}$$

By taking the difference between Eqs. 8.40 and 8.41 we get the relation between the fluid velocity v and the bias

$$2mv\,v_F = eV. \tag{8.42}$$

The fluid current is

$$I = env, \tag{8.43}$$

where n is the density. For a 1D non-interacting system the density is (from

Eq. 2.84)

$$n = \frac{2 k_F}{\pi} = \frac{2 m v_F}{\hbar \pi}. \qquad (8.44)$$

Replacing this density into Eq. 8.43 and using the relation 8.42 we finally get

$$I = e\, n v = \frac{2 e^2}{h} V, \qquad (8.45)$$

which gives the linear-response quantized conductance

$$G_0 = \frac{dI}{dV} = \frac{2 e^2}{h}, \qquad (8.46)$$

I have derived in Sec. 2.3.4 using the Kubo formalism, and in Sec. 3.3.4 using the Landauer approach.

The above result is from yet another derivation and suggests an alternative interpretation of the quantized conductance. The latter may be interpreted as the one associated with an ideal (i.e., $\eta = 0$) single-component liquid of charge e confined to flow in 1D. By "single-component" I mean a liquid with uniform velocity.

However, the basic assumption in all three derivations is that the electrons in the 1D channel get in and out at the infinite boundaries of the system without "reflecting" back, and that they acquire either a local distribution, or, in the present case, a local fluid velocity appropriate to the far boundaries. The latter may be interpreted as "basins" for the liquid. As already discussed in Sec. 2.3.4 and in Sec. 3.7.2, this conductance, and corresponding resistance, may be thus interpreted as the one an ideal electron liquid experiences when it adiabatically flows into an infinitely large basin filled with the same liquid.

When this happens, we lose information on the electron motion inside the basins (reservoirs) due to the dilution of degrees of freedom (Sec. 3.8.2). Therefore, we may equivalently interpret the quantized conductance as the one due to the *loss of information* at the boundary between the system we are interested in, and the basins (reservoirs) where we impose – or better yet, *measure* – a velocity field (or electrochemical potential) difference.

If we assume that only a fraction T of electrons is transmitted due to the presence of a barrier in the liquid, we can argue that, in the linear response, the current is an equal fraction of the current in the absence of the barrier, i.e., $I = envT$. The conductance is thus $G = T\, 2 e^2/h$ in accordance with the two-terminal result Eq. 3.3.4.

The multi-component liquid

Let us now assume that the liquid is formed by non-interacting particles with different velocities. We can then group all particles with a given velocity into a *component* of the liquid. This is equivalent to the separation of the scattering problem we have discussed in Sec. 3.3.3 into separate channels, each with a given velocity and density of states. The current in that approach was simply the sum of contributions from all channels.

In this hydrodynamic formalism we can do the same, where by channel we mean a single-component fluid where all particles (with a given density) share the same velocity. In the linear response, according to the above derivation, each fluid component contributes a conductance

$$G_i = \frac{2\,e^2}{h} T_i, \qquad (8.47)$$

with T_i the fraction of electrons that is transmitted due to the presence of a given barrier along the fluid flow.

Since, by assumption, all fluid components are *independent* of each other, namely they flow as if the other components do not exist (no friction between the different fluid layers), the total conductance is the sum over all components

$$G = \frac{2e^2}{h} \sum_i T_i, \qquad (8.48)$$

and the probabilities T_i can be interpreted as the eigenvalues of the matrix $\tau^\dagger \tau$ (see Eq. 3.211).

The above is clearly an idealization, since the different layers of the liquid do interact with each other via Coulomb forces, which create a viscous *drag* in the liquid. As discussed in Sec. 4.4 when interactions – beyond mean field – are present one cannot define independent channels, the concept of single-particle transmission breaks down, and Eq. 8.48 is not valid.

My goal for the following sections is to understand precisely the effect of viscosity on the dynamics of the electron liquid in nanojunctions.

8.4 Viscosity from Time-Dependent Current Density-Functional Theory

Before calculating the effect of viscosity on resistance, let us make a parallel between the hydrodynamic theory I have developed in the preceding sections and time-dependent current density-functional theory (TDCDFT). In Appendix F I introduce the basic theorem of this theory, which, given an initial condition, allows us to map the dynamics of the many-body problem into the dynamics of effective single particles (Kohn–Sham system).

Using this formulation one then solves the following set of effective single-particle equations (see Eq. F.1)

$$\left[i\hbar\frac{\partial}{\partial t} - \frac{1}{2m}\left(-i\hbar\nabla - \frac{e}{c}\mathbf{A}_{ext}(\mathbf{r},t) - \frac{e}{c}\mathbf{A}_{xc}(\mathbf{r},t)\right)^2 - V_{ext}(\mathbf{r},t)\right]\phi_k^{(KS)}(\mathbf{r},t) = 0, \tag{8.49}$$

with $V_{ext}(\mathbf{r},t)$ given in Eq. 8.33, and I have also included an arbitrary external vector potential $\mathbf{A}_{ext}(\mathbf{r},t)$.

The exchange-correlation vector potential $\mathbf{A}_{xc}(\mathbf{r},t)$ contains information on both the Pauli exclusion principle and on all other correlations effects. Its exact form is unknown. However, an approximate form has been evaluated within linear-response theory for the inhomogeneous electron liquid subject to a time-dependent perturbation (Vignale *et al.* 1997; Ullrich and Vignale, 2002). It has been shown that its time derivative gives rise to an *exchange-correlation field* $\mathbf{E}_{xc}$, which can be written in terms of the equilibrium density and velocity field in the following "hydrodynamic" form (see also Appendix F)

$$-e\,E_{xc,i}(\mathbf{r},t) = \frac{e}{c}\frac{\partial A_{xc,i}(\mathbf{r},t)}{\partial t} = -\nabla_i V_{xc}^{ALDA}(\mathbf{r},t) + \frac{1}{n(\mathbf{r})}\sum_j \frac{\partial \sigma_{xc,ij}(\mathbf{r},t)}{\partial r_j}, \tag{8.50}$$

where V_{xc}^{ALDA} is the exchange-correlation scalar potential in the adiabatic local density approximation (ALDA) (see Appendix E) and the exchange-correlation stress tensor is, in d dimensions ($d=2,3$)[18]

$$\sigma_{xc,ij}(\mathbf{r},t) = \int_{t_0}^t \left\{\tilde{\eta}(n(\mathbf{r}),t,t')\left[\nabla_j v_i(\mathbf{r},t') + \nabla_i v_j(\mathbf{r},t') - \frac{2}{d}\delta_{ij}\nabla\cdot\mathbf{v}(\mathbf{r},t')\right]\right.$$
$$\left. + \tilde{\zeta}(n(\mathbf{r}),t,t')\nabla\cdot\mathbf{v}(\mathbf{r},t')\delta_{ij}\right\}dt', \tag{8.51}$$

where t_0 is an arbitrary initial time we may take to $-\infty$.

The complex viscoelastic coefficients $\tilde{\eta}$ and $\tilde{\zeta}$ are related to the exchange-correlation kernel of the homogeneous electron liquid evaluated at the local electron density (see also discussion in Sec. 2.3.2). The coefficient $\tilde{\eta}(n,t,t')$ contains information on the *shear* properties of the electron liquid (like its shear viscosity and modulus), while the term $\tilde{\zeta}(n,t,t')$ on the *bulk* properties (like its bulk viscosity and modulus), which together with the pressure of

---

[18] Note that both the stress tensor 8.51 and the exchange-correlation force 8.50 are gauge invariant since they correspond to physical quantities. The ALDA force, $-\nabla_i V_{xc}^{ALDA}(\mathbf{r},t)$, may be alternatively written in terms of a vector potential (see Giuliani and Vignale 2005, p. 392-393).

8.4 Viscosity from Time-Dependent Current Density-Functional Theory

the liquid describes the response of the liquid to compression (cf. Eq. 8.30 and Eqs. K.1 and K.2 in Appendix K).

It is important to note that the field $\mathbf{E}_{xc}(\mathbf{r},t)$ is *not conservative*: it describes internal dissipation due to electron-electron interactions. In addition, it *does not vanish* in the d.c. limit of zero frequency: due to the dynamical behavior of the electrons, interactions that lead to viscosity cannot be eliminated at any frequency. This confirms once more a point I have made in Sec. 3.11: ground-state density-functional theory (DFT) *cannot capture* the full electron dynamics, even if we knew the exact ground-state functional. Therefore, the current one obtains using the scattering approach of Chapter 3 in combination with ground-state DFT *cannot* be equal to the true many-body current, even in the linear response.

Loss of memory approximation

Both viscoelastic coefficients $\tilde{\eta}$ and $\tilde{\zeta}$ may be non-local in time so that the stress tensor 8.51 includes *memory* effects, i.e., it depends on the full time evolution of the system, including initial conditions.

If we assume that memory effects can be neglected, namely the shear and bulk viscosities at a given time depend only on physical processes at that instant of time

$$\left. \begin{array}{l} \tilde{\eta}(n,t,t') \sim \tilde{\eta}(n,t')\delta(t-t') \\ \tilde{\zeta}(n,t,t') \sim \tilde{\zeta}(n,t')\delta(t-t') \end{array} \right\} \text{ memory-less approximation,} \quad (8.52)$$

then from Eq. 8.51 the stress tensor is

$$\sigma_{xc,ij}(\mathbf{r},t) = \tilde{\eta}(n,t)\left(\nabla_j v_i(\mathbf{r},t) + \nabla_i v_j(\mathbf{r},t) - \frac{2}{d}\delta_{ij}\nabla\cdot v(\mathbf{r},t)\right)$$
$$+ \tilde{\zeta}(n,t)\nabla\cdot\mathbf{v}(\mathbf{r},t')\delta_{ij}. \quad (8.53)$$

This is of the same form as the stress tensor 8.30 I have derived in the previous section without invoking linear-response theory, simply on the basis of a fast relaxation mechanism at the nanojunction. Therefore, we can interpret the quantity $-\nabla_i V_{xc}^{ALDA}(\mathbf{r},t)$ in Eq. 8.50 as the hydrodynamic force associated with the *exchange-correlation pressure*.

As anticipated, the bulk viscosity of the electron liquid is much smaller than the shear viscosity and can be thus neglected.[19] With this approximation, the stress tensor 8.53 is of the same form as the one entering the quantum Navier-Stokes equations 8.29.[20]

---

[19] More generally, one can show that the bulk viscosity $\zeta(n,t)$ vanishes in the limit of zero frequency, which is the one of interest here (Giuliani and Vignale, 2005).

[20] We will see in Sec. 8.4.2 that only its real part contributes to the d.c. conductance.

We have thus come full circle. We have shown that in nanostructures the stress tensor can be written in a form, Eq. 8.30, similar to the one entering the classical Navier–Stokes equations. But we also know from linear-response theory that the exchange-correlation effects can be represented in a similar form, Eq. 8.53. This form is likely to remain valid even beyond the linear response provided both the density and the velocity field are slowly varying functions of position (see Appendix F, and Giuliani and Vignale 2005).

8.4.1 Functional approximation, loss of information, and dissipative dynamics

As discussed in Sec. 8.2, the basic theorem of time-dependent current density-functional theory establishes that, for given initial conditions, the *exact* stress tensor $\mathbf{P}_{ij}$, Eq. 8.24, is a universal functional of the current density (see Appendinx F). This implies that the hierarchy of equations for the moments of the distribution function, Eq. 8.21, can be formally closed to all orders in the electron-electron interaction, and one could recover full information on the system dynamics (see also discussion in Sec. 2.8.3).

Consequently, the Navier–Stokes stress tensor in Eq. 8.29 can be seen as the first-order, non-trivial contribution to the *exact* stress tensor of the electron liquid (D'Agosta and Di Ventra, 2006).

If we knew the exact stress tensor, and the system were isolated – no external force present – the dynamics described by Eqs. 8.15 would be conservative, in the sense that the total energy of the system would remain constant in time. As I discussed in Sec. 2.8.6, a partially known Hamiltonian introduces irreversibility in the system dynamics. Therefore, by cutting off higher-order terms in the stress tensor we lose information on some correlation degrees of freedom, and the dynamics becomes irreversible, even though our original equations of motion are time-reversal invariant.

Here, irreversibility is also accompanied by dissipative dynamics: the viscosity of the liquid generates *energy dissipation* into the degrees of freedom we have left out by approximating the stress tensor.

From a practical point of view, if we want to obtain analytical results regarding the effects of viscosity on the transport properties of nanostructures we can use the Navier–Stokes equations 8.32 with the viscosity coefficient estimated from linear-response theory (see Appendix F). This is what I will do in the next section.

Instead, if we are interested in full-fledged first-principles calculations of transport in nanostructures, we can use, e.g., the micro-canonical picture described in the previous chapter, and solve the Kohn–Sham equations 8.49 with given initial conditions. (I will give later an example of such a calculation.) The theorems I proved in Chapter 7 guarantee that the more physical effects we include in the exchange-correlation functional – such as the viscosity – the closer we are to the true total many-body current at any given time.

Within the above approach one can thus study decoherence (Wijewardane and Ullrich, 2005; Kurzweil and Baer, 2006), energy loss (D'Agosta and Vignale, 2006) and other related electron-electron inelastic effects via the solution of effective single-particle equations. In the next section I will discuss the effect of viscosity on resistance in nanojunctions.

8.4.2 Effect of viscosity on resistance

Viscous resistance is by definition an inelastic electron-electron effect. Therefore, its estimate is far from trivial. However, from the preceding section we have seen that, in the context of time-dependent current density-functional theory, the viscous dynamics of the electron liquid can be formulated in the form of a non-conservative exchange-correlation "field" (Eq. 8.50), which can be written in terms of the density and current density.[21]

This field allows us to calculate the rate of energy dissipation associated with it (Sai et al., 2007a). Before proceeding, I want to stress that all considerations I make in this section pertain to a liquid in *laminar flow*. This is likely to be a good approximation at low currents, but may not hold for all currents and all nanostructures due to the possible onset of *turbulence* (see Sec. 8.5). I will briefly mention at the end of this discussion the consequences of turbulent flow on resistance.

Let us then work in the linear response, adopt the memory-less approximation 8.52 for the visco-elastic coefficients, and consider the d.c. transport limit. Let us also assume the system is in an ideal steady state with current I. We are interested only in the viscous (dynamical) contribution of the field 8.50. Let us call this field

$$E_i^{dyn}(\mathbf{r}) = -\frac{1}{e\,n(\mathbf{r})} \lim_{\omega \to 0} \nabla_j \sigma_{xc,ij}(\mathbf{r}, \omega), \qquad (8.54)$$

with (in three dimensions)

$$\lim_{\omega \to 0} \sigma_{xc,ij}(\mathbf{r}, \omega) = \eta(n) \left(\nabla_j v_i + \nabla_i v_j - \frac{2}{3} \delta_{ij} \nabla_k v_k \right), \qquad (8.55)$$

---

[21] Similar results have been found in Koentopp et al. (2006).

where $\eta(n) = \lim_{\omega \to 0} \operatorname{Re} \tilde{\eta}(n, \omega)$.

If **j** is the current density in the system, the rate of energy dissipation (the power) due to viscous effects is then (Sai *et al.*, 2007a)

$$P^{dyn} = \frac{dE}{dt} = e \sum_i \int d\mathbf{r}\, j_i(\mathbf{r})\, E_i^{dyn}(\mathbf{r})$$

$$= -\sum_{ij} \int d\mathbf{r}\, v_i\, \nabla_j \left[\eta(n) \left(\nabla_j v_i + \nabla_i v_j - \frac{2}{3}\delta_{ij} \nabla \cdot \mathbf{v} \right) \right],$$
(8.56)

where I have used the definition of velocity field 8.13.

Let us now assume that the viscosity varies weakly with position, $\eta(n) \simeq \eta$, so that we can take it out of the integral. This is not always a good approximation, especially at nanojunctions where the local density may vary substantially, e.g., molecular structures between metallic electrodes. However, it allows us to derive an analytical expression for the viscous resistance.

Integrating by parts the above expression and using the traceless property of the tensor $\sigma_{xc,ij}$, we can re-write Eq. 8.56 as (Exercise 8.4)

$$P^{dyn} = \frac{\eta}{2} \sum_{ij} \int d\mathbf{r} \left[\nabla_j v_i + \nabla_i v_j - \frac{2}{3}\delta_{ij} \nabla \cdot \mathbf{v} \right] \left[\nabla_j v_i + \nabla_i v_j - \frac{2}{3}\delta_{ij} \nabla \cdot \mathbf{v} \right],$$
(8.57)

which is similar to the corresponding quantity one obtains from classical hydrodynamics (see, e.g., Landau and Lifshitz, 1959b).

As a further simplifying approximation, let us suppose that the current density is constant, $|\mathbf{j}| = j$, and the velocity field is oriented only along the direction of current flow, that we take the x axis: $\mathbf{v} = n^{-1} j \hat{x}$, with $\hat{x}$ the unit vector in the x direction. Like the assumption of constant viscosity, this is also quite a strong approximation.

With this approximation Eq. 8.57 reduces to (Exercise 8.4)

$$P^{dyn} = \eta j^2 \int d\mathbf{r} \left[\frac{4}{3}(\nabla_x n^{-1})^2 + (\nabla_\perp n^{-1})^2 \right],$$
(8.58)

where $\perp$ represents the direction transverse to the velocity field.

Let us now write the total current $I = e j w$, where w is the cross section of the nanostructure. The resistance due to viscous effects is then

$$R^{dyn} = \frac{P^{dyn}}{I^2} = \frac{\eta}{e^2 w^2} \int d\mathbf{r} \left[\frac{4}{3}(\nabla_x n^{-1})^2 + (\nabla_\perp n^{-1})^2 \right],$$
(8.59)

which is a positive quantity and thus shows that viscous effects *increase* the resistance (Sai *et al.*, 2005). In addition, this resistance depends on the variations of the density at the nanostructure. The larger these variations, the bigger the contribution of viscous effects to the total resistance. This result can be understood quite easily by recalling that a local variation of the density affects the local velocity field and thus increases the viscous drag of the liquid.

From Eq. 8.59 we expect that this viscous resistance is small in all-metallic junctions where the density varies smoothly, while it may be a larger percentage of the total resistance in nanostructures made of a material different than that of the electrodes (e.g., molecular junctions) (Sai et al., 2005). However, despite the simple result embodied in Eq. 8.59 I need to make a few comments.

I have derived expression 8.59 within the linear response, and under the conditions – discussed in Appendix F – of validity of Eq. 8.50. I have also explicitly neglected viscosity variations, and any variation of the current density transverse to the direction of current flow.

In nanoscale structures the current density may vary rapidly in the transverse direction due to a decrease of velocity from the center of the channel to the edges of the conductor (see, e.g., Fig. 8.2). The transverse density and current density gradients can thus contribute significantly to the dissipation effects at nanojunctions.

In addition, the viscous resistance would be further enhanced if turbulence develops. As I will discuss in the next section, turbulent eddies may develop near the junction thus creating extremely large variations of the current density (D'Agosta and Di Ventra, 2006; Sai et al., 2007b). There is no analytical theory of turbulence. Therefore, a quantitative evaluation of the viscous resistance requires knowledge of the microscopic current and density distributions at the nanostructure, and the dissipation power – and associated resistance – can only be evaluated numerically for the case at hand.

8.5 Turbulent transport

Let us now turn to the type of solutions of Eqs. 8.32. We know from classical mechanics (Landau and Lifshitz, 1959b) that the solutions of Eqs. 8.32 describe many different regimes. The *non-linear regime* pertains to *turbulent flow* of the liquid, and is generally favored over the *linear* or *laminar* one.

By "favored" I mean that given a liquid confined by certain geometrical constraints, there exists a larger set of physical conditions (such as its density, current density, etc.) under which the turbulent solution exists compared to the laminar one.

The underlying physical reason for turbulence is that when the kinetic energy of the fluid in the direction of current flow (longitudinal kinetic energy) is much larger than its transverse component, the flow becomes unstable with respect to small perturbations. In this case, one would then observe a local velocity field that varies in space and time in an irregular way, and whose pattern is sensitive to initial conditions, much like the dynamics of chaotic systems.[22] In this regime, we would therefore expect the current to fluctuate in addition to the usual fluctuations due to shot noise and thermal noise I have discussed in Chapter 5.

I stress here that turbulence need not be *fully developed*. Fully developed turbulence corresponds to a completely ("uniformly") chaotic velocity field behavior in the whole system. This is the main topic of books on the subject, for the simple reason that in this case one can employ simple arguments – based on dimensional analysis – to extract general properties of a phenomenon that has eluded a satisfactory theoretical description since its discovery.

Indeed, in many systems – and in particular in nanoscale systems – turbulence, if formed, is unlikely to be fully developed. Instead, it originates at defects or geometrical constrictions, and its influence is limited to the proximity of such defects/constrictions. Notice also that the irregular turbulent current density pattern is not the same as that we expect in proximity to defects in an ideal steady state. This pattern is static, while turbulence is a *dynamical* phenomenon which does not even allow the definition of a steady state as that for which a finite autocorrelation time exists (Sec. 5.1): the system dynamics correlates in both space and time, generally according to power laws (Landau and Lifshitz, 1959b). In addition, as I will discuss in a moment, the departure from equilibrium is truly microscopic: in nanoscale systems turbulent eddies are of nanoscale dimensions.

On the other hand, the laminar solution shows a "regular" and "smooth" flow everywhere in space, and would not be so sensitive to initial conditions.

In classical fluid mechanics, in order to identify these regimes, it is customary to define a key quantity, the *Reynolds number Re*, as a non-dimensional quantity that can be constructed out of the physical parameters of the system, such as the density, the viscosity η, etc.

Let us then follow a similar convention in the quantum case (D'Agosta and Di Ventra, 2006). In $d > 1$ dimensions I give the following three equivalent

---

[22] Chaotic currents have been theoretically predicted and experimentally demonstrated (Zhang *et al.*, 1996), e.g., in two-dimensional GaAs/AlAs superlattices (Bulashenko and Bonilla, 1995), as well as in lower-dimensional superlattices (Zwolak *et al.*, 2003).

8.5 Turbulent transport

definitions of the Reynolds number for a system of linear dimension l,

$$Re = \frac{m n v l}{\eta} = \frac{m j l}{\eta} = \frac{m}{e} \frac{I l^{2-d}}{\eta}, \qquad (8.60)$$

where, given the current density j, I have used the approximate relation $I \approx e j l^{d-1}$ for the average total electrical current.[23]

The stable and stationary flow is typically laminar for small Re, while for large Re the flow is turbulent with a *critical* value R_{cr} – which is dependent only on the geometry of the structure – separating these two regimes (Landau and Lifshitz, 1959b). At, or near, the critical Re velocity eddies form in the fluid flow which superimpose on laminar flow. These eddies are initially stationary but, with increasing current, they diffuse in the system and new ones are generated until turbulence is fully developed. This occurs at Reynolds numbers typically much larger than R_{cr}. From the definition of Reynolds number 8.60 we immediately see that, given a geometrical constriction, turbulence is favored by increasing the average current in the system and/or by decreasing the viscosity.

Adiabatic vs. non-adiabatic potentials

In order to apply these concepts to nanostructures, and determine under which conditions the transition from laminar to turbulent flow occurs, we need to know the self-consistent confining potential V_{ext} that enters Eq. 8.32. Microscopically, this potential may be a complicated function of position – and possibly time – so that analytical solutions of the Navier–Stokes equations are generally not available.

In Landau and Lifshitz (1959b), several geometries are analyzed that are amenable to analytical solutions. In D'Agosta and Di Ventra (2006), two important cases are studied for the electron liquid: an *adiabatic* constriction (inset (a) of Fig. 8.1) and a *non-adiabatic* constriction (inset (b) of Fig. 8.1). For details of the calculations in 2D I refer the reader to D'Agosta and Di Ventra (2006). Here I just mention the main results.[24]

Adiabatic constrictions. The liquid forms a laminar *Poiseuille flow* (Landau and Lifshitz, 1959b). This flow is stable against small perturbations for

---

[23] From Eqs. 8.35 we easily see that the viscosity η has dimensions $[\eta] = [M/L^{d-2}T]$. For convenience, I am also assuming in definition 8.60 that the linear dimensions of the system are similar in all directions. Otherwise one simply replaces in 8.60, l^2 with $l_x l_y$ in 2D, and l^3 with $l_x l_y l_z$ in 3D.

[24] Note that these results are generally valid irrespective of the boundary conditions on the velocity field (D'Agosta and Di Ventra, 2006).

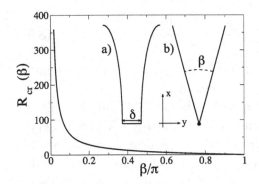

Fig. 8.1. Critical Reynolds number as a function of the angle β of the 2D geometry represented in the inset (b). Inset (a) is a schematic of an adiabatic constriction. Adapted with permission from D'Agosta and Di Ventra (2006).

almost all Re numbers.[25] The effect of viscosity on the resistance can then be estimated as described in Sec. 8.4.2.

Non-adiabatic constrictions. There is a critical Reynolds number which depends on the geometry of the constriction. In the simplest case represented in the inset (b) of Fig. 8.1 this critical number can be determined analytically (Landau and Lifshitz, 1959b; D'Agosta and Di Ventra, 2006) as a function of the opening angle of the constriction. It is plotted in Fig. 8.1 and separates the phase space in two regions: laminar flow for $Re < R_{cr}$, turbulent flow for $Re > R_{cr}$. If a turbulent regime is reached we cannot rely on a static calculation of conduction to obtain the resistance of the system, and a dynamical first-principles calculation – using, e.g., Eqs. 8.49 of time-dependent current density-functional theory – is necessary.

Since the microscopic self-consistent confining potential for electrons in many nanostructures is likely to be non-adiabatic, we expect turbulence to develop at sufficiently large currents and/or small values of viscosity.

Eddy size and energy dissipation

The dimension λ_0 of the smallest observable turbulent eddies can be estimated from dimensional analysis and is of the order of (Landau and Lifshitz, 1959b; D'Agosta and Di Ventra, 2006)

$$\lambda_0 \sim l \left(\frac{R_{cr}}{Re} \right)^{3/4} = l \left(\frac{I_{cr}}{I} \right)^{3/4}, \qquad (8.61)$$

where I_{cr} is the critical current, i.e., the current that corresponds to R_{cr}.

---

[25] Instabilities occur at a critical Reynolds number of about 10^4.

The length λ_0 is the smallest length scale over which energy is dissipated due to electron-electron friction. In fact, in the presence of turbulence, energy dissipation likely occurs via a *cascade process* (Landau and Lifshitz, 1959b), whereby the energy associated with the whole liquid – energy stored over a large length scale – is transferred to eddies of ever decreasing length scales, till the largest energy dissipation occurs at length scales of the order of λ_0.

For the non-adiabatic potential represented in inset (b) of Fig. 8.1 we easily see that λ_0 decreases rapidly with increasing angle. Let us estimate it for a 2D gold point contact. If we take as η the one calculated within linear-response theory (Eq. F.9 in Appendix F; $\eta/\hbar n \simeq 0.02$ for $r_s \simeq 3a_0$), $m \simeq 1m_e$ and a typical current of 10 μA, $Re \simeq 10$ from Eq. 8.60. In order to have turbulent flow, from Fig. 8.1 we see that we need to have an angle larger than $\pi/2$, i.e., a strongly non-adiabatic constriction.

In 3D the phase space for developing turbulence is larger than in 2D. We thus expect turbulence to develop at smaller Reynolds numbers in 3D nanostructures.

A numerical example

Let us illustrate the above results using the micro-canonical picture of transport of the previous chapter and time-dependent current density-functional theory. Figure 8.2 depicts the flow of electrons across a quasi-2D closed nanostructure, prepared as discussed in Sec. 7.3, in a range of biases between 0.02 V and 3 V. Electrons are compensated by a uniform background charge (jellium model with $r_s \simeq 3a_0$). Panels (a)–(d) correspond to the solution of the Eqs. 8.49 with the approximate exchange-correlation vector potential given in Eq. 8.50 in the memory-less assumption 8.52. Panels (e)–(h) correspond to the solution of the Navier–Stokes equations 8.35 for an incompressible liquid with the same viscosity, average density and velocity (Bushong et al., 2007a). Panel (a) has to be compared with panel (e); panel (b) with (f), and so on.

There are clearly some differences between the solutions of Eqs. 8.49 and the solutions of Eqs. 8.35. These differences are mostly due to the details of the charge configuration at the electrode-junction interface, in particular some degree of compressibility of the quantum liquid in the junction (see discussion in Sec. 7.3), which has been neglected in the solutions of Eqs. 8.35.

Nevertheless the similarity between the flow patterns within the two approaches is striking. At low biases, the flow is laminar and "smooth". At these biases the current density shows an almost perfect top–bottom sym-

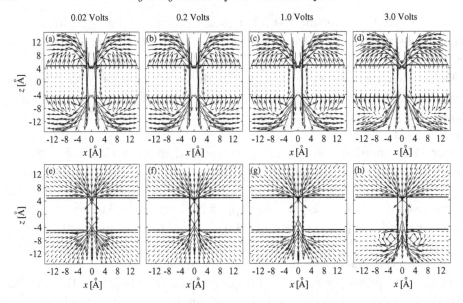

Fig. 8.2. Panels (a)–(d): Electron current density (arrows) for electrons moving from the top electrode to the bottom electrode across a nanojunction. The solid lines delimit the contour of the junction. The snapshots are taken at $t = 1.4$ fs from the initial time, for an initial bias of (a) 0.02 V, (b) 0.2 V, (c) 1.0 V and (d) 3.0 V. Panels (e)–(h): Velocity field solution of Eqs. (8.32) for a liquid with same velocity, density and viscosity as the quantum-mechanical one. Adapted with permission from Bushong et al. (2007a).

metry: the direction of the flow is symmetric with respect to the symmetry $z \to -z$.

With increasing current, however, a transition occurs: the symmetry $z \to -z$ of the current density breaks, and eddies start to appear in proximity to the junction. This is clearly evident, for instance, in Fig. 8.2(d) and (h). The outgoing current density in the bottom electrode has a more varied angular behavior, in contrast to the behavior in the top electrode, in which the electron liquid flows more uniformly toward the junction. As expected from the above analysis, these eddies are of atomic-scale dimension and are not static, but evolve in time and diffuse in the bottom electrode. Clearly, they would be different for different geometries of the junction/electrodes, and the density of the liquid.

Eventually, the left–right symmetry in the current density also breaks (see, Bushong et al., 2007a; Bushong et al., 2007b). An example of the current density in such a case is depicted on the front cover of this book for a 3D junction.[26]

---

[26] A possible experimental way to determine the transition from laminar to turbulent flow is

I finally stress that in these calculations, as in all the ones in this chapter, the viscosity is assumed to depend on the density only, and not on the current density (or bias). This is quite a strong approximation, and some bias dependence is expected to occur away from linear response. In this regime, the above numerical results can only represent a qualitative behavior of the electron flow.

8.6 Local electron heating

The internal friction of the electron liquid creates yet another phenomenon that is particularly important in nanostructures: *local electron heating*. This phenomenon is analogous to the local ionic heating I have discussed in Sec. 6.3 involving electron-ion interaction.

We can understand this effect similarly to the ionic heating. Due to the large current densities in nanojunctions compared to their bulk counterparts the electron-electron scattering rate increases locally in the junction (D'Agosta *et al.*, 2006). The *underlying Fermi sea* would thus "heat up" locally due to this increased scattering, via production of *electron-hole pairs* whose energy needs to be dissipated away from the nanostructure. By "locally" I mean both in the nanostructure and in its immediate neighborhood.[27]

Determining this effect via standard perturbation techniques seems hopeless: one would need to determine transition probabilities between many-body current-carrying states.

Instead, we can use the hydrodynamic picture we have developed here. In this picture this phenomenon can be understood as an increase of the velocity field of the liquid locally in the junction compared to its bulk value – the velocity of a liquid increases the more it is "squeezed" while its pressure decreases (cf. Bernoulli's equation 8.36).

This increased kinetic energy generates, via internal friction, heat in the junction, which eventually has to be dissipated away into the bulk electrodes. To this heating we can associate a *local electronic temperature*, $\theta_e(\mathbf{r}, t)$, in the sense discussed in Secs. 2.4 and 6.3: it is the *temperature ideally measured by a local probe at equilibrium with the electrons at the given position in space and at any given time*.[28]

> described in Bushong *et al.* (2007b), where it is suggested that the magnetic field created by the current shows an increased top-down asymmetry with increasing current, namely with increasing degree of turbulence.

[27] There are plenty of electrons at the contact between the nanostructure and the bulk electrodes.
[28] As always, while this resembles an operational definition, it is clearly not easy to realize experimentally.

This local electronic temperature would also affect the electron-phonon scattering rates, and consequently the bias dependence of the ionic temperature – which is relatively easier to measure than the temperature of the electrons. I want to determine both the bias dependence of the local electron heating, and its effect on ionic heating.

Let us start from simple considerations on the expected bias dependence of the electron heating. Assume first no electron-phonon scattering is present in the system and zero electronic temperature in the bulk electrodes, $\theta_0 = 0$. Let us also suppose no turbulence develops; otherwise this phenomenon would create fluctuations of the local temperature which are difficult to predict, and, in addition, energy dissipation may occur via the cascade process I have discussed in Sec. 8.5.

If V is the applied bias and R the resistance of the nanojunction, the power generated in the nanostructure due to exchange of energy between the current-carrying electrons and the underlying Fermi sea has to be a small fraction of the total power of the circuit V^2/R. Let us define this fraction as $P_e = \alpha(R)V^2/R$, with $\alpha(R)$ a positive constant – that may depend on R – to be determined from a microscopic theory.

At steady state this power has to balance the thermal current I_{th} (heat per unit time) carried away into the bulk electrodes by the electrons.[29] Let us assume that the thermal conductivity k follows a bulk law, i.e., it is related to the *specific heat* per unit volume measured at constant volume $C_V(\theta_e)$ through the Wiedemann–Franz law (see, e.g., Ashcroft and Mermin, 1975)

$$k = \frac{1}{3} v_F \lambda_e C_V(\theta_e), \qquad (8.62)$$

where θ_e is the electronic temperature, v_F is the Fermi velocity, and λ_e is the electron mean-free path. This law holds when the energy of each electron is, to a good approximation, conserved during scattering. I will justify this assumption in a moment (see Sec. 8.6.1).

At small temperatures the specific heat of an electron liquid is linearly dependent on temperature, $C_V(\theta_e) = \pi^2 k_B^2 n \theta_e / 2E_F$, where n is the density and E_F the Fermi energy. Using the relations $k_F^3 = 3\pi^2 n$ (valid for non-interacting electrons) and $E_F = m v_F^2/2$, we then get

$$k = \gamma \theta_e = \frac{k_F^2 k_B^2 \lambda_e}{9\hbar} \theta_e \qquad (8.63)$$

with k_F the Fermi momentum and k_B the Boltzmann constant.

---

[29] This process can be seen as the creation of electron-hole pairs that move away from the junction, eventually recombining in the bulk.

The thermal current, at small temperatures, is then $I_{th} \propto k\theta_e = \gamma\theta_e^2$. At steady state we have the condition

$$P_e = I_{th} \rightarrow \theta_e \propto V\sqrt{\frac{\alpha(R)}{\gamma}R}, \quad \text{zero background } \theta_0, \tag{8.64}$$

which predicts a linear increase of the electronic temperature with bias.[30] If the electrodes are at a finite temperature θ_0 then similar to the discussion that led us to Eq. 6.75 we have

$$P_e = I_{th}^{in} - I_{th}^{out} \rightarrow \theta_{eff} = (\theta_0^2 + \theta_e^2)^{1/2}, \quad \text{finite background } \theta_0. \tag{8.65}$$

8.6.1 Electron heat conduction

Before proceeding with the estimate of the temperature let us give a simple argument to justify the use of the bulk form, Eq. 8.63, of the electron heat conductivity in quasi-ballistic systems at low temperatures (and bias) and in the absence of electronic resonances and turbulence.[31]

Due to the small size of nanostructures and long electron inelastic mean free paths, we expect that, like the local energy of phonons, the local electron energy is dissipated away from the junction more efficiently via *elastic* scattering. Therefore, suppose electrons tunnel across the junction from the left electrode, which is in local equilibrium with a temperature θ_L and electrochemical potential μ_L, to the right electrode at temperature θ_R and electrochemical potential μ_R. If the transmission coefficient for electrons to move *elastically* from left to right is $T(E)$, the energy current they transport is of the form (cf. Eq. 6.83)

$$I_{th} \propto \int dE\, E\, T(E)\, [f_L(E, \theta_R) - f_R(E, \theta_L)] \tag{8.66}$$

where $f_{L(R)}$ is the local Fermi–Dirac distribution function of the left (right) electrode at its own local temperature and electrochemical potential (Eq. 1.80).

Let us set $\theta_R = 0$ and consider a small left electrode temperature $\theta_L \equiv \theta_e$ (cf. also discussion of the thermopower in Sec. 6.4). For small biases, in the absence of electronic resonances, we may take the transmission coefficient $T(E)$ to be independent of energy. Call this quantity $\mathcal{T} \equiv T(E)$.

---

[30] Note that, unlike the local ionic heating, there is no "critical" bias in the electron heating effect: current-carrying electrons can scatter the Fermi sea at any bias and create electron-hole pairs in the process.
[31] I am not aware of a complete theory of electron heat conduction in nanostructures.

From Eq. 8.66 we therefore see that $I_{th} \propto \mathcal{T}\theta_e^2$, i.e., the thermal conductivity must be of the form

$$k = \gamma \mathcal{T} \theta_e, \tag{8.67}$$

which is similar to the form 8.63 – with possibly a different coefficient γ – except that it explicitly contains the transmission coefficient. A larger value of $\mathcal{T}$ means a more efficient way for electrons to transport energy from one region to the other.

In the linear response, away from electronic resonances, and in the absence of turbulence we do not expect the coefficient γ to be much different from the bulk value appearing in Eq. 8.63.[32]

In what follows, therefore, I will assume k equal to

$$k = \frac{\mathcal{T} k_F^2 k_B^2 \lambda_e}{9\hbar} \theta_e, \tag{8.68}$$

with θ_e possibly dependent on position.

Clearly, for an accurate quantitative result one needs to determine all the above quantities – for each system – from first principles.

8.6.2 Hydrodynamics of heat transfer

We now need a microscopic theory to determine the coefficients entering Eq. 8.64. This theory needs to take into account the influence of electron-electron interactions on both the heat production and transport.

The hydrodynamic theory I have introduced in the previous sections is well suited for this problem. We know from classical fluid dynamics that the internal friction of a fluid gives rise to local entropy production, where the entropy in this non-equilibrium problem must be interpreted as the statistical entropy I have defined in Sec. 2.8. If there is a variation of temperature in the liquid, and the latter has a position- and time-dependent thermal conductivity $k(\mathbf{r}, t)$, then the balance equation of this entropy production is (Landau and Lifshitz, 1959b)

$$\boxed{\begin{aligned} n(\mathbf{r},t)\theta_e(r,t)D_t s(\mathbf{r},t) &= \sigma_{ij}(\mathbf{r},t)\partial_j v_i(\mathbf{r},t) \\ &\quad + \nabla \cdot [k(\mathbf{r},t)\nabla \theta_e(\mathbf{r},t)], \end{aligned}} \tag{8.69}$$

---

[32] Turbulence can produce local eddies in proximity to the junction where heat is generated nonlinearly at a high rate (Sec. 8.5). Electronic resonances produce multiple reflections inside the nanostructure (Sec. 3.7.1.1). Electron-hole pair production is thus enhanced with different high-order processes occurring at comparable rates. In both cases, the heat conductivity coefficient may deviate substantially from the bulk law.

where s is the local entropy per unit volume and σ_{ij} is the liquid stress tensor.[33] Equation 8.69 can also be derived from first principles for the electron liquid along the same arguments I have used to derive the quantum Navier–Stokes equations.[34]

Here I start from Eq. 8.69 in analogy to the classical case and study its steady-state solutions. The electron stress tensor is either Eq. 8.30, if we keep the bulk viscosity, or Eq. 8.29, if we neglect it.

Incompressible electron liquid

For a general nanostructure Eq. 8.69 needs to be solved numerically.[35] We can however estimate the order of magnitude of the heating effect in the case in which the electron liquid can be assumed incompressible.[36] We proceed along the following lines.

Let us postulate that the following relations are valid out of equilibrium as they are at (quasi-)equilibrium (Landau and Lifshitz, 1959b)

$$\frac{\partial s}{\partial t} = \left(\frac{\partial s}{\partial \theta_e}\right)_P \frac{\partial \theta_e}{\partial t} = \frac{c_P}{\theta_e}\frac{\partial \theta_e}{\partial t}, \qquad (8.70)$$

and

$$\nabla s = \left(\frac{\partial s}{\partial \theta_e}\right)_P \nabla \theta_e = \frac{c_P}{\theta_e}\nabla \theta_e, \qquad (8.71)$$

where the subscript P indicates that the derivative has to be done at constant pressure, and c_P is the electron specific heat at constant pressure.[37] Replacing 8.70 and 8.71 into 8.69 we then get the heat equation for an incompressible liquid,[38]

---

[33] As before, we are assuming that the system is in local thermal equilibrium, thus local thermodynamic quantities such as energy and entropy densities can be defined as I have done in Sec. 2.4. However, unlike the global equilibrium case, the statistical entropy defined here can both increase and decrease in time: the energy is transferred in and out of the local volumes where the entropy is defined.

[34] The derivation requires one to keep the third and fourth moments of the distribution 8.18, and neglect the fifth-order one. I refer the interested reader to Tokatly and Pankratov (1999), for such a derivation.

[35] See D'Agosta et al. (2006), for an analytical solution in the case of a quasi-adiabatic potential.

[36] As I have discussed previously, this is a good approximation for metallic quantum point contacts.

[37] The specific heat per unit volume is $C_P = n\,c_p$. At room, or lower, temperature, within corrections of the order of $(k_B \theta_e/E_F)^2$, we can approximate $C_P \approx C_V$.

[38] In the presence of temperature variations the density cannot be strictly constant. However, if the variations of the temperature are small and the velocity of the fluid is smaller than the Fermi velocity, the liquid can still be assumed incompressible.

$$\boxed{\sigma_{ij}(\mathbf{r})\partial_j v_i(\mathbf{r}) + \nabla \cdot [k(\mathbf{r},t)\nabla\theta_e(\mathbf{r},t)] = n\, c_P \left(\frac{\partial \theta_e}{\partial t} + \mathbf{v}(\mathbf{r},t)\cdot\nabla\theta_e(\mathbf{r},t)\right).}$$

(8.72)

Assuming that the temperature variations are maintained by external sources – in the present case the energy supplied by the battery and the "cooling" provided by the external environment – we can assume the temperature (and thermal conductivity) constant in time and solve the steady-state equation

$$\sigma_{ij}(\mathbf{r})\partial_j v_i(\mathbf{r}) + \nabla \cdot [k(\mathbf{r})\nabla\theta_e(\mathbf{r})] = n\, c_P\, \mathbf{v}(\mathbf{r}) \cdot \nabla\theta_e(\mathbf{r}). \qquad (8.73)$$

The first term on the left-hand side corresponds to heat generation due to the internal friction of the electron liquid. The second one on the left-hand side is the heat transfer (*diffusion*) due to the presence of temperature differences in the liquid, and would be present even in the absence of mechanical motion of the liquid.

Finally, the term on the right-hand side is the *advection* term and is related to heat transport due to the actual movement of the particles in the liquid. Equations 8.31, 8.32, and 8.73 constitute a complete set of equations to describe the charge and heat flow of an incompressible fluid at steady state.

Once again, the solution of Eq. 8.73 for an arbitrary external potential requires numerical integration. To obtain an analytical solution I proceed as follows.

Let us suppose, as we did in Sec. 8.4.2, that the velocity field does not vary appreciably in the directions transverse to current flow, which I take along the x axis. In addition, I take the viscosity to depend only on the density, and not on the local temperature.

The stress tensor is thus (from Eq. 8.30 without the bulk viscosity)

$$\sigma_{xx} = \left(2 - \frac{2}{d}\right)\eta\nabla_x v_x. \qquad (8.74)$$

Replacing Eq. 8.74 ($d=3$) and 8.68 into Eq. 8.73 we get[39]

$$\eta\frac{4}{3}\left(\frac{\partial v_x}{\partial x}\right)^2 + \frac{1}{2}\frac{k_F^2 k_B^2 \lambda_e}{9\hbar}\frac{\partial^2 \theta_e^2}{\partial x^2} = C_V\, v_x\frac{\partial \theta_e}{\partial x}. \qquad (8.75)$$

By assuming that both the first and second derivatives of the temperature

---

[39] I assume here $\mathcal{T} \doteq 1$ and $C_P = n\, c_P = C_V$. The inclusion of a finite transmission function requires specific knowledge of the velocity profile that leads to such transmission.

are appreciably different from zero within a length $L \ll \lambda_e$,[40] we see from the above equation that the ratio between the advection term and the diffusion one is of the order of $v_x L / v_F \lambda_e$. In quasi-ballistic nanoscale systems this ratio is much smaller than one.

I can therefore neglect the contribution of the advection term. The equation we need to solve is thus

$$\eta \frac{4}{3} \left(\frac{\partial v_x}{\partial x} \right)^2 + \frac{1}{2} \frac{k_F^2 k_B^2 \lambda_e}{9\hbar} \frac{\partial^2 \theta_e^2}{\partial x^2} = 0. \qquad (8.76)$$

The change of velocity with bias can be estimated as follows. Let us start from Bernoulli's equation 8.36 for the liquid slightly away from the junction in the x direction where the potential of the junction can be assumed zero,[41]

$$\frac{v_x^2}{2} + \frac{E_F}{m} = 2\frac{E_F}{m} + \frac{eV}{m}. \qquad (8.77)$$

If we write $v_x = v_F + \delta v_x$ and assume $\delta v_x \ll v_F$, from Eq. 8.77 we find $\delta v_x \approx eV/mv_F$. If L is again the length over which the velocity and temperature change most, we can linearize Eq. 8.76 as

$$\eta \frac{4}{3} \left(\frac{\delta v_x}{L} \right)^2 - \frac{k_F^2 k_B^2 \lambda_e}{9\hbar} \frac{\bar{\theta}_e^2}{L^2} = 0, \qquad (8.78)$$

where I have defined as $\bar{\theta}_e$ the average electronic temperature over the length L, and the minus sign is because the temperature increases inside the junction, thus its second derivative in space is negative. Solving Eq. 8.78 we then get

$$\bar{\theta}_e = V \frac{e}{\pi k_B} \sqrt{\frac{4\eta}{\hbar n k_F \lambda_e}}, \qquad (8.79)$$

where I have used the relation $k_F^3 = 3\pi^2 n$, with n the density of the liquid.

Equation F.8 in Appendix F shows that the ratio $\eta/\hbar n$ is a number dependent only on the "average distance" between electrons, $r_s a_0$. Call this ratio $c(r_s)$. Let us take $k_F \simeq 1$ Å^{-1} and $\lambda_e \simeq 1000$ Å, typical of metallic quantum point contacts.[42] From Eq. 8.79 we get

$$\bar{\theta}_e \approx V \beta \sqrt{c(r_s)}, \qquad (8.80)$$

with $\beta \sim 235$ K/V.

---

[40] That is equivalent to saying that most of the electron temperature variations occur in the nanostructure whose characteristic length is much smaller than the inelastic mean-free path.
[41] The factor of 2 in front of E_F on the right-hand side of Eq. 8.77 comes from the fact that we consider two Fermi gases, one on the right and one on the left of the junction, and we neglect the small interaction energy between the two.
[42] We are here implicitly assuming that both λ_e and the viscosity η do not vary appreciably in the bias range of interest.

In 3D, if we consider a gold point contact, $r_s \sim 3a_0$, $c(r_s) \sim 0.017$ (from Eq. F.8). We therefore get $\bar{\theta}_e \approx V \times 31$ [K/V]. At a bias of 0.1 V, the local electronic temperature is thus about 3 K. For comparison, the ionic one, at the same bias, is about 50 K (see Sec. 6.3.1 and Chen et al., 2003).

Recall, however, that the above is an oversimplified derivation meant to provide an order of magnitude of this phenomenon. The exact values of the electronic temperature clearly depend on the microscopic details of the junction and, to a large extent, on the exact form of the heat conductivity, Eq. 8.68, and the value of the inelastic mean-free path λ_e.

Let us finally note that, as in the case of localized phonon modes (see Sec. 6.3), if the generated electron-hole pairs are not able to diffuse away from the junction, then from the point of view of the electrons in the Fermi sea the current-carrying electrons are simply excited at an effective finite temperature of the order of eV/k_B. This is the amount of departure of the electron distribution from the equilibrium value, irrespective of the electron-electron interaction strength. In this case, much larger electronic temperatures may be obtained.

8.6.3 Effect of local electron heating on ionic heating

The discussion in the previous section neglects the correlated effect of electron-ion interaction. In particular, it does not take into account that part of the energy of the current-carrying electrons goes into the excitation of ionic vibrations. Since the latter effect generates a local ionic temperature, which, as discussed in Sec. 6.3, may be measured experimentally, the relation between local ionic temperature and local electron temperature may provide evidence of the electron heating effect.

If we consider the electron-phonon interaction, then each electron can either scatter with other electrons or with phonons. Since the source of energy is the same for both processes – the initial energy of the electron – the *ionic* temperature must be *lower* in the presence of the electron heating than without: part of the energy that would otherwise go to heat up the ions is "lost" into heating of the electrons.

Within the assumptions of the previous section, to first order we can assume the electron-electron and electron-phonon processes independent, and occurring with equal probability. Let us also suppose that, for a given bias, the local electron temperature in the absence of electron-phonon scattering, θ_e, is smaller than the ionic temperature in the absence of electron heating, θ_{ion}.

The energy lost by an electron due to electron heating is "seen" by the phonons as a sink of energy at the electron temperature θ_e. This energy however is "lost" by the phonons at *their own rate*. As done previously, let us assume a bulk lattice heat conduction law and start from zero background temperature for both electrons and ions. The energy per unit time not gained by the phonons due to electron heating is thus $-\alpha_{ph}\theta_e^4$, where α_{ph} is the lattice heat conductance coefficient (Sec. 6.3.1).

The balance between the power in and the power out at steady state is therefore (for $\theta_e < \theta_{ion}$)

$$\alpha_{ph}\theta_{eff}^4 = \alpha_{ph}\theta_{ion}^4 - \alpha_{ph}\theta_e^4 \rightarrow \theta_{eff} = (\theta_{ion}^4 - \theta_e^4)^{1/4}, \quad \text{zero background } \theta_0. \tag{8.81}$$

In terms of the bias, $\theta_{ion} = \gamma_{ph}\sqrt{V}$ and $\theta_e = \gamma_e V$, where γ_{ph} and γ_e need to be determined microscopically as discussed in the preceding section and in Sec. 6.3.

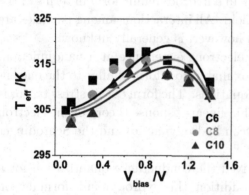

Fig. 8.3. Experimental local ionic temperature (symbols) as a function of bias for alkane-dithiol molecules of different lengths between bulk electrodes. Theoretical predictions are from Eq. 8.83 and are indicated as solid lines. Reprinted with permission from Huang et al. (2007).

Equation 8.81 is thus (for $V < (\gamma_{ph}/\gamma_e)^2$)

$$\theta_{eff} = (\gamma_{ph}^4 V^2 - \gamma_e^4 V^4)^{1/4}, \quad \text{zero background } \theta_0. \tag{8.82}$$

Finally, if we assume a finite background temperature θ_0 (for simplicity the same for electrons and ions), the effective ionic temperature is (see also discussion in Sec. 6.3)

$$\theta_{eff} = (\theta_0^4 + \gamma_{ph}^4 V^2 - \gamma_e^4 V^4)^{1/4}, \quad \text{finite background } \theta_0. \tag{8.83}$$

From the above we thus see that at low biases the ionic heating dominates, while at large biases electron heating increases in importance till it effectively "cools" the local ionic temperature of the nanojunction. In order to observe this phenomenon, the electron-phonon coefficient, γ_{ph}, and the corresponding electron-electron one, γ_e, must be of comparable magnitude. From the estimate in the previous section, we can argue that this is likely to occur in structures where ionic heating is small, as in insulating molecular structures.

A preliminary indication of this phenomenon has been reported experimentally in molecular structures (Huang et al., 2007). This is shown in Fig. 8.3 where the effective local ionic temperature as a function of bias for alkanedithiol molecules of different lengths has been extracted from the effective force necessary to break the junction. The temperature saturation and reversal as a function of bias is clearly visible.

Summary and open questions

In this chapter I have shown that the many-body time-dependent Schrödinger equation can be written equivalently in a hydrodynamic form in terms of the single-particle density and velocity field. All interactions among particles are lumped into a stress tensor, which, however, is generally unknown.

Due to the fast elastic collisions electrons experience at a nanojunction, the above stress tensor can be approximated to a form similar to the one entering the classical Navier–Stokes equations. The form of this stress tensor is also similar to the one obtained within linear-response theory of the electron liquid, making the connection between hydrodynamics and the Schrödinger equation even stronger.

Within this approach I have derived the conductance quantization for a charged non-viscous fluid in 1D. In addition, this hydrodynamic formulation suggests several phenomena with classical counterparts, such as turbulence of the electron liquid and local electron heating.

The latter in particular has consequences for the local ionic heating effect, namely it suggests that ions effectively "cool" at high biases due to "loss" of energy towards electron-electron interaction degrees of freedom.

All the results, however, are qualitative due to many factors. For one, the viscosity dependence on bias is unknown for nanostructures. Second, a microscopic theory of electron heat transport at nanojunctions is still lacking. In addition, it is not clear how heating is generated in the regime in which electron-electron and electron-phonon interactions are of the same magnitude. A theory that addresses the correlated dynamics of many electrons in interactions with ions is thus necessary.

Finally, due to its "economical" description in terms of single-particle properties such as the density and velocity field, the hydrodynamic formulation seems ideal to study poorly understood effects such as dephasing and memory effects in transport. Its generalization to spin-dependent phenomena would also contribute to our understanding of these issues in nanostructures.

Exercises

8.1 **The Madelung equations.** (i) Consider the de Broglie ansatz 8.3 for a single-particle wave-function. Insert this form into the Schrödinger equation 8.2. Show that this decouples into the two equations of motion 8.4 and 8.5. (ii) Prove that by defining the stress tensor 8.8 one obtains the Madelung equations 8.11. (iii) Show that the stress tensor 8.8 can be written as

$$P_{ij} = -\frac{\hbar^2}{4m}(\nabla^2 n)\delta_{ij} + \frac{\hbar^2}{4m}\frac{(\nabla_i n)(\nabla_j n)}{n}, \quad (E8.1)$$

where the first term on the right-hand side can be interpreted as the *pressure* and the second term as the *shear stress tensor*. Note, however, a substantial difference with the Navier–Stokes shear stress tensor 8.29. This latter one depends on the derivatives of *velocities*, not of densities.

8.2 **Vanishing of the quantum force in the single-particle case.** Show that for a single particle in an external potential the " hydrodynamic quantum force" associated with the "potential" 8.7 vanishes if $\sqrt{n}$ is an eigenstate of the Laplacian, namely

$$\nabla^2 \sqrt{n} = \lambda \sqrt{n}, \quad (E8.2)$$

with λ an arbitrary real number.

8.3 Consider a single particle in a non-magnetic external field. The expectation value of the paramagnetic current density operator over a state $\Psi(\mathbf{r},t)$ is

$$\mathbf{j}(\mathbf{r},t) = \frac{\hbar}{m}\text{Im}\left\{[\Psi(\mathbf{r},t)]^*\nabla\Psi(\mathbf{r},t)\right\}. \quad (E8.3)$$

Write this state using the de Broglie ansatz, Eq. 8.3. Show that the velocity field 8.13 is equivalent to the single-particle definition 8.6.

8.4 (i) Consider the rate of energy dissipation 8.56. Integrate this expression by parts, and use the fact that the tensor $\sigma_{xc,ij}$ (Eq. 8.55) is traceless to prove Eq. 8.57. (ii) Assume now the current density constant, $|\mathbf{j}| = j$, and the velocity field oriented along the direction of current flow x: $\mathbf{v} = n^{-1} j \hat{x}$. Show that in this case Eq. 8.57 reduces to Eq. 8.58.

Appendix A
A primer on second quantization

This appendix is meant to cover only the basic facts of *second quantization* that are relevant to this book, and it is thus not a comprehensive account of this formalism. The reader interested in an in-depth proof of all the statements presented here should consult books fully dedicated to many-body techniques (see, e.g., Fetter and Walecka, 1971; Mattuck, 1976; Mahan 1990).

Fermions

Let us consider a system of N *non-interacting* and *indistinguishable* particles that satisfy Fermi statistics (see Sec. 1.4.6). If $\{\phi_k(\mathbf{r})\}$ is a general complete set of orthonormal single-particle states – with k denoting any possible quantum number – the many-body wave-function describing this system is the *Slater determinant* of the $N \times N$ matrix built out of the single-particle states $\phi_{k_j}(\mathbf{r}_i)$ (with $1 \leq j, i \leq N$)

$$\Phi_{k_1,k_2,\ldots,k_N}(\mathbf{r}_1;\ldots;\mathbf{r}_N) = \frac{1}{\sqrt{N!}} \begin{vmatrix} \phi_{k_1}(\mathbf{r}_1) & \ldots & \phi_{k_1}(\mathbf{r}_N) \\ \vdots & & \vdots \\ \phi_{k_N}(\mathbf{r}_1) & \ldots & \phi_{k_N}(\mathbf{r}_N) \end{vmatrix}$$

$$\equiv \frac{1}{\sqrt{N!}} \det\left[\phi_{k_j}(\mathbf{r}_i)\right]. \qquad (A.1)$$

The set of all possible Slater determinants forms a complete basis set in the Hilbert space $\mathcal{H}_N$ of the N-particle system. This means that if the N particles interact with each other or with an external perturbation, a general many-body state of the system can be written as a linear combination of

Slater determinants

$$\Psi(\mathbf{r}_1;\ldots;\mathbf{r}_N) = \sum_{k_1,\ldots,k_N} C_{k_1,\ldots,k_N} \Phi_{k_1,k_2,\ldots,k_N}(\mathbf{r}_1;\ldots;\mathbf{r}_N), \quad (A.2)$$

with $C_{k_1,\ldots,k_N}$ arbitrary c-numbers.

Now, since the particles are indistinguishable it is much more convenient to use a representation in which the only information we care about is how many particles are in a specific single-particle state $\{\phi_k(\mathbf{r})\}$. If we denote with n_k the number of particles in state $\{\phi_k(\mathbf{r})\}$ (also called the *occupation number*) we can write in Dirac notation

$$\Phi_{k_1,k_2,\ldots,k_N}(\mathbf{r}_1;\ldots;\mathbf{r}_N) = \langle \mathbf{r}_1;\ldots;\mathbf{r}_N | n_{k_1},\ldots n_{k_N} \rangle, \quad (A.3)$$

with

$$\sum_{k_i} n_{k_i} = N. \quad (A.4)$$

More compactly we write

$$|\Phi_{k_1,k_2,\ldots,k_N}\rangle = |n_{k_1},\ldots,n_{k_N}\rangle, \quad (A.5)$$

which can be interpreted as the N-particle non-interacting state in which n_{k_1} particles are in the single-particle state $\{\phi_{k_1}(\mathbf{r})\}$, n_{k_2} particles in the state $\{\phi_{k_2}(\mathbf{r})\}$, and so on.

We are dealing now with a new class of states, which are called *Fock states*, and the corresponding space spanned by these states is called *Fock space*, which is an "extended" Hilbert space with variable number of particles.

Creation and annihilation operators

With this in mind, we may define operators on this space that either *create*, $\hat{a}^\dagger_{k_i}$, or *destroy*, $\hat{a}_{k_i}$, a particle in any given single-particle state $\{\phi_{k_i}(\mathbf{r})\}$. In particular, we may define the successive action of the creation operator $\hat{a}^\dagger_{k_i}$ on the *vacuum* state – the one with zero particles –

$$|0,0,\ldots,0\rangle \equiv |0\rangle \quad (A.6)$$

so that

$$|\Phi_{k_1,k_2,\ldots,k_N}\rangle = \hat{a}^\dagger_{k_1} \hat{a}^\dagger_{k_2} \ldots \hat{a}^\dagger_{k_3} |0\rangle. \quad (A.7)$$

Since we are dealing with fermions, a successive application of creation and destruction operators on the same state exchanges particles and thus the many-body wave-function has to change sign. We thus define the action of *creation* operators as

$$\hat{a}^\dagger_{k_i}|n_{k_1},\ldots,n_{k_i},\ldots\rangle = (-1)^{S_i}(1-n_{k_i})|n_{k_1},\ldots,n_{k_i}+1,\ldots\rangle \quad (A.8)$$

and the action of *annihilation* operators (Hermitian conjugate of the creation operators) as

$$\hat{a}_{k_i}|n_{k_1},\ldots,n_{k_i},\ldots\rangle = (-1)^{S_i} n_{k_i}|n_{k_1},\ldots,n_{k_i}-1,\ldots\rangle, \quad (A.9)$$

where the coefficient

$$S_i = n_{k_1} + \ldots + n_{k_{i-1}} = \sum_{l<i} n_{k_l} \quad (A.10)$$

takes into account the fact that each occupied single-particle state preceding the particle state i in $|n_{k_1},\ldots,n_{k_i},\ldots\rangle$ contributes a factor (-1) to the wave-function.

It is easy to prove from these definitions that

$$\hat{a}^\dagger_{k_i}\hat{a}_{k_i}|n_{k_1},\ldots,n_{k_i},\ldots\rangle = n_{k_i}|n_{k_1},\ldots,n_{k_i},\ldots\rangle, \quad (A.11)$$

namely the operator $\hat{a}^\dagger_{k_i}\hat{a}_{k_i}$ counts the number of particles in a given single-particle state. We then define the *occupation number operator* (or simply *number operator*)

$$\hat{n}_{k_i} = \hat{a}^\dagger_{k_i}\hat{a}_{k_i}, \quad (A.12)$$

and the *total number operator*

$$\hat{N} = \sum_{k_i} \hat{n}_{k_i} = \sum_{k_i} \hat{a}^\dagger_{k_i}\hat{a}_{k_i}, \quad (A.13)$$

which counts the total number of particles in a given many-body state. Similarly, we can prove that

$$\hat{a}_{k_i}\hat{a}^\dagger_{k_i}|n_{k_1},\ldots,n_{k_i},\ldots\rangle = (1-n_{k_i})|n_{k_1},\ldots,n_{k_i},\ldots\rangle, \quad (A.14)$$

so that

$$\hat{a}_{k_i}\hat{a}^\dagger_{k_i} = 1 - \hat{n}_{k_i}. \quad (A.15)$$

By summing A.15 and A.12 we get the relation

$$\hat{a}_{k_i}\hat{a}^\dagger_{k_i} + \hat{a}^\dagger_{k_i}\hat{a}_{k_i} = 1. \quad (A.16)$$

By using the definitions A.8 and A.9 one can also prove that for arbitrary states k_i and k_l

$$\{\hat{a}_{k_i},\hat{a}^\dagger_{k_l}\} = \hat{a}_{k_i}\hat{a}^\dagger_{k_l} + \hat{a}^\dagger_{k_l}\hat{a}_{k_i} = \delta_{k_i,k_l}, \quad (A.17)$$

and
$$\{\hat{a}_{k_i}, \hat{a}_{k_l}\} = \{\hat{a}^\dagger_{k_i}, \hat{a}^\dagger_{k_l}\} = 0. \tag{A.18}$$

The field operators

We can now work in the position representation and define the *field operators*

$$\psi^\dagger(\mathbf{r}) = \sum_k \langle \phi_k | \mathbf{r} \rangle \hat{a}^\dagger_k, \quad \text{creates particle at position } \mathbf{r} \tag{A.19}$$

and

$$\psi(\mathbf{r}) = \sum_k \langle \mathbf{r} | \phi_k \rangle \hat{a}_k, \quad \text{destroys particle at position } \mathbf{r}. \tag{A.20}$$

Using the commutation relations A.17 and A.18 one can easily prove that

$$\{\psi(\mathbf{r}), \psi^\dagger(\mathbf{r}')\} = \delta(\mathbf{r} - \mathbf{r}'), \tag{A.21}$$

and

$$\{\psi(\mathbf{r}), \psi(\mathbf{r}')\} = \{\psi^\dagger(\mathbf{r}), \psi^\dagger(\mathbf{r}')\} = 0. \tag{A.22}$$

Operators in second-quantized form

Here, I simply state the second-quantized form of the operators that are mainly used in this book. The demonstration of these results is straightforward but tedious, and I refer the interested reader to other books for a full derivation (see, e.g., Mattuck, 1976).

In terms of the field operators, the single-particle number density operator 1.24 can be written as

$$\hat{n}(\mathbf{r}) = \psi^\dagger(\mathbf{r})\psi(\mathbf{r}), \tag{A.23}$$

while the current density operator 1.22 is

$$\hat{\mathbf{j}}(\mathbf{r}, t) = \frac{\hbar}{2im} \lim_{\mathbf{r} \to \mathbf{r}'} (\nabla_\mathbf{r} - \nabla_{\mathbf{r}'}) \psi^\dagger(\mathbf{r}', t) \psi(\mathbf{r}, t), \tag{A.24}$$

where in the above expression the field operators are in the Heisenberg picture.

A Hamiltonian describing free electrons can be written as

$$\hat{H} = \sum_i \frac{\hat{p}_i}{2m} = \sum_k \frac{\hbar^2 k^2}{2m} \hat{a}_k^\dagger \hat{a}_k. \quad (A.25)$$

Finally, using the above field operators, we can write a one-body operator with position representation $U(\mathbf{r})$ as

$$\hat{U} = \int d\mathbf{r}\, \psi^\dagger(\mathbf{r})\, U(\mathbf{r})\, \psi(\mathbf{r}) = \int d\mathbf{r}\, U(\mathbf{r})\, \hat{n}(\mathbf{r}). \quad (A.26)$$

A two-body operator, with positional elements $W(\mathbf{r}, \mathbf{r}')$, can be written as

$$\hat{W} = \int d\mathbf{r} \int d\mathbf{r}'\, \psi^\dagger(\mathbf{r}) \psi^\dagger(\mathbf{r}')\, W(\mathbf{r}, \mathbf{r}')\, \psi(\mathbf{r}') \psi(\mathbf{r}). \quad (A.27)$$

Bosons

The above results can be easily extended to the case in which the particles follow Bose statistics. If

$$|\Psi_{N_1, N_2, \ldots}\rangle = |N_1, N_2 \ldots\rangle \quad (A.28)$$

is a general many-body wave-function of non-interacting bosons, with N_i the occupation of a single-particle boson state, we define the creation, $\hat{b}_j^\dagger$, and annihilation, $\hat{b}_j$, operators as

$$\hat{b}_j^\dagger |N_1, \ldots, N_j, \ldots\rangle = \sqrt{(N_j + 1)}\, |N_1, \ldots, N_j + 1, \ldots\rangle \quad (A.29)$$

$$\hat{b}_j |N_1, \ldots, N_i, \ldots\rangle = \sqrt{N_j}\, |N_1, \ldots, N_j - 1, \ldots\rangle. \quad (A.30)$$

It is then easy to prove that these operators satisfy the commutation relations

$$\left[b_j, b_{j'}^\dagger\right] = b_j b_{j'}^\dagger - b_j^\dagger b_{j'} = \delta_{j,j'} \quad (A.31)$$

and

$$\left[b_j, b_{j'}\right] = \left[b_j^\dagger, b_{j'}^\dagger\right] = 0. \quad (A.32)$$

Appendix B
The quantum BBGKY hierarchy

I derive here the BBGKY hierarchy generated from the equation of motion of the single-particle time-ordered Green's function for fermions

$$\mathbf{G}(\mathbf{r},t;\mathbf{r}',t') = -\frac{i}{\hbar}\langle \mathrm{T}\{\psi(\mathbf{r},t)\psi^\dagger(\mathbf{r}',t')\}\rangle. \qquad \text{(B.1)}$$

The field operators are written in the Heisenberg picture with respect to a general many-body Hamiltonian of the type

$$H = -\sum_i \frac{\hbar^2}{2m}\nabla_i^2 + \sum_{ij} w(\mathbf{r}_i - \mathbf{r}_j), \qquad \text{(B.2)}$$

with $w(\mathbf{r}_i - \mathbf{r}_j)$ a general two-body potential (such as the Coulomb potential) describing the interaction between any two particles.

The equation of motion of any operator $\hat{A}_H(t)$ in the Heisenberg picture is (Messiah, 1961)

$$i\hbar \frac{d\hat{A}_H(t)}{dt} = [\hat{A}_H(t), \hat{H}], \qquad \text{(B.3)}$$

where I have assumed that the operator $\hat{A}$ does not depend explicitly on time.

Using this equation for the field operator we get

$$i\hbar \frac{\partial \psi(\mathbf{r},t)}{\partial t} = -\frac{\hbar^2}{2m}\nabla^2 \psi(\mathbf{r},t) + \int d\mathbf{r}' w(\mathbf{r}-\mathbf{r}')\,\psi^\dagger(\mathbf{r}',t)\,\psi(\mathbf{r}',t)\,\psi(\mathbf{r},t), \qquad \text{(B.4)}$$

and its Hermitian conjugate

$$-i\hbar \frac{\partial \psi^\dagger(\mathbf{r},t)}{\partial t} = -\frac{\hbar^2}{2m}\nabla^2 \psi^\dagger(\mathbf{r},t) + \int d\mathbf{r}' w(\mathbf{r}-\mathbf{r}')\,\psi^\dagger(\mathbf{r},t)\,\psi^\dagger(\mathbf{r}',t)\,\psi(\mathbf{r}',t). \qquad \text{(B.5)}$$

By differentiating Eq. B.1 with respect to one of the two times we get

$$i\hbar \frac{\partial G(\mathbf{r},t;\mathbf{r}',t')}{\partial t} = \delta(t-t')\langle\{\psi(\mathbf{r},t),\psi^\dagger(\mathbf{r}',t)\}\rangle$$

$$+ i\hbar \left\langle T\left\{\frac{\partial \psi(\mathbf{r},t)}{\partial t}\psi^\dagger(\mathbf{r}',t')\right\}\right\rangle$$

$$= \delta(\mathbf{r}-\mathbf{r}')\delta(t-t') + \left\langle T\left\{\frac{\partial \psi(\mathbf{r},t)}{\partial t}\psi^\dagger(\mathbf{r}',t')\right\}\right\rangle, \quad (B.6)$$

where in the last equality I have used the commutation relation A.21.

By replacing Eq. B.4 into B.6 we obtain

$$\left(i\hbar\frac{\partial}{\partial t_1} + \frac{\hbar^2}{2m}\nabla_1^2\right) G(\mathbf{r}_1,t_1;\mathbf{r}'_1,t'_1) = \delta(\mathbf{r}_1-\mathbf{r}'_1)\delta(t_1-t'_1)$$

$$- i\int d\mathbf{r}\, w(\mathbf{r}_1-\mathbf{r}) G_2(\mathbf{r},t_1,\mathbf{r}_1,t_1;\mathbf{r},t_1^+,\mathbf{r}'_1,t'_1), \quad (B.7)$$

where the symbol t_1^+ means $\lim_{\epsilon \to 0^+}(t_1 + \epsilon)$, and $\mathbf{G}_2$ is the *two-particle* (or *four-time*) Green's function (with the compact notation $1 \equiv (\mathbf{r}_1,t_1)$)

$$G_2(1,2;2',1') = -\frac{1}{\hbar}\langle T\{\psi(1)\psi(2)\psi^\dagger(2')\psi^\dagger(1')\}\rangle. \quad (B.8)$$

By differentiating the two-particle Green's function B.8, we get an equation of motion containing the three-particle Green's function, whose equation of motion depends on the four-particle Green's function, and so on.

By defining the n-particle Green's function as

$$G_n(1,2,\ldots,n;n',\ldots,2',1')$$
$$= \frac{(-i)^n}{\hbar}\langle T\{\psi(1)\psi(2)\ldots\psi(n)\psi^\dagger(n')\ldots\psi^\dagger(2')\psi^\dagger(1')\}\rangle, \quad (B.9)$$

we get the general relation between the $\mathbf{G}_n$ and the $\mathbf{G}_{n-1}$ Green's functions ($n \geq 2$)

$$\left(i\hbar\frac{\partial}{\partial t_i} + \frac{\hbar^2}{2m}\nabla_i^2\right) G_n(1,\ldots,n;n',\ldots,1')$$

$$= \delta(\mathbf{r}_1-\mathbf{r}'_1)\delta(t_1-t'_1) + \hbar\sum_{j=2}^n (-1)^{i+j}\delta(\mathbf{r}_i-\mathbf{r}'_j)\delta(t_i-t'_j)$$

$$\times G_{n-1}(1,\ldots,i-1,i+1,\ldots,n;n',\ldots,(j+1)',(j-1)',\ldots,1')$$

$$- i\int d\mathbf{r}\, w(\mathbf{r}_j-\mathbf{r})\, G_{n+1}(\mathbf{r}t_i,1,\ldots,n;n',\ldots,1',\mathbf{r}\,t_i^+), \quad (B.10)$$

where $\mathbf{G}_{n-1}(1,\ldots,i-1,i+1,\ldots,n;n',\ldots,(j-1)',(j+1)',\ldots,1')$ is the $(n-1)$ Green's function where the coordinates i and j have been omitted.

The above relation represents the BBGKY hierarchy. It can be trivially generalized in the presence of an additional single-particle potential $U(\mathbf{r})$ in the Hamiltonian B.2, by subtracting this potential in the parenthesis on the left-hand side of Eq. B.10.

Appendix C
The Lindblad equation

Let us consider a closed system whose degrees of freedom can be divided into two *distinguishable* sets – call them S and B, e.g., electrons and phonons – and we are interested in the dynamics of only one of the two, say, the electrons. Call the set of degrees of freedom S the "system". These two sets of degrees of freedom are mutually interacting, but do not exchange particles, namely the number of particles of S is fixed.

Let us also suppose that the other set of degrees of freedom B is so large that we are not interested in its microscopic dynamics, or it is simply impossible to calculate. For both mathematical and physical reasons, by "large" I mean *infinitely large*. This set of degrees of freedom then acts as an environment for the system S (Sec. 1.2).

Given an initial condition, the dynamics of $S + B$ is *reversible*, so that it is generated by a *group* of unitary operators $U(t)$ (Eq. 1.18) on the Hilbert space of $S + B$, that depends on one parameter: the time t. On the other hand, if we follow only the degrees of freedom of S, by considering the environment B infinite (hence with an infinite Poincaré recurrence time – Sec. 1.2.1), we impose a preferential direction of time because, due to the interaction of S with B, during time evolution some correlations in the system S are "lost" into the degrees of freedom of B without the possibility to recover them (see Sec. 2.8).

We have thus transformed the original, reversible closed system $S + B$ dynamics into an *irreversible* description of the reduced dynamics of S. We therefore expect that, unlike the group of unitary transformations $U(t)$, there exists in the present case – under appropriate conditions to be given – a *semigroup* of "maps" (operators) that describes the dynamics of S. By "semigroup" I mean one that generates a continuous dynamics of the reduced statistical operator $\hat{\rho}_S$, but only along one direction of time.[1] In particular, we expect this dynamics to be generated by an equation of motion of the type

$$\frac{d\hat{\rho}_S(t)}{dt} = \mathcal{L}\hat{\rho}_S(t). \tag{C.1}$$

In this appendix, I will first formulate a general theorem due to Lindblad (Lindblad, 1976) which, under specific conditions, rigorously shows what *form* the above super-operator $\mathcal{L}$ must have (super-operator because it maps an operator into another). Second, I will derive, from first principles, the same result so that a more physical understanding of Eq. C.1 can be obtained.

C.1 The Lindblad theorem

Assume that the dynamical semigroup generated by Eq. C.1 also preserves the three fundamental properties of the statistical operator $\hat{\rho}_S$

(i) Hermiticity: all probabilities are real,
(ii) Conservation of trace: $\text{Tr}\{\hat{\rho}_S(t)\} = 1$,
(iii) Positivity: $\langle \Psi|\hat{\rho}_S(t)|\Psi\rangle \geq 0$, $\forall\ |\Psi\rangle$ in the Hilbert space. This guarantees that the probabilities associated with all possible states are non-negative.

In addition, under the condition that the map $\hat{\rho}_S(t=0) \to \hat{\rho}_S(t)$ is linear, the Lindblad theorem states that the super-operator $\mathcal{L}$ must have a specific form, so that Eq. C.1 reads

$$\partial_t \hat{\rho}_S(t) = -\frac{i}{\hbar}\left[\hat{H}_S, \hat{\rho}_S(t)\right] + \sum_\alpha \left[-\frac{1}{2}\hat{V}_\alpha^\dagger \hat{V}_\alpha \hat{\rho}_S(t) - \frac{1}{2}\hat{\rho}_S(t)\hat{V}_\alpha^\dagger \hat{V}_\alpha + \hat{V}_\alpha \hat{\rho}_S(t)\hat{V}_\alpha^\dagger\right], \tag{C.2}$$

with $\hat{H}_S$ an Hermitian operator, and $\{V_\alpha\}$ a set of (Lindblad) operators not necessarily Hermitian. However, both $\hat{H}_S$ and the bath operators must be *time independent*.[2]

The terms $\hat{V}_\alpha \hat{\rho}_S(t)\hat{V}_\alpha^\dagger$ in the Lindblad equation C.2 create "quantum jumps" between states of the system, so that they can be associated with

---

[1] In mathematical terms a semigroup does not have the inverse-element property of a group, and hence does not allow a unique invertibility of the system dynamics. If the environment B is finite (but large), we can still describe a "locally irreversible" dynamics of the subsystem S so long as the time scales associated with the latter are much longer than any other correlation time of the subsystem B. In this case, however, there is no *exact* dynamical semigroup in the mathematical sense, since we can always invert the system dynamics (albeit at very long times).

[2] An example of Lindblad operators and the use of the Lindblad equation C.2 was given in Sec. 2.2.1 when discussing relaxation and dephasing.

quantum fluctuations, while the other terms, $\frac{1}{2}\hat{V}_\alpha^\dagger \hat{V}_\alpha \hat{\rho}_S(t)$ and $\frac{1}{2}\hat{\rho}_S(t)\hat{V}_\alpha^\dagger \hat{V}_\alpha$, compensate for such fluctuations – to preserve properties (i) and (ii) of the statistical operator – and can be thus interpreted as *dissipative* terms (see also discussion in Sec. G.1 of Appendix G).[3]

Equation C.2 is invariant under the transformations

$$\hat{H}_S \to \hat{H}_S - \frac{i}{2}\sum_\alpha \left(c_\alpha \hat{V}_\alpha - c_\alpha^* \hat{V}_\alpha^\dagger \right); \quad \hat{V}_\alpha \to \sum_\beta C_{\alpha\beta} \hat{V}_\beta + c_\alpha \hat{1}, \quad \text{(C.3)}$$

with $\{C\}_{\alpha\beta}$ a unitary matrix, and $\{c_\alpha\}$ a set of arbitrary c-numbers.[4]

Note that the operators $\hat{H}_S$, $\{V_\alpha\}$ need to be independent of time for the proof of the theorem to be valid.[5] However, time dependence of these operators does not mean that an equation of the form C.2 cannot be derived.[6]

The class of equations C.2 with time-dependent operators are generally called *quantum master equations*.[7] It is easy to prove that they preserve properties (i) and (ii) of the statistical operators.

However, these quantum master equations may *not* guarantee property (iii), namely there may be solutions of Eq. C.2 for arbitrary time-dependent Hamiltonians and/or bath operators that are not positive, and thus do not correspond to physical states. Positivity may be lost at specific times, and this may also depend on the initial conditions: some initial conditions may lead to unphysical states, while others may still be proper states despite time-dependent operators (see, e.g., Maniscalco *at al.*, 2004). This means that for time-dependent Hamiltonians and/or bath operators – namely, whenever the dynamical semigroup property is violated – one needs to check that condition (iii) is satisfied for the physical problem at hand.

This issue does not appear if one starts from the *stochastic* Schrödinger equation G.1 that I will discuss in Appendix G. For given Hamiltonian and bath operators, the statistical operator one calculates from this stochastic equation, by definition, always corresponds to a well-defined physical state.

---

[3] Note that the energy quantum jumps induced by the bath may be very long range in space, in the sense that its influence may be "felt" in the *whole* system.

[4] This implies that the operators V_α can always be chosen traceless.

[5] A dynamical semigroup cannot be strictly defined for time-dependent Hamiltonians.

[6] In fact, the derivation of Eq. C.2 with time-dependent Hamiltonian and Lindblad operators rests on the general decomposition of a super-operator $\mathcal{F}(\hat{\rho})$ in *Kraus operators* $\hat{E}_\mu$

$$\mathcal{F}(\hat{\rho}) = \sum_\mu \hat{E}_\mu \hat{\rho} \hat{E}_\mu^\dagger,$$

where $\sum_\mu \hat{E}_\mu^\dagger \hat{E}_\mu = \hat{1}$ (see, e.g., Breuer and Petruccione, 2002).

[7] Time-dependent bath operators that are local in time may be interpreted as some sort of non-Markovian dynamics, whereby the interaction of the bath with the system changes in time, but it carries information only at the time at which the statistical operator is evaluated, and not on its past dynamics (Paz and Zurek, 1999).

C.2 Derivation of the Lindblad equation

There are several ways to derive the Lindblad equation C.1 from first principles (see, e.g., van Kampen, 1992). Here I follow one due to Nakajima (Nakajima, 1958; see also Zwanzig, 1960).

Let us write the time-independent Hamiltonian of the system plus the environment degrees of freedom as

$$\hat{H}_{tot} = \hat{H}_S \otimes \hat{1}_B + \hat{1}_S \otimes \hat{H}_B + \alpha \hat{H}_{int}, \quad (C.4)$$

where $\hat{1}_B$ is the identity operator acting on the Hilbert space of the environment $\mathcal{H}_B$, $\hat{1}_S$ is the identity operator on the Hilbert space of the system $\mathcal{H}_S$, and H_{int} describes the interaction between our system and the environment. I have also introduced a small parameter α to stress that I consider this interaction *weak* so that some sort of perturbation theory on the interaction can be performed. The Hamiltonians $\hat{H}_{tot}$ and H_{int} are operators acting on the tensor product Hilbert space $\mathcal{H}_{tot} = \mathcal{H}_S \otimes \mathcal{H}_B$.

The statistical operator of the total system $\hat{\rho}_{tot}(t)$ follows the Liouville–von Neumann equation 1.60

$$\frac{d\hat{\rho}_{tot}(t)}{dt} = -\frac{i}{\hbar}\left[\hat{H}_{tot}, \hat{\rho}_{tot}(t)\right] \equiv \mathcal{L}_{tot}\hat{\rho}_{tot}(t) = (\mathcal{L}_S + \mathcal{L}_B + \alpha \mathcal{L}_{int})\hat{\rho}_{tot}, \quad (C.5)$$

where I have introduced super-operators for each Hamiltonian appearing in Eq. C.4. A formal solution of the above equation is (if $\mathcal{L}_{tot}$ is time-independent)

$$\hat{\rho}_{tot}(t) = e^{\mathcal{L}_{tot}t}\hat{\rho}(t = t_0). \quad (C.6)$$

I am interested in the dynamics of the *reduced* statistical operator $\hat{\rho}_S(t)$

$$\hat{\rho}_S(t) = \text{Tr}_B\{\hat{\rho}_{tot}(t)\}, \quad (C.7)$$

where Tr_B means that we are tracing out all degrees of freedom of the environment B. As discussed in Sec. 1.4.5, this operator does not generally obey a closed equation of motion of the type C.1. Let us then proceed as follows.

The projection method

Let us suppose the two subsystems S and B are *uncorrelated* at the initial time t_0 (and earlier times as well)

$$\hat{\rho}_{tot}(t_0) = \hat{\rho}_S(t_0) \otimes \hat{\rho}_B(t_0), \quad (C.8)$$

where the two statistical operators $\hat{\rho}_S(t_0)$ and $\hat{\rho}_B(t_0)$ act on their respective Hilbert spaces.

C.2 Derivation of the Lindblad equation

Let us now define a *super-projector* P (a time-independent linear operator acting on $\mathcal{H}_{tot}$, with $P^2 = P$) such that I can divide the total statistical operator in two parts[8]

$$\hat{\rho}_{tot} = P\hat{\rho}_{tot} + (1-P)\hat{\rho}_{tot} \equiv \hat{\rho}_1 + \hat{\rho}_2. \tag{C.9}$$

For a general operator $\hat{X}$ on the composite Hilbert space $\mathcal{H}_{tot}$ I define the action of this operator as follows[9]

$$P\hat{X} = \hat{\rho}_B(t_0)\mathrm{Tr}_B\{\hat{X}\}, \tag{C.10}$$

which for the total statistical operator is simply

$$P\hat{\rho}_{tot}(t) = \hat{\rho}_B(t_0)\mathrm{Tr}_B\{\hat{\rho}_{tot}(t)\} \equiv \hat{\rho}_B(t_0)\hat{\rho}_S(t), \tag{C.11}$$

namely it projects into the subspace S without otherwise affecting the environment degrees of freedom. By construction, it thus commutes with both $\mathcal{L}_S$ and $\mathcal{L}_B$ and its action on $\mathcal{L}_B$ is zero

$$P\mathcal{L}_S - \mathcal{L}_S P = 0, \qquad P\mathcal{L}_B = \mathcal{L}_B P = 0. \tag{C.12}$$

Energy renormalization due to the bath

In addition, let us assume that the action of the interaction Hamiltonian averages to zero over the environment degrees of freedom

$$\mathrm{Tr}_B\{\hat{H}_{int}\hat{\rho}_B(t_0)\} = 0. \tag{C.13}$$

In terms of the super-projector P this is equivalent to

$$P\mathcal{L}_{int}P = 0, \tag{C.14}$$

or explicitly in terms of the statistical operators

$$\left[\mathrm{Tr}_B\{\hat{H}_{int}\hat{\rho}_B(t_0)\}\hat{\rho}_S(t) - \hat{\rho}_S(t)\mathrm{Tr}_B\{\hat{H}_{int}\hat{\rho}_B(t_0)\}\right] = 0. \tag{C.15}$$

Assumption C.13 is reasonable for homogeneous interactions with the environment. If it is not satisfied, we can always absorb the average of the interaction Hamiltonian $\hat{H}_{int}$ over the equilibrium state of the environment into the Hamiltonian $\hat{H}_S$: it simply contributes to the unitary evolution of our system S. This is equivalent to saying that the eigenvalues of the system Hamiltonian are *renormalized* by the interaction with the bath.

---

[8] By definition C.9, $\hat{\rho}_2$ is not actually a statistical operator in the proper sense, but simply the difference between statistical operators.

[9] This is one particular choice. For another type of super-projector see, e.g., Zwanzig, 1960.

Memory kernels

The Liouville–von Neumann equation C.5 can then be written in terms of two coupled equations for $\hat{\rho}_1$ and $\hat{\rho}_2$

$$P\frac{d\hat{\rho}_{tot}(t)}{dt} = \frac{d\hat{\rho}_1(t)}{dt} = \mathcal{L}_S\hat{\rho}_1(t) + \alpha P\mathcal{L}_{int}\hat{\rho}_2(t), \tag{C.16}$$

$$(1-P)\frac{d\hat{\rho}_{tot}(t)}{dt} = \frac{d\hat{\rho}_2(t)}{dt} = (1-P)\mathcal{L}_{tot}\hat{\rho}_2(t) + \alpha(1-P)\mathcal{L}_{int}\hat{\rho}_1(t). \tag{C.17}$$

The formal solution of Eq. C.17 is

$$\hat{\rho}_2(t) = \alpha \int_{t_0}^{t} d\tau\, e^{\tau(1-P)\mathcal{L}_{tot}}(1-P)\mathcal{L}_{int}\hat{\rho}_1(t-\tau)$$

$$= \alpha \int_{t_0}^{t} d\tau\, e^{\tau(1-P)\mathcal{L}_{tot}}\mathcal{L}_{int}\hat{\rho}_1(t-\tau), \tag{C.18}$$

where in the last equality I have used Eq. C.14.

By substituting C.18 into Eq. C.16 gives (using also the definition C.11 of super-projector)

$$\frac{d\hat{\rho}_S(t)}{dt} = \mathcal{L}_S\hat{\rho}_S(t) + \alpha^2 \int_{t_0}^{t} d\tau\, F(\tau)\hat{\rho}_S(t-\tau), \tag{C.19}$$

where

$$F(\tau) = \mathrm{Tr}_B\{\mathcal{L}_{int} e^{\tau(1-P)\mathcal{L}_{tot}}\mathcal{L}_{int}\hat{\rho}_B(t_0)\} \tag{C.20}$$

is sometimes called the *memory kernel*.

Equation C.19 is still exact, namely it is an equivalent form of the Liouville–von Neumann equation C.5. It is not yet in the Lindblad form C.1, since one needs the full *history* of the dynamics of the reduced statistical operator $\hat{\rho}_S$.

We can simplify it by assuming that there exists an autocorrelation time τ_c (see definition in Sec. 5.1) such that

$$F(\tau) \simeq 0, \quad \text{for } \tau > \tau_c. \tag{C.21}$$

For this to be true the environment B must have a *dense* energy spectrum so that its internal correlations cannot affect its dynamics in the long-time limit.

This approximation allows us to extend the integral range in Eq. C.19 from t to ∞, when $t > \tau_c$. In addition, since P projects out the action of both $\mathcal{L}_S$ and $\mathcal{L}_B$ we have

$$P\mathcal{L}_{tot} = \alpha P\mathcal{L}_{int}. \tag{C.22}$$

C.2 Derivation of the Lindblad equation

If the interaction $\hat{H}_{int}$ is weak we can formally expand the exponential appearing in the function $F(\tau)$ and retain only terms of order α^2 in Eq. C.19. We then get[10]

$$\frac{d\hat{\rho}_S(t)}{dt} = \mathcal{L}_S \hat{\rho}_S(t) + \alpha^2 \int_0^\infty d\tau \left[\text{Tr}_B \{ \mathcal{L}_{int} e^{\tau(\mathcal{L}_S + \mathcal{L}_B)} \mathcal{L}_{int} \hat{\rho}_B(t_0) \} \right] \hat{\rho}_S(t - \tau). \quad (C.23)$$

Markov approximation

This equation still contains memory. Let us then call τ_S the typical time scale dictated by the unitary evolution of the system S (namely, caused by $\mathcal{L}_S$) over which the state of the system S changes even in the absence of the environment. If

$$\tau_c \ll \tau_S \quad (C.24)$$

we can make the *Markov approximation*[11]

$$\hat{\rho}_S(t - \tau) \simeq \hat{\rho}_S(t), \quad (C.25)$$

so that no history (memory) of the statistical operator is required to determine its time evolution. This is equivalent to saying that when the time evolution of our system is long compared to the time necessary for the environment to "forget" quantum correlations, the system dynamics is *local* in time.

We may then neglect the action of $\mathcal{L}_S$ in Eq. C.23, so that the latter becomes

$$\frac{d\hat{\rho}_S(t)}{dt} = \left\{ \mathcal{L}_S + \alpha^2 \int_0^\infty d\tau \left[\text{Tr}_B \{ \mathcal{L}_{int} e^{\tau \mathcal{L}_B} \mathcal{L}_{int} \hat{\rho}_B(t_0) \} \right] \right\} \hat{\rho}_S(t). \quad (C.26)$$

The above is the desired equation of motion – which is in the Lindblad form C.1 – for the reduced statistical operator $\hat{\rho}_S(t)$.

Let us write the interaction Hamiltonian as[12]

$$\hat{H}_{int} = \hat{V}_S \otimes \hat{V}_B, \quad (C.27)$$

---

[10] The next term in the series contains a time integral of range τ_c. Therefore, this term is of order $\alpha^3 \tau_c^3$. "Weakness" of $\hat{H}_{int}$ then strictly means $\alpha \tau_c \ll 1$.
[11] The relative error one makes is thus of order τ_c / τ_S.
[12] We could write $\hat{H}_{int}$ as the sum of product operators. This would give rise to a summation in Eq. C.28.

namely the product between an operator acting on the system and an operator acting on the environment. Equation C.26 then becomes

$$\frac{d\hat{\rho}_S(t)}{dt} = \mathcal{L}_S \hat{\rho}_S(t) - \frac{\alpha^2}{2\hbar} \int_0^\infty d\tau \left[\text{Tr}_B\{\hat{V}_B e^{\tau \mathcal{L}_B} \hat{V}_B \hat{\rho}_B(t_0)\} \hat{V}_S [\hat{V}_S, \hat{\rho}_S(t)] \right.$$
$$\left. - \text{Tr}_B\{\hat{V}_B e^{\tau \mathcal{L}_B} \hat{V}_B \hat{\rho}_B(t_0)\} [\hat{V}_S, \hat{\rho}_S(t)] \hat{V}_S \right], \qquad (C.28)$$

which can be written in operator form as in Eq. C.2.

C.3 Steady-state solutions

The steady-state solutions, $\hat{\rho}_S^{ss}$, of the Lindblad equation C.2 are defined as

$$\frac{d\hat{\rho}_S^{ss}}{dt} = 0 \quad \Longrightarrow \quad \mathcal{L} \hat{\rho}_S^{ss} = 0. \qquad (C.29)$$

This is equivalent to finding the eigenstates of the super-operator $\mathcal{L}$ with zero eigenvalue

$$(\mathcal{L} - \lambda \hat{1}) \hat{\rho}_S^{ss} = 0, \qquad \lambda = 0, \qquad \text{stationary solutions.} \qquad (C.30)$$

For an arbitrary $\mathcal{L}$ – not necessarily time-independent – the above equation may have no solution or many solutions, namely several steady-state statistical operators in the long-time limit. All other non-zero eigenvalues, $\lambda \neq 0$, correspond to non-stationary states.

Appendix D
Ground-state Density-Functional Theory

I summarize here the concepts of density-functional theory (DFT) as applied to the ground state of many-electron systems. The reader interested in a comprehensive description of ground-state DFT is urged to consult the book by Parr and Young (1989).

D.1 The Hohenberg–Kohn theorem

Let us start from an N-electron Hamiltonian

$$\hat{\mathbf{H}} = \hat{\mathbf{T}} + \hat{\mathbf{W}} + \hat{\mathbf{V}}_{ext}, \qquad (D.1)$$

where $\hat{\mathbf{T}}$ is the kinetic operator and $\hat{\mathbf{W}}$ is the electron-electron interaction operator. The operator $\hat{\mathbf{V}}_{ext}$ describes a local static external potential (like the electron-ion potential in Eq. 6.2) and can be written in terms of the density operator 1.22 as (see Eq. A.26)

$$\hat{\mathbf{V}}_{ext} = \int d\mathbf{r}\, V_{ext}(\mathbf{r})\, \hat{n}(\mathbf{r}). \qquad (D.2)$$

Let us then take an arbitrary many-body wave-function $|\Psi\rangle$, antisymmetric in the exchange of orbital and spin coordinates of two electrons and construct the density (see Sec. 2.8.3)

$$n(\mathbf{r}) = N \sum_{s_1 \ldots s_N} \int d\mathbf{r}_2 \ldots d\mathbf{r}_N\, |\Psi(\mathbf{r}, s_1; \mathbf{r}_2, s_2; \ldots; \mathbf{r}_N, s_N)|^2, \qquad (D.3)$$

where I have indicated with s_i the spin coordinate of electron i. This density satisfies the condition

$$\int d\mathbf{r}\, n(\mathbf{r}) = N, \qquad (D.4)$$

where the integral is over all space.

Let us now assume that for a given $V_{ext}(\mathbf{r})$ we have found a density $n(\mathbf{r})$, satisfying D.4, which corresponds to the *ground state* of Hamiltonian D.1. Such a density is called *V-representable*.[1] Changing the external potential, clearly changes the density.

On the other hand, the Hohenberg–Kohn theorem states that (Hohenberg and Kohn, 1964)

Theorem: *Two external potentials, which differ by more than a constant, cannot give the same ground-state density,*

thus establishing a one-to-one correspondence between the external potential and the ground-state density.

For a proof of this theorem I refer the reader to the original paper (Hohenberg and Kohn, 1964) or the book by Parr and Young (1989). The above theorem has tremendous consequences since it implies that, if one knows the ground-state density, then, in principle, one can determine the scalar external potential that determines that density, hence the Hamiltonian D.1, from which the ground-state many-body wave-function can be obtained, and, therefore, *all* ground-state properties of the N-electron system (see also Sec. 2.8.3).

D.2 The Kohn–Sham equations

From the above theorem we also know that the total energy of the system in its ground state is a *functional* of the density

$$E[n] = \langle \Psi_{gs}|\hat{H}|\Psi_{gs}\rangle = \langle \Psi_{gs}|\hat{\mathbf{T}}+\hat{\mathbf{W}}|\Psi_{gs}\rangle + \langle \Psi_{gs}|\hat{\mathbf{V}}_{ext}|\Psi_{gs}\rangle$$
$$\equiv F[n] + \int d\mathbf{r}\, V_{ext}(\mathbf{r})\, n(\mathbf{r}), \qquad (D.5)$$

where $|\Psi_{gs}\rangle$ is a ground-state many-body wave-function, and $n(\mathbf{r})$ is the corresponding density. The functional $F[n]$ is the *same* for all electron systems and we make the following *ansatz*

$$F[n] = T_s[n] + \frac{e^2}{2}\int d\mathbf{r}\int d\mathbf{r}'\frac{n(\mathbf{r})n(\mathbf{r}')}{|\mathbf{r}-\mathbf{r}'|} + E_{xc}[n]$$
$$\equiv T_s[n] + E_H[n] + E_{xc}[n], \qquad (D.6)$$

where $T_s[n]$ is the kinetic energy of a non-interacting system with the *same* ground-state density $n(\mathbf{r})$ as the interacting one. The second term in Eq. D.6

---

[1] Note that, if the ground state is degenerate, given an external potential $V_{ext}(\mathbf{r})$, we can find more than one ground-state density.

D.2 The Kohn–Sham equations

is the Hartree energy, and the last term is called the *exchange-correlation energy functional*, and contains information on the exchange interaction between the electrons and *all* other correlation effects of the many-electron system.[2]

The *functional derivative* of the Hartree part of D.6 with respect to the density gives the Hartree potential energy 3.11

$$V_H(\mathbf{r}) = e^2 \int d\mathbf{r}' \frac{n(\mathbf{r}')}{|\mathbf{r} - \mathbf{r}'|} = \frac{\delta E_H[n]}{\delta n(\mathbf{r})}. \tag{D.7}$$

Similarly, we define the *exchange-correlation potential* as the functional derivative

$$V_{xc}(\mathbf{r}) \equiv \frac{\delta E_{xc}[n]}{\delta n(\mathbf{r})}. \tag{D.8}$$

Finally, by minimizing the functional

$$E_{KS}[n] = T_s[n] + E_H[n] + E_{xc}[n] + \int d\mathbf{r}\, V_{ext}(\mathbf{r})\, n(\mathbf{r}) \tag{D.9}$$

with respect to the density subject to constraint D.4 we find the Kohn–Sham (KS) equations

$$\boxed{\left[-\frac{\hbar^2}{2m}\nabla^2 + V_H(\mathbf{r}) + V_{xc}(\mathbf{r}) + V_{ext}(\mathbf{r})\right]\phi_k^{KS}(\mathbf{r}, s) = \epsilon_k \phi_k^{KS}(\mathbf{r}, s),} \tag{D.10}$$

that correspond to the solution of the Schrödinger equation of *auxiliary* non-interacting electrons in the presence of the potential $V_{KS} \equiv V_H(\mathbf{r}) + V_{xc}(\mathbf{r}) + V_{ext}(\mathbf{r})$.

Solving the above equations yields the wave-functions $\phi_k^{KS}(\mathbf{r}, s)$, from which the ground-state density is

$$n(\mathbf{r}) = \sum_s \sum_{k=1}^{N} |\phi_k^{KS}(\mathbf{r}, s)|^2, \tag{D.11}$$

that is equivalent to the density of a single *Slater determinant* (Eq. A.1)

$$|\Psi^{KS}\rangle = \frac{1}{\sqrt{N!}} \det\left[|\phi_{k_i}^{KS}\rangle\right] \tag{D.12}$$

constructed from the N lowest-occupied KS wave-functions.[3]

---

[2] Physically, this energy corresponds to the Coulomb interaction between any electron and the *exchange-correlation hole* that accompanies it. This hole is due to the repulsive electron-electron interaction which "depletes" charge density around that given electron.

[3] Like the single-particle KS states, this Slater determinant is just an "operational" tool, and generally *not* a good approximation of the many-body wave-function.

D.3 Generalization to grand-canonical equilibrium

The Hohenberg–Kohn theorem can be generalized to the case in which the system is in grand-canonical equilibrium with a reservoir specified by a chemical potential $\bar{\mu}$ and temperature θ. In this case the theorem states (Mermin, 1965)

Theorem: *Given a many-particle system in grand-canonical equilibrium with a reservoir at chemical potential $\bar{\mu}$ and temperature θ, two external potentials, which differ by more than a constant, cannot give the same ensemble-averaged density,*

where the ensemble-averaged density is

$$n(\mathbf{r}) = \langle \hat{n}(\mathbf{r}) \rangle = \operatorname{Tr}\left\{\hat{\rho}_G^{eq}\, \hat{n}(\mathbf{r})\right\}, \tag{D.13}$$

with $\hat{\rho}_G^{eq}$ the grand-canonical statistical operator 1.76. The expectation value of any observable can then be written, in principle, as a functional of the ensemble-averaged density D.13.

Note also that any approximation to the exact functional makes the Kohn-Sham Hamiltonian *stochastic*, i.e., a different Hamiltonian for every element of the statistical ensemble (Sec. 1.4.4).

D.4 The local density approximation and beyond

In order to solve Eqs D.10 one needs to know the exact exchange-correlation functional. This is unknown and approximations are thus necessary. The most commonly used one is known as the *local density approximation* (LDA). It consists in approximating the exchange-correlation functional $E_{xc}[n]$ as

$$E_{xc}^{LDA}[n] = \int d\mathbf{r}\, n(\mathbf{r})\, \epsilon_{xc}\left(n(\mathbf{r})\right), \tag{D.14}$$

where $\epsilon_{xc}\left(n(\mathbf{r})\right)$ is the exchange-correlation energy per particle of the *homogeneous* electron liquid with density n. The LDA exchange-correlation potential at position $\mathbf{r}$ is then

$$V_{xc}^{LDA}(\mathbf{r}) \equiv \left.\frac{d\left[n\,\epsilon_{xc}(n)\right]}{dn}\right|_{n(\mathbf{r})}, \tag{D.15}$$

that of a homogeneous electron liquid that has density n at that position. Approximate formulas for the exchange-correlation energy per particle of the homogeneous electron liquid have been proposed (Perdew and Zunger, 1981) based on interpolation of Monte Carlo numerical results (Ceperley and Alder, 1980).

D.4 The local density approximation and beyond

Several other schemes have been suggested to improve on the LDA. One, known as *generalized gradient approximation* (GGA), assumes a form for the exchange-correlation energy functional where the gradient of the density of the electron liquid at any point in space appears explicitly (see, e.g., Perdew *et al.*, 1996). For isolated systems these functionals improve the binding energy compared to LDA. However, I am not aware of a systematic study of the differences between these functionals and LDA on static transport problems.

In practice, given an approximate form for the exchange-correlation scalar potential, Eqs. D.10 are solved iteratively until self-consistency – in the density and/or potential – is achieved within a given numerical accuracy. For instance, one starts from a guess for the density $n(\mathbf{r})$, calculates the potential that enters the KS equations D.10, solves the latter with given boundary conditions – in the case of steady-state transport with scattering boundary conditions, Chapter 3 – and calculates a new density via Eq. D.11.

Appendix E
Time-Dependent DFT

I extend the concepts of ground-state DFT to the time-dependent domain, namely for many-body systems out of equilibrium. A comprehensive review of time-dependent DFT (TDDFT) can be found in, e.g., Marques *et al.* (2006).

E.1 The Runge–Gross theorem

Let us start from an N-electron Hamiltonian

$$\hat{H}(t) = \hat{T} + \hat{W} + \hat{V}_{ext}(t) \tag{E.1}$$

where $\hat{T}$ is the kinetic operator and $\hat{W}$ is the electron-electron interaction operator. The operator $\hat{V}_{ext}(t)$ is a time-dependent local external potential

$$\hat{V}_{ext}(t) = \int d\mathbf{r}\, V_{ext}(\mathbf{r}, t)\, \hat{n}(\mathbf{r}). \tag{E.2}$$

Let the system described by Hamiltonian E.1 evolve from an arbitrary initial condition $|\Psi_0\rangle$ at time t_0. Call $n(\mathbf{r}, t)$ the density that evolves from the initial state $|\Psi_0\rangle$ under the action of potential $V_{ext}(\mathbf{r}, t)$, and call $n'(\mathbf{r}, t)$ the density evolving from the *same* initial state $|\Psi_0\rangle$ under the action of potential $V'_{ext}(\mathbf{r}, t)$. The Runge–Gross theorem (Runge and Gross, 1984) states that

> **Theorem:** *If two external potentials differ by more than a time-dependent (space-independent) constant*
>
> $$V_{ext}(\mathbf{r}, t) \neq V'_{ext}(\mathbf{r}, t) + c(t), \tag{E.3}$$
>
> *then the corresponding densities, $n(\mathbf{r}, t)$ and $n'(\mathbf{r}, t)$, are different.*

For a proof of this theorem I refer the reader to the original paper (Runge and Gross, 1984; see also van Leeuwen, 1999). I make here only some remarks on the conditions necessary for its proof. First, the potentials $V_{ext}(\mathbf{r},t)$ and $V'_{ext}(\mathbf{r},t)$ need to be expandable in Taylor series about the initial time t_0. Second, the density has to go to zero at infinity – which is the case for a finite closed system with bound electrons – or it is the density of an infinite but periodic system. The theorem does not exclude the possibility that two different potentials produce the *same* density if the time evolution occurs from two *different* initial conditions.[1]

Finally, the current density is not generally V-representable: the mapping between current density and external scalar potential is not invertible (D'Agosta and Vignale, 2005). This is easy to understand: at any given time and point in space a scalar potential provides just one value, while the current density is a vector. Therefore, the KS current density in TDDFT is *not* guaranteed to be exact; only its divergence via the continuity equation 1.11.

E.2 The time-dependent Kohn–Sham equations

From the above theorem one can formulate the time-dependent version of the Kohn–Sham equations

$$\left[i\hbar\frac{\partial}{\partial t} + \frac{\hbar^2}{2m}\nabla^2 - V_H(\mathbf{r},t) - V_{xc}(\mathbf{r},t) - V_{ext}(\mathbf{r},t)\right]\phi_k^{KS}(\mathbf{r},s,t) = 0, \quad \text{(E.4)}$$

where $V_H(\mathbf{r},t)$ and $V_{xc}(\mathbf{r},t)$ are the time-dependent generalizations of the corresponding potentials D.7 and D.8, respectively, of the static case. The density is then

$$n(\mathbf{r},t) = n^{KS}(\mathbf{r},t) \equiv \sum_s \sum_{k=1}^{N} |\phi_k^{KS}(\mathbf{r},s,t)|^2. \quad \text{(E.5)}$$

E.3 The adiabatic local density approximation

The time-dependent version of the KS equations poses further requirements on the construction of the exchange-correlation scalar potential. The potential $V_{xc}(\mathbf{r},t)$ depends on the *history* of the system – included initial condition – so that it cannot be written simply as the functional derivative with respect to the density.

---

[1] For a discussion of these conditions see Giuliani and Vignale (2005).

It is nonetheless common to approximate it as the ground-state exchange-correlation potential D.15 evaluated at the *instantaneous* density $n(\mathbf{r},t)$. This is known as the *adiabatic local density approximation* (ALDA):

$$V_{xc}^{ALDA}(\mathbf{r},t) = \left.\frac{d\left[n\,\epsilon_{xc}(n)\right]}{dn}\right|_{n(\mathbf{r},t)}. \tag{E.6}$$

Equations E.4 are generally solved the following way. Assign an initial condition. This is typically (but not necessarily) the solution of the ground-state KS equations for some Hamiltonian (e.g., the Hamiltonian in Eqs. E.4 without the time-dependent part in the potentials). These equations are then solved numerically by time-evolving the initial KS states, where at every time step one determines the density from which the Hartree potential and the exchange-correlation potential can be determined.

Due to numerical instabilities, typical time evolutions that can be achieved with small numbers of atoms and/or electrons are of the order of tens of femtoseconds. A small time scale, but enough to establish a quasi-steady state in a nanoscale system (see Sec. 7.6).

In order to improve beyond the ALDA one has to work with time-dependent *current* DFT (Appendix F).

Appendix F
Time-Dependent Current DFT

In this appendix I outline the main tenets of time-dependent current DFT (TD-CDFT). A pedagogical and extended account of this approach can be found in the book by Giuliani and Vignale (2005).

F.1 The current density as the main variable

In Appendix E I have hinted at the problem of obtaining a local representation of the exchange-correlation scalar potential V_{xc} (the one appearing in the Kohn–Sham equations E.4) in terms of the density. Without going into details, I just mention that the physical reason why such a representation does not exist for an inhomogeneous electron liquid is related to the strong non-local (in space) functional dependence of this potential on the density.

A way out of this problem is to realize that due to gauge invariance a time-dependent scalar potential $V(\mathbf{r}, t)$ can always be represented by a longitudinal vector potential $\mathbf{A}(\mathbf{r}, t)$.[1] We also know that the density is related to the current density via the continuity equation 1.11. Putting these two facts together, we therefore see that the response of the density operator to an external scalar potential can be expressed as the response of the *current density* operator to an external vector potential, i.e., we can reformulate time-dependent density-functional theory in terms of the current density instead of the density. It turns out that a local representation of the exchange-correlation *vector potential* in terms of the current density can be derived.

---

[1] By longitudinal I mean that it has a finite divergence and zero curl. The relation between the longitudinal vector and scalar potentials is

$$\mathbf{A}(\mathbf{r}, t) = \frac{c}{e} \int_{t_0}^{t} \nabla V(\mathbf{r}, t') dt'.$$

Using the current density as the main variable we can then formulate the following theorem of TD-CDFT due to Ghosh and Dhara (Ghosh and Dara, 1988)

Theorem: *Given an initial quantum state $|\Psi_0\rangle$, two external vector potentials $\mathbf{A}_{ext}(\mathbf{r},t)$ and $\mathbf{A}'_{ext}(\mathbf{r},t)$ that produce the same current density $\mathbf{j}(\mathbf{r},t)$, must necessarily coincide, up to a gauge transformation.*

The given current density is then said to be **A**-*representable*.[2] It is also obvious why – unlike in TDDFT – the map between the current density and vector potentials is generally invertible: they are both vectors.

The above theorem also guarantees that the expectation value of any physical observable can be in principle written as a functional of the current density and the given initial condition.

In addition, one can choose to work with an auxiliary (Kohn–Sham) system in which the interactions among particles have been switched off and solve the TD-CDFT Kohn–Sham equations

$$\left[i\hbar\frac{\partial}{\partial t} - \frac{1}{2m}\left(-i\hbar\nabla - \frac{e}{c}\mathbf{A}_{ext}(\mathbf{r},t) - \frac{e}{c}\mathbf{A}_{xc}(\mathbf{r},t)\right)^2 - V_{ext}(\mathbf{r}) - V_H(\mathbf{r},t)\right]\phi_k^{KS}(\mathbf{r},t) = 0, \quad (\text{F.1})$$

where, due to gauge invariance, all time dependence of the external potential has been included in the vector potential $\mathbf{A}_{ext}(\mathbf{r},t)$,[3] and $\mathbf{A}_{xc}(\mathbf{r},t)$ is an exchange-correlation vector potential that is a functional of the initial condition $|\Psi_0\rangle$ and the current density $\mathbf{j}(\mathbf{r},t')$ at all *previous* times $t' \leq t$. This functional describes all quantum correlations in the system and is dependent on its *history*.

F.2 The exchange-correlation electric field

Within the above formulation it was shown that an approximate local representation of $\mathbf{A}_{xc}(\mathbf{r},t)$ can be obtained (Vignale and Kohn, 1996).

To be precise, let us consider an external field with frequency ω and wave-vector k. Let us call v_F and k_F the local (in space) Fermi velocity and wave-vector of the electron liquid, respectively, and $q = |\nabla n_0/n_0|$ the rate

---

[2] For a different, and more recent, proof of the theorem see Vignale (2004).
[3] For a static electric field, $\mathbf{A}(\mathbf{r},t)$ can be chosen identically zero, and $V_{ext}(\mathbf{r})$ is simply the static electron-ion interaction.

F.2 The exchange-correlation electric field

of spatial variation of its ground-state density n_0. Under the assumptions that $q, k \ll \omega/v_F, k_F$, one can show, using linear-response theory, that the exchange-correlation vector potential $\mathbf{A}_{xc}(\mathbf{r}, t)$ is local in space (albeit non-local in time) and can be written as (Vignale et al., 1997)

$$\frac{e}{c}\frac{\partial \mathbf{A}_{xc,i}(\mathbf{r}, t)}{\partial t} \equiv -e\mathbf{E}_{xc}(\mathbf{r}, t) = -\nabla_i V_{xc}^{ALDA}(\mathbf{r}, t) + \frac{1}{n(\mathbf{r})}\sum_j \frac{\partial \sigma_{xc,ij}(\mathbf{r}, t)}{\partial r_j}, \tag{F.2}$$

where $V_{xc}^{ALDA}(\mathbf{r}, t)$ is given in Eq. E.6 and $\sigma_{xc,ij}(\mathbf{r}, t)$ is a stress tensor that has the following *hydrodynamic* form in d dimensions

$$\sigma_{xc,ij}(\mathbf{r}, t) = \int_{t_0}^{t}\left\{\tilde{\eta}(n(\mathbf{r}), t, t')\left[\nabla_j v_i(\mathbf{r}, t') + \nabla_i v_j(\mathbf{r}, t') - \frac{2}{d}\delta_{i,j}\nabla_k v_k(\mathbf{r}, t')\right]\right.$$
$$\left. + \tilde{\zeta}(n(\mathbf{r}), t, t')\nabla \cdot \mathbf{v}(\mathbf{r}, t')\delta_{ij}\right\}dt'. \tag{F.3}$$

The complex coefficients $\tilde{\eta}$ and $\tilde{\zeta}$ represent, respectively, the shear and bulk viscosity of the electron liquid. With the assumption that the external perturbation is periodic in time we can Fourier transform Eq. F.2 to get

$$-e\mathbf{E}_{xc}(\mathbf{r}, \omega) = -\nabla_i V_{xc}^{ALDA}(\mathbf{r}, \omega) + \frac{1}{n(\mathbf{r})}\sum_j \frac{\partial \sigma_{xc,ij}(\mathbf{r}, \omega)}{\partial r_j}, \tag{F.4}$$

with

$$\sigma_{xc,ij}(\mathbf{r}, \omega) = \tilde{\eta}(n, \omega)\left[\nabla_j v_i + \nabla_i v_j - \frac{2}{d}\delta_{i,j}\nabla_k v_k\right] + \tilde{\zeta}(n, \omega)\nabla \cdot \mathbf{v}\delta_{ij}. \tag{F.5}$$

The functions $\tilde{\eta}(n, \omega)$ and $\tilde{\zeta}(n, \omega)$ are generally complex. Their full expressions in terms of the exchange-correlation kernel of the homogeneous electron gas can be found in Giuliani and Vignale (2005). Below I report approximate formulas for the real part of the viscosity coefficient in the zero-frequency limit, the only quantity that is of interest in most of this book.[4]

Alternatively, one can choose to work in a different gauge and solve the equations

$$\left[i\hbar\frac{\partial}{\partial t} - \frac{1}{2m}\left(-i\hbar\nabla - \frac{e}{c}\mathbf{A}_{ext}(\mathbf{r}, t) - \frac{e}{c}\mathbf{A}_{xc}^{dyn}(\mathbf{r}, t)\right)^2\right.$$
$$\left. - V_{ext}(\mathbf{r}) - V_H(\mathbf{r}, t) - V_{xc}^{ALDA}(\mathbf{r}, t)\right]\phi_k^{KS}(\mathbf{r}, t) = 0, \tag{F.6}$$

with

$$\mathbf{A}_{xc,i}^{dyn}(\mathbf{r}, t) = \mathbf{A}_{xc,i}^{dyn}(\mathbf{r}, t_0) + \frac{c}{e}\int_{t_0}^{t}dt'\frac{1}{n(\mathbf{r}, t')}\sum_j \frac{\partial \sigma_{xc,ij}(\mathbf{r}, t')}{\partial r_j}. \tag{F.7}$$

---

[4] In this limit only the shear viscosity is finite, the bulk viscosity being zero.

Note that in the above I have included the time dependence of the density as well. This is beyond the linear-response result expressed in Eq. F.2. However, it is reasonable to think that Eq. F.2 is valid beyond linear response provided both the density and the velocity field are slowly varying function of space so that non-linear terms, involving three-particle (and higher order) Green's functions, may be neglected (see discussion in Sec. 8.2.1 and Giuliani and Vignale, 2005).

Equations F.1 (or F.6) are solved similarly to Eqs. E.4 of TDDFT. One starts with an initial condition for the density – and typically from an initial condition with zero velocity field so that $\mathbf{A}^{dyn}_{xc,i}(\mathbf{r}, t_0) = 0$, and then evolves the initial KS states in time.

There is however a major difference with the standard TDDFT equations: the presence of an exchange-correlation vector potential that depends on the *history* of the dynamics in a non-trivial way.

Even if one uses the memory-less approximation 8.52, and thus the stress tensor F.3 can be approximated as in Eq. 8.53, one needs to determine the exchange-correlation vector potential at a time t as in Eq. F.7. This represents an extra numerical burden compared to the solution of the TDDFT equations E.4.

F.3 Approximate formulas for the viscosity

In the zero-frequency limit (or more precisely for $\hbar\omega \ll E_F$) the viscosity $\eta = \lim_{\omega \to 0} \text{Re}\,\tilde{\eta}(\omega)$ can be approximated in three and two dimensions as (Conti and Vignale, 1999)[5]

$$\eta_{3D} = \frac{\hbar n}{60(r_s/a_0)^{-3/2} + 80(r_s/a_0)^{-1} - 40(r_s/a_0)^{-2/3} + 62(r_s/a_0)^{-1/3}}, \quad (\text{F.8})$$

and

$$\eta_{2D} = \left[\left(\frac{(r_s/a_0)^2}{6\pi} \ln\sqrt{\frac{2a_0}{er_s}} + 0.25(r_s/a_0)^2\right)^{-1} + 21(r_s/a_0)^{-2} \right.$$
$$\left. + 23(r_s/a_0)^{-1/2} + 13\right]^{-1} \hbar n, \quad (\text{F.9})$$

where e is the Nepero number, n the density, a_0 the Bohr radius, and $r_s = (3/4\pi n)^{1/3}$ in 3D, and $r_s = (1/\pi n)^{1/2}$ in 2D.

---

[5] Note that
$$\hbar n = \frac{1.6 \times 10^{-4}}{(r_s/a_0)^3} \text{ Pa s.}$$

F.3 Approximate formulas for the viscosity

These are the expressions used in Chapter 8 to estimate the viscosity. Note, however, that these formulas have been derived from a fit to numerical results based on mode-mode coupling theory (Nifosi et al., 1998), and may not describe well the viscosity of the electron liquid at low densities.

Indeed, the *exact* low-density limit of the viscosity of the electron liquid is not known. Due to strong correlation effects in the low-density electron liquid, where the electron system is close to crystallization, it is reasonable to suspect that the relative viscosity η/n might increase well above the values one obtains from Eqs. F.8 and F.9 (Sai et al., 2007a).

In addition, the presence of a nanostructure introduces an effective cut-off for the transfer of momentum between adjacent layers of the electron liquid, which is at the origin of the viscosity (see Appendix K). This effective cut-off may increase the viscosity beyond the value obtained from Eqs. F.8 and F.9.

Appendix G
Stochastic Time-Dependent Current DFT

In this appendix I extend the theorems of dynamical density functional theories to the case in which the system is not closed, namely it is in dynamical interaction with a bath, or a set of baths. This theory goes under the name of stochastic time-dependent current density-functional theory (Di Ventra and D'Agosta, 2007).

G.1 The stochastic Schrödinger equation

Since in density-functional theory Hamiltonians depend on the density and/or current density, they are in general different for every element of a statistical ensemble. This precludes us from writing a closed equation of motion for the statistical operator (see discussion in Sec. 1.4.4). In addition, we have discussed in Appendix C that for time-dependent Hamiltonians and/or bath operators positivity of the density operator is not guaranteed by the quantum master equation C.2. This means that one could obtain solutions of Eq. C.2 that do not correspond to physical states.

To avoid these pitfalls we therefore start from the following *stochastic* Schrödinger equation (see, e.g., van Kampen, 1992)[1]

$$\partial_t |\Psi(t)\rangle = -\frac{i}{\hbar} \hat{H}(t)|\Psi(t)\rangle - \frac{1}{2}\hat{V}^\dagger \hat{V}|\Psi(t)\rangle + \ell(t)\hat{V}|\Psi(t)\rangle, \qquad \text{(G.1)}$$

where $\hat{V}$ is an operator – possibly dependent on time – describing the interaction of the bath with the many-body system. Here $\hat{H}(t)$ is an Hermitian

---

[1] From a numerical point of view the advantage of the stochastic Schrödinger equation compared to an equation of motion for the density matrix is that the former requires the knowledge of $N-1$ components of the state vector (the -1 comes from the normalization constraint), while the latter $(N+2)(N-1)/2$ elements of the density matrix (taking into account the constraints of hermiticity and unit trace). However, several runs, each one corresponding to different realizations of the stochastic process $\ell(t)$, are necessary to compute ensemble averages with the stochastic Schrödinger equation.

operator, not necessarily equal to the Hamiltonian of the system in the *absence* of the bath. Indeed, it may also describe the *renormalization* of the states of the system due to the interaction with the bath (see discussion in Appendix C).[2] Given an initial condition, Eq. G.1 always defines a proper physical state, whether the Hamiltonian is stochastic or time-dependent, or both (and also if the operator $\hat{V}$ is time-dependent).

For simplicity we will take it to be the Hamiltonian of an interacting N-electron system subject to an external field with vector potential $\mathbf{A}_{ext}$

$$\hat{H}(t) = \frac{1}{2m}\sum_i \left[\hat{p}_i - \frac{e}{c}\mathbf{A}_{ext}(\hat{r}_i,t)\right]^2 + \frac{1}{2}\sum_{i\neq j} U(\hat{r}_i - \hat{r}_j), \qquad (G.2)$$

with $U(\hat{r}_i - \hat{r}_j)$ a potential describing two-particle interactions (e.g., the Coulomb potential 2.96).[3]

Without loss of generality the stochastic process, $\ell(t)$, is chosen such that it has both zero ensemble average and δ-autocorrelation, i.e.,

$$\overline{\ell(t)} = 0; \qquad \overline{\ell(t)\ell(t')} = \delta(t-t'), \qquad (G.3)$$

where the symbol $\overline{\cdots}$ indicates the average over a statistical ensemble of identical systems all prepared in some initial state $|\Psi_0\rangle$. The following discussion will be valid even if the initial state is not pure, namely, it represents a given macro-state of the system.

The last term on the right-hand side of Eq. G.1 describes precisely the *fluctuations* induced by the bath; the second term is the compensating *dissipative* part.

We can generalize the above to more than one bath, each described by an operator $\hat{V}_\alpha$, by solving

$$\partial_t|\Psi(t)\rangle = \left[-\frac{i}{\hbar}\hat{H}(t) - \frac{1}{2}\sum_\alpha \hat{V}_\alpha^\dagger \hat{V}_\alpha + \sum_\alpha \ell_\alpha(t)\hat{V}_\alpha\right]|\Psi(t)\rangle, \qquad (G.4)$$

with

$$\overline{\ell_\alpha(t)} = 0; \qquad \overline{\ell_\alpha(t)\ell_\beta(t')} = \delta_{\alpha\beta}\delta(t-t'). \qquad (G.5)$$

In what follows, I will consider only one bath operator. However, all conclusions apply to an arbitrary number of baths.

---

[2] Note that Eq. G.1 is a *linear* stochastic Schrödinger equation. Stochastic Schrödinger equations that contain *non-linear* terms may be derived which are equivalent to the linear equation G.1 (see, e.g., Collet and Gardiner, 1984). The difference between the linear and non-linear type of equations is in the interpretation of the stochastic process. In the former case it is interpreted as an "input" type of noise, e.g., due to fluctuations of the environment. In the second case, it is considered as the "output" fluctuations of the system dynamics induced by the presence of the environment. The theorem of stochastic TD-CDFT (Sec. G.3) applies to both cases.

[3] Here I am working in a gauge in which the dynamical external scalar potential is zero at any time.

The stochastic Schrödinger equation G.1 preserves the ensemble-averaged orthonormality of any two states. This is easy to show by expanding Eq. G.1 in a small interval of time Δt[4]

$$|\Psi(t+\Delta t)\rangle = \left[\hat{1} + \int_t^{t+\Delta t} dt' \left(-\frac{i}{\hbar}\hat{H}(t') - \frac{1}{2}\hat{V}^\dagger\hat{V} + \ell(t')\hat{V}\right)\right] |\Psi(t)\rangle. \quad (G.6)$$

The ensemble-averaged norm is thus preserved in time:[5]

$$\overline{\langle\Psi(t+\Delta t)|\Psi(t+\Delta t)\rangle} = \overline{\langle\Psi(t)|\Psi(t)\rangle}. \quad (G.7)$$

In general, given any two orthogonal states $|\Psi_\alpha(t)\rangle$ and $|\Psi_\beta(t)\rangle$

$$\overline{\langle\Psi_\alpha(t)|\Psi_\beta(t)\rangle} = \delta_{\alpha\beta}. \quad (G.8)$$

G.2 Derivation of the quantum master equation

The above results are equivalent to the formalism of statistical operators I have introduced in Sec. 1.4, provided the Hamiltonian does not depend on microscopic degrees of freedom, such as the density or current density. Indeed, as in Eq. 7.13, we define the statistical operator[6]

$$\hat{\rho}(t) = \overline{|\Psi(t)\rangle\langle\Psi(t)|} \equiv \sum_i p_i(t) |\Psi_i(t)\rangle\langle\Psi_i(t)|, \quad (G.9)$$

with $\sum_i p_i(t) = 1$, so that (from Eq. G.7)

$$\overline{\langle\Psi(t)|\Psi(t)\rangle} = \text{Tr}\{\hat{\rho}(t)\} = 1. \quad (G.10)$$

By construction, this statistical operator is Hermitian, positive, and its trace is conserved at all times.

Consider now a general operator $\hat{O}(t)$ in the Heisenberg picture (see Sec. 1.4), and define the following stochastic "evolution" operator

$$\hat{U}(t+\Delta t, t) = \left[\hat{1} + \int_t^{t+\Delta t} dt' \left(-\frac{i}{\hbar}\hat{H}(t') - \frac{1}{2}\hat{V}^\dagger\hat{V} + \ell(t')\hat{V}\right)\right], \quad (G.11)$$

---

[4] Note that due to the presence of the stochastic term $\ell(t)$ Eq. G.1 is a *stochastic differential equation*. Therefore, the "standard" calculus does not apply, and one has to use stochastic methods. In this appendix, I follow the Itô calculus (see, e.g., van Kampen, 1992).

[5] One may impose the normalization of the state vector for *every* realization of the stochastic process at every instant of time. This leads to a non-linear stochastic Schrödinger equation which provides an equivalent physical description as Eq. G.1 (see, e.g., Goetsch et al., 1995).

[6] In this definition I am assuming for simplicity that the initial state of the system is pure. If the initial state of the system is mixed with macro-state $\{|\Psi_0^n\rangle, p_n\}$, then definition G.9 of statistical operator includes an extra summation

$$\hat{\rho}(t) = \sum_n p_n \overline{|\Psi^n(t)\rangle\langle\Psi^n(t)|},$$

where $|\Psi^n(t)\rangle \equiv \{|\Psi_i^n(t)\rangle\}$ is the ensemble of state vectors corresponding to the initial condition $|\Psi_0^n\rangle$.

G.2 Derivation of the quantum master equation

where Δt is a small interval of time.

The expectation value of the operator $\hat{O}(t)$ at a time $t + \Delta t$ is then

$$\langle \hat{O}(t+\Delta t)\rangle = \langle \Psi_0|\hat{U}^\dagger(t+\Delta t, t)\hat{O}(t)\hat{U}^\dagger(t+\Delta t, t)|\Psi_0\rangle. \quad (G.12)$$

By carrying out the calculations explicitly and averaging over the stochastic process according to Eq. G.3 we find, in the limit $\Delta t \to 0$, the ensemble-averaged time evolution

$$\partial_t \overline{\langle \hat{O}(t)\rangle} = \frac{i}{\hbar}\overline{\left\langle \left[\hat{H}(t), \hat{O}(t)\right]\right\rangle} - \frac{1}{2}\overline{\langle \hat{V}^\dagger \hat{V}\hat{O}(t)\rangle} - \frac{1}{2}\overline{\langle \hat{O}(t)\hat{V}^\dagger\hat{V}\rangle} + \overline{\langle \hat{V}^\dagger \hat{O}(t)\hat{V}\rangle}.$$

$$= \frac{i}{\hbar}\left\langle \left[\hat{H}(t), \overline{\hat{O}(t)}\right]\right\rangle - \frac{1}{2}\langle \hat{V}^\dagger\hat{V}\overline{\hat{O}(t)}\rangle - \frac{1}{2}\langle \overline{\hat{O}(t)}\hat{V}^\dagger\hat{V}\rangle + \langle \hat{V}^\dagger \overline{\hat{O}(t)}\hat{V}\rangle, \quad (G.13)$$

where in the second equality I have assumed that the Hamiltonian and the bath operator do not depend explicitly on the wave-functions (or density), and the symbol $\langle \cdots \rangle$ means $\langle \Psi_0|\cdots|\Psi_0\rangle$, the average over the initial state.

Using the definition G.9 of statistical operator, I can rewrite Eq. G.13 as

$$\partial_t \text{Tr}\{\hat{\rho}(t)\hat{O}\} = \frac{i}{\hbar}\text{Tr}\{\hat{\rho}(t)[\hat{H}(t), \hat{O}]\}$$

$$+ \text{Tr}\left\{\hat{\rho}(t)\left(-\frac{1}{2}\hat{V}^\dagger\hat{V}\hat{O} - \frac{1}{2}\hat{O}\hat{V}^\dagger\hat{V} + \hat{V}^\dagger\hat{O}\hat{V}\right)\right\}$$

$$= -\frac{i}{\hbar}\text{Tr}\{[\hat{H}(t), \hat{\rho}(t)]\hat{O}\}$$

$$+ \text{Tr}\left\{\left(-\frac{1}{2}\hat{V}^\dagger\hat{V}\hat{\rho}(t) - \frac{1}{2}\hat{\rho}(t)\hat{V}^\dagger\hat{V} + \hat{V}\hat{\rho}(t)\hat{V}^\dagger\right)\hat{O}\right\}, \quad (G.14)$$

where I have used the cyclic property of the trace (Eq. E1.3 in Exercise 1.1), and the operator $\hat{O}$ is now in the Schrödinger picture.

Since the above result must be true for an arbitrary operator $\hat{O}$, we conclude that the statistical operator satisfies the following equation of motion

$$\partial_t \hat{\rho}(t) = -\frac{i}{\hbar}\left[\hat{H}(t), \hat{\rho}(t)\right] - \frac{1}{2}\hat{V}^\dagger\hat{V}\hat{\rho}(t) - \frac{1}{2}\hat{\rho}(t)\hat{V}^\dagger\hat{V} + \hat{V}\hat{\rho}(t)\hat{V}^\dagger, \quad (G.15)$$

which is of the Lindblad form C.2, but with a time-dependent Hamiltonian.

If the Hamiltonian does not depend on the wave-functions, the physics described by the stochastic Schrödinger equation G.1 is thus the same as that described by a quantum master equation, with the additional benefit that the statistical operator derived from the stochastic Schrödinger equantion G.1 is, by definition, positive at all times.[7]

Using the number density and current density operators (Eqs. 1.24 and 1.25, respectively) the ensemble-averaged density is

$$\overline{n(\mathbf{r},t)} \equiv \overline{\langle \hat{n}(\mathbf{r},t) \rangle} = \text{Tr}\{\hat{\rho}(t)\,\hat{n}(\mathbf{r})\}, \qquad (\text{G.16})$$

and the ensemble-averaged current density

$$\overline{\mathbf{j}(\mathbf{r},t)} \equiv \overline{\langle \hat{\mathbf{j}}(\mathbf{r},t) \rangle} = \text{Tr}\{\hat{\rho}(t)\,\hat{\mathbf{j}}(\mathbf{r})\}. \qquad (\text{G.17})$$

From the first equality of Eq. G.13, we get the equation of motion for the ensemble-averaged density

$$\frac{\partial \overline{n(\mathbf{r},t)}}{\partial t} = -\nabla \cdot \overline{\mathbf{j}(\mathbf{r},t)} + \left\langle \hat{V}^\dagger \hat{n} \hat{V} - \frac{1}{2}\hat{V}^\dagger \hat{V} \hat{n} - \frac{1}{2}\hat{n}\hat{V}^\dagger \hat{V} \right\rangle. \qquad (\text{G.18})$$

This is the generalization of the continuity equation 1.11 for a system open to an environment. The last term on the right-hand side of Eq. G.18 is identically zero for bath operators that are local in space (Frensley, 1990). Most transport theories satisfy this requirement since the action that a true bath does on the system is derived from microscopic mechanisms (e.g., inelastic processes) which are generally local (Frensley, 1990).

If this is not the case, then this term represents *instantaneous* transfer of charge between disconnected – and possibly macroscopically far away – regions of the system without the need of mechanical motion (represented by the first term on the right-hand side of Eq. G.18). This instantaneous "action at a distance" is reminiscent of the postulate of wave-packet reduction (Sec. 1.3.3) whereby the system may change its state in a non-unitary way upon measurement.

Here, it is the result of the memory-less approximation that underlies the stochastic Schrödinger equation G.1. By assuming that the bath correlation times are much shorter than the times associated with the dynamics of the system (in fact, in the Markov approximation these correlation times are assumed zero), we have lost information on the microscopic interaction mechanisms at time scales of the order of the correlation times of the bath (see also discussion after Eq. C.25 in Appendix C). In other words, we have coarse-grained the time evolution of our system, and we are therefore unable to follow its dynamics on time scales smaller than this time resolution (see, e.g., Gebauer and Car, 2004).

---

[7] We also say that the stochastic Schrödinger equantion G.1 defines a *stochastic unraveling* of the corresponding quantum master equation.

G.3 The theorem of Stochastic TD-CDFT

We can now formulate the following theorem (Di Ventra and D'Agosta, 2007)

Theorem: *Given an initial state $\hat{\rho}(t = t_0)$, and a set of bath operators $\{\hat{V}_\alpha\}$, two external vector potentials $\mathbf{A}_{ext}(\mathbf{r}, t)$ and $\mathbf{A}'_{ext}(\mathbf{r}, t)$ that produce the same ensemble-averaged current density $\overline{\mathbf{j}(\mathbf{r}, t)}$ must necessarily coincide, up to a gauge transformation.*

For a proof of the theorem I refer the reader to the original paper (Di Ventra and D'Agosta, 2007). Here, I mention the important consequence that any ensemble-averaged current density that is *interacting* **A**-representable is also *non-interacting* **A**-representable. This means that an ensemble-averaged current density that can be obtained from a vector potential acting on the true many-body system can be obtained from a *suitable* vector potential acting on a non-interacting system (for given initial condition and set of bath operators).

Let us write this "auxiliary" Kohn–Sham vector potential as

$$\mathbf{A}_{eff}(\mathbf{r}, t) = \mathbf{A}_{ext}(\mathbf{r}, t) + \mathbf{A}_{xc}\left[\overline{\mathbf{j}(\mathbf{r}, t')}, \hat{\rho}(t = t_0), \{\hat{V}_\alpha\}\right], \quad (G.19)$$

with the exchange-correlation component a functional of the initial state, the ensemble-averaged current density at times $t' \leq t$ (it depends on the history of the system), and on the bath operators $\{\hat{V}_\alpha\}$. The role of the exchange-correlation vector potential is precisely to provide the correct ensemble-averaged current density at any given time.

This shows that the dynamics of the many-particle system can be mapped into the dynamics of an auxiliary Kohn–Sham Slater determinant evolving according to[8]

$$\partial_t |\Psi^{KS}(t)\rangle = -\frac{i}{\hbar} \hat{H}^{KS}(t) |\Psi^{KS}(t)\rangle - \frac{1}{2} \hat{V}^\dagger \hat{V} |\Psi^{KS}(t)\rangle + \ell(t) \hat{V} |\Psi^{KS}(t)\rangle, \quad (G.20)$$

where

$$\hat{H}^{KS}(t) = \sum_i \left\{ \frac{1}{2m}\left[\hat{p}_i - \frac{e}{c}\mathbf{A}_{ext}(\hat{r}_i, t) - \frac{e}{c}\mathbf{A}_{xc}(\hat{r}_i, t)\right]^2 + V_{ext}(\hat{r}_i) + V_H(\hat{r}_i, t) \right\}, \quad (G.21)$$

---

[8] For a single bath operator. To generalize to many operators one simply sums their dynamics as in Eq. G.4.

where here I have explicitly added a possible scalar external potential $V_{ext}(\hat{r}_i)$ (e.g., the electron-ion potential), and I have isolated the Hartree potential $V_H(\hat{r}_i, t)$.

The above also shows that for a general operator $\hat{V}$ acting on the many-body system one can "decouple" its quantum correlations, but not necessarily the statistical correlations induced by the presence of the bath.

It is only when the bath operator $\hat{V}$ acts on single-particle states or depends on the single-particle density only (call it $\hat{V}_{sp}$) that Eq. G.20 can be written as

$$\partial_t |\phi_k^{KS}(t)\rangle = -\frac{i}{\hbar} \hat{H}^{KS}(t)|\phi_k^{KS}(t)\rangle - \frac{1}{2}\hat{V}_{sp}^\dagger \hat{V}_{sp}|\phi_k^{KS}(t)\rangle + \ell(t)\hat{V}_{sp}|\phi_k^{KS}(t)\rangle, \tag{G.22}$$

with $\hat{H}^{KS}(t)$ a single Kohn–Sham Hamiltonian appearing in Eq. G.21.

Finally, any approximation to the functionals appearing in the Kohn–Sham Hamiltonian G.21 makes the latter stochastic (Sec. 1.4.4). For instance, the commonly used *ansatz* that isolates the Hartree potential $V_H(\mathbf{r}, t)$ from the exchange-correlation vector potential, as in Eq. G.21, can be made assuming that the density appearing in it is the density of *each* element of the ensemble, or the ensemble-averaged density. This is because in general

$$\int d\mathbf{r} \int d\mathbf{r}' \frac{\overline{\langle \hat{n}(\mathbf{r})\rangle \langle \hat{n}(\mathbf{r}')\rangle}}{|\mathbf{r}-\mathbf{r}'|} \neq \int d\mathbf{r} \int d\mathbf{r}' \frac{\overline{\langle \hat{n}(\mathbf{r})\rangle}\, \overline{\langle \hat{n}(\mathbf{r}')\rangle}}{|\mathbf{r}-\mathbf{r}'|}. \tag{G.23}$$

From a physical point of view, the first choice is the most natural since every element of the ensemble has its own density. This, however, precludes us from writing a closed equation of motion for the statistical operator, as discussed in Sec. 1.4.4, while this case is intrinsically built in the solution of Eq. G.20, thus adding no extra computational complexity to it.

If we choose the second possibility, one needs to include in the exchange-correlation also the statistical correlations of the direct Coulomb interaction at different points in space. These correlations may be very large, and possibly much larger than the Coulomb interaction between the average densities (right-hand side of Eq. G.23).

In actual calculations, one may apply, as a starting point, available approximations for $\mathbf{A}_{xc}$, like the ones introduced in Appendix F. The calculations are carried out as described in Appendix F, with the difference that one has now to average over the different dynamical realizations of the stochastic process G.3.

Appendix H
Inelastic corrections to current and shot noise

I give here explicit (and approximate) formulas for the first-order correction to the wave-functions due to electron-phonon scattering in the weak backscattering approximation. I then write the corresponding corrections to the current and shot noise.

Up to first order, the correction to the wave-function due to electron-phonon interaction is

$$|\Phi_E^{L(R)}; N_j\rangle = |\Psi_E^{L(R)}; N_j\rangle + |\delta\Psi_E^{L(R)}; N_j\rangle, \tag{H.1}$$

where according to standard first-order perturbation theory the correction term is

$$|\delta\Psi_E^\alpha; N_j\rangle = \lim_{\epsilon \to 0^+} \sum_{\alpha'=L,R} \sum_{M_{j'}} \int dE' D_{E'}^{\alpha'} \frac{\langle \Psi_{E'}^{\alpha'}; M_{j'}|\hat{H}_{e-ph}|\Psi_E^\alpha; N_j\rangle |\Psi_{E'}^{\alpha'}; M_{j'}\rangle}{\varepsilon(E, N_j) - \varepsilon(E', M_{j'}) + i\epsilon}, \tag{H.2}$$

with $\varepsilon(E, N_j) = E + N_j \hbar \omega_j$ the energy of state $|\Psi_E^\alpha; N_j\rangle$.

One can solve the above integral using the relation

$$\lim_{\epsilon \to 0^+} \int \frac{F(z)}{z + i\epsilon} = P \int \frac{F(z)}{z} - i\pi F(0), \tag{H.3}$$

where P indicates the Cauchy principal value.

In the weak backscattering case one can approximate the principal value with $-i\pi F(0)$ so that the correction to the wave-function can be written as

$$|\delta\Psi_E^\alpha; N_j\rangle = (B_{j,1}^\alpha + B_{j,3}^\alpha)|\Psi_{E-\hbar\omega_j}^{\alpha'}; N_j + 1\rangle + (B_{j,2}^\alpha + B_{j,4}^\alpha)|\Psi_{E+\hbar\omega_j}^{\alpha'}; N_j - 1\rangle, \tag{H.4}$$

where $B_{j,1}^\alpha$, $B_{j,2}^\alpha$, $B_{j,3}^\alpha$ and $B_{j,4}^\alpha$ are the amplitudes corresponding to the diagrams depicted in Fig. 6.2.

For instance, for $\left|\delta\Psi_E^R; N_j\right\rangle$ the coefficients are[1]

$$B_{j,1}^R = -\mathrm{i}2\pi\sqrt{\frac{\hbar}{2\omega_j}} \sum_i A_{i,j} J_{E-\hbar\omega_j,E}^{i,LR} D_{E-\hbar\omega_j}^L \sqrt{(1+N_j) f_E^R (1 - f_{E-\hbar\omega_j}^L)},$$

$$B_{j,2}^R = -\mathrm{i}2\pi\sqrt{\frac{\hbar}{2\omega_j}} \sum_i A_{i,j} J_{E+\hbar\omega_j,E}^{i,LR} D_{E+\hbar\omega_j}^L \sqrt{N_j f_E^R (1 - f_{E+\hbar\omega_j}^L)},$$

$$B_{j,3}^R = \mathrm{i}2\pi\sqrt{\frac{\hbar}{2\omega_j}} \sum_i A_{i,j} J_{E-\hbar\omega_j,E}^{i,RL} D_{E-\hbar\omega_j}^L \sqrt{(1+N_j) f_E^L (1 - f_{E-\hbar\omega_j}^R)},$$

$$B_{j,4}^R = \mathrm{i}2\pi\sqrt{\frac{\hbar}{2\omega_j}} \sum_i A_{i,j} J_{E+\hbar\omega_j,E}^{i,RL} D_{E+\hbar\omega_j}^L \sqrt{N_j f_E^L (1 - f_{E+\hbar\omega_j}^R)}. \quad (\text{H.5})$$

Similarly, the coefficients in $\left|\delta\Psi_E^L; N_j\right\rangle$ have the forms $B_{j,k}^L = B_{j,k}^R (L \rightleftharpoons R)$, where $k = 1, \ldots, 4$; the notation $(L \rightleftharpoons R)$ means interchange of labels R and L.

Let us now calculate the inelastic contribution to the current from these wave-functions.[2] Consistent with the approximations I have made so far, I calculate this contribution only for the case of small bias, but still larger than the first few vibrational modes of the nanostructure. I write this contribution for the case in which the electronic temperature is zero ($\theta = 0$).

Let us define the current operator (Sec. 5.2)

$$\hat{I}(x) = \frac{e\hbar}{2mi} \int d\mathbf{r}_\perp \left[\hat{\psi}^\dagger(\mathbf{r}) \left(\frac{\partial}{\partial x} \hat{\psi}(\mathbf{r})\right) - \left(\frac{\partial}{\partial x} \hat{\psi}^\dagger(\mathbf{r})\right) \hat{\psi}(\mathbf{r}) \right], \quad (\text{H.6})$$

where now the field operator is given by Eq. 6.27 with the single-particle electronic states as modified in H.1. We then calculate the average of the current operator over the electronic degrees of freedom using the properties 5.24. Assuming that the electrochemical potential on the left, μ_L, is higher than the electrochemical potential on the right, μ_R, then, at $\theta = 0$, due to the Pauli exclusion principle only the processes (c) and (d) in Fig. 6.2 contribute.

At $\theta = 0$ (but finite ionic temperature), the current has thus the elastic contribution I_{el}

$$I_{\text{el}} = \frac{e\hbar}{2mi} \int_{\mu_R}^{\mu_L} dE \int d\mathbf{r}_\perp \tilde{I}_{E,E}^{RR}, \quad (\text{H.7})$$

---

[1] There are also the terms proportional to $\sqrt{f_E^\alpha (1 - f_{E\pm\hbar\omega_j}^\alpha)}$ that have a smaller amplitude at low temperatures and small phonon frequencies.

[2] This neglects the interference terms 6.34 between the zero-order and the second-order correction to the wave-functions.

which is precisely the average current 5.28 I have calculated in Sec. 5.2 with $\tilde{I}_{E,E}^{\alpha\alpha}$ defined in Eq. 5.27. The inelastic contribution I_{in} is instead[3]

$$I_{in} = -\frac{e\hbar}{2mi} \int_{\mu_R}^{\mu_L} dE \int d\mathbf{r}_\perp \tilde{I}_{E,E}^{RR} \sum_j \left(|B_{j,1}^R|^2 + |B_{j,2}^R|^2 \right). \tag{H.8}$$

The sum of the above two terms thus gives the total current

$$I = I_{el} + I_{in} = \frac{e\hbar}{2mi} \int_{\mu_R}^{\mu_L} dE \int d\mathbf{r}_\perp \tilde{I}_{E,E}^{RR} \left[1 - \sum_j \left(|B_{j,1}^R|^2 + |B_{j,2}^R|^2 \right) \right]. \tag{H.9}$$

Since only the modulus squared of the coefficients $B_{j,k}^R$ appears in Eq. H.8, the current, under weak backscattering conditions, is *reduced* by inelastic effects.

Inelastic effects on shot noise

One can follow similar arguments and find, within the same approximations, the corresponding inelastic correction to shot noise. In this case one finds Chen and Di Ventra (2005)

$$S_{RR} = \frac{e^2\hbar^3}{2m^2} \int_{\mu_L}^{\mu_R} dE \left| \int d\mathbf{r}_\perp \tilde{I}_{E,E}^{LR} \right|^2 \left[1 + \sum_{j;k=1,2} |B_{j,k}^R \cdot B_{j,k}^{L*}|^2 \right], \tag{H.10}$$

which instead shows an *increase* of shot noise due to inelastic scattering with respect to the elastic contribution (see Eq. 5.33). This increase can be interpreted as increased correlations between states of different directionality introduced by the electron-phonon coupling.

---

[3] In Eqs. H.8 and H.10 the coefficients $B_{j,k}$ are the ones in H.5 but *without* the terms $\sqrt{f_E^\alpha(1-f_E^\beta)}$. In addition, these equations have been simplified by using $\tilde{I}_{E\pm\hbar\omega_j,E\pm\hbar\omega_j}^{RR} \simeq \tilde{I}_{E,E}^{RR}$, valid for energies close to the electrochemical potentials. (The electrochemical potentials have energies of eVs, to be compared with the typical energy of a few meVs of phonons in nanoscale junctions.)

Appendix I
Hydrodynamic form of the Schrödinger equation

I discuss some steps necessary to show that the time-dependent many-body Schrödinger equation 1.16 can be written in hydrodynamic form.[1]

Let us start from the general Hamiltonian operator

$$\hat{H} = \hat{T} + \hat{W} + \hat{V}_{ext} \qquad (I.1)$$

describing electrons interacting via Coulomb forces (described by the operator $\hat{W}$) subject to a time-dependent external potential represented by the operator $\hat{V}_{ext}$.[2] The operator $\hat{T}$ describes the kinetic energy of the particles. Using the relations A.26 and A.27 we can write these operators in second-quantization form so that the Hamiltonian I.1 can be written as[3]

$$\begin{aligned}\hat{H} = &- \int d\mathbf{r}\, \psi^\dagger(\mathbf{r}) \frac{\hbar^2 \nabla^2}{2m} \psi(\mathbf{r}) \\ &+ \frac{1}{2} \int d\mathbf{r} \int d\mathbf{r}'\, \psi^\dagger(\mathbf{r})\psi^\dagger(\mathbf{r}') \frac{1}{|\mathbf{r}-\mathbf{r}'|} \psi(\mathbf{r}')\psi(\mathbf{r}) \\ &+ \int d\mathbf{r}\, \psi^\dagger(\mathbf{r}) V_{ext}(\mathbf{r},t) \psi(\mathbf{r}), \end{aligned} \qquad (I.2)$$

where the field operators have been defined in Eqs. A.19 and A.20. Using Eq. B.3 we find the following equation of motion for the field operator in the Heisenberg picture

$$\begin{aligned}i\hbar \frac{\partial}{\partial t}\psi(\mathbf{r},t) = &-\frac{\hbar^2 \nabla^2}{2m} \psi(\mathbf{r},t) + \int d\mathbf{r}'\, \psi^\dagger(\mathbf{r}',t) \frac{1}{|\mathbf{r}-\mathbf{r}'|} \psi(\mathbf{r}',t)\psi(\mathbf{r},t) \\ &+ V_{ext}(\mathbf{r},t)\psi(\mathbf{r},t). \end{aligned} \qquad (I.3)$$

---

[1] This derivation follows closely the one in Tokatly (2005), see also Martin and Schwinger (1959).
[2] Here the external potential may be the electron-ion potential plus any other possible external potential.
[3] The kinetic energy is a one-body operator, the Coulomb interaction is a two-body operator, and we assume, as we did in Sec. 4.3.1, that the external field can be described by a one-body potential.

The equation of motion for the density operator, Eq. A.23,

$$\hat{n}(\mathbf{r},t) \equiv \psi^\dagger(\mathbf{r},t)\psi(\mathbf{r},t), \tag{I.4}$$

follows similarly from Eq. B.3. Using Eq. I.3 (and the hermiticity of the Hamiltonian) we get

$$\begin{aligned}
i\hbar\frac{\partial \hat{n}(\mathbf{r},t)}{\partial t} &= i\hbar\left[\frac{\partial \psi^\dagger(\mathbf{r}',t)}{\partial t}\psi(\mathbf{r}',t) + \psi^\dagger(\mathbf{r}',t)\frac{\partial \psi(\mathbf{r}',t)}{\partial t}\right] \\
&= -\frac{\hbar^2}{2m}\left\{\psi^\dagger(\mathbf{r},t)\nabla^2\psi(\mathbf{r},t) - [\nabla^2\psi^\dagger(\mathbf{r},t)]\psi(\mathbf{r},t)\right\} \\
&= -i\hbar\frac{\hbar}{2mi}\nabla\cdot\left\{\psi^\dagger(\mathbf{r},t)\nabla\psi(\mathbf{r},t) - [\nabla\psi^\dagger(\mathbf{r},t)]\psi(\mathbf{r},t)\right\} \\
&= -i\hbar\nabla\cdot\hat{\mathbf{j}}(\mathbf{r},t), \tag{I.5}
\end{aligned}$$

where in the last equality I have used the definition A.24 of current density operator.

If we now take the average of Eq. I.5 (with respect to the statistical operator of the system) we recover the continuity equation 1.11. By defining the *fluid velocity*

$$\mathbf{v}(\mathbf{r},t) = \frac{\mathbf{j}(\mathbf{r},t)}{n(\mathbf{r},t)} = \frac{\langle\hat{\mathbf{j}}(\mathbf{r},t)\rangle}{\langle\hat{n}(\mathbf{r},t)\rangle}, \tag{I.6}$$

we can rewrite the continuity Eq. 1.11 as

$$D_t n(\mathbf{r},t) + n(\mathbf{r},t)\nabla\cdot\mathbf{v}(\mathbf{r},t) = 0, \tag{I.7}$$

where the convective derivative is given in Eq. 8.10.

Following similar arguments, i.e., by differentiating in time the current density operator A.24, one can show that the expectation value of the current density operator satisfies the following "Newtonian" equation of motion

$$m\frac{\partial \mathbf{j}(\mathbf{r},t)}{\partial t} + \mathbf{F}^{kin}(\mathbf{r},t) + \mathbf{F}^{e-e}(\mathbf{r},t) + n(\mathbf{r},t)\nabla_\mathbf{r} V_{ext}(\mathbf{r},t) = 0, \tag{I.8}$$

where I have defined the *force densities* due to the kinetic part of the Hamiltonian

$$\begin{aligned}
\mathbf{F}_i^{kin}(\mathbf{r},t) &= \frac{\hbar^2}{2m}\sum_j \frac{\partial}{\partial r_j}\left\langle \nabla_i\psi^\dagger\nabla_j\psi + \nabla_j\psi^\dagger\nabla_i\psi - \frac{\delta_{ij}}{2}\nabla^2(\psi^\dagger\psi)\right\rangle \\
&\equiv \sum_j \frac{\partial}{\partial r_j}\tilde{T}_{ij}(\mathbf{r},t) \equiv \nabla_j\tilde{T}_{ij}(\mathbf{r},t), \tag{I.9}
\end{aligned}$$

and due to the Coulomb interaction among electrons

$$F_i^{e-e}(\mathbf{r},t) = \int d\mathbf{r}' \nabla_i \frac{1}{|\mathbf{r}-\mathbf{r}'|} \left\langle \psi^\dagger(\mathbf{r},t)\psi^\dagger(\mathbf{r}',t)\psi(\mathbf{r}',t)\psi(\mathbf{r},t) \right\rangle. \quad (\text{I.10})$$

In Eq. I.9 I have also defined the *second-rank tensor* $\tilde{\mathbf{T}}_{ij}$

$$\tilde{\mathbf{T}}_{ij} = \frac{\hbar^2}{2m} \left\langle \nabla_i \psi^\dagger \nabla_j \psi + \nabla_j \psi^\dagger \nabla_i \psi - \frac{\delta_{ij}}{2} \nabla^2(\psi^\dagger \psi) \right\rangle, \quad (\text{I.11})$$

and, in the last equality of Eq. I.9, I have used the convention that repeated indexes indicate summation over the three spatial coordinates.

Equation I.8 can be interpreted as the second law of mechanics applied to the electron liquid: all forces, internal ($\mathbf{F}^{e-e}$ and $\mathbf{F}^{kin}$) and external ($-n\nabla V_{ext}$) to the fluid, produce a change in velocity.

Assuming that the integration volume in I.10 extends beyond the range of the interactions, $\mathbf{F}_i^{e-e}(\mathbf{r},t)$ can also be written as the divergence of a second-rank tensor[4]

$$F_i^{e-e}(\mathbf{r},t) = \sum_j \frac{\partial}{\partial r_j} \mathbf{W}_{ij}(\mathbf{r},t) \equiv \nabla_j \mathbf{W}_{ij}(\mathbf{r},t), \quad (\text{I.12})$$

where

$$\mathbf{W}_{i,j}(\mathbf{r},t) = -\frac{1}{2} \int d\mathbf{r}' \frac{r'_i r'_j}{|\mathbf{r}'|} \frac{\partial |\mathbf{r}'|^{-1}}{\partial |\mathbf{r}'|}$$
$$\times \int_0^1 d\lambda \, \langle \psi^\dagger(\mathbf{r}+\lambda\mathbf{r}',t)\psi^\dagger(\mathbf{r}-(1-\lambda)\mathbf{r}',t)$$
$$\times \psi(\mathbf{r}-(1-\lambda)\mathbf{r}',t)\psi(\mathbf{r}+\lambda\mathbf{r}',t)\rangle, \quad (\text{I.13})$$

with λ a positive parameter that defines a line which connects two interacting particles.

Let us now use the definition of convective derivative 8.10 and rewrite the kinetic stress tensor as

$$\mathbf{T}_{ij} = \tilde{\mathbf{T}}_{ij}(\mathbf{r},t) - m\,n(\mathbf{r},t)\,v_i(\mathbf{r},t)\,v_j(\mathbf{r},t). \quad (\text{I.14})$$

Let us define the total (two-particle) stress tensor

$$\mathbf{P}_{ij}(\mathbf{r},t) = \mathbf{W}_{ij}(\mathbf{r},t) + \mathbf{T}_{ij}(\mathbf{r},t). \quad (\text{I.15})$$

We can then rewrite Eq. I.8 as

$$m n(\mathbf{r},t) D_t v_j(\mathbf{r},t) + \nabla_i \mathbf{P}_{ij}(\mathbf{r},t) + n(\mathbf{r},t) \nabla_j V_{ext}(\mathbf{r},t) = 0. \quad (\text{I.16})$$

---

[4] Due to screening effects and for a finite and closed system this statement is generally true, i.e. one can always find a volume large enough that the interactions are negligibly small at the surface enclosing this volume.

Solving the coupled equations I.7 and I.16 with given initial conditions is therefore equivalent to solving for the many-body Schrödinger equation 1.16. We have thus proven the equivalence

$$i\hbar \frac{d|\Psi(t)\rangle}{dt} = \mathbf{H}|\Psi(t)\rangle$$
$$\Updownarrow$$
$$\begin{cases} D_t n(\mathbf{r},t) + n(\mathbf{r},t)\nabla \cdot \mathbf{v}(\mathbf{r},t) = 0 \\ mn(\mathbf{r},t)D_t \mathbf{v}_j(\mathbf{r},t) + \nabla_i \mathbf{P}_{ij}(\mathbf{r},t) + n(\mathbf{r},t)\nabla_j V_{ext}(\mathbf{r},t) = 0, \end{cases}$$

(I.17)

where the reverse relation (from the hydrodynamic equations of motion to the Schrödinger equation) is guaranteed by the theorems of dynamical density-functional theories (see discussion in Sec. 8.2).

In order to solve I.17 one needs to derive an equation of motion for the *two-particle* density matrix $G_2 = \langle \psi^\dagger(\mathbf{r})\psi^\dagger(\mathbf{r}')\psi(\mathbf{r}')\psi(\mathbf{r})\rangle$, which enters in the definition of the interaction part of the stress tensor, Eq. I.13. This equation of motion can be derived from the Heisenberg equation of motion of the particle field operators, Eq. I.3. Carrying out the calculation explicitly, one would find that this equation of motion contains the three-particle density matrix $G_3 = \langle \psi^\dagger(\mathbf{r})\psi^\dagger(\mathbf{r}',t)\psi^\dagger(\mathbf{r}'')\psi(\mathbf{r}'')\psi(\mathbf{r}')\psi(\mathbf{r})\rangle$, i.e., one needs the stress tensor involving interactions among three particles (call this $\mathbf{P}^{(3)}_{ijk}$). In turn, the equation of motion for the three-particle density matrix (stress tensor) contains the four-particle density matrix (stress tensor) and so forth, thus generating the BBGKY infinite hierarchy of nested equations explicitly written in Appendix B, making the problem practically unsolvable.

Appendix J
Equation of motion for the stress tensor

I derive here the equation of motion for the stress tensor $\mathbf{P}_{ij}$, Eq. 8.21.

Let us start from the Boltzmann equation 4.155 for the non-equilibrium distribution function f

$$\frac{\partial f(\mathbf{r},\mathbf{p},t)}{\partial t} + \frac{\mathbf{p}}{m} \cdot \nabla_\mathbf{r} f(\mathbf{r},\mathbf{p},t) - \nabla_\mathbf{r} V_{ext}(\mathbf{r},t) \cdot \nabla_\mathbf{p} f(\mathbf{r},\mathbf{p},t) = \mathbf{I}[f], \quad (J.1)$$

for an arbitrary external potential $V_{ext}(\mathbf{r},t)$. The number density and current density are related to the distribution function as

$$n(\mathbf{r},t) = \sum_\mathbf{p} f(\mathbf{r},\mathbf{p},t) \quad (J.2)$$

and

$$\mathbf{j}(\mathbf{r},t) = \sum_\mathbf{p} \frac{\mathbf{p}}{m} f(\mathbf{r},\mathbf{p},t). \quad (J.3)$$

Let us move to a reference frame co-moving with the liquid (called the *Lagrangian reference frame*) via the transformation (Tokatly and Pankratov, 1999)

$$\mathbf{p} \to \mathbf{p} + m\mathbf{v}, \quad (J.4)$$

where $\mathbf{v}$ is the velocity of each point $\mathbf{r}$ in the liquid.[1] Call $f^L(\mathbf{r},\mathbf{p},t) = f(\mathbf{r},\mathbf{p}+m\mathbf{v},t)$ the distribution function in this reference frame. We have

$$\sum_\mathbf{p} f^L(\mathbf{r},\mathbf{p},t) = \sum_\mathbf{p} f(\mathbf{r},\mathbf{p}+m\mathbf{v},t) = \sum_{\mathbf{p}'} f(\mathbf{r},\mathbf{p}',t) = n(\mathbf{r},t), \quad (J.5)$$

---

[1] This is a local *non-inertial* reference frame.

and
$$\sum_{\mathbf{p}} \frac{\mathbf{p}}{m} f^L(\mathbf{r}, \mathbf{p}, t) = \sum_{\mathbf{p}} \frac{\mathbf{p}}{m} f(\mathbf{r}, \mathbf{p} + m\mathbf{v}, t) = \sum_{\mathbf{p}'} \left(\frac{\mathbf{p}'}{m} - \mathbf{v} \right) f(\mathbf{r}, \mathbf{p}', t) = 0, \tag{J.6}$$

i.e., the density is the same as in the laboratory frame, while the current density is zero in the Lagrangian frame.

The Boltzmann equation J.1 in the Lagrangian frame is

$$D_t f^L + \frac{\mathbf{p}}{m} \cdot \nabla_{\mathbf{r}} f^L - e \nabla V_{ext} \cdot \nabla_{\mathbf{p}} f^L - \nabla \cdot \mathbf{v} \, \mathbf{p} \cdot \nabla_{\mathbf{p}} f^L - m D_t \mathbf{v} \cdot \nabla_{\mathbf{p}} f^L = \mathbf{I}[f^L]. \tag{J.7}$$

If we integrate Eq. J.7 over all momenta $\mathbf{p}$ we obtain

$$D_t n(\mathbf{r}, t) + n(\mathbf{r}, t) \nabla \cdot \mathbf{v}(\mathbf{r}, t) = \sum_{\mathbf{p}} \mathbf{I}[f^L] = 0, \tag{J.8}$$

where the last condition is set by the conservation of number of particles (see Exercise 2.12). Equation J.8 is the continuity equation.

Let us now write Eq. J.7 in index notation (summation over repeated indexes is implied), multiply it by the products of momentum components $p_i p_j$ and sum over all the momenta. We then get

$$\sum_{\mathbf{p}} p_i p_j D_t f^L + \sum_{\mathbf{p}} \frac{p_i p_j p_k}{m} \nabla_k f^L - e \nabla_k V_{ext} \sum_{\mathbf{p}} p_i p_j \nabla_{p_k} f^L$$
$$- \sum_{\mathbf{p}} p_i p_j p_k \nabla_k v_l \nabla_{p_l} f^L - \sum_{\mathbf{p}} p_i p_j (m D_t v_k) \nabla_{p_k} f^L = \sum_{\mathbf{p}} p_j p_j \mathbf{I}[f^L]. \tag{J.9}$$

Let us define
$$\mathbf{P}_{ij} = \frac{1}{m} \sum_{\mathbf{p}} p_i p_j f^L, \tag{J.10}$$

and
$$\mathbf{P}^{(3)}_{ijk} = \frac{1}{m^2} \sum_{\mathbf{p}} p_i p_j p_k f^L, \tag{J.11}$$

the second and third moment of the distribution, respectively. Equation J.9 can then be written as

$$D_t \mathbf{P}_{ij} + \mathbf{P}_{ij} \nabla \cdot \mathbf{v} + \mathbf{P}_{ik} \nabla_k v_j + \mathbf{P}_{kj} \nabla_k v_i + \nabla_k \mathbf{P}^{(3)}_{ijk} = \frac{1}{m} \sum_{\mathbf{p}} p_i p_j \mathbf{I}[f], \tag{J.12}$$

which is Eq. 8.21. One can then prove that an equation of the same form exists for the stress tensor in the laboratory frame by inverting the transformation J.4.

Let us now write the stress tensor as

$$\mathbf{P}_{ij}(\mathbf{r},t) = P(\mathbf{r},t)\delta_{ij} - \boldsymbol{\pi}_{ij}(\mathbf{r},t), \tag{J.13}$$

with $\text{Tr}\{\boldsymbol{\pi}_{ij}(\mathbf{r},t)\} = 0$. We introduce J.13 into J.12 and neglect the term $\nabla_k \mathbf{P}^{(3)}_{ijk}$. We get

$$D_t P \delta_{ij} + [\nabla_k v_i P \delta_{kj} + \nabla_k v_j P \delta_{ik} + \nabla_i v_i P \delta_{ij}]$$
$$- D_t \boldsymbol{\pi}_{ij} - [\nabla_k v_i \boldsymbol{\pi}_{kj} + \nabla_k v_j \boldsymbol{\pi}_{ik} + \nabla_i v_i \boldsymbol{\pi}_{ij}]$$
$$= \frac{1}{m} \sum_{\mathbf{p}} p_i p_j \mathbf{I}[f]. \tag{J.14}$$

By taking the trace of Eq. J.14 and using the fact that $\boldsymbol{\pi}_{ij}(\mathbf{r},t)$ is traceless we get in d dimensions

$$D_t P + \left(1 + \frac{2}{d}\right) P \nabla_k v_k - \frac{2}{d} \boldsymbol{\pi}_{kj} \nabla_k v_j = 0, \tag{J.15}$$

which is precisely Eq. 8.26. By introducing J.15 into J.14 we finally get

$$-D_t \boldsymbol{\pi}_{ij} - \left[\nabla_k v_i \boldsymbol{\pi}_{jk} + \nabla_k v_j \boldsymbol{\pi}_{ik} + \nabla_k v_k \boldsymbol{\pi}_{ij} - \frac{2}{d} \boldsymbol{\pi}_{kl} \nabla_k v_l \delta_{ij}\right]$$
$$+ P\left[\nabla_k v_i \delta_{kj} + \nabla_k v_i \delta_{ik} - \frac{2}{d}\delta_{ij}\nabla_k v_k\right] = \frac{1}{m}\sum_{\mathbf{p}} p_i p_j \mathbf{I}[f]. \tag{J.16}$$

This is equivalent to 8.27.

The derivation of the quantum Navier–Stokes equations then follows as explained in Sec. 8.2.1.

Appendix K
Cut-off of the viscosity divergence

In this appendix I give a simple argument to show that the presence of a nanostructure cuts off the typical divergence of the electron viscosity at zero frequency and zero temperature (Abrikosov and Khalatnikov, 1959).[1] At finite frequencies the shear $\tilde{\mu}$ and bulk $\tilde{B}$ moduli of a liquid are complex functions (Landau and Lifshitz, 1959)

$$\tilde{\mu}(\omega) = \mu(\omega) - i\omega\eta(\omega), \quad \text{complex shear modulus,} \quad (K.1)$$

and

$$\tilde{B}(\omega) = B(\omega) - i\omega\zeta(\omega), \quad \text{complex bulk modulus,} \quad (K.2)$$

where η is the shear viscosity, ζ the bulk viscosity, $\mu(\omega)$ is the shear modulus and $B(\omega)$ is the bulk modulus, and $\mu(\omega \to 0) = 0$.

Since $\mu(\omega)$ and $-\omega\eta(\omega)$ are the real and imaginary parts, respectively, of the same response function $\tilde{\mu}(\omega)$, they are related via the Kramers–Krönig relation

$$\mu(\omega) = \mu(\omega \to \infty) - 2\frac{P}{\pi}\int_0^\infty d\omega' \frac{\omega'^2}{\omega'^2 - \omega^2} \eta(\omega'), \quad (K.3)$$

where P indicates the Cauchy principal value. In the limit of $\omega \to 0$ the above equation reads

$$\mu(\omega \to \infty) \equiv \mu_\infty = 2\frac{P}{\pi}\int_0^\infty d\omega'\, \eta(\omega') \simeq \frac{\eta(\omega \to 0)}{\tau}, \quad (K.4)$$

where τ is the total scattering time, and the last step is justified by the fact that the major contribution to the integral comes from frequencies within a range of the order of $1/\tau$.

---

[1] This is the order of the limits to obtain this divergence: first one sets the frequency to zero, and then the temperature to zero. The opposite order – first the temperature goes to zero and then the frequency vanishes – does not lead to a viscosity divergence. In this regime, the viscosity η tends to a constant and rather small value.

In the absence of the nanostructure and for a translationally invariant system, τ diverges as $1/\theta_e^2$, where θ_e is the electronic temperature. From K.4 we then see that the viscosity diverges at zero frequency and zero temperature (Abrikosov and Khalatnikov, 1959).

Physically this is due to the fact that for a translationally invariant system at zero temperature the quasi-particles close to the Fermi energy are very long-lived and can transport momentum – and thus friction of every "adjacent" layer of the liquid – over very large distances, thus making the liquid move as a whole (similar to a solid behavior).

On the other hand, the nanostructure introduces a finite lifetime τ_c (independent of temperature) due to the elastic scattering of electrons at the junction (see Sec. 7.6) which cuts off the divergence of the zero-frequency viscosity $\eta(\omega \to 0)$ via the relation (from K.4)

$$\eta(\omega \to 0) = \mu_\infty \tau_c, \tag{K.5}$$

as if a finite temperature is present. The presence of the nanostructure does not allow quasi-particles to transport momentum to every layer of the liquid over all distances. This result also shows that the zero-frequency viscosity is structure-dependent and may be much larger than the one obtained from Eqs. F.8 and F.9 of Appendix F.

Appendix L
Bernoulli's equation

Here, I derive Bernoulli's equation from the Navier–Stokes equations 8.35. Let us use the identity

$$\mathbf{v} \cdot \nabla \mathbf{v} = \frac{1}{2}\nabla(\mathbf{v} \cdot \mathbf{v}) - \mathbf{v} \times (\nabla \times \mathbf{v}), \tag{L.1}$$

and replace it into 8.35. We then get

$$\frac{\partial \mathbf{v}}{\partial t} + \nabla\left(\frac{\mathbf{v} \cdot \mathbf{v}}{2} + \frac{P}{n} + \frac{V_{ext}}{m}\right) - \eta \nabla^2 \mathbf{v} = \mathbf{v} \times (\nabla \times \mathbf{v}). \tag{L.2}$$

For an inviscid liquid ($\eta = 0$) at steady state we get

$$\nabla\left(\frac{\mathbf{v} \cdot \mathbf{v}}{2} + \frac{P}{n} + \frac{V_{ext}}{m}\right) = \mathbf{v} \times (\nabla \times \mathbf{v}). \tag{L.3}$$

A *streamline* is defined as the line such that its tangent at any point gives the direction of the velocity at that point (Landau and Lifshitz, 1959). If we denote with l the unit length along a streamline, the projection of Eq. L.3 onto a streamline gives

$$\frac{\partial}{\partial l}\left(\frac{\mathbf{v} \cdot \mathbf{v}}{2} + \frac{P}{n} + \frac{V_{ext}}{m}\right) = 0, \tag{L.4}$$

since $\mathbf{v} \times (\nabla \times \mathbf{v})$ is orthogonal to $\mathbf{v}$ and hence its projection on the streamline is zero. Equation L.4 finally gives Bernoulli's equation

$$\frac{\mathbf{v} \cdot \mathbf{v}}{2} + \frac{P}{n} + \frac{V_{ext}}{m} = \text{constant}. \tag{L.5}$$

References

Abrikosov, A.A. and I.M. Khalatnikov, Rep. Prog. Phys. **22**, 329 (1959).
Agraït, N., C. Untiedt, G. Rubio-Bollinger, and S. Vieira Phys. Rev. Lett. **88**, 216803 (2002).
Agraït, N., A. Levy Yeyati, and J.M. van Ruitenbeek, Phys. Rep. **377**, 81 (2003).
Altshuler, B.L., A.G. Aronov, and D.E. Khmelnitsky, J. Phys. C, **15**, 7367, (1982).
Ashcroft, N.W., and N.D. Mermin, *Solid State Physics*, (Saunders College Publishing, 1975).
Averin D.V., and K.K. Likharev, J. Low Temp. Phys. **62**, 345 (1986).
Balian, R., *From Microphysics to Macrophysics*, (Springer-Verlag, Berlin, 1991).
Baranger, H.U., and A.D. Stone, Phys. Rev. B **40**, 8169 (1989).
Bardeen, J., Phys. Rev. Lett. **6**, 57 (1961).
Benoit, A.D., S. Washburn, C.P. Umbach, R. B. Laibowitz, and R. A. Webb, Phys. Rev. Lett. **57**, 1765 (1986).
Blanter, Ya. M., and M. Büttiker, Phys. Rep. **336**, 1 (2000).
Bloch, F., Z. Phys. **81**, 363 (1933).
Bogachek, E.N., A.G. Scherbakov, and U. Landman, Phys. Rev. B **54**, R11094 (1996).
Boltzmann, L.W., Ber. Wien. Akad. **66**, 275 (1872).
Bomze, Yu., G. Gershon, D. Shovkun, L.S. Levitov, and M. Reznikov, Phys. Rev. Lett. **95**, 176601 (2005).
Bransden, B.H., and C.J. Joachain, *Physics of Atoms and Molecules* (John Wiley & Sons, New York, 1983).
Breuer, H.-P., and F. Petruccione, *The Theory of Open Quantum Systems*, (Oxford University Press, Oxford, 2002).
Bulashenko, O.M., and L.L. Bonilla, Phys. Rev. B **52** 7849 (1995).
Bushong, N., and M. Di Ventra, cond-mat/07110762.
Bushong, N., N. Sai, and M. Di Ventra, Nano Lett. **5**, 2569 (2005).
Bushong, N., J. Gamble, and M. Di Ventra, Nano Lett. **7**, 1789 (2007a).
Bushong, N., Y.V. Pershin, and M. Di Ventra, Phys. Rev. Lett. **99**, 226802 (2007b).
Butcher, P.N., J. Phys. Cond. Matt. **2**, 4869 (1990).
Büttiker, M., Phys. Rev. Lett. **57**, 1761 (1986).
Büttiker, M., IBM J. Res. Dev. **32**, 316 (1988).
Büttiker, M., Y. Imry, and R. Landauer, Phys. Lett. **96A**, 365 (1983).
Büttiker, M., Y. Imry, R. Landauer, and S. Pinhas, Phys. Rev. B **31**, 6207 (1985).

Caroli, C., R. Combescot, P. Nozieres, D. Saint-James, J. Phys. C: Solid St. Phys. **4**, 916 (1971).
Casati, G. and I. Guarnieri, Phys. Rev. Lett. **42**, 1579 (1979).
Ceperley, D.M., and B.J. Alder, Phys. Rev. Lett. **45**, 566 (1980).
Chen, Y.-C., and M. Di Ventra, Phys. Rev. B **67**, 153304 (2003).
Chen, Y.-C., and M. Di Ventra, Phys. Rev. Lett. **95**, 166802 (2005).
Chen, Y.-C., M. Zwolak, and M. Di Ventra, Nano Lett. **3**, 1691 (2003).
Chen, Y.-C., M. Zwolak, and M. Di Ventra, Nano Lett. **5**, 621 (2005).
Cheng, C.-L., J. S. Evans, and T. V. Voorhis, Phys. Rev. B **74**, 155112 (2006).
Cini, M., Phys. Rev. B **22**, 5887 (1980).
Collet, M.J., and C.W. Gardiner, Phys. Rev. A **30**, 1386 (1984).
Conti, S., and G. Vignale, Phys. Rev. B **60**, 7966 (1999).
Cui, X., M. Freitag, R. Martel, L. Brus, and Ph. Avouris, Nano Lett. **3**, 783 (2003).
D'Agosta, R., and M. Di Ventra, J. Phys. Cond. Matt. **18**, 11059 (2006).
D'Agosta, R., and G. Vignale, Phys. Rev B **71**, 245103 (2005).
D'Agosta, R., and G. Vignale, Phys. Rev. Lett. **96**, 016405 (2006).
D'Agosta, R., N. Sai, and M. Di Ventra, Nano Lett. **6**, 2935 (2006).
Dalibard, J., Y. Castin, and K. Mølmer, Phys. Rev. Lett. **68**, 580 (1992).
D'Amico, I., and G. Vignale, Phys. Rev. B **65**, 085109 (2002).
Damle, P.S., A. W. Ghosh, and S. Datta, Phys. Rev. B **64**, 201403 (2001).
Datta, S., *Electronic Transport in Mesoscopic Systems* (Cambridge University Press, Cambridge, 1995).
Di Ventra, M., and R. D'Agosta, Phys. Rev. Lett. **98**, 226403 (2007).
Di Ventra, M., and N.D. Lang, Phys. Rev. B **65**, 045402 (2002).
Di Ventra, M., and S.T. Pantelides, Phys. Rev. B **61**, 16207 (2000).
Di Ventra, M., and T.N. Todorov, J. Phys. Cond. Matt. **16**, 8025 (2004).
Di Ventra, M., S.T. Pantelides, and N.D. Lang, Phys. Rev. Lett. **84**, 979 (2000).
Di Ventra, M., S.T. Pantelides, and N. D. Lang, Phys. Rev. Lett. **88**, 046801 (2002).
Di Ventra, M., S. Evoy, and J.R. Heflin, *Introduction to Nanoscale Science and Technology* (Springer, New York, 2004a).
Di Ventra, M., Y.-C. Chen, and T.N. Todorov, Phys. Rev. Lett. **92**, 176803 (2004b).
Drude, P., Annalen der Physik **1**, 566 (1900).
Dutta, P., and P.M. Horn, Rev. Mod. Phys. **53**, 497 (1981).
Engquist, L. and P.W. Anderson, Phys. Rev. B **24**, 1151 (1981).
Entin-Wohlman, O., C. Hartzstein, and Y. Imry, Phys. Rev. B **34**, 921 (1986).
Evans, D.J., E.G.D. Cohen, G.P. Morriss, Phys. Rev. Lett. **71**, 2401 (1993).
Ferry, D.K., and S.M. Goodnick, *Transport in Nanostructures* (Cambridge University Press, Cambridge, 1997).
Fetter, A.L., and J.D. Walecka, *Quantum Theory of Many Particle Systems* (McGraw-Hill, Boston, 1971).
Feynmann, R.P., Phys. Rev. **56**, 340 (1939).
Fisher, D.S. and P.A. Lee, Phys. Rev. B **23**, 6851 (1981).
Frenkel, J., Phys. Rev. **36**, 1604 (1930).
Frensley, W.R., Rev. Mod. Phys. **62**, 745 (1990).
Fröhlich, H., Physica **37**, 215 (1967).
Galperin, M., M.A. Ratner, and A. Nitzan, J. Chem. Phys. **121**, 11965 (2004).
Gaspard, P., Prog. Th. Phys. Supplement **165**, 33 (2006).
Gebauer, R. and R. Car, Phys. Rev. B **70**, 125324 (2004).
Ghosh, S.W., and B.M. Deb, Phys. Rep. **92**, 1 (1982).
Ghosh, S.W., and A.K. Dhara, Phys. Rev. A **38**, 1149 (1988).

Giuliani, G.F., and G. Vignale, *Quantum Theory of the Electron Liquid* (Cambridge University Press, Cambridge, 2005).
Goldstein, H., *Classical Mechanics* (Addison Wesley, Massachusetts, 1950).
Goetsch, P., R. Graham, and F. Haake, Phys. Rev. A **51**, 136 (1995).
Gross, E.K.U., D. Mearns, and L.N. Oliveira, Phys. Rev. Lett. **61**, 1518 (1988).
Gurvitz, S.A., Phys. Rev. B **57**, 6602 (1998).
Harrison, W.A., *Electronic Structure and the Properties of Solids* (Dover, New York, 1989).
Haug, H., and A.P. Jauho, *Quantum Kinetics in Transport and Optics of Semiconductors* (Springer, Berlin, 1996).
Hellmann, H., *Einfhrung in die Quantenchemie* (Franz Deuticke, Leipzig, 1937), Sec. 54.
Hohenberg, P., and W. Kohn, Phys. Rev. **136**, B864 (1964).
Huang, Z.F., B.Q. Xu, Y.C. Chen, M. Di Ventra, and N.J. Tao, Nano Lett. **6**, 1240 (2006).
Huang, Z.F., F. Chen, R. D'Agosta, P.A. Bennett, M. Di Ventra, and N.J. Tao, Nature Nanotechnology **2**, 698 (2007).
Iannaccone, G., G. Lombardi, M. Macucci, and B. Pellegrini, Phys. Rev. Lett. **80**, 1054 (1998).
Imry, Y., in *Directions in Condensed Matter Physics*, G. Grinstein and G. Mazenko eds. (World Scientific Press, Singapore, 1986).
Jackson, J.D., *Classical Electrodynamics* (John Wiley & Sons, New York, 1975).
Jaynes, E.T., Phys. Rev. A **4**, 747 (1971).
Johnson, M., and R.H. Silsbee, Phys. Rev. Lett. **55**, 1790 (1985).
Kadanoff, L.P., and G. Baym, *Quantum Statistical Mechanics* (Benjamin, New York, 1962).
Kamenev, A., and W. Kohn, Phys. Rev. B **63**, 155304 (2001).
Keldysh, L.V., *Zh. Eksp. Teor. Fiz.* **47**, 1515 (1964) [Sov. Phys.-JETP **47**, 804 (1978)].
Khinchine, A.I., *Statistical Mechanics* (Dover Publication, New York, 1949).
Kikkawa, J.M., and D.D. Awshalom, Nature **397**, 139 (1999).
Kinoshita, T., T. Wenger, and D.S. Weiss, Nature **440**, 900 (2006).
Koentopp, M., K. Burke, and F. Evers, Phys. Rev. B **73**, 121403 (2006).
Kohn, W., and L.J. Sham, Phys. Rev. **140**, A1133 (1965).
Kondo, J., Prog. Theor. Phys. **32**, 37 (1964).
Kubo, R., J. Phys. Soc. Japan **12**, 570 (1959).
Kubo, R., M. Toda, and N. Hashitsume, *Statistical Physics II: Non-equilibrium Statistical Mechanics* (Springer-Verlag, Berlin, 1985).
Kumar, P., and R.S. Sorbello, Thin Solid Films **25**, 25 (1975).
Kurth, S., *et al.*, Phys. Rev. B **72**, 035308 (2005).
Kurzweil, Y., and R. Baer, Phys. Rev. B **73**, 075413 (2006).
Lagerqvist, J., Y.-C. Chen, and M. Di Ventra, Nanotechnology **15**, S459 (2004).
Landau, L.D., and E.M. Lifshitz, *Statistical Physics, Course of Theoretical Physics*, Vol. 5 (Pergamon Press, 1959a).
Landau, L.D., and E.M. Lifshitz, *Fluid Mechanics, Course of Theoretical Physics*, Vol. 6 (Pergamon Press, 1959b).
Landauer, R., IBM J. Res. Dev. **1**, 223 (1957).
Landauer, R., J. Phys. Cond. Matt. **1**, 8099 (1989).
Landauer, R., and J.W.F. Woo, Phys. Rev. B **10**, 1266 (1974).
Lang, N.D., Phys. Rev. B **45**, 13599 (1992).

Lang, N.D., Phys. Rev. B **52**, 5335 (1995).
Lee, H., L.S. Levitov, A.Yu. Yakovets, Phys. Rev. B **51**, 4079 (1995).
Levitov, L.S., and G.B. Lesovik, JETP Lett. **58**, 230 (1993).
Lévy, L.P., G. Dolan, J. Dunsmuir, and H. Bouchiat, Phys. Rev. Lett. **64**, 2074 (1990).
Lindblad, G., Commun. Math. Phys. **48**, 119 (1976).
Ludolph, B., and J.M. van Ruitenbeek, Phys. Rev. B **59**, 12290 (1999).
Luttinger, J.M., J. Math. Phys. **4**, 1154 (1963).
Madelung, E., Z. Phys. **40**, 332 (1926).
Mahan, G., *Many-Particle Physics* (Plenum, New York, 1990).
Mal'shukov, A.G., and C.S. Chu, Phys. Rev. Lett. **97**, 076601 (2006).
Maniscalco, S. *et al.*, J. Opt. B: Quantum Semiclass. Opt. **6**, S98 (2004).
Marques, M.A.L., C.A. Ullrich, F. Nogueira, A. Rubio, K. Burke, and E.K.U. Gross, in *Time-Dependent Density Functional Theory, Lecture Notes in Physics*, Vol. 706 (Springer, Berlin, 2006).
Martin, P.C., and J. Schwinger, Phys. Rev. **115**, 1342 (1959).
Mattuck, R.D., *A Guide to Feynman Diagrams in the Many-Body Problem* (Dover, New York, 1976).
Meir, Y., and N. Wingreen, Phys. Rev. Lett. **68**, 2512 (1992).
Meir, Y., N. Wingreen, and P.A. Lee, Phys. Rev. Lett. **66**, 3048 (1991).
Mermin, N.D., Phys. Rev. A **137**, 1441 (1965).
Messiah, A., *Quantum Mechanics* (North-Holland, New York, 1961).
Migdal, A.B., Sov. Phys. JETP **7**, 996 (1958).
Montgomery, M.J., J. Hoekstra, T.N. Todorov, and A. Sutton, J. Phys. Cond. Matt. **15**, 731 (2003).
Muller, C.J., J.M. van Ruitenbeek, and L.J. de Jongh, Physica C **191**, 485 (1992).
Nakajima, S., Prog. Theor. Phys. **20**, 948 (1958).
Newton, R.G., *Scattering Theory of Waves and Particles* (McGraw-Hill, New York, 1966).
Nifosi, R., S. Conti and M.P. Tosi, Phys. Rev. B, **58**, 12758 (1998).
Oseledec, V. I., Trudy. Mosk. Mat. Obsc. [Moscow Math. Soc.] **19**, 197 (1968).
Palacios, J.J., A.J. Prez-Jimenez, E. Louis, E. SanFabian, and J.A. Verges, Phys. Rev. B **66**, 035322 (2002).
Park, H., J. Park, A. Lim, E Anderson, A. Alivisatos, P.L. McEuen, Nature **407**, 57 (2000).
Park, J., A.N. Pasupathy, J.I. Goldsmith, *et al.* Nature **417**, 722 (2002).
Parr, R.G., and W. Young, *Density Functional Theory of Atoms and Molecules* (Oxford University Press, New York, 1989).
Pathria, R.K., *Statistical Mechanics* (Pergamon Press, Oxford, 1972).
Patton, K.R., and M.R. Geller, Phys. Rev. B. **64**, 155320 (2001).
Paulsson, M. and S. Datta, Phys. Rev. B **67**, 241403 (2003).
Paz, J.P. and W.H. Zurek, in *Proceedings of the 72nd Les Houches Summer School: Coherent Atomic Matter Waves*, R. Kaiser, C. Westbrook, and F. David, Eds. (Springer-Verlag, 1999).
Perdew, J.P., and A. Zunger, Phys. Rev. B **23**, 5048 (1981).
Perdew, J.P., K. Burke, and M. Ernzerhof, Phys. Rev. Lett. **77**, 3865 (1996).
Pershin, Y.V., and M. Di Ventra, Phys. Rev. B **75**, 193301 (2007).
Pichard, J.L., M. Sanquer, K. Slevin, and P. Debray, Phys. Rev. Lett. **65**, 1812 (1990).

Pollock, D.D., *Thermoelectricity: Theory, Thermometry, Tool* (American Society for Testing and Materials, Philadelphia, PA, 1985).
Prigogine, I., *Introduction to the Thermodynamics of Irreversible Processes* (Wiley, New York, 1967).
Pulay, P., in *Modern Theoretical Chemistry*, H.F. Schaefer Ed. (Plenum, New York, 1977), Vol. 4.
Rammer, J., and H. Smith, Rev. Mod. Phys. **58**, 323 (1986).
Reddy, P., *et al.* Science **315**, 1568 (2007).
Reichl, L.E., *A Modern Course in Statistical Physics* (John Wiley & Sons, New York, 1998).
Rego, L.G.C., and G. Kirczenow, Phys. Rev. Lett. **81**, 232 (1998).
Ross, S.M., *Stochastic Processes*, second edition (John Wiley & Sons, New York, 1996).
Runge, E., and E.K.U. Gross, Phys. Rev. Lett. **52**, 997 (1984).
Sai, N., M. Zwolak, G. Vignale, and M. Di Ventra, Phys. Rev. Lett. **94**, 186810 (2005).
Sai, N., M. Zwolak, G. Vignale, and M. Di Ventra, Phys. Rev. Lett. **98**, 259702 (2007a).
Sai, N., N. Bushong, R. Hatcher, and M. Di Ventra Phys. Rev. B **75**, 115410 (2007b).
Saito, R., G. Dresselhaus, M.S. Dresselhaus, *Physical Properties of Carbon Nanotubes*, (World Scientific, Singapore, 1998).
Sakurai, J.J., *Advanced Quantum Mechanics* (Addison Wesley, New York, 1967).
Schottky, W., Annalen der Physik **57**, 541 (1918).
Schwab, K., J.L. Arlett, J.M. Worlock, M.L. Roukes, Physica E **9**, 60 (2001).
Schwinger, J., Phys. Rev. **72**, 742 (1947).
Segal, D., A. Nitzan, J. Chem. Phys. **117**, 3915 (2002).
Sivan, U., and Y. Imry, Phys. Rev. B **33**, 551 (1986).
Sorbello, R.S., in *Solid State Physics*, H. Ehrenreich and F. Spaepen Eds. (Academic Press, New York, 1998), Vol. **51**.
Stefanucci, G., and C.-O. Almbladh, Phys. Rev. B **69**, 195318 (2004).
Stern, A., Y. Aharonov, and Y. Imry, Phys. Rev. A **41**, 3436 (1990).
Todorov, T.N., Phil. Mag. B **77**, 965 (1998).
Tokatly, I.V., Phys. Rev. B **71**, 165104 (2005).
Tokatly, I.V., and O. Pankratov, Phys. Rev. B **60**, 15550 (1999).
Tsutsui, M., S. Kurokawa, A. Sakai, Appl. Phys. Lett. **90**, 133121 (2007).
Ullrich, C.A., and G. Vignale, Phys. Rev. B **65**, 245102 (2002).
van den Brom, H.E., and J.M. van Ruitenbeek, Phys. Rev. Lett. **82**, 1526 (1999).
van Kampen, N.G., *Stochastic Processes in Physics and Chemistry* (Elsevier, Amsterdam, 1992).
van Leeuwen, R., Phys. Rev. Lett. **52**, 997 (1999).
van Wees, B.J., H. van Houten, C.W.J. Beenakker, *et al.* Phys. Rev. Lett. **60**, 848 (1988).
Vignale, G., Phys. Rev. B **70**, 201102(R) (2004).
Vignale, G., and W. Kohn, Phys. Rev. Lett. **77**, 2037 (1996).
Vignale, G., C.A. Ullrich, and S. Conti, Phys. Rev. Lett. **79**, 4878 (1997).
Wallstrom, T.C., Phys. Rev. A **49**, 1613 (1994).
Wang, Z., *et al.*, Science **317**, 787 (2007).
Wharam, D., T.J. Thornton, R. Newbury, *et al.* J. Phys. C: Solid State Phys. **21**, L209 (1988).

Wijewardane, H.O., and C.A. Ullrich, Phys. Rev. Lett. **95**, 086401 (2005).
Xue, Y., S. Datta, and M.A. Ratner, J. Chem. Phys. **115**, 4292 (2001).
Yang, Z., and M. Di Ventra, Phys. Rev. B **67**, 161311(R) (2003).
Yang, Z., A. Tackett, and M. Di Ventra, Phys. Rev. B **66**, 041405 (2002).
Yang, Z., N.D. Lang, and M. Di Ventra, Appl. Phys. Lett. **82**, 1938 (2003).
Yang, Z., M. Chshiev, M. Zwolak, Y.-C. Chen, and M. Di Ventra, Phys. Rev. B **71**, 041402 (2005).
Yao, J., Y.-C. Chen, M. Di Ventra, and Z.Q. Yang, Phys. Rev. B **73**, 233407 (2006).
Yu, Z.G., and M. E. Flatté, Phys. Rev. B **66**, 201202 (2002).
Yu, L.H., and D. Natelson, Nano Lett. **4**, 79 (2004).
Zhang, Y., J. Kastrup, R. Klann, K.H. Ploog, and H.T. Grahn, Phys. Rev. Lett. **77**, 3001 (1996).
Zurek, W.H., Phys. Today **44**, 36 (1991).
Žutić, I., J. Fabian, and S. Das Sarma, Rev. Mod. Phys. **76**, 323 (2004).
Zwanzig, R., J. Chem. Phys. **33**, 1338 (1960).
Zwolak, M., and M. Di Ventra, Rev. Mod. Phys. **80**, 141 (2008).
Zwolak, M., and M. Di Ventra, Appl. Phys. Lett. **81**, 925 (2002).
Zwolak, M., D. Ferguson, and M. Di Ventra, Phys. Rev. B **67**, 081303 (2003).

Index

advection, 408
Aharonov–Bohm effect, 192, 381
analytic continuation, 304
anharmonic effects, 320
apparatus, measurement, 7, 15, 32, 93
approach
   Landauer, 35
   Langevin, 278
   partition-free, 214, 346
   partitioning, 108, 146, 149, 154, 213
approximation
   adiabatic, 53, 213, 282
     in presence of current, 283
   Born, 54, 136, 303
     self-consistent, 303
   conserving, 225
   generalized gradient, 435
   local density, 434
     adiabatic, 357, 358, 392, 437
   loss of memory, 393, 401
   Markov, 28, 429
   mean-field, 60, 65, 70, 73, 89, 107, 137, 166, 230
   non-crossing, 300
   quasi-particle, 229
   relaxation-time, 81, 82, 98, 255, 371, 384
   single-particle, 283
attractor, 367
average
   macroscopic, 70
   planar, 70
Avogadro's number, 8

basin of attraction, 367
bath, 7
   phonon, 25, 44, 369
BBGKY hierarchy, 81, 84, 108, 218, 356, 420, 457
bias, 3, 39, 70, 103, 128, 213, 348, 358
Boltzmann constant, 8, 404
Boltzmann equation, 35, 49, 83
   classical, 79, 80
   quantum, 235, 251, 254, 382

Born–Oppenheimer
   approximation, see approximation, adiabatic
   surface, 282
bosons, 83, 219, 223, 419
   shot noise for, 263
   time-ordering operator, 287
branch cut, 141
brownian motion, 65
bubble diagram, 304

capacitance, 6, 245, 247, 260, 278, 348
channels, 110, 117
   closed, 177
   incoherent, see independent
   independent, 47, 110, 111
   open, 177
characteristic polynomial, 174
charge
   bound, 5, 356
   free, 5, 356
charge neutrality, 369
coherence, 22
   length, 48
   long-range, 62
   orbital, 76
   spin, 76
collision, 77
   elastic, 43
   inelastic, 43
   integral, 80, 255
compressibility, isothermal, 65
conductance, 39
   differential, 130
   four-probe, 181, 375
   heat, 315, 411
   multi-channel, 182
   multi-probe, 185, 348
   quantized, 130, 388
   single-channel, 179
   two-probe, 128, 130
conductivity, 39
   Drude, 42

Index

frequency-dependent, 95
frequency-dependent, 56
sum-rule, 42
tensor, 40
conductor, ohmic, 40
constant of the motion, 363
constriction
  adiabatic, 399
  non-adiabatic, 400
correlations, 22, 27, 32, 47, 59, 63, 78, 84, 107, 110, 209, 216, 237, 266, 423, 443, 453
Coulomb blockade
  non-equilibrium Green's function approach to, 248
  othodox picture, 245
coupling
  electron-phonon, 285, 290, 291
  spin-orbit, 75, 190
cross-correlation, 270
cumulant, 274
current
  alternating, 348
  chaotic, 398
  confined, 14
  density, 1, 4, 90, 458
  diamagnetic, 55
  diffusion, 72
  direct, 53, 348, 384, 393
  displacement, 276
  drift, 72
  fluctuations, see noise
  inelastic, 290
    from NEGF, 296
    from perturbation theory, 291
  localized, 5, 356
  magnetization, 5, 356
  moments of, 261
  non-variational properties of, 199
  paramagnetic, 55
  persistent, 14
  polarization, 5, 356
  spontaneously fluctuating, 63
  total, 4, 6, 7, 123, 351–356

de Broglie ansatz, 378, 380, 413
decoherence, 22, 44, 46
density
  ground-state, 57
  number, 4, 89, 458
  of states, 66, 123, 205, 226
    local, 291, 294, 322
  phase, 77, 79
  probability, 78, 81, 118, 228
  reduced, 26, 78, 84
density-functional theory, 24, 57, 89, 92, 351
  stochastic time-dependent current, 5, 57, 351, 356, 444
  time-dependent, 57, 351
  time-dependent current, 90, 354
dephasing, see decoherence
diffusion, 69, 408

coefficient, 65, 66, 73
response coefficient, 73
dilution
  energy, 9
  geometrical, 181, 390
  of degrees of freedom, 112
dimensional analysis, 398, 400
dipole
  local resistivity, 114, 333
  local resistivity spin, 197
disorder, see entropy
distribution
  Bernoulli, 274
  binomial, 274
  gaussian, 275
  Lorentzian, 258
  Poisson, 264, 265
  statistical, 261
distribution function
  Bose–Einstein, 31
  Fermi–Dirac, 31, 77, 415
  local equilibrium, 111, 126, 374
  non-equilibrium, 78
dots, quantum, 2, 245
drag
  spin Coulomb, 75
  viscous, 391, 397
drift, 69
Drude
  model, 35, 39
  peak, 95
dyadic, 36
dynamics
  dissipative, 380
  irreversible, 9, 90
  non-Markovian, 425

eigenchannel basis, 162, 272
eigendifferential, 14, 119, 336
Einstein relation, 65
electro-migration, see force, current-induced
electron-hole pair, 303, 403
electron-pumping effect, 316
electrons
  core, 283
  independent, see approximation, mean-field
  left-moving, 112
  right-moving, 111
  valence, 283
electrostatic potential, 374
energy dissipation, 96, 314, 394, 400
  rate of, 396
ensemble, 4, 19, 20, 24, 25, 92, 94, 261
  canonical, 30, 63, 87
  grand-canonical, 29
  micro-canonical, 64
entanglement, see correlations
enthalpy, 388
entropy
  canonical, see entropy, thermodynamic
  concavity, 92, 99, 357

flux, 91, 99
  local, 71, 407
  micro-canonical, 86
  production, 99, 367, 406
    rate, 91, 99
  statistical, 8
    classical, 85
    Kubo, 91
    of measurement, 93
    of open systems, 90
    of pure states, 87
    of stochastic Hamiltonians, 92
    quantum, 86
  thermodynamic, 68, 87, 98
environment, 7, 423
equation
  Bernoulli, 388, 463
  Boltzmann, 458
  continuity, 6, 62, 75, 225, 361, 378
  drift-diffusion, 73
  Dyson
    interacting, 222
    non-interacting, 137
  Hartree–Fock, 24, 166
  Keldysh, 236
  Langevin, 278
  Lindblad, 28, 44, 424
  Liouville, 97
  Liouville–von Neumann, 23, 51, 211, 426
  Lippmann–Schwinger
    time-dependent, 132
    time-independent, 140
  Schrödinger, 12, 51, 211
    stochastic, 28, 354, 444
  stochastic differential, 278, 446
equilibrium
  global, 50, 51, 212
  local, 49, 71, 80, 83, 101
    approach to, 82
  quasi-static, 69
  thermal, see equilibrium, thermodynamic
  thermodynamic, 31, 69, 74, 219, 320, 407
ergodicity, 64, 262
Euler equations, 387
evolution
  non-unitary, 28, 152, 336
  unitary, 12, 213, 423, 427
exchange-correlation
  field, 392, 440
  functional, 433
  hole, 433
  kernel, 59
  pressure, 393
  scalar potential, 392, 433, 437
  vector potential, 59, 392, 439, 449

Fano factor, 273
Fermi
  energy, 31, 66, 69, 117
  momentum, 66
  sea, 403
  statistics, see distribution function
  velocity, 48, 66
  wave-vector, 117
ferromagnet, 197
field
  electric, 72, 333
    inhomogeneous, 35
    uniform, 57
  magnetic, 109, 156, 190
  velocity, 376, 378, 380
field operator, 218, 267, 287, 289, 418
flow
  irrotational, 380
  laminar, 387, 395, 397
  turbulent, 260, 387, 397
fluctuations, 6, 258
  spectral density of, 265
  spontaneous, 63
  thermal, 4
fluid, see liquid
Fock
  space, 286, 416
  states, 416
force
  and topology, 341
  conservative character, 340
  current-induced
    elastic, 329
    inelastic, 329
  direct, 330
  from linear response, 332
  out of equilibrium, 335
  Pulay, 334, 345
  sign, 333
  wind, 330
formalism
  Keldysh, 209
  Kubo, 35, 50
  linear-response, 35, 50
  non-equilibrium Green's function, 209
Fourier transform, 42, 117, 132, 140–142, 222, 223, 225, 239, 240, 253
friction, see viscosity
Friedel oscillations, 146, 374
function
  autocorrelation, 62, 261
    voltage, 279
  characteristic, 274
  correlation, 97, 224, 278
    current-current, 62
  distribution, 77
  Heaviside, 31, 117
  moment generating, 274
  sample, 261
  spectral, 225, 228
  functional, 24, 80, 211, 221, 245, 432
    derivative, 433

g-factor, 191
gas
  Fermi, 48, 266, 409

Index

degenerate, 31, 66
 non-degenerate, 32, 63, 74, 76
 ideal, 30, 48, 63, 73, 74, 78
gauge, 52, 56
 invariance, 13, 96, 192, 392, 439
Green's function
 advanced, 133, 223
 anti-time-ordered, 232
 contour-ordered, 217, 231
 equation of motion, 220
 equilibrium, 217
 free, 134
 greater, 224
 lesser, 224
 principal-value, 204
 relation between retarded and advanced, 134
 retarded, 133, 223
 series expansion, 136
 three-particle, 220
 time-ordered, 218
 two-particle, 220
group, 423

Hamilton equations, 80, 83
Hamiltonian
 electron-ion, 281
 Hartree, 108, 231
 Kohn–Sham, 355
  ground-state, 109
  stochastic, 434
 random, see stochastic
 stochastic, 24, 92
 transfer, see tunneling Hamiltonian
heat dissipation, see energy dissipation
heat transfer, 406
heating
 local electron, 403, 410
 local ionic, 280, 312, 410
 vs current-induced forces, 343
Helmholtz, free energy, 68
hypothesis, ergodic, 64, 264

idempotency, 38
information, loss of, 9, 32, 84–86, 94, 357, 390, 394
 in Kubo approach, 91
 in Landauer approach, 107
 with stochastic Hamiltonians, 92
information, maximum, 18, 21
insulator, 67
interaction
 electron-electron, 2, 47, 110, 356, 369
 electron-ion, see interaction, electron-phonon
 electron-phonon, 110, 280, 286–288, 290, 365, 369, 410, 451
 phonon-phonon, 286
invariance, time-reversal, 83, 122
Itô calculus, 446

jellium model, 338, 341, 358, 401

Kohn–Sham
 equations, 57, 432
  time-dependent, 437
 functional, see exchange-correlation functional
 orbitals, 58
Kondo effect, 210
Kramers–Krönig relation, 461

leads, 105, 108, 112, 115, 116
lifetime, electron-phonon, 307
Lindbladian, 28
liquid
 electron, 376
 Fermi, 228
 incompressible, 388, 401, 407
 Luttinger, 210
 viscous, 376
localization length, 175
Lyapunov exponent, 176

Madelung equations, 379
magnetization, 5
magnetoresistance, 194
mass, effective, 41, 95, 299
material
 diamagnetic, 5
 ferromagnetic, 190
 paramagnetic, 5
matrix
 T, 157
 S, 167
  unitarity of, 162, 174, 205
 density, 21
  N-particle, 89
  Kohn–Sham, 59
  single-particle, 89
 pseudo-unitary, 173, 206
 random, 176
 reduced density, 103
 transfer, 167
mean-free path, 48
 inelastic, 48, 290, 294
measurement
 complete, 19, 32
 continuous, 32
 current, 15–17, 19, 93
 ideal, 16
 outcome of, 32, 94
memory, 27, 28, 83, 171, 368
 kernels, 428
mesoscopic, 1, 36
micro-canonical approach, 35, 346
 conductance, 374
micro-reversibility, 84
mobility, 74
modes, transverse, see channels
molecule, 112
 alkanethiol, 323
 octanedithiol, 313
moment

474 Index

dipole, 5
magnetic, 5

nanotube, 2, 112, 118
Navier–Stokes equations, 382, 387
noise, 6
   $1/f$, 258
   external, 258
   input, 445
   internal, 258
   Johnson–Nyquist, see thermal
   output, 445
   Poisson, 264
   shot, 260, 398, 453
   spectrum, 265
   sub-poissonian, 273
   super-poissonian, 273
   telegraph, 258
   thermal, 260, 263, 275, 277–279, 398
   turbulent, 260
normal mode, 285

observables
   commuting, 12, 18
   compatible, 18
occupation number, 31, 248, 266, 275, 416
Ohm's law, 39
Onsager relation, 196, 206
operator
   current, 13, 120
   current density, 13
   density, see operator, statistical
   density of states, 143
   diamagnetic current density, 13
   equilibrium statistical, 29
   Kraus, 425
   Lindblad, 28, 424
   linear, 427
   number density, 13
   occupation number, 417
   paramagnetic current density, 13
   Pauli, 75
   position, 13
   projection, 16
   scattering, see collision, integral
   statistical, 19–21, 93
      properties of, 37
      reduced, 26, 426
      time-evolution, 23
      truncation of, 33
   super-, 28, 29, 104, 424, 430
   time-evolution, 12
      series expansion, 217
   time-ordering, 12
   velocity, 13
optical lattices, 368

particle flux, conservation, 122, 125, 162, 165, 187, 189
particles
   indistinguishable, 9, 30, 77, 415

   interacting, 77
   non-interacting, 30, 79, 415
path
   evolutionary, 363
   semiclassical, 41
phase, 43
phonon, 25, 29, 44, 83
   absorption, 293, 307
   dressed, 288
   emission, 293, 306
      spontaneous, 294
   in presence of current, 340
   localized, 308, 316
   propagator, 288
      free, 287, 308
   subsystem, 284
picture
   Heisenberg, 21
   interaction, 53
   Schrödinger, 21
polarization, 5, 288
   spin, 75, 195, 197
positivity, statistical operator, 21, 54, 355, 424
potential
   chemical, 29, 68, 72
   electrochemical, 70, 103
      local, 179
      spin-dependent, 72
   electrostatic, see Hartree
   Hartree, 70, 109, 128, 438
   non-local, 193
pressure, 385–387, 407, 413
principal value, 204
principle
   exclusion, see principle, Pauli
   initial-state maximum entropy, 367
   maximum entropy, 367
   Pauli, 59, 66, 83, 84, 354
   Prigogine, 367
probability
   inelastic transition, 292
   reflection, 121, 267
   transmission, 122, 267
probe, 128
   capacitative, 128
   floating, 188
   invasive, 129, 185, 195
   local, 313, 403
   non-invasive, 128, 173, 179
   voltage, 43, 69, 185, 188, 324
process
   cascade, 401
   dephasing, 43, 110, 171, 186, 187, 189
   inelastic, 280, 284, 370
   irreversible, 84
   normal, 320, 369
   stationary, 262
   stochastic, 261, 278, 445
      δ-correlated, 354
   Umklapp, 319, 320, 369
projection method, 426

Index

projector, super-, 427
propagator, *see* Green's function
pseudopotential, 283, 330, 332, 338

quantum master equation, 28, 425, 446
   stochastic unraveling of the, 448
quantum of conductance, 67
quantum point contact, 1, 131

reference frame
   co-moving, *see* reference frame, Lagrangian
   Lagrangian, 383, 458
   non-inertial, 458
reflection amplitude, 158
regime
   ballistic, 67, 178
   hydrodynamic, *see* equilibrium, local
   localization, 176
   ohmic, 67, 177
   quasi-ballistic, 178
relaxation
   energy, 44
   momentum, 43, 50, 369, 370
   spin, 75
renormalization, energy, 28, 105, 148, 151, 229, 299, 427
reservoir, 7, 67, 69, 103, 105, 111, 129, 166, 178, 186, 202
   phonon, 289
resistance, 6, 39, 110, 348
   definition of, 42
   dynamical, *see* resistance, viscous
   quantized, 67
   viscous, 395, 396
resistivity, 39
resolution of the identity, 143
resonance, 128, 148
response, 39, 115
   function, 55
      current-current, 55, 58
   linear, 40, 50, 52, 392
Reynolds number, 398
   critical, 399

scattering
   theory, 132
   boundary conditions, 106
   elastic, 109, 249, 405
   electron-electron, *see* interaction
   electron-phonon, *see* interaction
   inelastic, 283, 290
      selection rules, 296
   rate
      electron-phonon, 291
   solutions, 106
   theory, 102, 106
screening, 129
   length, 114
   magnetic, 5
   partial, 115
Seebeck coefficient, *see* thermopower

self-consistency, 70, 199
self-energy, 136
   electron-phonon
      equilibrium, 301
      non-equilibrium, 307
   greater, 234
   irreducible, 221
   lesser, 234
   operator, 149
semiconductor, 197
   doped, 74, 299
   non-degenerate, 74
semigroup, 423
similarity transformation, 175
single-electron tunneling, *see* Coulomb blockade
Slater determinant, 415
Sommerfeld expansion, 326
specific heat, 404, 407
spectral representation, 142
spectrum
   continuum, 14
   dense, 8, 44, 103
   discrete, 14
spin
   accumulation, 197
   degeneracy, 66, 67
   electron, 75, 190
   flip, 75
   majority, 72, 197
   minority, 72, 197
   nuclear, 75, 190
   polarization, 72
spin valve, 193
spinor, 191
state
   evanescent, 118
   incoherent, 23
   incoming, 138
   lifetime, 137, 148, 151, 229
      elastic, 166
      inelastic, 299
   macro-, 20, 21, 25, 38, 102
   micro-, 17, 19–22, 26, 32–34, 38, 107
   mixed, *see* macro-
   mixing, 47
   outgoing, 139
   pure, *see* micro-
   quasi-steady, 347, 360
      global stability, 364
      variational definition, 360
   square-integrable, 119
   stationary, *see* steady
   steady, 7, 104, 121, 200, 203, 346
      and initial conditions, 364
      approach to, 369
   uncorrelated, 27, 112
statistics, counting, 262, 265, 274
Stosszahlansatz, 84
streamline, 463
subbands, 116

superconductor, 62
superlattice, 398
symmetry breaking, spontaneous, 84
symmetry, time-reversal, see invariance, time-reversal
system
  chaotic, 398
  closed, 7, 23, 51, 107, 211
  finite, 8
  homogeneous, 57
  infinite, 9, 106
  inhomogeneous, 40
  isolated, 8
  isotropic, 66
  Kohn–Sham, 58, 351
  macroscopic, 1, 46
  mesoscopic, 178
  non-interacting, 228
  open, 28, 32, 102
  strongly correlated, 228
  translationally invariant, 369
  weakly interacting, 228

temperature
  Debye, 319
  local electronic, 286, 403
  local ionic, 286, 313
tensor, 40, 42
  shear stress, 413
  stress, 378, 381
    Navier–Stokes, 385
theorem
  H-, 86, 98
  central limit, 275
  fluctuation, 88
  fluctuation-dissipation, 60, 65, 97, 226, 257
  Hellmann–Feynman, 334, 345
  Hohenberg–Kohn, 432
  Lindblad, 424
  Liouville, 79
  Migdal, 303
  optical, 206
  Oseledec, 176
  Poincaré recurrence, 8
  Runge–Gross, 436
  total current, 351–356
  variational, 200
  Wiener–Khinchin, 265
thermopower, 323
  dynamical effects, 327
  transient response, 327
time
  autocorrelation, 103, 105, 398, 428
    and steady state, 262
    definition of, 262
  coherence, 43
  correlation, see time, autocorrelation
  decoherence, 46
  elastic relaxation, 43
  inelastic relaxation, 43
  momentum relaxation, 49

Poincaré recurrence, 8, 349
relaxation, 41, 81, 90, 95, 230
transit, 2
trace, 20, 36
transition probability, 81
transitions, inelastic, 280, 292, 319
transmission amplitude, 159
transmission coefficient, 124
transport
  incoherent, 43
  phase coherent, 43, 110
  spin-dependent, 75, 190
  viewpoints, 35
tunneling
  Hamiltonian, 206
  resonant, 169, 244
  sequential, 171
  single-electron, see Coulomb blockade
turbulence, 260, 395, 406
  eddies, 400
  fully developed, 398

ultra-cold gases, 368

variable, extensive, 9
vector potential, 56, 379, 392, 445
  longitudinal, 439
velocity
  drift, 4, 74
  thermal, 4
vertex corrections, 302
virtual process, 304
viscosity, 198
  bulk, 386, 441, 461
  shear, 376, 386, 441, 461
    approximate formulas, 442
voltmeter, see probe, voltage

wave-packet, 105, 114, 119, 370
  collapse, 16
  postulate of reduction, 16
Wiedemann–Franz law, 404
wires, atomic, 2

Printed in the United States
By Bookmasters